Communications in Computer and Information Science   1964

Rationale
The CCIS series is devoted to the publication of proceedings of computer science conferences. Its aim is to efficiently disseminate original research results in informatics in printed and electronic form. While the focus is on publication of peer-reviewed full papers presenting mature work, inclusion of reviewed short papers reporting on work in progress is welcome, too. Besides globally relevant meetings with internationally representative program committees guaranteeing a strict peer-reviewing and paper selection process, conferences run by societies or of high regional or national relevance are also considered for publication.

Topics
The topical scope of CCIS spans the entire spectrum of informatics ranging from foundational topics in the theory of computing to information and communications science and technology and a broad variety of interdisciplinary application fields.

Information for Volume Editors and Authors
Publication in CCIS is free of charge. No royalties are paid, however, we offer registered conference participants temporary free access to the online version of the conference proceedings on SpringerLink (http://link.springer.com) by means of an http referrer from the conference website and/or a number of complimentary printed copies, as specified in the official acceptance email of the event.

CCIS proceedings can be published in time for distribution at conferences or as post-proceedings, and delivered in the form of printed books and/or electronically as USBs and/or e-content licenses for accessing proceedings at SpringerLink. Furthermore, CCIS proceedings are included in the CCIS electronic book series hosted in the SpringerLink digital library at http://link.springer.com/bookseries/7899. Conferences publishing in CCIS are allowed to use Online Conference Service (OCS) for managing the whole proceedings lifecycle (from submission and reviewing to preparing for publication) free of charge.

Publication process
The language of publication is exclusively English. Authors publishing in CCIS have to sign the Springer CCIS copyright transfer form, however, they are free to use their material published in CCIS for substantially changed, more elaborate subsequent publications elsewhere. For the preparation of the camera-ready papers/files, authors have to strictly adhere to the Springer CCIS Authors' Instructions and are strongly encouraged to use the CCIS LaTeX style files or templates.

Abstracting/Indexing
CCIS is abstracted/indexed in DBLP, Google Scholar, EI-Compendex, Mathematical Reviews, SCImago, Scopus. CCIS volumes are also submitted for the inclusion in ISI Proceedings.

How to start
To start the evaluation of your proposal for inclusion in the CCIS series, please send an e-mail to ccis@springer.com.

Biao Luo · Long Cheng · Zheng-Guang Wu ·
Hongyi Li · Chaojie Li
Editors

Neural Information Processing

30th International Conference, ICONIP 2023
Changsha, China, November 20–23, 2023
Proceedings, Part X

 Springer

Editors
Biao Luo ⓘ
School of Automation
Central South University
Changsha, China

Zheng-Guang Wu ⓘ
Institute of Cyber-Systems and Control
Zhejiang University
Hangzhou, China

Chaojie Li ⓘ
School of Electrical Engineering
and Telecommunications
UNSW Sydney
Sydney, NSW, Australia

Long Cheng ⓘ
Institute of Automation
Chinese Academy of Sciences
Beijing, China

Hongyi Li ⓘ
School of Automation
Guangdong University of Technology
Guangzhou, China

ISSN 1865-0929   ISSN 1865-0937 (electronic)
Communications in Computer and Information Science
ISBN 978-981-99-8140-3   ISBN 978-981-99-8141-0 (eBook)
https://doi.org/10.1007/978-981-99-8141-0

Preface

Welcome to the 30th International Conference on Neural Information Processing (ICONIP2023) of the Asia-Pacific Neural Network Society (APNNS), held in Changsha, China, November 20–23, 2023.

The mission of the Asia-Pacific Neural Network Society is to promote active interactions among researchers, scientists, and industry professionals who are working in neural networks and related fields in the Asia-Pacific region. APNNS has Governing Board Members from 13 countries/regions – Australia, China, Hong Kong, India, Japan, Malaysia, New Zealand, Singapore, South Korea, Qatar, Taiwan, Thailand, and Turkey. The society's flagship annual conference is the International Conference of Neural Information Processing (ICONIP). The ICONIP conference aims to provide a leading international forum for researchers, scientists, and industry professionals who are working in neuroscience, neural networks, deep learning, and related fields to share their new ideas, progress, and achievements.

ICONIP2023 received 1274 papers, of which 394 papers were accepted for publication in Communications in Computer and Information Science (CCIS), representing an acceptance rate of 30.93% and reflecting the increasingly high quality of research in neural networks and related areas. The conference focused on four main areas, i.e., "Theory and Algorithms", "Cognitive Neurosciences", "Human-Centered Computing", and "Applications". All the submissions were rigorously reviewed by the conference Program Committee (PC), comprising 258 PC members, and they ensured that every paper had at least two high-quality single-blind reviews. In fact, 5270 reviews were provided by 2145 reviewers. On average, each paper received 4.14 reviews.

We would like to take this opportunity to thank all the authors for submitting their papers to our conference, and our great appreciation goes to the Program Committee members and the reviewers who devoted their time and effort to our rigorous peer-review process; their insightful reviews and timely feedback ensured the high quality of the papers accepted for publication. We hope you enjoyed the research program at the conference.

October 2023

Biao Luo
Long Cheng
Zheng-Guang Wu
Hongyi Li
Chaojie Li

Organization

Honorary Chair

Weihua Gui   Central South University, China

Advisory Chairs

Jonathan Chan   King Mongkut's University of Technology
                              Thonburi, Thailand
Zeng-Guang Hou   Chinese Academy of Sciences, China
Nikola Kasabov   Auckland University of Technology, New Zealand
Derong Liu   Southern University of Science and Technology,
                              China
Seiichi Ozawa   Kobe University, Japan
Kevin Wong   Murdoch University, Australia

General Chairs

Tingwen Huang   Texas A&M University at Qatar, Qatar
Chunhua Yang   Central South University, China

Program Chairs

Biao Luo   Central South University, China
Long Cheng   Chinese Academy of Sciences, China
Zheng-Guang Wu   Zhejiang University, China
Hongyi Li   Guangdong University of Technology, China
Chaojie Li   University of New South Wales, Australia

Technical Chairs

Xing He   Southwest University, China
Keke Huang   Central South University, China
Huaqing Li   Southwest University, China
Qi Zhou   Guangdong University of Technology, China

Local Arrangement Chairs

Wenfeng Hu	Central South University, China
Bei Sun	Central South University, China

Finance Chairs

Fanbiao Li	Central South University, China
Hayaru Shouno	University of Electro-Communications, Japan
Xiaojun Zhou	Central South University, China

Special Session Chairs

Hongjing Liang	University of Electronic Science and Technology, China
Paul S. Pang	Federation University, Australia
Qiankun Song	Chongqing Jiaotong University, China
Lin Xiao	Hunan Normal University, China

Tutorial Chairs

Min Liu	Hunan University, China
M. Tanveer	Indian Institute of Technology Indore, India
Guanghui Wen	Southeast University, China

Publicity Chairs

Sabri Arik	Istanbul University-Cerrahpaşa, Turkey
Sung-Bae Cho	Yonsei University, South Korea
Maryam Doborjeh	Auckland University of Technology, New Zealand
El-Sayed M. El-Alfy	King Fahd University of Petroleum and Minerals, Saudi Arabia
Ashish Ghosh	Indian Statistical Institute, India
Chuandong Li	Southwest University, China
Weng Kin Lai	Tunku Abdul Rahman University of Management & Technology, Malaysia
Chu Kiong Loo	University of Malaya, Malaysia
Qinmin Yang	Zhejiang University, China
Zhigang Zeng	Huazhong University of Science and Technology, China

Publication Chairs

Zhiwen Chen	Central South University, China
Andrew Chi-Sing Leung	City University of Hong Kong, China
Xin Wang	Southwest University, China
Xiaofeng Yuan	Central South University, China

Secretaries

Yun Feng	Hunan University, China
Bingchuan Wang	Central South University, China

Webmasters

Tianmeng Hu	Central South University, China
Xianzhe Liu	Xiangtan University, China

Program Committee

Rohit Agarwal	UiT The Arctic University of Norway, Norway
Hasin Ahmed	Gauhati University, India
Harith Al-Sahaf	Victoria University of Wellington, New Zealand
Brad Alexander	University of Adelaide, Australia
Mashaan Alshammari	Independent Researcher, Saudi Arabia
Sabri Arik	Istanbul University, Turkey
Ravneet Singh Arora	Block Inc., USA
Zeyar Aung	Khalifa University of Science and Technology, UAE
Monowar Bhuyan	Umeå University, Sweden
Jingguo Bi	Beijing University of Posts and Telecommunications, China
Xu Bin	Northwestern Polytechnical University, China
Marcin Blachnik	Silesian University of Technology, Poland
Paul Black	Federation University, Australia
Anoop C. S.	Govt. Engineering College, India
Ning Cai	Beijing University of Posts and Telecommunications, China
Siripinyo Chantamunee	Walailak University, Thailand
Hangjun Che	City University of Hong Kong, China

Wei-Wei Che	Qingdao University, China
Huabin Chen	Nanchang University, China
Jinpeng Chen	Beijing University of Posts & Telecommunications, China
Ke-Jia Chen	Nanjing University of Posts and Telecommunications, China
Lv Chen	Shandong Normal University, China
Qiuyuan Chen	Tencent Technology, China
Wei-Neng Chen	South China University of Technology, China
Yufei Chen	Tongji University, China
Long Cheng	Institute of Automation, China
Yongli Cheng	Fuzhou University, China
Sung-Bae Cho	Yonsei University, South Korea
Ruikai Cui	Australian National University, Australia
Jianhua Dai	Hunan Normal University, China
Tao Dai	Tsinghua University, China
Yuxin Ding	Harbin Institute of Technology, China
Bo Dong	Xi'an Jiaotong University, China
Shanling Dong	Zhejiang University, China
Sidong Feng	Monash University, Australia
Yuming Feng	Chongqing Three Gorges University, China
Yun Feng	Hunan University, China
Junjie Fu	Southeast University, China
Yanggeng Fu	Fuzhou University, China
Ninnart Fuengfusin	Kyushu Institute of Technology, Japan
Thippa Reddy Gadekallu	VIT University, India
Ruobin Gao	Nanyang Technological University, Singapore
Tom Gedeon	Curtin University, Australia
Kam Meng Goh	Tunku Abdul Rahman University of Management and Technology, Malaysia
Zbigniew Gomolka	University of Rzeszow, Poland
Shengrong Gong	Changshu Institute of Technology, China
Xiaodong Gu	Fudan University, China
Zhihao Gu	Shanghai Jiao Tong University, China
Changlu Guo	Budapest University of Technology and Economics, Hungary
Weixin Han	Northwestern Polytechnical University, China
Xing He	Southwest University, China
Akira Hirose	University of Tokyo, Japan
Yin Hongwei	Huzhou Normal University, China
Md Zakir Hossain	Curtin University, Australia
Zengguang Hou	Chinese Academy of Sciences, China

Lu Hu	Jiangsu University, China
Zeke Zexi Hu	University of Sydney, Australia
He Huang	Soochow University, China
Junjian Huang	Chongqing University of Education, China
Kaizhu Huang	Duke Kunshan University, China
David Iclanzan	Sapientia University, Romania
Radu Tudor Ionescu	University of Bucharest, Romania
Asim Iqbal	Cornell University, USA
Syed Islam	Edith Cowan University, Australia
Kazunori Iwata	Hiroshima City University, Japan
Junkai Ji	Shenzhen University, China
Yi Ji	Soochow University, China
Canghong Jin	Zhejiang University, China
Xiaoyang Kang	Fudan University, China
Mutsumi Kimura	Ryukoku University, Japan
Masahiro Kohjima	NTT, Japan
Damian Kordos	Rzeszow University of Technology, Poland
Marek Kraft	Poznań University of Technology, Poland
Lov Kumar	NIT Kurukshetra, India
Weng Kin Lai	Tunku Abdul Rahman University of Management & Technology, Malaysia
Xinyi Le	Shanghai Jiao Tong University, China
Bin Li	University of Science and Technology of China, China
Hongfei Li	Xinjiang University, China
Houcheng Li	Chinese Academy of Sciences, China
Huaqing Li	Southwest University, China
Jianfeng Li	Southwest University, China
Jun Li	Nanjing Normal University, China
Kan Li	Beijing Institute of Technology, China
Peifeng Li	Soochow University, China
Wenye Li	Chinese University of Hong Kong, China
Xiangyu Li	Beijing Jiaotong University, China
Yantao Li	Chongqing University, China
Yaoman Li	Chinese University of Hong Kong, China
Yinlin Li	Chinese Academy of Sciences, China
Yuan Li	Academy of Military Science, China
Yun Li	Nanjing University of Posts and Telecommunications, China
Zhidong Li	University of Technology Sydney, Australia
Zhixin Li	Guangxi Normal University, China
Zhongyi Li	Beihang University, China

Ziqiang Li	University of Tokyo, Japan
Xianghong Lin	Northwest Normal University, China
Yang Lin	University of Sydney, Australia
Huawen Liu	Zhejiang Normal University, China
Jian-Wei Liu	China University of Petroleum, China
Jun Liu	Chengdu University of Information Technology, China
Junxiu Liu	Guangxi Normal University, China
Tommy Liu	Australian National University, Australia
Wen Liu	Chinese University of Hong Kong, China
Yan Liu	Taikang Insurance Group, China
Yang Liu	Guangdong University of Technology, China
Yaozhong Liu	Australian National University, Australia
Yong Liu	Heilongjiang University, China
Yubao Liu	Sun Yat-sen University, China
Yunlong Liu	Xiamen University, China
Zhe Liu	Jiangsu University, China
Zhen Liu	Chinese Academy of Sciences, China
Zhi-Yong Liu	Chinese Academy of Sciences, China
Ma Lizhuang	Shanghai Jiao Tong University, China
Chu-Kiong Loo	University of Malaya, Malaysia
Vasco Lopes	Universidade da Beira Interior, Portugal
Hongtao Lu	Shanghai Jiao Tong University, China
Wenpeng Lu	Qilu University of Technology, China
Biao Luo	Central South University, China
Ye Luo	Tongji University, China
Jiancheng Lv	Sichuan University, China
Yuezu Lv	Beijing Institute of Technology, China
Huifang Ma	Northwest Normal University, China
Jinwen Ma	Peking University, China
Jyoti Maggu	Thapar Institute of Engineering and Technology Patiala, India
Adnan Mahmood	Macquarie University, Australia
Mufti Mahmud	University of Padova, Italy
Krishanu Maity	Indian Institute of Technology Patna, India
Srimanta Mandal	DA-IICT, India
Wang Manning	Fudan University, China
Piotr Milczarski	Lodz University of Technology, Poland
Malek Mouhoub	University of Regina, Canada
Nankun Mu	Chongqing University, China
Wenlong Ni	Jiangxi Normal University, China
Anupiya Nugaliyadde	Murdoch University, Australia

Toshiaki Omori	Kobe University, Japan
Babatunde Onasanya	University of Ibadan, Nigeria
Manisha Padala	Indian Institute of Science, India
Sarbani Palit	Indian Statistical Institute, India
Paul Pang	Federation University, Australia
Rasmita Panigrahi	Giet University, India
Kitsuchart Pasupa	King Mongkut's Institute of Technology Ladkrabang, Thailand
Dipanjyoti Paul	Ohio State University, USA
Hu Peng	Jiujiang University, China
Kebin Peng	University of Texas at San Antonio, USA
Dawid Połap	Silesian University of Technology, Poland
Zhong Qian	Soochow University, China
Sitian Qin	Harbin Institute of Technology at Weihai, China
Toshimichi Saito	Hosei University, Japan
Fumiaki Saitoh	Chiba Institute of Technology, Japan
Naoyuki Sato	Future University Hakodate, Japan
Chandni Saxena	Chinese University of Hong Kong, China
Jiaxing Shang	Chongqing University, China
Lin Shang	Nanjing University, China
Jie Shao	University of Science and Technology of China, China
Yin Sheng	Huazhong University of Science and Technology, China
Liu Sheng-Lan	Dalian University of Technology, China
Hayaru Shouno	University of Electro-Communications, Japan
Gautam Srivastava	Brandon University, Canada
Jianbo Su	Shanghai Jiao Tong University, China
Jianhua Su	Institute of Automation, China
Xiangdong Su	Inner Mongolia University, China
Daiki Suehiro	Kyushu University, Japan
Basem Suleiman	University of New South Wales, Australia
Ning Sun	Shandong Normal University, China
Shiliang Sun	East China Normal University, China
Chunyu Tan	Anhui University, China
Gouhei Tanaka	University of Tokyo, Japan
Maolin Tang	Queensland University of Technology, Australia
Shu Tian	University of Science and Technology Beijing, China
Shikui Tu	Shanghai Jiao Tong University, China
Nancy Victor	Vellore Institute of Technology, India
Petra Vidnerová	Institute of Computer Science, Czech Republic

Shanchuan Wan	University of Tokyo, Japan
Tao Wan	Beihang University, China
Ying Wan	Southeast University, China
Bangjun Wang	Soochow University, China
Hao Wang	Shanghai University, China
Huamin Wang	Southwest University, China
Hui Wang	Nanchang Institute of Technology, China
Huiwei Wang	Southwest University, China
Jianzong Wang	Ping An Technology, China
Lei Wang	National University of Defense Technology, China
Lin Wang	University of Jinan, China
Shi Lin Wang	Shanghai Jiao Tong University, China
Wei Wang	Shenzhen MSU-BIT University, China
Weiqun Wang	Chinese Academy of Sciences, China
Xiaoyu Wang	Tokyo Institute of Technology, Japan
Xin Wang	Southwest University, China
Xin Wang	Southwest University, China
Yan Wang	Chinese Academy of Sciences, China
Yan Wang	Sichuan University, China
Yonghua Wang	Guangdong University of Technology, China
Yongyu Wang	JD Logistics, China
Zhenhua Wang	Northwest A&F University, China
Zi-Peng Wang	Beijing University of Technology, China
Hongxi Wei	Inner Mongolia University, China
Guanghui Wen	Southeast University, China
Guoguang Wen	Beijing Jiaotong University, China
Ka-Chun Wong	City University of Hong Kong, China
Anna Wróblewska	Warsaw University of Technology, Poland
Fengge Wu	Institute of Software, Chinese Academy of Sciences, China
Ji Wu	Tsinghua University, China
Wei Wu	Inner Mongolia University, China
Yue Wu	Shanghai Jiao Tong University, China
Likun Xia	Capital Normal University, China
Lin Xiao	Hunan Normal University, China
Qiang Xiao	Huazhong University of Science and Technology, China
Hao Xiong	Macquarie University, Australia
Dongpo Xu	Northeast Normal University, China
Hua Xu	Tsinghua University, China
Jianhua Xu	Nanjing Normal University, China

Xinyue Xu	Hong Kong University of Science and Technology, China
Yong Xu	Beijing Institute of Technology, China
Ngo Xuan Bach	Posts and Telecommunications Institute of Technology, Vietnam
Hao Xue	University of New South Wales, Australia
Yang Xujun	Chongqing Jiaotong University, China
Haitian Yang	Chinese Academy of Sciences, China
Jie Yang	Shanghai Jiao Tong University, China
Minghao Yang	Chinese Academy of Sciences, China
Peipei Yang	Chinese Academy of Science, China
Zhiyuan Yang	City University of Hong Kong, China
Wangshu Yao	Soochow University, China
Ming Yin	Guangdong University of Technology, China
Qiang Yu	Tianjin University, China
Wenxin Yu	Southwest University of Science and Technology, China
Yun-Hao Yuan	Yangzhou University, China
Xiaodong Yue	Shanghai University, China
Paweł Zawistowski	Warsaw University of Technology, Poland
Hui Zeng	Southwest University of Science and Technology, China
Wang Zengyunwang	Hunan First Normal University, China
Daren Zha	Institute of Information Engineering, China
Zhi-Hui Zhan	South China University of Technology, China
Baojie Zhang	Chongqing Three Gorges University, China
Canlong Zhang	Guangxi Normal University, China
Guixuan Zhang	Chinese Academy of Science, China
Jianming Zhang	Changsha University of Science and Technology, China
Li Zhang	Soochow University, China
Wei Zhang	Southwest University, China
Wenbing Zhang	Yangzhou University, China
Xiang Zhang	National University of Defense Technology, China
Xiaofang Zhang	Soochow University, China
Xiaowang Zhang	Tianjin University, China
Xinglong Zhang	National University of Defense Technology, China
Dongdong Zhao	Wuhan University of Technology, China
Xiang Zhao	National University of Defense Technology, China
Xu Zhao	Shanghai Jiao Tong University, China

Contents – Part X

Human Centred Computing

Paper Recommendation with Multi-view Knowledge-Aware Attentive
Network 3
  Yuzhi Chen, Pengjun Zhai, and Yu Fang

Biological Tissue Sections Instance Segmentation Based on Active
Learning 16
  Yanan lv, Haoze Jia, Haoran Chen, Xi Chen, Guodong Sun, and Hua Han

Research on Automatic Segmentation Algorithm of Brain Tumor Image
Based on Multi-sequence Self-supervised Fusion in Complex Scenes 28
  Guiqiang Zhang, Jianting Shi, Wenqiang Liu, Guifu Zhang,
  and Yuanhan He

An Effective Morphological Analysis Framework of Intracranial Artery
in 3D Digital Subtraction Angiography 50
  Haining Zhao, Tao Wang, Shiqi Liu, Xiaoliang Xie, Xiaohu Zhou,
  Zengguang Hou, Liqun Jiao, Yan Ma, Ye Li, Jichang Luo, Jia Dong,
  and Bairu Zhang

Effective Domain Adaptation for Robust Dysarthric Speech Recognition 62
  Shanhu Wang, Jing Zhao, and Shiliang Sun

TCNet: Texture and Contour-Aware Model for Bone Marrow Smear
Region of Interest Selection 74
  Chengliang Wang, Jian Chen, Xing Wu, Zailin Yang, Longrong Ran,
  and Yao Liu

Diff-Writer: A Diffusion Model-Based Stylized Online Handwritten
Chinese Character Generator 86
  Min-Si Ren, Yan-Ming Zhang, Qiu-Feng Wang, Fei Yin,
  and Cheng-Lin Liu

EEG Epileptic Seizure Classification Using Hybrid Time-Frequency
Attention Deep Network 101
  Yunfei Tian, Chunyu Tan, Qiaoyun Wu, and Yun Zhou

A Feature Pyramid Fusion Network Based on Dynamic Perception
Transformer for Retinal OCT Biomarker Image Segmentation 114
Xiaoming Liu and Yuanzhe Ding

LDW-RS Loss: Label Density-Weighted Loss with Ranking Similarity
Regularization for Imbalanced Deep Fetal Brain Age Regression 125
Yang Liu, Siru Wang, Wei Xia, Aaron Fenster, Haitao Gan, and Ran Zhou

Segment Anything Model for Semi-supervised Medical Image
Segmentation via Selecting Reliable Pseudo-labels 138
*Ning Li, Lianjin Xiong, Wei Qiu, Yudong Pan, Yiqian Luo,
and Yangsong Zhang*

Aided Diagnosis of Autism Spectrum Disorder Based on a Mixed Neural
Network Model 150
*Yiqian Luo, Ning Li, Yudong Pan, Wei Qiu, Lianjin Xiong,
and Yangsong Zhang*

A Supervised Spatio-Temporal Contrastive Learning Framework
with Optimal Skeleton Subgraph Topology for Human Action Recognition 162
*Zelin Deng, Hao Zhou, Wei Ouyang, Pei He, Song Yun, Qiang Tang,
and Li Yu*

Multi-scale Feature Fusion Neural Network for Accurate Prediction
of Drug-Target Interactions 176
Zhibo Yang, Binhao Bai, Jinyu Long, Ping Wei, and Junli Li

GoatPose: A Lightweight and Efficient Network with Attention Mechanism ... 189
Yaxuan Sun, Annan Wang, and Shengxi Wu

Sign Language Recognition for Low Resource Languages Using Few Shot
Learning 203
*Kaveesh Charuka, Sandareka Wickramanayake,
Thanuja D. Ambegoda, Pasan Madhushan, and Dineth Wijesooriya*

T Cell Receptor Protein Sequences and Sparse Coding: A Novel Approach
to Cancer Classification 215
*Zahra Tayebi, Sarwan Ali, Prakash Chourasia, Taslim Murad,
and Murray Patterson*

Weakly Supervised Temporal Action Localization Through Segment
Contrastive Learning 228
Zihao Jiang and Yidong Li

S-CGRU: An Efficient Model for Pedestrian Trajectory Prediction   244
Zhenwei Xu, Qing Yu, Wushouer Slamu, Yaoyong Zhou, and Zhida Liu

Prior-Enhanced Network for Image-Based PM2.5 Estimation
from Imbalanced Data Distribution   260
Xueqing Fang, Zhan Li, Bin Yuan, Xinrui Wang, Zekai Jiang,
Jianliang Zeng, and Qingliang Chen

Dynamic Data Augmentation via Monte-Carlo Tree Search for Prostate
MRI Segmentation   272
Xinyue Xu, Yuhan Hsi, Haonan Wang, and Xiaomeng Li

Language Guided Graph Transformer for Skeleton Action Recognition   283
Libo Weng, Weidong Lou, and Fei Gao

A Federated Multi-stage Light-Weight Vision Transformer for Respiratory
Disease Detection   300
Pranab Sahoo, Saksham Kumar Sharma, Sriparna Saha,
and Samrat Mondal

Curiosity Enhanced Bayesian Personalized Ranking for Recommender
Systems   312
Yaoming Deng, Qiqi Ding, Xin Wu, and Yi Cai

Modeling Online Adaptive Navigation in Virtual Environments Based
on PID Control   325
Yuyang Wang, Jean-Rémy Chardonnet, and Frédéric Merienne

Lip Reading Using Temporal Adaptive Module   347
Jian Huang, Lianwei Teng, Yewei Xiao, Aosu Zhu, and Xuanming Liu

AudioFormer: Channel Audio Encoder Based on Multi-granularity
Features   357
Jialin Wang, Yunfeng Xu, Borui Miao, and Shaojie Zhao

A Context Aware Lung Cancer Survival Prediction Network by Using
Whole Slide Images   374
Xinyu Liu, Yicheng Wang, and Ye Luo

A Novel Approach for Improved Pedestrian Walking Speed Prediction:
Exploiting Proximity Correlation   387
Xiaohe Chen, Zhiyong Tao, Mei Wang, and Yuanzhen Zhou

MView-DTI: A Multi-view Feature Fusion-Based Approach
for Drug-Target Protein Interaction Prediction 400
  Jiahui Wen, Haitao Gan, Zhi Yang, Ming Shi, and Ji Wang

User Multi-preferences Fusion for Conversational Recommender Systems 412
  Yi Zhang, Dongming Zhao, Bo Wang, Kun Huang, Ruifang He,
  and Yuexian Hou

Debiasing Medication Recommendation with Counterfactual Analysis 426
  Pei Tang, Chunping Ouyang, and Yongbin Liu

Early Detection of Depression and Alcoholism Disorders by EEG Signal 439
  Hesam Akbari and Wael Korani

Unleash the Capabilities of the Vision-Language Pre-training Model
in Gaze Object Prediction 453
  Dazhi Chen and Gang Gou

A Two-Stage Network for Segmentation of Vertebrae and Intervertebral
Discs: Integration of Efficient Local-Global Fusion Using 3D Transformer
and 2D CNN 467
  Zhiqiang Li, Xiaogen Zhou, and Tong Tong

Integrating Multi-view Feature Extraction and Fuzzy Rank-Based
Ensemble for Accurate HIV-1 Protease Cleavage Site Prediction 480
  Susmita Palmal, Sriparna Saha, and Somanath Tripathy

KSHFS: Research on Drug-Drug Interaction Prediction Based
on Knowledge Subgraph and High-Order Feature-Aware Structure 493
  Nana Wang, Qian Gao, and Jun Fan

Self-supervised-Enhanced Dual Hierarchical Graph Convolution Network
for Social Recommendation 507
  Yixing Guo, Weimin Li, Jingchao Wang, and Shaohua Li

Dynamical Graph Echo State Networks with Snapshot Merging
for Spreading Process Classification 523
  Ziqiang Li, Kantaro Fujiwara, and Gouhei Tanaka

Trajectory Prediction with Contrastive Pre-training and Social Rank
Fine-Tuning 535
  Chenyou Fan, Haiqi Jiang, Aimin Huang, and Junjie Hu

Three-Dimensional Medical Image Fusion with Deformable
Cross-Attention 551
   Lin Liu, Xinxin Fan, Chulong Zhang, Jingjing Dai, Yaoqin Xie,
   and Xiaokun Liang

Handling Class Imbalance in Forecasting Parkinson's Disease
Wearing-off with Fitness Tracker Dataset 564
   John Noel Victorino, Sozo Inoue, and Tomohiro Shibata

Real-Time Instance Segmentation and Tip Detection for Neuroendoscopic
Surgical Instruments 579
   Rihui Song, Silu Guo, Ni Liu, Yehua Ling, Jin Gong, and Kai Huang

Spatial Gene Expression Prediction Using Hierarchical Sparse Attention 594
   Cui Chen, Zuping Zhang, and Panrui Tang

Author Index 607

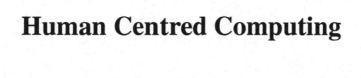

Human Centred Computing

Paper Recommendation with Multi-view Knowledge-Aware Attentive Network

Yuzhi Chen, Pengjun Zhai$^{(\boxtimes)}$, and Yu Fang

Tongji University, Shanghai 201804, China
{2132906,1810369,fangyu}@tongji.edu.cn

Abstract. The paper recommendation system aims to recommend potential papers of interest to users from massive data. Many efforts introduced knowledge graphs to solve problems such as data sparsity faced by traditional recommendation methods and used GNN-based techniques to mine the features of users and papers. However, existing work has not emphasized the quality of the knowledge graph construction, and has not optimized the modeling method from the scenario of paper recommendation, which makes the quality of recommendation results have room for improvement. In this paper, we proposed a Multi-View Knowledge-aware Attentive Network (MVKAN). Specifically, we first designed a knowledge graph construction method based on keynote extraction for better recommendation assistance. We then designed mechanisms for aggregation and propagation of graph attention from three views: the connectivity importance of entities, user's time preferences, and short-cut links to users based on tag similarity. This helps to model the representation of users and papers more effectively. Results from the experiments show that our model outperforms the baselines.

Keywords: Paper recommendation · Knowledge graph · Graph attention network

1 Introduction

With the exponential growth in literature, traditional paper searches faced a severe information overload problem. The paper recommendation system [2] can efficiently find papers that match the researcher's interests in massive data. Traditional content-based and collaborative filtering methods suffer from sparse interaction data, cold start, and poor interpretability. Therefore, many works mitigated these problems by introducing auxiliary information such as knowledge graph (KG) [1]. Such approaches typically used graph neural network techniques (GNNs) to encode representations of nodes [3], aggregating and propagating the features of user and paper nodes in the KG structure. Examples include RippleNet [4], which defined interest propagation networks, KGCN [6], which applied graph convolutional neural networks to KG, KGAT [5], which combined graph attention mechanisms in KG, and so on [1, 3].

Under the specific task of KG-enhanced paper recommendation, the first step is to build a domain-specific KG based on the papers [7, 16, 18, 19], which is constructed to

B. Luo et al. (Eds.): ICONIP 2023, CCIS 1964, pp. 3–15, 2024.
https://doi.org/10.1007/978-981-99-8141-0_1

explore potential associated paths on the side of papers. Then, the high-order connectivity [5] are captured by coming from the KG. In this paper, we have the following points that distinguish us from previous work:

Quality Knowledge Graph Construction. Many works [7, 13–15] used unsupervised keyword extraction algorithms or topic models to obtain keywords from the abstract of a paper. These keywords are used to query external knowledge bases to build the KG. However, the keywords extracted in this way do not represent the core information of the paper properly. This may blindly link to off-topic worthless triplets, thereby 'diluting' the meaningful high-order connectivity in the constructed KG.

Optimization of Information Aggregation Mechanisms to Suit Paper Recommendation Scenarios

Connectivity Importance of Entities. Previous work [5, 6, 8] did not consider the connectivity of entities when aggregating neighborhood information. For example, KGAT calculated attention weights based only on the distance between entities in the KG embedding (KGE) space [12]. Figure 1 shows an example where P_1 will be biased towards the less info-dense region A_2 during training. Instead, P_1 should prioritize information from A_1, as it implies more messaging paths to user and paper nodes.

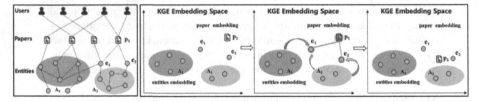

Fig. 1. An example showing the impact of different connectivity importance of neighbors

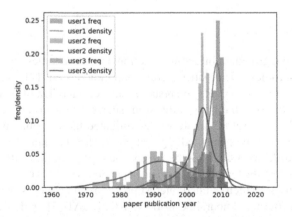

Fig. 2. Different time preference distributions exist for different users

Distribution of Users' Time Preference for Papers. We found that most users had a distinct time preference in their paper interactions which can be shown in Fig. 2. In this paper, we call this the locality of time preferences: papers corresponding to a specific interest of the user are contracted within a certain time window.

Enriching the Semantic Links on the User Side with Additional Information. Restricted by privacy policies, most datasets remove information about users. As a result, most previous work has given up on mining potential linking relationships on the user side and instead focused only on extending the KG on the paper side [2].

Due to the above problems, we propose MVKAN, a paper recommendation framework based on knowledge graph construction and a Multi-View Knowledge-aware Attentive Network. Our main contributions include:

1. We define and extract the keynote pattern of a paper to construct KG, and design experiments to verify that it can better support downstream recommendation tasks.
2. We redesigned the graph attention mechanism to guide the computation of graph attention from multiple views such as the connectivity importance of entities and the temporal preferences of users.
3. We capture the similarity of users' interests from tags and abstract them as short-cut links on the user side. This allows users to merge information from similar others, thus 'short-circuiting' the capturing of high-order connectivity.

2 Problem Formulation

User-Paper Interaction Bipartite Graph. The paper recommendation involves two main sides, the set of users $U = \{u_1, \ldots, u_n\}$ and the set of papers $P = \{p_1, \ldots, p_n\}$. We model the user-paper interaction as a bipartite graph $G_1 = \{(u, y_{up}, p) | u \in U, p \in P)\}$, when $y_{up} = 1$ indicates a *interact* edge between user node u and paper node p, otherwise $y_{up} = 0$.

Keynote Knowledge Graph. We extract the core information of the paper from the abstract, organize it into keynote triplets (which will be defined in detail in Scct. 3.1), and link it to an external knowledge base to build a knowledge graph. We refer to this as the keynote knowledge graph $G_2 = \{V, R, K\}$, where $V = \{v_1, \ldots, v_n\}$ is the set of entities, $R = \{r_1, \ldots, r_n\}$ is the set of relations, and K is the set of knowledge, where the knowledge is represented by the symbolic triplet (h, r, t), where $h, t \in V, r \in R$.

Collaborative Knowledge Graph (CKG). We will construct the paper-entity alignment set A in Sect. 3.1 and use it to link G_1 and G_2, thus gathering into a collaborative knowledge graph $G = \{G_1, A, G_2\}$, which integrates users' interactions and knowledge derived from the paper side [5].

Top-K Paper Recommendation Task. Our goal is to generate a Top-K list $P_u^k = \{p_1, \ldots, p_k\}$ of paper recommendations for $u \in U$. We need to train the model to learn a functional mapping $F(u, p | \Theta, G, u \in U, p \in P)$ from U to P, where Θ is all trainable parameters. F will be used to predict the probability $\hat{y}_{u,p} \in [0, 1]$ that u will click on p. Sorted by $\hat{y}_{u,p}$, the model can compute P_u^k for each user.

3 The Proposed Framework

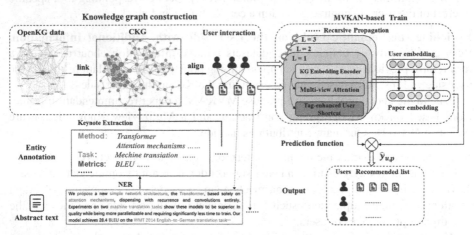

Fig. 3. Overview of the proposed framework

Figure 3 shows the framework of our method, which consist of two main components: (1) Knowledge graph construction. It describes the details of constructing G_2 based on keynote extraction and how it can be merged with G_1 into a CKG. (2) The Multi-View Knowledge-aware Attentive layer (MVKAN layer). It is also composed of three parts. The first is a KGE-based encoder, followed by a multi-view graph attention mechanism and short-cut links for users based on tag similarity. We will then describe how to combine the output of multiple MVKAN layers to calculate the final representation of the user and paper, thus for the paper recommendation.

3.1 Knowledge Graph Construction

Keynote Extraction. Inspired by [27], We summarized 5 triplets patterns in Table 1 to model the keynote of a paper, which is more representative of the paper and the points of interest that users focus on. The keynote relates to 4 types of entities: Task (subject area, task proposition, etc.), Method (the method used/proposed), Measure (metrics, benchmark), and Material (environment, dataset, equipment).

The core of the task is to capture the semantics corresponding to the above triplets in the abstract text and extract the tail entities. We have summarized trigger words in Table 1 by reading extensive papers. With trigger words and their extended word forms, it is possible to converge the scope of the keynote extraction to the key sentences, from which entities can be extracted to populate the predefined triplets.

We transformed the entity extraction into a NER task [25]. We first applied BIO to annotate parts of the abstract data, using four types of entities as labels. We then train the BERT [21] + LSTM + CRF model for automatic annotation and organize them into predefined triplets. They will be linked to XLORE [20] for further semantic additions and finally complemented into G_2. For papers with few extracted entities or linking failure,

Table 1. Keynote triplets definitions and Partial trigger word prototypes

Triplets	Trigger words
(Paper, review, Task)	review, survey, summarize, overview
(Paper, solved, Task)	solve, problem, focus, describe, present
(Paper, design, Method)	evaluate, model, benchmark, dataset
(Paper, used, Measure)	use, introduce, build, propose, design
(Paper, based, Material)	...

we also extracted concept terms from other sentences in the abstract using unsupervised methods as additional linking aids.

Collaborative Knowledge Graph Construction. We define paper-entity alignment set $A = \{(p, align, v)|p \in P, v \in G_2\}$. For the unsupervised method, v is the entity matched by unsupervised keywords, and *align* is the keyword mapping relation; for the keynote extraction, v is the extracted keynote entity, and *align* is the relations in the keynote triplets we defined. By linking G_1 and G_2 through A, we can obtain the collaborative knowledge graph $G = \{(h, r, t)|(h, t) \in V', r \in R'\}$, where $V' = V \cup U \cup P$ and $R' = R \cup \{interact\} \cup \{align\}$. To ensure the quality of G, we filtered the entities and relations with frequencies less than 10 and removed the associated triplets. We will also compare the impact of the KG constructed by traditional methods (unsupervised extraction only) on recommendation results in Sect. 4.3.

3.2 Multi-view Knowledge-Aware Attentive Layers

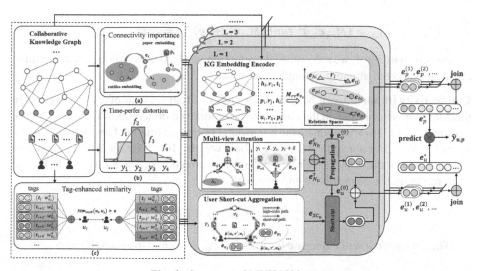

Fig. 4. Structure of MVKAN layers

Next, we will present how to combine the constructed G with MVKAN layers in Fig. 4. This is mainly achieved by the following modules:

KGE-Based Encoder. We use a KGE-based encoder [12] to encode the structural information of G into the representation e_v of entity v. It maps the symbolic triplet (h, r, t) to a distributed vector pair (e_h, e_r, e_t). The embedding model along the translational distance idea [12] usually sets the scoring function $g(h, r, t)$ for (h, r, t). Thereby, the loss function can be obtained according to the paired loss pattern:

$$L_{KGE} = \sum_{(h,r,t)\in K} \sum_{(h',r,t')\in K'} \varphi[\gamma + g(h, r, t) - g(h', r, t')]_+ \tag{1}$$

where K' is the negative samples set; φ is the transition function and γ is the hyperparameters of the KGE model. We use the TransD [22] for encoding:

$$e_h = M_{rh}e_{h_s} = \left(e_{r_p}e_{h_p}^\top + I\right)e_{h_s}, e_t = M_{rt}e_{t_s} = \left(e_{r_p}e_{t_p}^\top + I\right)e_{t_s} \tag{2}$$

$$g(h, r, t) = -\|e_h + e_r - e_t\|_2^2 \tag{3}$$

This means that each object v in a triplet is represented by the semantic vector e_{v_s} and the projection vector e_{v_p}. I is the unit matrix. For entity h, whose 1-hop neighborhood horizons [6] is $N_h = \{t|(h, r, t) \in K\}$, the information e_{N_h} passed from it can be obtained as follows:

$$e_{N_h} = \sum_{(h,r,t)\in N_h} \text{softmax}\left(\pi_{h,t}^r\right) e_t = \sum_{(h,r,t)\in N_h} \tilde{\pi}_{h,t}^r e_t \tag{4}$$

where $\tilde{\pi}_{h,t}^r$ is obtained by normalizing $\pi_{h,t}^r$ in N_h. $\pi_{h,t}^r$ is the attention weight [5], which measures the importance of neighbors only by the distance in the KGE space:

$$\pi_{h,t}^r = (M_{rh}e_h)^\top tanh(M_{rt}e_t + e_r) \xrightarrow{simplified} e_h \cdot e_t \tag{5}$$

As shown in Fig. 4, we redesigned the attention mechanism to guide the computation and aggregation of attention through three views:

Connectivity Importance of Neighboring Nodes. From the view of entities, we design the attention factor based on the connectivity importance of neighboring nodes. We use random walk [23] to approximate the k-hop neighborhood horizon of entity h (we set k = 4), called $N_h^{rw_k}$. For a neighbor t_i, its PageRank weight [16] $PR(t_i)$ in $N_h^{rw_k}$ is calculated as a measure for the importance of the potential information region it links for h. We use $c_h^{t_i}$ to represent the connectivity importance of node t_i in $N_h^{rw_k}$:

$$c_h^{t_i} = PR_{N_h^{rw_k}}(t_i), t_i \in N_h \tag{6}$$

Thereby, node h will aggregate information $e_{N_h}^c$ from its 1-hop neighbors:

$$e_{N_h}^c = \sum_{(h,r,t_i)\in N_h} \text{softmax}(c_h^{t_i}\pi_{h,t_i}^r)e_{t_i} = \sum_{(h,r,t_i)\in N_h} \tilde{\pi}_{ch,t_i}^r e_{t_i} \tag{7}$$

Locality of User's Time Preference. Inspired by [24], we completed the publication year of papers in the dataset from Google Scholar. For a paper-year series $Y = \{(p_1, y_1), \ldots, (p_n, y_n)\}$ in the interaction of user u, we define the factor $\tau_u^{p_i}$ to reflect the time preference of u's reading interest on paper p_i:

$$\tau_u^{p_i} = 1 + \frac{\sum_{y_m \in Y_h} f_{y_m} + \varepsilon}{\sum_{(p_i, y_n) \in Y} f_{y_n}} \tag{8}$$

where $Y_h = \left[max(min_y, y_i - \delta), min(max_y, y_i + \delta) \right]$ is the time preference locality domain of p_i; f_{y_i} represents the frequency of y_i, δ is the locality threshold we defined, and ε is the smoothing factor. It degenerates to a uniform weight when δ increases to infinity or the time distribution tends to be average. Here we take $\delta = 1$, $\varepsilon = 1$. For user u, $\tau_u^{p_i}$ will be involved in the computation of the information $e_{N_u}^\tau$ from N_u:

$$e_{N_u}^\tau = \sum_{(u,r,p_i) \in N_u} \text{softmax}(\tau_u^{p_i} \pi_{u,p_i}^r) e_{p_i} = \sum_{(u,r,p_i) \in N_u} \tilde{\pi}_{\tau u, p_i}^r e_{p_i} \tag{9}$$

Short-Cut Links for Users Based on Tag Similarity. Unlike the tags of commodity recommendation (e.g., *funny, tasty, expensive*), the tags assigned to papers by users (e.g., *bayesian-network, python-tools*) naturally imply the user's interest or motivation, not just the perceptual rating. Therefore, we can mine the user's interest and potential connections from the tags. We refer to text similarity to calculate the semantic similarity between users' tag lists. All tags typed by user u form the series $S_u = \{(t_1, f_1), \ldots, (t_n, f_n)\}$, where t_n is the nth tag and f_n is the frequency of t_n in the tag list of u. For all tags, the distribution $D = \{(t_1, d_1), \ldots, (t_n, d_n)\}$ can be obtained relative to the user set U. d_n is the number of times that t_n appears in the series of tags for different users. Thus, the weight of t_i for u can be represented based on TF-IDF:

$$w_u^i = \frac{f_i}{\sum_{k=1}^n f_i} \times log\left(\frac{m}{d_i + 1}\right) \tag{10}$$

Then we can obtain series $S_u' = \{(t_1, w_u^1), \ldots, (t_n, w_u^n)\}$, where w_u^n measures the importance of t_n on the interest composition of u. Assuming that S_u' is the "text" of u's reading interest expanded in the tag dimension, we use the pre-trained model [21] to encode the vector v_i for each tag, and represent the user vector v_u as a weighted average of S_u'. We use Cosine similarity $\theta(u_i, u_j)$ to represents the similarity of the interests between u_i and u_j in the tag dimension. When two users have similar interests on θ, they can provide potential recommendation value to each other. As shown in Fig. 4(c), we create a virtual two-way edge r' for users whose similarity is above the threshold α. Then we define the tag similarity horizon $N_{u_i}^{SC} = \{u_j | \theta(u_i, u_j) \geq \alpha\}$ for user u_i. Essentially, this is equal to opening a short-cut link pathway, which can quickly spread the information $e_{SC_{u_i}}$ passed by potential semantic neighbors for u_i:

$$e_{SC_{u_i}} = \sum_{u_j \in N_{u_i}^{SC}} \tilde{\mu}\left(u_i, r', u_j\right) e_{u_j} = \sum_{u_j \in N_{u_i}^{SC}} \text{softmax}(\theta(u_i, u_j)) e_{u_j} \tag{11}$$

3.3 Information Aggregation and High-Order Propagation

In particular, for user u, we need to aggregate the structural information $e_{N_u}^{\tau}$ from the paper side (G_2) and the short-cut information e_{SC_u} from the user side during training:

$$e_{N_u} = f_{agg}\left(e_{N_u}^{\tau}, e_{SC_u}\right) = \text{LeakyReLU}\left(W\left(e_{N_u}^{\tau} + e_{SC_u}\right)\right) \tag{12}$$

where $W \in R^{d' \times d}$ is the learnable weight matrix to distill the effective information for aggregation and d is the transformation size. Thereby, for any entity $v \in V'$ in CKG, we use the Bi-Interaction [5] aggregator to obtain its 1-layer representation $e_v^{(1)}$:

$$
\begin{aligned}
e_v^{(1)} &= f_{Bi-Interaction}\left(e_v^{(0)}, e_{N_v}^{(0)}\right) \\
&= \text{LeakyReLU}\left(W_a\left(e_v^{(0)} + e_{N_v}^{(0)}\right)\right) + \text{LeakyReLU}\left(W_m\left(e_v^{(0)} \odot e_{N_v}^{(0)}\right)\right)
\end{aligned} \tag{13}
$$

here $e_{N_v}^{(0)}$ is the neighbor domain information obtained by entity v at the initial layer of aggregation, $W_a, W_m \in R^{d' \times d}$ is the trainable weight matrix, and $\odot$ is the element-wise product. We use a recursive approach [5, 6] to propagate high-order connectivity information and obtain the representation of entities in different layers by aggregating information L times. For an entity v in the graph, its l-layer is represented as:

$$e_v^{(l)} = f_{Bi-Interaction}\left(e_v^{(l-1)}, e_{N_v}^{(l-1)}\right) \tag{14}$$

where $e_{N_v}^{(l)}$ of layer l depends on $e_v^{(l-1)}$ and $e_{N_v}^{(l-1)}$ of layer $l-1$, which allows the model to gradually aggregate information of the remote nodes on the multi-hop path. It is worth noting that the short-cut links we created play a " reduced order" in this process. It helps the model learn higher-order connectivity information by exploring potential short-cut channels.

3.4 Prediction and Optimization

After aggregating the L-layer MVKAN information, the model will output the representations of user u and paper p at different layers. We need to join them into final representations as inputs for the model predictions:

$$e_u^* = f_{join}\left(e_u^{(0)}, e_u^{(1)}, \ldots, e_u^{(L)}\right), e_p^* = f_{join}\left(e_p^{(0)}, e_p^{(1)}, \ldots, e_p^{(L)}\right) \tag{15}$$

Here we use the concatenation operation [5] to compute e_u^* and e_u^*. After that, the probability that u is interested in p is calculated using inner product:

$$\hat{y}_{u,p} = e_u^{*\top} \cdot e_p^* \tag{16}$$

For the user-paper interaction, we use BPR [9] to build the loss function as follows:

$$L_{UP} = \sum\nolimits_{(u,p,p') \in O} -\ln \sigma\left(\hat{y}_{u,p} - \hat{y}_{u,p'}\right) \tag{17}$$

where $\sigma(\cdot)$ is the sigmoid function and O defined as: $O = \{(u, p, p')|(u, p) \in O^+, (u, p') \in O^-\}$. O^+ is the train set and O^- is the negative sample set. We generate negative samples from a random sample of papers that users do not interact with. Finally, the following joint loss function will be used to optimize the entire model:

$$L_{total} = L_{KGE} + L_{UP} + \lambda \|\Theta\|_2^2 \tag{18}$$

the last term of the above equation is used to prevent overfitting.

4 Experiments

4.1 Experiment Settings

Dataset. Citeulike-a is a real dataset from the paper-sharing site CiteULike [26]. The dataset contains data on papers, user-paper interactions, and tags collected from CiteULike. We randomly divide the dataset into train, valid, and test sets in the ratio of 7:1:2. Table 2 shows the statistics of our dataset. We named the CKG constructed using unsupervised extraction of keywords as citeUnsup and the CKG built by keynote extraction as citeKExtra. (To be fair, We try to make both sides equal in size).

Table 2. The statistics of datasets.

Dataset	Knowledge Graph			User-Item Interaction				
	entities	relations	triplets	Users	Items	Inters	Tags	Sparsity
citeUnsup	42331	26	303912	5551	16980	204987	17951	99.78%
citeKNote	43528	29	325703					

Metrics and Parameter Setting. We selected the common metrics Precision, Recall, and NDCG to evaluate the proposed model in Table 3. In the KGE, the embedding dimension is set to 100 for entities and 64 for relations. We set the batch size to 2048 for KGE and 1024 for BPR loss. For short-cut links, we take the similarity threshold $\alpha = 0.7$. All data distributions or calculations used for training are obtained from the train set only to prevent data leakage. We train L_{KGE} and L_{UP} alternately, using mini-batch Adam Optimizer and Xavier initializer.

4.2 Experiment Results and Analysis

The previous recommendation algorithms can be divided into two categories: not using knowledge graph and using knowledge graph enhancement. We select models from these two categories in recent years as baseline. Table 3 shows the results of the comparison experiments and we can draw the following conclusions:

Table 3. Performance Comparison.

Models		Prec@K		Recall@K		NDCG@K	
		K = 20	K = 40	K = 20	K = 40	K = 20	K = 40
Non-KG	BPRMF [9]	0.0994	0.0735	0.1889	0.2705	0.1765	0.2050
	NFM [10]	0.1164	0.0837	0.2296	0.3170	0.2157	0.2461
	TGCN [8]	0.1298	0.1064	0.2456	0.3427	0.2291	0.2693
KG-enhanced	CKE [11]	0.1192	0.0863	0.2318	0.3243	0.2184	0.2506
	CFKG [17]	0.1101	0.0818	0.2214	0.3058	0.2103	0.2421
	RippleNet [4]	0.1254	0.0985	0.2373	0.3305	0.2253	0.2622
	KGAT [5]	0.1282	0.1002	0.2397	0.3384	0.2272	0.2669
Proposed	MVKAN	**0.1344**	**0.1127**	**0.2485**	**0.3497**	**0.2356**	**0.2781**

- Our proposed model outperforms the baselines in all metrics. RippleNet and KGAT outperform CKE and CFKG, which shows the importance of mining multiple information in CKG. The results also show that the design of the MVKAN layer can model the representation of users and paper nodes more effectively than the interest propagation model of RippleNet and the indiscriminate attention mechanism of KGAT, thus enhancing the relevance of the recommendation results.
- In most cases, KG-enhanced models outperform non-KG models, but this is not absolute. This challenges the ability of the model to extract and mine information from the KG. TGCN is the SOTA of tag-awar recommendation models, which outperforms CKE, CFKG, and RippleNet in all three metrics. However, since TGCN only constructs the tag-user-item graph without integrating other effective knowledge to upgrade into a knowledge graph, it misses potential higher-order connectivity links, and thus it performs lower than MVKAN.

4.3 Comparison of Ablation Experiments

Comparison of the Effect of KG Construction Methods. We experimented separately to compare the effects of the CKGs constructed by the two approaches on the recommendation results. In addition, we design the high-order connectivity metric HO_u^k. It represents the probability of visiting other user nodes by performing a k-step random walk in $N_u^{rw_k}$ (Sect. 3.2) from the node u. The average value $\overline{HO^k}$ is obtained for all user nodes, which can reflect the effective high-order connectivity in the CKG.

The results are given in Table 4. From the high-order connectivity metrics we defined, citeKExtra significantly outperforms citeUnsup, thus verifying that using the keynote extraction approach to construct the KG can obtain more effective high-order connectivity paths. This also verified the performance improvement of the recommendation result metrics from the side and enhanced the interpretability of the model.

Effect of Different Components on MVKAN. To verify the impact of different components in MVKAN, we set up 3 variants of MVKAN for experiments on the citeTExtra.

Table 4. Performance comparison between different KG construction.

Dataset	HO^k		Prec@20	Recall@20	NDCG@20
	k = 3	k = 4			
citeUnsup	0.4439	0.4067	0.1292	0.2413	0.2308
citeKExtra	0.4923	0.4652	0.1344	0.2485	0.2356

w/o CI means removing the neighbor connectivity importance attention, w/o TP means removing the time preference attention mechanism, and w/o SC removes the short-cut mechanism. The experimental results are shown in Table 5.

The results show a decrease in performance for all three variants compared to MVKAN, which verifies the validity of all parts of the model. Among them, the short-cut mechanism contributes the most to the performance because it helps to capture more potential high-order connectivity. And the lowest contribution is made by time preference attention, possibly because the distribution of time preferences of some users does not follow the locality, but tends to be average or dispersed, thus degrading to undifferentiated weights.

Table 5. Effect of MVKAN components

Variants	Prec@20	Recall@20	NDCG@20
w/o CI	0.1305	0.2246	0.2318
w/o TP	0.1319	0.2463	0.2331
w/o SC	0.1292	0.2412	0.2292

5 Conclusion and Future Work

In this paper, we construct a collaborative knowledge graph based on the keynote of papers. We design a knowledge-aware attention framework for the paper recommendation, MVKAN, which expands the graph attention and information aggregation mechanism in terms of the connectivity importance of neighboring nodes, the time-preference locality of users, and the short-cut link of users. Experiments on public datasets show that our model can effectively improve the quality of recommendation. Finally, we analyze the impact of each component in MVKAN through ablation experiments. In future work, we will further refine the accuracy of graph modeling and explore the impact of hard-negative sample mining algorithms on model performance.

Acknowledgments. This research was funded by the National Key Research and Development Program of China (No. 2019YFB2101600).

References

1. Guo, Q., Zhuang, F., Qin, C., et al.: A survey on knowledge graph-based recommender systems. IEEE Trans. Knowl. Data Eng. **34**(8), 3549–3568 (2020)
2. Kreutz, C.K., Schenkel, R.: Scientific paper recommendation systems: a literature review of recent publications. Int. J. Digit. Libr. **23**(4), 335–369 (2022)
3. Gao, Y., Li, Y.F., Lin, Y., et al.: Deep learning on knowledge graph for recommender system: a survey. arXiv preprint arXiv:2004.00387 (2020)
4. Wang, H., Zhang, F., Wang, J., et al.: RippleNet: propagating user preferences on the knowledge graph for recommender systems. In: Proceedings of the 27th ACM International Conference on Information and Knowledge Management, pp. 417–426 (2018)
5. Wang, X., He, X., Cao, Y., et al:. KGAT: knowledge graph attention network for recommendation. In: Proceedings of the 25th ACM SIGKDD International Conference on Knowledge Discovery & Data Mining, pp. 950–958 (2019)
6. Wang, H., Zhao, M., Xie, X., et al.: Knowledge graph convolutional networks for recommender systems. In: The World Wide Web Conference, pp. 3307–3313 (2019)
7. Ali, Z., Qi, G., Muhammad, K., et al.: Paper recommendation based on heterogeneous network embedding. Knowl. Based Syst. **210**, 106438 (2020)
8. Chen, B., Guo, W., Tang, R., et al.: TGCN: tag graph convolutional network for tag-aware recommendation. In: Proceedings of the 29th ACM International Conference on Information & Knowledge Management, pp. 155–164 (2020)
9. Rendle, S., Freudenthaler, C., Gantner, Z., et al.: BPR: Bayesian personalized ranking from implicit feedback. arXiv preprint arXiv:1205.2618 (2012)
10. He, X., Chua, T.S.: Neural factorization machines for sparse predictive analytics. In: Proceedings of the 40th International ACM SIGIR conference on Research and Development in Information Retrieval, pp. 355–364 (2017)
11. Zhang, F., Yuan, N.J., Lian, D., et al.: Collaborative knowledge base embedding for recommender systems. In: Proceedings of the 22nd ACM SIGKDD International Conference on Knowledge Discovery and Data Mining, pp. 353–362 (2016)
12. Wang, Q., Mao, Z., Wang, B., et al.: Knowledge graph embedding: a survey of approaches and applications. IEEE Trans. Knowl. Data Eng. **29**(12), 2724–2743 (2017)
13. Renuka, S., Raj Kiran, G.S.S., Rohit, P.: An unsupervised content-based article recommendation system using natural language processing. In: Jeena Jacob, I., Kolandapalayam Shanmugam, S., Piramuthu, S., Falkowski-Gilski, P. (eds.) Data Intelligence and Cognitive Informatics. AIS, pp. 165–180. Springer, Singapore (2021). https://doi.org/10.1007/978-981-15-8530-2_13
14. Wang, X., Xu, H., et al.: Scholarly paper recommendation via related path analysis in knowledge graph. In: International Conference on Service Science, pp. 36–43. IEEE (2020)
15. Tang, H., Liu, B., Qian, J.: Content-based and knowledge graph-based paper recommendation: Exploring user preferences with the knowledge graphs for scientific paper recommendation. Concur. Comput. Pract. Exp. **33**(13), e6227 (2021)
16. Manrique, R., Marino, O.: Knowledge graph-based weighting strategies for a scholarly paper recommendation scenario. KaRS@ RecSys **7**, 1–4 (2018)
17. Ai, Q., Azizi, V., Chen, X., et al.: Learning heterogeneous knowledge base embeddings for explainable recommendation. Algorithms **11**(9), 137 (2018)
18. Wang, B., Weng, Z., Wang, Y.: A novel paper recommendation method empowered by knowledge graph: for research beginners. arXiv preprint arXiv:2103.08819 (2021)
19. Li, X., Chen, Y., Pettit, B., et al.: Personalised reranking of paper recommendations using paper content and user behavior. ACM Trans. Inf. Syst. (TOIS) **37**(3), 1–23 (2019)

20. Jin, H., Li, C., Zhang, J., et al.: XLORE2: large-scale cross-lingual knowledge graph construction and application. Data Intell. **1**(1), 77–98 (2019)
21. Liu, X., Yin, D., Zheng, J., et al.: OAG-BERT: towards a unified backbone language model for academic knowledge services. In: Proceedings of the 28th ACM SIGKDD Conference on Knowledge Discovery and Data Mining, pp. 3418–3428 (2022)
22. Ji, G., He, S., Xu, L., et al.: Knowledge graph embedding via dynamic mapping matrix. In: Proceedings of the 53rd Annual Meeting of the Association for Computational Linguistics and the 7th International Joint Conference on Natural Language Processing (Volume 1: Long Papers), pp. 687–696 (2015)
23. Perozzi, B., Al-Rfou, R., Skiena, S.: DeepWalk: online learning of social representations. In: Proceedings of the 20th ACM SIGKDD International Conference on Knowledge Discovery and Data Mining, pp. 701–710 (2014)
24. Hao, L., Liu, S., Pan, L.: Paper recommendation based on author-paper interest and graph structure. In: 2021 IEEE 24th International Conference on Computer Supported Cooperative Work in Design (CSCWD), pp. 256–261. IEEE (2021)
25. Li, J., Sun, A., Han, J., et al.: A survey on deep learning for named entity recognition. IEEE Trans. Knowl. Data Eng. **34**(1), 50–70 (2020)
26. Wang, H., Chen, B., Li, W.J.: Collaborative topic regression with social regularization for tag recommendation. In: 23th International Joint Conference on Artificial Intelligence (2013)
27. Wu, J., Zhu, X., Zhang, C., Hu, Z.: Event-centric tourism knowledge graph—a case study of Hainan. In: Li, G., Shen, H., Yuan, Y., Wang, X., Liu, H., Zhao, X. (eds.) KSEM 2020. LNCS, vol. 12274, pp. 3–15. Springer, Cham (2020). https://doi.org/10.1007/978-3-030-551 30-8_1

Biological Tissue Sections Instance Segmentation Based on Active Learning

Yanan lv[1,2], Haoze Jia[1,2], Haoran Chen[2,3], Xi Chen[1(✉)], Guodong Sun[1], and Hua Han[3,4(✉)]

[1] Institute of Automation, Chinese Academy of Sciences, Beijing 100190, China
xi.chen@ia.ac.cn
[2] School of Artificial Intelligence, University of Chinese Academy of Sciences, Beijing 100190, China
[3] State Key Laboratory of Multimodal Artificial Intelligence Systems, Institute of Automation, Chinese Academy of Sciences, Beijing 100190, China
hua.han@ia.ac.cn
[4] School of Future Technology, University of Chinese Academy of Sciences, Beijing 100190, China

Abstract. Precisely identifying the locations of biological tissue sections on the wafer is the basis for microscopy imaging. However, the sections made of different biological tissues are different in shape. Therefore, the instance segmentation network trained in the existing dataset may not be suitable for detecting new sections, and the cost of making the new dataset is high. Therefore, this paper proposes an active learning algorithm for biological tissue section instance segmentation. The algorithm can achieve better results with only a few images for training when facing the new segmentation task of biological tissue sections. The algorithm adds a loss prediction module on the instance segmentation network, weights the uncertainty of the instance segmentation mask by the posterior category probability, and finally calculates the value of the sample. Then, we select the sample with the most significant value as the training set, so we can only label a small number of samples, and the network can achieve the expected performance. The algorithm is robust to different shapes of tissue sections and can be applied to various complex scenes to segment tissue sections automatically. Furthermore, experiments show that only labeling 30% samples of the whole training set makes the network achieve the expected performance.

Keywords: Instance Segmentation · biological tissue section · Active Learning

1 Introduction

Serial section electron microscopy [1, 2] is widely used to study biological tissues at the microscopic scale [3–5]. In the imaging process of the serial sectioning microscope, the wafer with the slices attached is first imaged to obtain the navigation map of the wafer. As shown in Fig. 1a, each wafer can contain hundreds of tissue slices. Then mark the location of each slice on the navigation map, and calculate the mapping between the pixel

B. Luo et al. (Eds.): ICONIP 2023, CCIS 1964, pp. 16–27, 2024.
https://doi.org/10.1007/978-981-99-8141-0_2

coordinates of the wafer navigation map and the physical coordinates of the microscope stage. Through this mapping, the corresponding relationship between the pixels of the navigation map and the physical coordinate system of the stage is established. Finally, the absolute coordinates of each slice on the wafer on the stage are obtained.

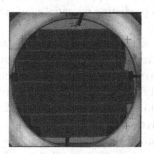 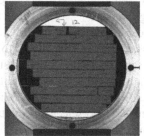 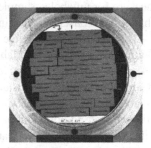

(a) mouse cortical tissue sections (b) zebrafish brain tissue sections (c) mouse testicular tissue sections

Fig. 1. The wafers of different biological tissue sections. (a) Mouse cortical tissue sections, (b) zebrafish brain tissue sections, (c) mouse testicular tissue sections.

With the continuous development of microscopic imaging technology and the gradual deepening of scientific research, researchers have an increasing demand for the observation of microscopic and mesoscopic structures, and the volume of observation samples is also increasing [6–8]. Automatic and accurate segmentation of sections on the wafer can effectively speed up the microscope imaging process. Our work in this paper is an improvement and continuation of the work of Sun et al. [9]. Sun's Frequency-Aware Instance Segmentation framework (FANet) uses two-dimensional discrete cosine transform and a multi-frequency channel attention mechanism to realize the instance segmentation of sections on the wafer.

However, the sections on wafers from different biological tissues are quite different (Fig. 1), and the instance segmentation network based on deep learning is primarily data-driven, relying on the support of its vast data set from the application field. Therefore, instance segmentation networks trained on existing tissue section datasets may not be suitable for new biological tissue section segmentation. When performing automatic segmentation tasks for new biological tissue sections, the dataset will likely need to be recreated. Active learning [10–12] uses particular strategies to find the most valuable data in the sample data that has yet to be labeled and then incorporates the data into the training set, to obtain the expected model processing effect with less training data.

The research on active learning can be divided into uncertainty-based and distribution-based methods. The method based on uncertainty is done by measuring the uncertainty of the model's predictions on the unlabeled data points using various metrics, then selecting the most informative unlabeled data and adding them to the training set so that the neural network can quickly learn helpful information. Lewis et al. [13, 14] employed the posterior probability of prediction category as a criterion to select informative instances for active learning. Specifically, samples that exhibit similar posterior probabilities across different categories, with minimal discrimination among them, were sampled and added to the training set. Roth et al. [15] leveraged the difference between the posterior probabilities of the top two predicted categories with the highest prediction probability as a measurement method. When the two categories with the highest posterior probabilities are closely situated, the model is deemed insufficient to learn and discriminate such sample types, thus warranting their inclusion in the training set. Joshi et al. [16], Luo et al. [17], and Settles et al. [18] used entropy as an uncertainty measure for unlabeled samples. Settles et al. [19] adopted predicting potential gradient changes as a basis for selecting samples.

Distribution-based active learning selects data points with uncertain distributional characteristics from an unlabeled dataset. The goal is to reduce the labeling effort by selecting the most informative samples based on their distributional properties. Nguyen et al. [20] proposed a clustering algorithm to estimate the underlying distribution of unlabeled samples. Meanwhile, Elhamifar et al. [21], Guo et al., and Yang et al. [22] utilized discrete optimization methods to select informative samples based on the distributional characteristics learned from the data. Mac et al. [23] and Hasan et al. [24] took the context-aware method to select samples that can represent the global sample distribution by the distance between the samples and their surrounding samples. Sener et al. [25] defined the active learning method as the problem of core set selection. Yoo et al. [10] added a loss prediction module to predict the loss of the main network when inferring uncertain samples. They selected the sample with the greatest loss to be labeled and added to the training set.

In the present research, active learning is mainly focused on image classification [26–30] and object detection [31, 32] tasks, and few are used in the research of instance segmentation. The algorithm proposed in this paper introduces the loss prediction module to FANet. Simultaneously, the instance segmentation mask uncertainty weighted by the posterior category probability is calculated. Finally, the loss predicted by the loss prediction module is combined with the weighted mask uncertainty as the final value of samples, achieving the expected performance of the instance segmentation network with only a small number of images.

2 Methodology

To address the challenge encountered by instance segmentation networks in adapting to novel data and necessitating a substantial amount of manual effort to reconstruct datasets, we propose an active learning algorithm for biological tissue sections instance segmentation. The algorithm seeks to select the most effective samples for training instance segmentation networks, requiring only a small number of images for training to achieve the expected results. The active learning algorithm proposed in this paper is improved from two aspects. One is to add a loss prediction module to the main instance segmentation network to predict the loss of the main network. The second is to calculate the uncertainty of the instance segmentation mask by the posterior category probability. Finally, through the amalgamation of loss and uncertainty indicators, selectively extracting samples from the sample pool that have yet to be comprehensively learned by the network is executed, followed by their classification and assimilation into the training set for network refinement. Ultimately achieve the goal of making the instance segmentation network achieve the expected performance when only a small number of samples are used for training.

2.1 Loss Prediction Module

In the course of training, samples that have not been sufficiently learned by the model will yield output values that deviate substantially from the ground truth, leading to a pronounced increase in the associated loss. The structure of the loss prediction module is based on the method of the paper [10]. And the architecture of the loss prediction is shown in Fig. 2.

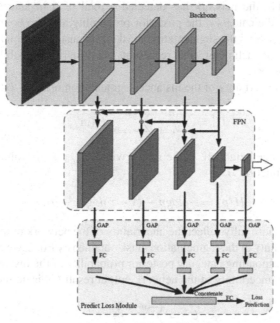

Fig. 2. The architecture of the loss prediction. Where GAP is Global Average Pooing, FC is Fully Connected.

The loss prediction module takes the features in the middle block of the main network as input, and outputs the prediction loss $\hat{l}$, the loss prediction module takes the training loss l of the main network as the true value, and calculates its training loss as $L_{loss} = MSE(\hat{l}, l)$. Finally, the training loss of the whole network is defined as

$$L = l + \lambda \cdot L_{loss}\left(\hat{l}, l\right) \tag{1}$$

where λ is a positive constant to balance the loss of main network and loss prediction module.

2.2 Uncertainty of Instance Segmentation Result

The uncertainty of the network output can reflect the network learning on the input samples, and a higher uncertainty indicates that the network has no confidence in the input and has not fully learned. In this paper, the quantification of network output uncertainty is derived through the confluence of segmentation mask and classification results, and by incorporating the utilization of instance segmentation mask uncertainty, which is weighted by category posterior probability, as a valuable metric for gauging sample uncertainty, the instance segmentation network is enabled to select appropriate samples effectively.

During the inference of the instance segmentation network, given an input image $x \in R^{H \times W}$, the network outputs results as $y = \{(y_{cls}^i, y_{prob}^i, y_{bbox}^i, y_{mask}^i)\}, i \in \{1, 2, \ldots, n\}$, where n means that the network has detected a total of n instances in the sample, $y_{cls}, y_{prob}, y_{bbox}$ is the category, the posterior probability and the bounding box of the instance. y_{mask} represents the segmentation mask of the instance, which is a normalized grayscale image, and each pixel value represents the posterior probability that the pixel belongs to the category.

We define the uncertainty of the instance segmentation mask as U^i

$$U^i = \sum_{u=0}^{H-1} \sum_{v=0}^{W-1} H(m_{u,v}^i) \tag{2}$$

where H is the image height, W is the image width, $m_{u,v}^i$ is the value of y_{mask}^i on the pixel (u, v). The function H is

$$H(p) = -plogp - (1-p)log(1-p). \tag{3}$$

In an effort to more fully reflect the uncertainty of the network output results, on the basis of the uncertainty of the segmentation mask obtained by Eq. 2, we consider comprehensively the corresponding category posterior probability of the instance segmentation result. Finally, the uncertainty of the network output result U^y is defined.

$$U^y = \sum_{i=0}^{n-1} \frac{1}{p^i} \cdot U^i \tag{4}$$

2.3 Active Learning Strategy for Instance Segmentation

Following the acquisition of the output generated by the loss prediction module and the uncertainty attributed to the instance segmentation result, the value of the input image x is subsequently derived, denoted as V_x.

$$V_x = \hat{l} + \eta \cdot U^y \tag{5}$$

where $\hat{l}$ is prediction loss of the loss prediction module, U^y is the uncertainty of the network output result, and η is a positive constant.

The process of the active learning strategy for instance segmentation network is as follows:

Algorithm: active learning strategy for instance segmentation

Input: S_n^0 (Sample pool with n unlabeled samples)

$\quad\quad\quad T_k^0$ (Initial training set with k labeled samples)

$\quad\quad\quad \Theta_{main}^0$ (Instance Segmentation network model)

$\quad\quad\quad \Theta_{loss}^0$ (Loss prediction module)

Initialization: $i \leftarrow 0$

While: the network achieves expected performance or the cost of labeling samples reaches threshold

$\quad\quad$ 1. using training set $T_{k\times(i+1)}^i$ to train Θ_{main}^i and Θ_{main}^i until them converge

$\quad\quad$ 2. Θ_{main}^i and Θ_{loss}^i infer all samples in $S_{n-k\times i}^i$ and obtain the value of all the samples

$V = \{V_0, V_1, ..., V_{n-k\times i}\}$

$\quad\quad$ 3. Sorting V in descending order

$\quad\quad$ 4. Label the first k samples from V, add them to the training set $T_{k\times(i+1)}^i$, and remove them from $S_{n-k\times i}^i$

$\quad\quad$ 5. $i \leftarrow i + 1$

Output: Θ_{main}^i

3 Experimental Results

3.1 Dataset

To train and evaluate the algorithm, we propose a dataset biological tissue sections of wafer as the benchmark. The wafer image of mouse testicular tissue was selected for the experiment, as shown in Fig. 1c. The sections were divided into three categories: incomplete section (Fig. 3a), broken section (Fig. 3b) and normal section (Fig. 3c). Crop the wafer images to (1024 × 1024), with 10% overlap between the sub images. After labeling the sub images with Labelme [33], it is divided into training set and validation set. The final training set contains 972 sub images, and the validation set contains 647 sub images.

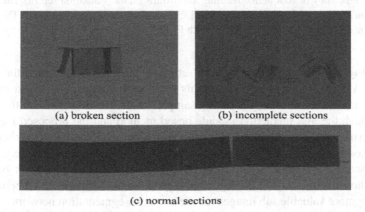

(a) broken section   (b) incomplete sections

(c) normal sections

Fig. 3. Different categories of mouse testicular tissue sections in dataset. (a) Incomplete section, (b) broken sections, (c) normal section.

3.2 Results

To evaluate the efficacy of the proposed algorithm, we have compared the result with the baseline method (random select), the entropy-based method [17] and the loss prediction module-based method [10]. Experiments are implemented in mmdetection [34]. The models are trained using stochastic gradient descent (SGD), all models are trained with one Tesla P40 GPU.

In the experiment, 972 sub images in the whole training set are used as the sample pool, and 6% of them are randomly selected as the initial training set to train the network. After network training converges, inference is performed on all sub images in the sample pool. Then, arrange all the sub images in descending order according to their value, and the top 6% sub images are selected to join the training set, form a new training set to continue training the network, and loop until the termination condition is reached. Finally, after each training session, the results of the instance segmentation network on the validation set are shown in Fig. 4. We use the average precision (AP) as the evaluation index in the experiment.

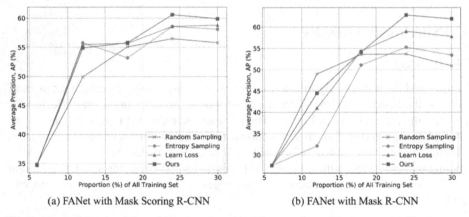

(a) FANet with Mask Scoring R-CNN   (b) FANet with Mask R-CNN

Fig. 4. The efficiency of four active learning algorithms on the validation set. (a) The instance segmentation network is Mask Scoring R-CNN [35] with FANet as its backbone, (b) The instance segmentation network is Mask R-CNN [36] with FANet as its backbone.

From Fig. 4, since the initial training set of all methods is the same, the average precision (AP) of all comparison algorithms on the validation set is almost the same. Setting the method of randomly selecting samples as the baseline method, the entropy based method, the loss prediction module based method and the proposed method all outperform the baseline method. Significantly, the proposed method performs better than other methods, especially when the training set gradually increases. Both experiments on Mask Scoring R-CNN [35] and Mask R-CNN [36] with FANet as their backbone network, show that the active learning method proposed in this paper has a better effect in selecting more valuable sub images for the instance segmentation network.

When active learning method is used to train the network, theoretically the ideal performance that the network can achieve is the result of training with all data in the entire sample pool. Table 1 shows the ideal performance and the results of the instance segmentation network after each training session. It can be seen that the method proposed in this paper only uses less than 30.0% sub images of the entire training set to train the instance segmentation network, and its average precision (AP) on the validation set is close to ideal performance. Only labeling 30.0% of the dataset can make the instance segmentation network achieve more than 91.0% performance, which indicates that the active learning algorithm proposed in this paper can effectively select sub images, and greatly saves the cost of human time and effort on data labeling.

Table 1. Comparison for average precision (AP) of four active learning methods on different proportions of all Training Set.

Method		AP (%) on Proportion of All Training Set					
		6	12	18	24	30	100
Mask Scoring R-CNN [35] with FANet	Random	34.9	49.9	55.1	56.5	55.8	63.5
	Entropy [17]	34.8	55.8	53.2	58.6	58.1	
	Learn Loss [10]	34.7	55.5	55.7	58.6	58.8	
	ours	34.8	54.9	55.8	60.6	59.9	
Mask R-CNN [36] with FANet	Random	27.6	49.0	53.6	53.7	50.9	67.7
	Entropy [17]	27.5	32.1	51.1	55.3	53.4	
	Learn Loss [10]	27.5	41.0	54.4	59.0	57.8	
	ours	27.5	44.5	54.2	62.8	61.9	

The results of each active learning method applied to instance segmentation networks on the validation set are shown in Fig. 5. It can be seen that whether it is based on Mask Scoring R-CNN or Mask R-CNN with FANet as its backbone, the network trained using the active learning method proposed in this paper can perform better on the tissue section. In the case of stick sections, the instance segmentation network is challenging to distinguish the boundaries between different sections, resulting in the spread of segmentation masks. This indicates that the network cannot learn enough information during training. The active learning algorithm proposed in this paper makes the training sets contain more information so the network can perform better.

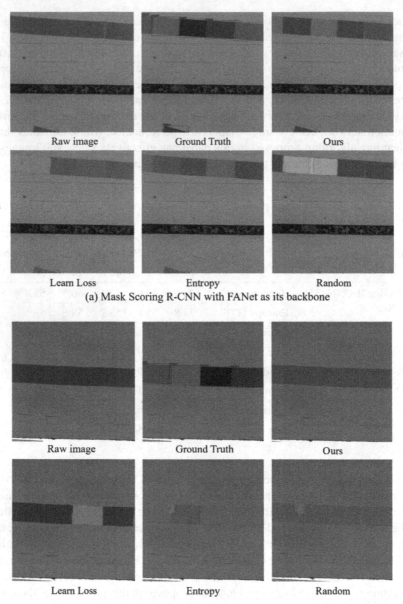

(a) Mask Scoring R-CNN with FANet as its backbone

(b) Mask R-CNN with FANet as its backbone

Fig. 5. The results of each active learning method applied to instance segmentation network on the validation set. (a) Mask Scoring R-CNN with FANet as its backbone, (b) Mask R-CNN with FANet as its backbone.

4 Conclusion

The neural network based on deep learning is mainly data-driven, and its excellent performance in application tasks cannot be separated from the support of an extensive data set. Due to the differences in sections made from different biological tissues, the instance segmentation network may not be suitable for new section detect. When conducting new automatic segmentation tasks on new biological tissue sections, remaking the dataset and retraining or fine-tuning the network using the new dataset may be necessary. However, the production of data sets requires a lot of human time and effort.

In the paper, we propose an active learning algorithm for biological tissue sections instance segmentation. The algorithm improves its effectiveness through two aspects. One is adding a loss prediction module on the instance segmentation network to predict a loss for the input network samples, thereby obtaining samples that are more helpful for network training. The second is weighting the uncertainty of the instance segmentation mask by the posterior category probability, and finally determining the most valuable images for learning. By selecting the sample with the most significant value as the training set, we can only label a small number of samples, and the network can achieve the expected performance.

Experimental results show that the algorithm proposed in this paper performs significantly better than other comparative algorithms on the validation set, especially when the training set is gradually increasing. And we only label 30.0% samples of the total training set to train the instance segmentation network, its average precision on the validation set is close to the ideal performance.

In conclusion, the active learning algorithm for biological tissue sections instance segmentation proposed in this paper can achieve better results with only a few images for training when facing the new segmentation task of biological tissue sections.

Funding. This work has been supported by the Scientific research instrument and equipment development project of Chinese Academy of Sciences [YJKYYQ20210022 to H.H.].

References

1. Harris, K.M., Perry, E., Bourne, J., et al.: Uniform serial sectioning for transmission electronmicroscopy. J. Neurosci. **26**(47), 12101–12103 (2006)
2. Hayworth, K.J., Morgan, J.L., Schalek, R., et al.: Imaging ATUM ultrathin section libraries with wafer mapper: a multi-scale approach to EM reconstruction of neural circuits. Front. Neural Circuits **8**, 68 (2014)
3. Shapson-Coe, A., Januszewski, M., Berger, D.R., et al.: A connectomic study of a petascale fragment of human cerebral cortex. BioRxiv (2021)
4. Li, P.H., Lindsey, L.F., Januszewski, M., et al.: Automated reconstruction of a serial-section EM drosophila brain with flood-filling networks and local realignment. Microsc. Micro Anal. **25**(S2), 1364–1365 (2019)
5. Vishwanathan, A., Ramirez, A.D., Wu, J., et al.: Modularity and neural coding from a brainstem synaptic wiring diagram. BioRxiv (2021)
6. Hildebrand, D.G.C., Cicconet, M., Torres, R.M., et al.: Whole-brain serial-section electron microscopy in larval zebrafish. Nature **545**(7654), 345–349 (2017)

7. Yin, W., Brittain, D., Borseth, J., et al.: A petascale automated imaging pipeline for mapping neuronal circuits with high-throughput transmission electron microscopy. Nat. Commun. **11**(1), 1–12 (2020)
8. Larsen, N.Y., Li, X., Tan, X., et al.: Cellular 3D-reconstruction and analysis in the human cerebral cortex using automatic serial sections. Commun. Biol. **4**(1), 1–15 (2021)
9. Sun, G., Wang, Z., Li, G., Han, H.: Robust frequency-aware instance segmentation for serial tissue sections. In: Wallraven, C., Liu, Q., Nagahara, H. (eds.) ACPR 2021. LNCS, vol. 13188, pp. 379–389. Springer, Cham (2022). https://doi.org/10.1007/978-3-031-02375-0_28
10. Yoo, D., Kweon, I.S.: Learning loss for active learning. In: Proceedings of the IEEE/CVF Conference on Computer Vision and Pattern Recognition, pp. 93–102 (2019)
11. Sinha, S., Ebrahimi, S., Darrell, T.: Variational adversarial active learning. In: Proceedings of the IEEE/CVF International Conference on Computer Vision, pp. 5972–5981 (2019)
12. Tran, T., Do, T.T., Reid, I., et al.: Bayesian generative active deep learning. In: International Conference on Machine Learning, pp. 6295–6304. PMLR (2019)
13. Lewis, D.D., Catlett, J.: Heterogeneous uncertainty sampling for supervised learning. In: Machine Learning Proceedings 1994, pp. 148–156. Elsevier (1994)
14. Lewis, D.D., Gale, W.A.: A sequential algorithm for training text classifiers. In: Croft, B.W., van Rijsbergen, C.J. (eds.) SIGIR 1994, pp. 3–12. Springer, London (1994). https://doi.org/10.1007/978-1-4471-2099-5_1
15. Roth, D., Small, K.: Margin-based active learning for structured output spaces. In: Fürnkranz, J., Scheffer, T., Spiliopoulou, M. (eds.) ECML 2006. LNCS, vol. 4212, pp. 413–424. Springer, Heidelberg (2006). https://doi.org/10.1007/11871842_40
16. Joshi, A.J., Porikli, F., Papanikolopoulos, N.: Multi-class active learning for image classification. In: 2009 IEEE Conference on Computer Vision and Pattern Recognition, pp. 2372–2379. IEEE (2009)
17. Luo, W., Schwing, A., Urtasun, R.: Latent structured active learning. In: Advances in Neural Information Processing Systems, 26 (2013)
18. Settles, B., Craven, M.: An analysis of active learning strategies for sequence labeling tasks. In: Proceedings of the 2008 Conference on Empirical Methods in Natural Language Processing, pp. 1070–1079 (2008)
19. Settles, B., Craven, M., Ray, S.: Multiple-instance active learning. In: Advances in Neural Information Processing Systems, 20 (2007)
20. Nguyen, H.T., Smeulders, A.: Active learning using pre-clustering. In: Proceedings of the Twenty-First International Conference on Machine Learning, p. 79 (2004)
21. Elhamifar, E., Sapiro, G., Yang, A., et al.: A convex optimization framework for active learning. In: Proceedings of the IEEE International Conference on Computer Vision, pp. 209–216 (2013)
22. Yang, B., Bender, G., Le, Q.V., et al.: CondConv: conditionally parameterized convolutions for efficient inference. In: Advances in Neural Information Processing Systems, 32 (2019)
23. Mac Aodha, O., Campbell, N.D., Kautz, J., et al.: Hierarchical subquery evaluation for active learning on a graph. In: Proceedings of the IEEE Conference on Computer Vision and Pattern Recognition, pp. 564–571 (2014)
24. Hasan, M., Roy-Chowdhury, A.K.: Context aware active learning of activity recognition models. In: Proceedings of the IEEE International Conference on Computer Vision, pp. 4543–4551 (2015)
25. Sener, O., Savarese, S.: Active learning for convolutional neural networks: a core-set approach. arXiv preprint arXiv:1708.00489 (2017)
26. Tang, Y.P., Huang, S.J.: Self-paced active learning: query the right thing at the right time. In: Proceedings of the AAAI Conference on Artificial Intelligence, vol. 33, pp. 5117–5124 (2019)

27. Beluch, W.H., Genewein, T., Nürnberger, A., et al.: The power of ensembles for active learning in image classification. In: Proceedings of the IEEE Conference on Computer Vision and Pattern Recognition, pp. 9368–9377 (2018)
28. Lin, L., Wang, K., Meng, D., et al.: Active self-paced learning for cost-effective and progressive face identification. IEEE Trans. Pattern Anal. Mach. Intell. **40**(1), 7–19 (2017)
29. Liu, Z.Y., Huang, S.J.: Active sampling for open-set classification without initial annotation. In: Proceedings of the AAAI Conference on Artificial Intelligence, vol. 33, pp. 4416–4423 (2019)
30. Liu, Z.Y., Li, S.Y., Chen, S., et al.: Uncertainty aware graph Gaussian process for semi-supervised learning. In: Proceedings of the AAAI Conference on Artificial Intelligence, vol. 34, pp. 4957–4964 (2020)
31. Aghdam, H.H., Gonzalez-Garcia, A., van de Weijer, J., et al.: Active learning for deep detection neural networks. In: Proceedings of the IEEE/CVF International Conference on Computer Vision, pp. 3672–3680 (2019)
32. Zhang, B., Li, L., Yang, S., et al.: State-relabeling adversarial active learning. In: Proceedings of the IEEE/CVF Conference on Computer Vision and Pattern Recognition, pp. 8756–8765 (2020)
33. Wada, K.: Labelme: Image Polygonal Annotation with Python. https://github.com/wkentaro/labelme. https://doi.org/10.5281/zenodo.5711226
34. Chen, K., Wang, J., Pang, J., et al.: Mmdetection: Open MMLab detection toolbox and benchmark. arXiv preprint arXiv:1906.07155 (2019)
35. Huang, Z., Huang, L., Gong, Y., et al.: Mask scoring R-CNN. In: Proceedings of the IEEE/CVF Conference on Computer Vision and Pattern Recognition, pp. 6409–6418 (2019)
36. He, K., Gkioxari, G., Dollár, P., et al.: Mask R-CNN. In: Proceedings of the IEEE International Conference on Computer Vision, pp. 2961–2969 (2017)

Research on Automatic Segmentation Algorithm of Brain Tumor Image Based on Multi-sequence Self-supervised Fusion in Complex Scenes

Guiqiang Zhang[1]([✉]), Jianting Shi[2], Wenqiang Liu[3]([✉]), Guifu Zhang[4], and Yuanhan He[4]

[1] Anhui Institute of Information Engineering, Wuhu 241009, China
zgq1451296146@163.com

[2] College of Computer and Software Engineering, Heilongjiang University of Science and Technology, Harbin 150020, China

[3] School of Information and Electronic Engineering, Zhejiang Gongshang University, Hangzhou 310018, China
lwq@zjgsu.edu.cn

[4] State Grid Anhui Ultra High Voltage Company, Hefei 230041, China

Abstract. Brain tumors play a crucial role in medical diagnosis and treatment planning. Extracting tumor information from MRI images is essential but can be challenging due to the limitations and intricacy of manual delineation. This paper presents a brain tumor image segmentation framework that addresses these challenges by leveraging multiple sequence information. The framework consists of encoder, decoder, and data fusion modules. The encoder incorporates Bi-ConvLSTM and Transformer models, enabling comprehensive utilization of both local and global details in each sequence. The decoder module employs a lightweight MLP architecture. Additionally, we propose a data fusion module that integrates self-supervised multi-sequence segmentation results. This module learns the weights of each sequence prediction result in an end-to-end manner, ensuring robust fusion results. Experimental validation on the BRATS 2018 dataset demonstrates the excellent performance of the proposed automatic segmentation framework for brain tumor images. Comparative analysis with other multi-sequence fusion segmentation models reveals that our framework achieves the highest Dice score in each region. To provide a more comprehensive background, it is important to highlight the significance of brain tumors in medical diagnosis and treatment planning. Brain tumors can have serious implications for patients, affecting their overall health and well-being. Accurate

This research was funded by Scientific Research Fund of Zhejiang Provincial Education Department, Grant/Award Number: Y202147323, National Natural Science Foundation of China under Grant NSFC-61803148, basic scientific research business cost project of Heilongjiang provincial undergraduate universities in 2020 (YJSCX2020-212HKD) and basic scientific research business cost project of Heilongjiang provincial undergraduate universities in 2022 (2022-KYYWF-0565).

B. Luo et al. (Eds.): ICONIP 2023, CCIS 1964, pp. 28–49, 2024.
https://doi.org/10.1007/978-981-99-8141-0_3

segmentation of brain tumors from MRI images is crucial for assessing tumor size, location, and characteristics, which in turn informs treatment decisions and prognosis. Currently, manual delineation of brain tumors from MRI images is a time-consuming and labor-intensive process prone to inter-observer variability. Automating this segmentation task using advanced image processing techniques can significantly improve efficiency and reliability.

Keywords: Medical Image Segmentation · brain tumor · Multiple sequences · self-supervised fusion · Trans-former

1 Introduction

In recent years, the rapid development of various imaging technologies has revolutionized the field of medical imaging, providing a crucial reference for clinical diagnosis and advancing scientific research. The effective utilization of medical images begins with the precise segmentation of tissues. Segmentation involves extracting the region of interest from the entire image, enabling doctors to focus on diagnostically significant areas while disregarding irrelevant regions. Accurate segmentation plays a pivotal role in facilitating doctors' analysis of medical conditions and reducing the risk of misdiagnosis. However, the current practice of manually delineating each slice in an MRI head image heavily relies on the subjective judgment of doctors and is a time-consuming task. Moreover, the repetitive nature of this high-intensity work of ten leads to fatigue induced errors.

To overcome these challenges, the field of computer vision has witnessed the emergence of automatic segmentation techniques for medical images, attracting significant attention from researchers. Existing segmentation methods encompass both traditional vision algorithms and cutting-edge deep learning-based approaches. Traditional image processing methods for segmentation rely on hand-crafted features such as geometry, grayscale, and texture. However, in recent years, deep learning has emerged as a dominant paradigm, surpassing traditional methods in terms of segmentation performance. Notably, deep learning-based brain tumor segmentation methods can be categorized into three main categories:

CNN-based models: Convolutional Neural Networks (CNNs) have become a cornerstone in medical image segmentation, with various architectures designed specifically for this task. These architectures include single-channel CNNs [1], multi-channel CNNs [2], and cascade frameworks [3]. Among them, the UNet architecture [4] has gained wide recognition for its effectiveness in preserving detailed information by incorporating skip connections and shorter paths. RNN-based models: Recurrent Neural Networks (RNNs), particularly the Long Short-Term Memory (LSTM) network, have been successfully applied to image classification and segmentation by Graves et al. [5]. Building upon this foundation, Le et al. [6] combined CNNs with RNNs to tackle the challenging task of tumor segmentation.

Transformer-based models: The recent integration of Transformers into medical image segmentation has sparked significant advancements in the field. TransUNet [7] stands as the pioneering model that successfully combines the power of Transformers and CNNs for medical image segmentation. Swin-UNet [8], on the other hand, represents the first pure Transformer architecture for medical image segmentation, leveraging the strengths of Swin Transformers in both the encoder and decoder stages [9]. Within the realm of 3D MR brain tumor image segmentation, TransBTS [10] innovatively embeds Transformers into the U-Net architecture by connecting the Transformer encoder to the bottom, enabling the capture of long-distance feature dependencies within the global receptive field. Furthermore, Jia et al. [11] modified TransBTS, introducing BiTr-UNet that incorporates Transformer encoders at two different scales. Additionally, NVIDIA proposed UNETR [12], a model specifically designed for processing 3D medical images. UNETR utilizes a 12-layer Transformer encoder as a feature extractor, seamlessly integrating it with the decoder strategy of UNet through progressive upsampling and skip connections for effective feature fusion and size recovery.

Despite the remarkable achievements of existing methods in brain tumor segmentation, there exists a considerable opportunity for further improvement. Most existing models are built upon a single architecture, which limits their capacity to extract nested structural features of brain tumors. Moreover, current Transformer based methods often overlook the potential benefits of extracting distinct feature information from different sequences commonly present in brain tumor images. Addressing these challenges, this paper proposes a novel multi-sequence feature self-supervised fusion segmentation network (MSSFN) that synergistically integrates the strengths of CNNs, LSTMs, and Transformers.

The structure of this paper is as follows: the first section provides an insightful introduction and establishes the research background; the second section presents the novel brain tumor segmentation framework proposed in this paper, showcasing its rational design and innovative features; the third section meticulously details the experimental setup, conducted experiments, and insightful analysis of the obtained results; finally, the fourth section presents a comprehensive summary of the entire paper, highlighting the key findings and potential avenues for future research.

2   Methods

2.1   Motivation for Model Design

The majority of existing medical image segmentation models tackle the multi-task segmentation of multiple tumor tissues by designing deeper network layers. This approach decouples complex segmentation tasks into several subtasks, which are then cascaded together. However, a drawback of this cascaded network is that the entire training process is not end-to-end, leading to reduced efficiency in both training and inference. Given that the brain mainly consists of soft tissues such as gray matter, white matter, cerebrospinal fluid, cerebellum, and brainstem, there is significant morpho logical variation among these tissues. Although decomposing the multi-classification task into subtasks reduces the overall task complexity, achieving local optima for each subtask does not guarantee achieving the global optimum.

To address these challenges, we propose a parallel end-to-end model. This model independently performs multi-task learning on different sequences, converting the multi-region tumor sub-segmentation problems into multiple parallel segmentation tasks. For feature extraction from each sequence, we utilize a feature extractor that combines CNN, RNN, and Transformer. This approach helps to focus on different regions of interest in various modal data, thereby alleviating the problem of class im-balance. Finally, we incorporate the idea of ensemble learning to effectively fuse the output results from different model sub-modules, overcoming incomplete output results due to noise or other factors during the segmentation process and enhancing the robustness of the segmentation results.

2.2 Multi-sequence Self-supervised Fusion Segmentation Network

Our proposed multi-sequence feature adaptive fusion segmentation network aims to leverage the complementary information from multiple MR images of the same patient with different modalities. The network architecture, as illustrated in Fig. 1, follows a U-Net-like encoding-decoding structure.

The encoder component of the network plays a crucial role in feature extraction and incorporates ConvLSTM and Transformer modules. The ConvLSTM module is utilized to capture spatial dependencies by combining convolutional operations with LSTM, allowing the network to leverage the local detail information from each sequence. On the other hand, the Transformer module is employed to capture long-range dependencies and capture global information. This combination enables the network to effectively utilize both local and global information from the multi-sequence inputs.

Moving to the decoder component, we adopt a lightweight MLP module. The MLP module is responsible for decoding the extracted features into segmentation results. By using a lightweight design, we strike a balance between model complexity and segmentation performance.

Additionally, we introduce an unsupervised segmentation result quality evaluation network. This network is employed to assess the quality of the segmentation results obtained from different sequences. It enables the network to dynamically update and learn the weights associated with each sequence segmentation result. This self-supervised learning approach allows the fusion process to be optimized and results in more accurate and robust final segmentation outputs.

2.3 Backbone: Bi-ConvLSTM

MRI brain tumor images are three-dimensional images. To make full use of the three-dimensional data and reduce information loss, it is necessary to design a suitable backbone. In this study, we employ a module called Bi-ConvLSTM, which connects the convolutional kernel and LSTM, as the backbone. The convolutional kernel allows for the utilization of two-dimensional image information from each slice, while LSTM captures information across slices. In the standard ConvLSTM, only forward dependencies are considered, but the reverse dependencies also contain valuable information. Therefore, in this paper, both

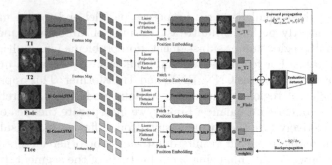

Fig. 1. The structure diagram of Multi-sequence self-supervised fusion segmentation network.

the forward and reverse parameters are taken as inputs, and the formula is as follows:

$$Y_t = \tanh\left(W_y^h * \boldsymbol{h}_t + W_y^{\bar{h}} * \overleftarrow{h}_t + b\right) \tag{1}$$

where $\boldsymbol{h}_t$ and $\overleftarrow{h}_t$ are the hidden state tensors of the forward and reverse states, b is the bias, Y_t represents the final output, using the hyperbolic tangent function to combine the output of the forward and reverse states in a nonlinear manner. The framework of Bi-ConvLSTM is shown in Fig. 2.

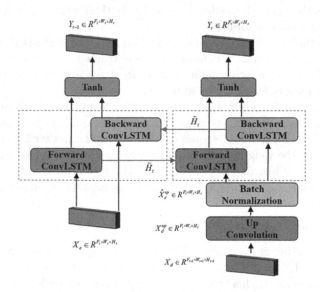

Fig. 2. Bi-ConvLSTM structure diagram.

2.4 Neck: Transformer+MLP

The Transformer architecture has shown great success in natural language processing tasks. In this paper, we leverage the attention mechanism of Transformer and apply it to the brain tumor image segmentation task as the Neck part of our overall model. The main idea is to transform the feature map into a "sequence" similar to text data by dividing it into smaller blocks. For this purpose, we incorporate the Trans-former and MLP model as the Neck component of our overall model architecture. The Transformer module is responsible for capturing global information and enhancing segmentation accuracy, while the MLP module is utilized to decode the feature maps into segmentation results.

At the core of the Transformer architecture lies the self-attention mechanism, which calculates the semantic relationships between different blocks within a feature map. In our approach, we employ a self-attention module with Multi-head, as illustrated in Fig. 3, to enable the model to capture and integrate both local and global dependencies effectively.

By incorporating the Transformer+MLP Neck into our segmentation framework, we enable the model to leverage the power of the attention mechanism in capturing long-range dependencies and semantic associations within the brain tumor images. This allows for more accurate and precise segmentation results, improving the overall performance of our proposed framework.

Figure 3 illustrates the self-attention module with Multi-head used in our approach. This module calculates the attention weights for each position within the feature map, enabling the model to focus on relevant regions and extract meaningful features for accurate segmentation.

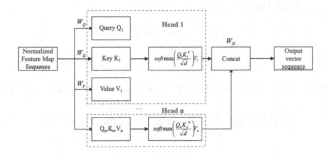

Fig. 3. Multi-head attention module.

Multi-head attention mechanism helps to learn features of different representation spaces. Assuming that the Transformer module has H attention heads, then for a set of layer normalized feature map sequence $X = [x_1, x_2, \ldots, x_n]$, each attention head of the self-attention module will calculate its Query matrix, Key matrix and Value matrix for each set of sequence x_i.

The component vectors of these three matrices are obtained by multiplying the parameter-trainable weight matrix and the normalized block sequence elements. The purpose is to increase the potential feature extraction ability of the network for the feature map slice sequence, namely:

$$q_i^{(h')} = W_i^{(h')^{(Q)}} x_i, k_i^{(h')} = W_i^{(h')^{(K)}} x_i, v_i^{(h')} = W_i^{(h')^{(V)}} x_i \qquad (2)$$

The query vector q, key vector k, and value vector v of each feature map block are actually linear maps of the original feature map slice vectors. After each attention head obtains the corresponding q, k, and v vectors of each block, the inner product of the key vector and the query vector is used to calculate its corresponding attention coefficient. Similar to the commonly used cosine similarity algorithm, the Transformer's attention mechanism can be considered mathematically as a measure of the similarity between two vectors, and its expression is:

$$\alpha_{i,j}^{(h')} = \frac{k_j^T \cdot q_i}{\sqrt{d_k}} \qquad (3)$$

where, $k_j = k_j^{(h')}$, $q_i = q_i^{(h')}$, d_k is the scale factor (that is, the dimensions of q_i and k_j).

Next, each row of the attention matrix is exponentially normalized using the softmax function:

$$\hat{\alpha}_{i,j}^{(h')} = \frac{\exp\left\{\alpha_{i,j}^{(h')}\right\}}{\sum_{k=1}^n \exp\left\{\alpha_{i,j}^{(h')}\right\}} \qquad (4)$$

Use the normalized attention coefficient to weight and sum the value vectors of each block, so that the attention vector is obtained as:

$$b_i^{(h')} = \sum_{j=1}^n \hat{\alpha}_{i,j}^{(h')} v_j^{(h')} \qquad (5)$$

where, $b_i^{(h')}$ represents the i-th output vector of the h-th attention head of the Transformer.

Equations (3) and (5) can also be written in matrix form:

$$B^{(h')} = \text{soft max}\left(\frac{\left(K^{(h')}\right)^T Q^{(h')}}{\sqrt{d_k}}\right) V^{(h')} \qquad (6)$$

Finally, splice the output vectors of all H headers in order and multiply them by an output matrix W_O to obtain the output vector corresponding to the i-th input block vector x_i:

$$b_i = W_O \begin{bmatrix} b_i^{(1)} \\ b_i^{(2)} \\ \dots \\ b_i^{(H)} \end{bmatrix} \qquad (7)$$

Assuming that the input of the l-th encoder is z_{l-1}, the input-output relationship of the first $L-1$ encoders can be expressed as:

$$z_l' = z_{l-1} + LN\left[SA\left(z_{l-1}\right)\right]$$
$$z_l = z_l' + LN\left[MLP\left(z_l'\right)\right] \tag{8}$$

This paper considers that the receptive field range of the Transformer encoder is large, so the image segmentation model based on the Transformer structure does not need to use a heavyweight decoder, and only needs to use a simple MLP structure as a decoder to achieve excellent results. In the decoder part of this paper, the MLP layer is used to form the upsampling module, and the output features of the downsampling module are spliced with the output features of the upsampling module through skip connections to restore the loss during downsampling. Finally, the fully connected layer is used to output pixel-level result.

2.5   Head: Self-supervised Multi-sequence Integration

This section will introduce the proposed self-supervised multi-sequence segmentation result integration method, which can learn the weight parameters of each sequence prediction result end-to-end, so that the final fusion result can be optimal.

In order to effectively evaluate the quality of the predicted probabilities output by the segmentation model, we need to choose a good evaluation index that can effectively represent the gap between the segmentation result and the real label. We choose the Jensen-Shannon (JS) divergence to measure the difference between our predictions and the true labels:

$$JS\left(d\,\|\,g\right) = \frac{1}{2}KL\left(d\left\|\frac{d+g}{2}\right.\right) + \frac{1}{2}KL\left(g\left\|\frac{d+g}{2}\right.\right) \tag{9}$$

where, d represents the segmentation result after degradation, that is, the incomplete network prediction output relative to the real label; g represents the real label of the segmentation result.

In order to obtain the evaluation result of the prediction probability of the segmentation network, according to formula 10, given the input image I, the final prediction result $y(I)$ is obtained by weighting and summing the prediction results obtained by the segmentation network. Superimpose the fused result with the original image data to get $[y(I); I]$, input it to the quality evaluation network, and finally calculate the predicted value Q.

$$Q = h([y(I); I])$$
$$= h\left(\left[\sum_{i=1}^{3}\sum_{j=1}^{D}w_{ij}f_{ij}(I); I\right]\right) \tag{10}$$

Due to the relatively simple task of evaluating the quality of segmentation results, we chose AlexNet [14] with fewer parameters. The network is mainly composed of convolutional layer, maximum pooling layer, and Relu activation function. At the end of the network we use a global average pooling layer (GAP) to obtain compressed features of the same size as the number of feature channels. By using the GAP layer, the network can be allowed to input images of any size, which can ensure that we will not be affected by the size limit of the network output after data preprocessing. After the GAP layer, we use a fully connected layer, and finally use the Sigmoid function to get the final prediction score Q, $Q \in [0, 1]$.

Due to the use of different sequences of MRI input into the segmentation network, four different segmentation results are obtained, which are then uniformly converted into the form of the axial plane. In order to fuse the segmentation results of different sequences, we set up 4 different sets of learnable weight parameters $w\_T1, w\_T2, w\_Flair, w\_T1ce$, whose size is $1 * 1 * D$, Where D represents the number of 2D slices contained in the depth direction of the 3D image. Since the input image of the evaluation network is a 2D image, for each slice of the 3D image, we have a corresponding learnable weight multiplied by it, and the output results of different viewing angles are weighted and summed to obtain the final fusion result. Our goal is that the score Q of the final fusion result in the quality evaluation network is as low as possible, that is, the JS divergence between the prediction result and the real manual label is as low as possible, indicating that the fusion result is as close as possible to the real segmentation result.

Given the input original segmentation image I, 4 different segmentation result prediction probabilities $f_i : f_1, f_2, f_3$ are obtained through four parallel model branches, where $f_i(I) \in [0, 1]$, so the result after the final integration is $y(I) = \sum_{i=1}^{3} \sum_{j=1}^{D} w_{ij} f_{ij}(I)$. w_{ij} satisfies $w_{ij} \geq 0$. For the self-supervised evaluation network, the purpose is to let the network evaluate the quality of the output and fusion results of the segmentation network without the help of manual segmentation labels, and obtain the quality evaluation value, as shown in formula (10). Therefore, for each final image segmentation result, its final optimization goal for weight coefficients can be expressed as follows:

$$w_{ij}^{*} = \arg\min_{w_{ij}} h\left(\left[\sum_{i=1}^{3} \sum_{j=1}^{D} w_{ij} f_{ij}(I); I\right]\right) \tag{11}$$

During the training process, the segmentation results of each sequence in the same group of images are accumulated through the quality evaluation network to obtain the predicted value, and used as the overall loss function for its weight update. The back-propagation gradient update calculation formula is as follows:

$$\nabla_{w_{ij}} \text{loss} = \frac{\partial h([y(I); I])}{\partial w_{ij}} \tag{12}$$

3   Experiment and Results

3.1   Dataset

IFor our experimental evaluation and validation, we employed the BraTS2018 dataset, which is a widely used brain tumor MRI dataset [15+source]. The BraTS2018 dataset consists of a comprehensive collection of multimodal glioma MR data, encompassing a total of 285 sets. Among these, there are 210 sets of High-Grade Glioma (HGG) data and 75 sets of Low-Grade Glioma (LGG) data. Each set of data in this multimodal dataset includes four sequences: T1-weighted (t1), T2-weighted (t2), Fluid-Attenuated Inversion Recovery (FLAIR), and T1-weighted post-contrast (t1ce).

To perform accurate segmentation, it is necessary to divide the dataset into three distinct regions: the whole tumor (WT), the enhancing tumor (ET), and the tumor core (TC). Each of these regions represents a specific aspect of the tumor and requires individual segmentation. The BraTS2018 dataset provides corresponding annotations for these three regions, enabling the evaluation and comparison of segmentation performance.

By leveraging this diverse and well-annotated dataset, our goal is to assess the effectiveness and performance of our proposed multi-sequence self-supervised fusion segmentation network. The BraTS2018 dataset serves as a valuable resource for training and evaluating our model, as it offers a comprehensive range of brain tumor im-ages along with corresponding ground truth annotations. This dataset provides the necessary information to train our model and evaluate its segmentation capabilities accurately.

Using the BraTS2018 dataset, we can thoroughly evaluate our model's ability to segment brain tumor regions across multiple sequences. The dataset's richness and diversity allow us to analyze the model's performance under various scenarios and provide insights into its strengths and limitations. By conducting experiments on this dataset, we aim to demonstrate the effectiveness and robustness of our proposed segmentation framework in accurately delineating brain tumor regions.

3.2   Experimental Setup

Data Preprocessing. To alleviate the problem of positive and negative class imbalance in medical images, we first obtain image bounding boxes based on the minimum boundaries of valid pixels in MR images, and then use the bounding boxes to uniformly crop images across all modalities. In the training process, we randomly cut a certain size of image block from the 3D image obtained after cutting as the training input of the segmentation network. We also limit the number of voxels of each category in the input image block to ensure that it is higher than a certain threshold. This will reduce the impact of too many background pixel values on the training process to a certain extent, and improve the efficiency of the training process.

Evaluating Indicator. This article uses Dice Similarity Coefficient (DSC) and Hausdorff Distance (HD) to evaluate model segmentation performance.

DSC measures the overall overlap between predictions and labels, where recall and false positives are equally important. For each of the three tumor regions, we calculate the output of the model $P \in \{0, 1\}$ and the corresponding label $T \in \{0, 1\}$, and the formula of DSC is expressed as follows:

$$\mathrm{Dice}(P, T) = \frac{|P_1 \wedge T_1|}{(|P_1| + |T_1|)/2} \tag{13}$$

where $\wedge$ represents the logical AND operator, and P_1 and T_1 represent the part of the prediction result and the pixel value of 1 in the label.

Hausdorff distance is used to measure the coincidence degree of the prediction result and the segmentation label in the segmentation boundary part. Hausdorff distance calculates the maximum value of the least squares distance between all points on a given volume surface and all surface points on another volume surface, and the two-way Hausdorff distance takes the maximum value of two one-way distances, and its calculation formula is expressed as:

$$\mathrm{Haus}(P, T) = \max \left\{ \sup_{p \in \partial P_1} \inf_{t \in \partial T_1} d(p, t), \sup_{t \in \partial T} \inf_{p \in \partial P} d(t, p) \right\} \tag{14}$$

where P_1 represents the part with a predicted value of 1 in the output result, ∂P_1 represents its segmentation boundary, T_1 is the corresponding label, ∂T_1 represents the segmentation boundary in the corresponding label, p and t are the boundary points in the prediction result and the corresponding label, respectively, $d(p, t)$ Denotes the least squares distance from the boundary P_1 to the boundary T_1.

Training Environment. In experiments, we implemented the proposed MSSFN using the Pytorch framework and trained it using NVIDIA GTX 1080Ti GPU. Due to GPU memory constraints, the size of training image patches is $96 \times 96 \times 96$ voxels, and we set the batch size to 1. In order to prevent the useful voxels of a certain category in the input image block from being zero during the sampling process and reduce the impact of class imbalance on the model, we filter the cropped image blocks during the sampling process. The image blocks whose number of useful voxels in each category is less than the threshold of 2000 voxels are removed. We trained the model for 200 epochs.

Baselines. This paper uses BiTr-UNet [11], Swin-UNet [8], Cascaded network [16], TumorGAN [17], Multiple model ensemble [18], NestedFormer [19] as comparison models. Among them, BiTr-UNet, Swin-UNet, and Cascaded network can only process a single sequence of data, while TumorGAN, Multiple model ensemble, and NestedFormer can process multiple sequence data at the same time as MSSFN.

3.3   Results

The evaluation scores of the segmentation results under the BraTS 2018 validation set are shown in Table 1. The results clearly indicate the superiority of the proposed MSSFN (Multi-Sequence Segmentation Fusion Network) in brain tumor segmentation.

Upon analyzing the results in Table 1, it is evident that the average performance of the segmentation results obtained by fusing different sequences is significantly better than that of applying a single sequence. This observation highlights the importance of multi-sequence fusion in improving the accuracy and robustness of the segmentation results. By fusing the segmentation results from multiple sequences, the MSSFN method effectively eliminates false positive voxels and enhances the overall segmentation accuracy.

Furthermore, the MSSFN model demonstrates a substantial improvement in the Dice score for the enhanced tumor region compared to the best-performing single-sequence model. The Dice score, which measures the overlap between the predicted and ground truth segmentation masks, serves as a key indicator of segmentation performance. The nearly 8% higher Dice score achieved by our model for the enhanced tumor region signifies the effectiveness of our proposed approach in accurately delineating tumor boundaries.

Moreover, when comparing with other multi-sequence fusion segmentation models, our proposed MSSFN consistently achieves the best Dice score in each region after fusing the segmentation results of four different sequences. This outcome emphasizes the superiority of the self-supervised multi-sequence ensemble approach introduced in this study. By leveraging the complementary information from multiple sequences, MSSFN effectively captures diverse tumor characteristics, resulting in more accurate and reliable segmentation outcomes.

The remarkable performance of MSSFN on the BraTS 2018 validation set demonstrates its efficacy and potential for clinical application. The ability to generate precise and reliable tumor segmentations is of paramount importance in clinical decision-making and treatment planning. By providing doctors with accurate and detailed tumor segmentation results, MSSFN empowers them to make well-informed diagnoses and treatment decisions, ultimately improving patient outcomes.

The findings from this evaluation validate the effectiveness and superiority of the MSSFN method in brain tumor segmentation. The fusion of multi-sequence segmentation results, coupled with the self-supervised learning approach, contributes to the robustness and accuracy of the segmentation outcomes. These results highlight the significant advancements achieved in the field of medical image segmentation, bringing us closer to the goal of efficient and accurate tumor analysis.

Table 1. Comparison of segmentation results of different models on the BraTS 2018 dataset validation set.

Model		Dice		
		ET	WT	TC
single sequence	BiTr-UNet	0.6837	0.8476	0.6783
	Swin-UNet	0.7194	0.8924	0.7864
	Cascaded network	0.7395	0.8862	0.8239
Multiple sequences	TumorGAN	0.7483	0.8864	0.7938
	Multiple model ensemble	0.7374	0.9013	0.7674
	NestedFormer	0.7822	0.8995	0.8375
	MSSFN(ours)	0.7984	0.9002	0.8411

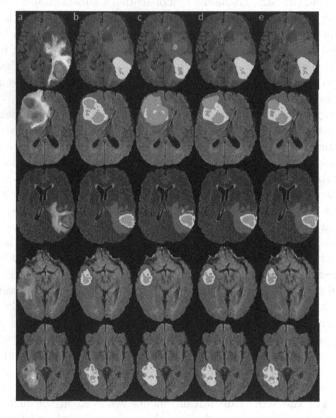

Fig. 4. Visual comparison of brain tumor segmentation results.

Figure 4 is a schematic diagram of the segmentation results. From top to bottom, the comparison of the segmentation results of 5 patients is shown, and each column from left to right represents: a) The original image (for the convenience of display, only the Flair sequence images are selected as the original image for comparison), b) the real image of manual segmentation, c) the predictive segmentation image of Cascaded network, d) the predictive segmentation image of Nested Former model, e) the predictive segmentation image of MSSFN model. The corresponding relationship between colors and labels in the figure is: NETC (green area), ED (red area), ET (yellow area), WT (green area+red area+yellow area), TC (green area+yellow area).

It can be seen that the easiest to distinguish and the best segmentation performance is WT, and the segmentation results when using three different models are relatively similar. For TC and ET, it is relatively difficult to distinguish, especially for ET, which has poor contrast ratio and scattered distribution. The segmentation maps obtained by the MSSFN model are the closest to the real segmentation results on WT, TC, and ET.

3.4 Ablation Experiment

The effectiveness of the MSSFN model structure design is further evaluated through ablation experiments. The structure of MSSFN is characterized by the encoder based on Bi-ConvLSTM and transformer and the decoder based on MLP, and the overall structure is similar to U-Net. Therefore, this section further evaluates the superiority of the encoder and decoder design in this paper by replacing the encoder and decoder of MSSFN with the corresponding components of U-Net. Specifically, the model of changing the encoder based on Bi-ConvLSTM and transformer to a fully convolutional encoder is denoted as MSSFN\E, and the model of changing an MLP-based decoder to a fully convolutional decoder is denoted as MSSFN\E d. The results of the ablation experiments are shown in Table 2.

Table 2. Results of ablation experiments.

Method	Dice			Hausdorff Dist.		
	ET	WT	TC	ET	WT	TC
MSSFN\E	0.7253	0.8749	0.7923	5.2948	7.7885	11.9521
MSSFN\D	0.7920	0.8974	**0.8474**	4.9193	7.0525	**11.2184**
MSSFN	**0.7984**	**0.9002**	0.8411	**4.8875**	**7.0428**	11.2294

From the results in Table 2, we can see that the segmentation result of MSSFN model is better than that of changing the encoder to U-Net encoder. It means that the Bi-ConvLSTM and transformer proposed in this paper have better feature extraction capabilities, and some obvious false positive voxels can be eliminated in the segmentation results, making the final segmentation results more robust and accurate. In the segmentation of enhanced tumor (ET) and tumor core (TC) parts with few voxels and complex boundaries, the encoders using Bi-ConvLSTM and transformer can improve the Dice score by 10.1% and 6.2%, respectively. Except that the performance of the tumor core (TC) part is slightly lower than that of the CNN-based decoder, the overall performance of the model in this paper is still better. Our model achieves desirable performance with a simpler decoder architecture.

4   Comparison with Existing Methods

4.1   Traditional Image Processing Methods

The traditional image processing method has some applications in the field of brain tumor segmentation. These methods mainly rely on hand-designed feature extraction and image processing algorithms to achieve segmentation by calculating similarities or differences between different features in the image. In the brain tumor segmentation task, the commonly used features include geometric features, gray features and texture features.

Geometric features primarily describe the shape and size of the objects in the image, which is important for distinguishing between tumors and normal tissue. The gray scale feature reflects the brightness and contrast of the image, which is useful for distinguishing the boundaries between the tumor and the surrounding tissue. Texture features describe the detail and texture in the image, which is very helpful for identifying the internal structure and morphology of the tumor.

The advantage of traditional method is that its algorithm is relatively simple and easy to understand and implement. For example, threshold segmentation is a common image processing method that sets a threshold to distinguish the foreground and background in an image. This method is simple and effective, but it is very sensitive to threshold selection. In addition, edge detection methods, such as Sobel, Canny, etc., are also widely used in image segmentation. They identify objects by detecting edges in the image. These methods usually have fast computing speed and small model size, which are suitable for some resource-limited application scenarios.

However, traditional image processing methods have some limitations in brain tumor segmentation. First, hand-designed features often fail to fully capture the complexity and variety of brain tumor images. Due to differences in the shape, size and location of brain tumors, hand-designed features are often difficult to adapt to different conditions, leading to inaccurate segmentation results. Secondly, traditional methods usually rely on the prior knowledge of domain experts and manual adjustment of parameters, which makes the adaptability of these methods limited and cannot flexibly adapt to the needs of different data sets and application scenarios. In addition, the performance of traditional methods is usually limited by the ability of feature selection and representation, which can not fully explore the potential information in brain tumor images.

In order to overcome these limitations, some traditional methods try to introduce more complex features, such as waveform based features, statistics based features, etc., to improve segmentation image processing methods. The traditional image processing method has some applications in the field of brain tumor segmentation. These methods mainly rely on hand-designed feature extraction and image processing algorithms to achieve segmentation by calculating similarities or differences between different features in the image. In the brain tumor segmentation task, the commonly used features include geometric features, gray features and texture features. Geometric features primarily describe the shape and size of the objects in the image, which is important for distinguishing between tumors and normal tissue. The gray scale feature reflects the brightness and contrast of the image, which is useful for distinguishing the boundaries between the tumor and the surrounding tissue. Texture features describe the detail and texture in the image, which is very helpful for identifying the internal structure and morphology of the tumor.

The advantage of traditional method is that its algorithm is relatively simple and easy to understand and implement. For example, threshold segmentation is a common image processing method that sets a threshold to distinguish the foreground and background in an image. This method is simple and effective, but it is very sensitive to threshold selection. In addition, edge detection methods, such as Sobel, Canny, etc., are also widely used in image segmentation. They identify objects by detecting edges in the image. These methods usually have fast computing speed and small model size, which are suitable for some resource-limited application scenarios.

However, traditional image processing methods have some limitations in brain tumor segmentation. First, hand-designed features often fail to fully capture the complexity and variety of brain tumor images. Due to differences in the shape, size and location of brain tumors, hand-designed features are often difficult to adapt to different conditions, leading to inaccurate segmentation results. Secondly, traditional methods usually rely on the prior knowledge of domain experts and manual adjustment of parameters, which makes the adaptability of these methods limited and cannot flexibly adapt to the needs of different data sets and application scenarios. In addition, the performance of traditional methods is usually limited by the ability of feature selection and representation, which can not fully explore the potential information in brain tumor images.

4.2   Methods Based on Deep Learning

In recent years, deep learning-based brain tumor segmentation methods have made remarkable progress. These methods use deep neural networks for end-to-end learning of images without manual feature design. Among them, convolutional neural network (CNN) is one of the most commonly used models, such as U-Net, SegNet and FCN. These methods can extract semantic information from low level to high level by stacking convolutional layers and pooling layers, and can segment brain tumor regions accurately. In addition, some methods use attention mechanisms, jump linking and other techniques to further improve segmentation performance. However, the traditional CNN model has some limitations in processing multi-modal images and global context information.

With the development of deep learning, some new network structures and models have been proposed and applied to brain tumor segmentation. For example, by introducing deconvolution layer, the full convolutional network (FCN) can accept the input image of any size and output the segmentation result of the same size as the input image, so as to realize the end-to-end segmentation of the image at the pixel level. In addition, by adding skip connections to the traditional convolutional network, U-Net model enables the network to better retain the details of the image, thus improving the accuracy of segmentation. Another example is the SegNet model, which establishes the corresponding relationship between encoder and decoder, so that the decoder can use the encoder's feature map to carry out more accurate feature recovery, thus improving the accuracy of segmentation.

Although deep learning methods have made remarkable achievements in brain tumor segmentation, there are still some challenges. First of all, deep learning methods usually require a large amount of annotated data for training, and it is often difficult to obtain a large amount of annotated data in practical applications. Secondly, the training process of deep learning methods usually requires a large amount of computing resources and time, which may be unacceptable in some resource-limited scenarios. In addition, deep learning methods are less explanatory, which can be a problem in some application scenarios that require interpretation. Therefore, how to solve these challenges and further improve the performance of deep learning methods in brain tumor segmentation is an important direction of future research.

4.3 Our Proposed Method

Compared with traditional image processing methods and methods based on deep learning, our proposed brain tumor image segmentation framework has the following advantages and improvements:

Firstly, our framework uses the information of multi-modal brain tumor images for segmentation, integrates the features of different modal images, and can capture different features of brain tumors more comprehensively, thus improving the accuracy of segmentation. Traditional methods and methods based on deep learning often only use the information of a single mode and cannot fully explore the potential information of multi-mode images.

Secondly, we introduced Bi-ConvLSTM and Transformer modules as components of the encoder. Bi-ConvLSTM improves feature representation by utilizing both local details and global context information. The Transformer module captures global information and handles long distance dependencies. This combination can better extract the features of brain tumor images and make the segmentation results more robust.

In addition, we propose a self-supervised multi-mode fusion segmentation method that can automatically learn the weight parameters of each mode prediction result to optimize the final fusion result. Compared with the method of manually setting weights, our method is more adaptive and flexible.

Finally, experiments on the BraTS2018 data set prove that our method achieves excellent performance on multiple evaluation indexes, and can obtain better segmentation results compared with the existing multi-modal fusion segmentation model. The brain tumor image segmentation framework shows obvious advantages compared with traditional image processing methods and deep learn-based methods. By taking full advantage of multi-modal information and introducing Bi-ConvLSTM and Transformer modules, our approach is able to segment brain tumor regions more accurately and with better robustness and adaptability. These results validate the validity and innovation of our proposed method, and provide new directions and ideas for further research and application in the field of brain tumor image processing.

To further elaborate on the proposed method, we have designed our framework with a focus on adaptability and flexibility. Our method is not limited to a specific type of brain tumor or a specific set of imaging modalities. Instead, it is capable of handling a wide range of brain tumor types and imaging modalities, making it a versatile tool for brain tumor segmentation.

Moreover, the use of Bi-ConvLSTM and Transformer modules in our framework allows for a more efficient and effective feature extraction process. The Bi-ConvLSTM module is capable of capturing both local details and global context information from the images, while the Transformer module is designed to handle long distance dependencies in the data. This combination of modules enables our framework to extract a richer set of features from the images, leading to more accurate and robust segmentation results.

In terms of the self-supervised multi-mode fusion segmentation method, it is designed to optimize the final fusion result by automatically learning the weight parameters of each mode prediction result. This approach is more adaptive and flexible compared to methods that rely on manually setting weights, as it allows the system to adjust the weight parameters based on the specific characteristics of the data.

The effectiveness of our proposed method is demonstrated through experiments on the BraTS2018 data set. The results of these experiments show that our method outperforms existing multi-modal fusion segmentation models in terms of multiple evaluation indexes. This validates the effectiveness and innovation of our proposed method, and highlights its potential for further research and application in the field of brain tumor image processing.

5 Conclusions

In this paper, we have made significant advancements in brain tumor segmentation by leveraging the multi-sequence information of brain tumor images and integrating Bi-ConvLSTM and Transformer models. Our proposed encoder structure, combining Bi-ConvLSTM and Transformer, effectively captures both local details and global context from the images. Compared to existing multimodal fusion methods for brain tumor segmentation, our self-supervised multi-sequence integration method demonstrates superior performance. It not only achieves a balanced prediction across each modality but also attains the highest performance on the BraTS2018 dataset.

The MSSFN method introduced in this study holds substantial potential value and promising applications in clinical practice. By providing more accurate and detailed tumor segmentation results, MSSFN can assist doctors in making well-informed diagnoses and treatment decisions. The automated and high-precision tumor segmentation achieved by MSSFN significantly reduces the workload for doctors and other healthcare professionals involved in brain tumor analysis. Additionally, it paves the way for further research in the field of brain tumor image processing, enabling researchers to explore new avenues and develop more advanced techniques.

The findings of this study contribute to the growing body of knowledge in the field of brain tumor segmentation and bring us closer to the goal of efficient and accurate tumor analysis. However, there are still opportunities for future improvements and extensions. For instance, incorporating more advanced deep learning techniques or exploring additional imaging modalities could further enhance the performance and versatility of the proposed segmentation framework. Furthermore, expanding the evaluation to larger and more diverse datasets would provide a more comprehensive assessment of the proposed method's generalizability and robustness.

In summary, the advancements achieved in this study underscore the potential of multi-sequence integration and the fusion of Bi-ConvLSTM and Transformer models in improving brain tumor segmentation. The proposed MSSFN method holds promise for practical implementation, offering valuable support to medical professionals and re-searchers in the field of brain tumor analysis. Through ongoing research and devel-opment, we can continue to enhance the accuracy, efficiency, and clinical applicability of brain tumor segmentation techniques, ultimately benefiting patients and advancing the field of medical image analysis.

The proposed MSSFN method has demonstrated its effectiveness and superiority in brain tumor segmentation. The integration of Bi-ConvLSTM and Transformer models in the encoder structure allows for a more comprehensive and accurate representation of brain tumor images, capturing both local details and global context. This is a significant improvement over existing methods, which often fail to fully capture the complex and diverse features of brain tumor images. The self-supervised multi-sequence integration method, a key component of MSSFN, has shown to outperform existing multimodal fusion methods in terms of prediction balance across each modality and overall performance on the BraTS2018 dataset.

The potential applications of MSSFN in clinical practice are promising. The method's ability to provide more accurate and detailed tumor segmentation results can greatly assist doctors in making informed diagnoses and treatment decisions. This not only improves the quality of patient care but also reduces the workload of healthcare professionals involved in brain tumor analysis. The high precision and automation of the MSSFN method also make it a valuable tool for further research in the field of brain tumor image processing. It opens up new avenues for the development of more ad-vanced techniques and the exploration of new research directions.

While the findings of this study contribute significantly to the field of brain tumor segmentation, there are still opportunities for further improvements and extensions. The proposed segmentation framework could be enhanced by incorporating more advanced deep learning techniques or exploring additional imaging modalities. Moreover, evaluating the method on larger and more diverse datasets would provide a more comprehensive assessment of its generalizability and robustness.

6   Outlook and Prospects Outlook

In this paper, we have presented the MSSFN (Multi-Sequence Segmentation Fusion Network) method, which demonstrates significant advancements in the field of medical image segmentation, particularly in the context of brain tumor analysis. By leveraging the multi-sequence information of brain tumor images and integrating Bi-ConvLSTM and Transformer models, we have achieved improved segmentation accuracy and robustness compared to existing methods.

The proposed MSSFN method holds great promise for further advancements and applications in the field of medical image segmentation. We have identified several areas for future improvements and developments to enhance the performance and versatility of MSSFN.

Firstly, we propose to focus on the improved network architecture and model de-sign of MSSFN. By exploring more efficient and accurate segmentation algorithms, such as introducing sophisticated attention mechanisms, multi-scale feature fusion, and adaptive learning methods, we can further enhance the network performance. Con-tinuously refining and optimizing the model architecture will enable MSSFN to achieve even higher levels of accuracy and efficiency in brain tumor segmentation.

Moreover, we see potential in expanding the application of MSSFN to other medical imaging tasks beyond brain tumor segmentation. Fields such as cardiac image segmentation, lung lesion detection, and breast cancer screening can benefit from the high-precision segmentation capability of MSSFN. Adapting and fine-tuning the model for different medical imaging applications will contribute to improved diagnoses and treatment planning in various healthcare domains.

Furthermore, the integration of MSSFN with clinical information and decision support systems can enhance its value and impact in clinical practice. By combining patients' clinical data, genomics information, and image segmentation results, MSSFN can assist doctors in making more accurate diagnoses, personalized treatment recommendations, and prognosis assessments. This integration will lead to more comprehensive and personalized healthcare services, ultimately improving patient outcomes.

To further validate the performance and robustness of MSSFN, we propose con-ducting validation and application studies using large-scale multi-center datasets. Verifying MSSFN's suitability across different institutions and imaging devices will enhance its reliability and feasibility in real-world clinical practice. Involving multiple centers and diverse patient populations will allow for a thorough evaluation of the generalizability and effectiveness of MSSFN.

Additionally, the fusion of multi-modal image information holds great potential for advancing medical image segmentation. While MSSFN currently focuses on mul-ti-sequence brain tumor image segmentation, future developments can involve the segmentation of multi-modal images. Incorporating various modalities such as mag-netic resonance imaging, positron emission tomography, and computed tomography can provide comprehensive and multi-dimensional disease information, enabling more accurate diagnoses and treatment decisions. Leveraging the complementary infor-mation from different modalities will further improve the precision and reliability of medical image segmentation.

Encouraging the open sourcing of MSSFN's code and model parameters and promoting collaboration within the research community will foster innovation and drive advancements in the field of medical image segmentation. By sharing resources, insights, and expertise, researchers can collectively contribute to the improvement and optimization of MSSFN. Open source initiatives and collaborative efforts will accelerate the development and adoption of state-of-the-art segmentation techniques, benefiting the wider healthcare community.

In clinical applications, MSSFN holds significant potential value. By providing fast and accurate image segmentation results, MSSFN can assist doctors in making more precise diagnoses and treatment decisions, reducing human errors and subjectivity. It can enhance surgical navigation accuracy, support surgical planning, and facilitate treatment evaluation. Additionally, MSSFN provides powerful tools for researchers to explore disease pathologies and influencing factors, enabling a deeper understanding of diseases and paving the way for advancements in medical research.

References

1. Kamnitsas, K., Ledig, C., Newcombe, V.F., Simpson, J.P., Kane, A.D., Menon, D.K., Rueckert, D., Glocker, B.: Efficient multi-scale 3D CNN with fully connected CRF for accurate brain lesion segmentation. Med. Image Anal. **36**, 61–78 (2017)
2. Su, C.-H., Chung, P.-C., Lin, S.-F., Tsai, H.-W., Yang, T.-L., Su, Y.-C.: Multi-scale attention convolutional network for Masson stained bile duct segmentation from liver pathology images. Sensors **22**, 2679 (2022)
3. Fu, X., Zeng, D., Huang, Y., Zhang, X.P., Ding, X.: A weighted variational model for simultaneous reflectance and illumination estimation. In Proceedings of the 2016 IEEE Conference on Computer Vision and Pattern Recognition (CVPR), Las Vegas, NV, USA, 27–30 June 2016
4. Ronneberger, O., Fischer, P., Brox, T.: U-net: convolutional networks for biomedical image segmentation. In: Navab, N., Hornegger, J., Wells, W.M., Frangi, A.F. (eds.) MICCAI 2015. LNCS, vol. 9351, pp. 234–241. Springer, Cham (2015). https://doi.org/10.1007/978-3-319-24574-4_28
5. Byeon, W., Breuel, T.M., Raue, F., Liwicki, M.: Scene labeling with LSTM recurrent neural networks. In: Proceedings of the 2015 IEEE Conference on Computer Vision and Pattern Recognition (CVPR), Boston, MA, USA, pp. 3547–3555, 8–10 June 2015
6. Le, T.H.N., Gummadi, R., Savvides, M.: Deep recurrent level set for segmenting brain tumors. In: Frangi, A.F., Schnabel, J.A., Davatzikos, C., Alberola-López, C., Fichtinger, G. (eds.) MICCAI 2018. LNCS, vol. 11072, pp. 646–653. Springer, Cham (2018). https://doi.org/10.1007/978-3-030-00931-1_74

7. Chen, J.N., et al.:TransUNet: transformers make strong encoders for medical image segmentation. arXiv, arXiv:2102.04306, https://arxiv.org/abs/2102.04306 (2021)
8. Cao, H., et al.: Swin-Unet: unet-like pure transformer for medical image segmentation. arXiv arXiv:2105.05537, https://arxiv.org/abs/2105.05537 (2021)
9. Liu, Z., et al.: Swin transformer: hierarchical vision transformer using shifted windows. In: Proceedings of the 2021 IEEE/CVF International Conference on Computer Vision (ICCV), Montreal, QC, Canada, pp. 9992–10002, 10–17 October 2021
10. Wang, W., Chen, C., Ding, M., Yu, H., Zha, S., Li, J.: TransBTS: multimodal brain tumor segmentation using transformer. In: de Bruijne, M., et al. (eds.) MICCAI 2021. LNCS, vol. 12901, pp. 109–119. Springer, Cham (2021). https://doi.org/10.1007/978-3-030-87193-2_11
11. Jia, Q.; Shu, H. BiTr-Unet: a CNN-transformer combined network for MRI brain tumor segmentation. arXiv arXiv:2109.12271 (2021)
12. Hatamizadeh, A., et al.: UNETR: transformers for 3d medical image segmentation. In: Proceedings of the IEEE/CVF Winter Conference on Applications of Computer Vision, Wai-koloa, HI, USA, vol. 4–8, pp. 574–584 (2022)
13. Song, H., Wang, W., Zhao, S., Shen, J., Lam, K.: Pyramid dilated deeper ConvLSTM for video salient object detection. In: Proceedings of the European Conference on Computer Vision (ECCV), Munich, Germany, 8–14 September 2018
14. Krizhevsky, A., Sutskever, I., Hinton, G.E.: ImageNet classification with deep convolutional neural networks. In: Proceedings of the Neural Information Processing System (NIPS), Harrahs and Harveys, Lake Tahoe, NV, USA, Vol. 2, pp. 1097–1105, 3–8 December 2012
15. Menze, B.H., et al.: The multimodal brain tumor image segmentation benchmark (BRATS). IEEE Trans. Med. Imaging **34**, 1993–2024 (2015)
16. Wang, G., Li, W., Ourselin, S., Vercauteren, T.: Automatic brain tumor segmentation using cascaded anisotropic convolutional neural networks. In: Proceedings of the International MICCAI Brainlesion Workshop, Quebec City, QC, Canada, 10–14 September 2017
17. Li, Q., Yu, Z., Wang, Y., Zheng, H.: TumorGAN: a multi-modal data augmentation framework for brain tumor segmentation. Sensors **20**, 4203 (2020)
18. Kamnitsas, K., et al.: Ensembles of multiple models and architectures for robust brain tumour segmentation. arXiv arXiv:1711.01468 (2017)
19. Xing, Z., Yu, L., Wan, L., Han, T., Zhu, L.: NestedFormer: nested modality-aware transformer for brain tumor segmentation. In: Proceedings of the International MICCAI Brainlesion Workshop, Singapore, vol. 18–22, pp. 273–283 (2022)

An Effective Morphological Analysis Framework of Intracranial Artery in 3D Digital Subtraction Angiography

Haining Zhao[1,2], Tao Wang[3], Shiqi Liu[1(✉)], Xiaoliang Xie[1(✉)], Xiaohu Zhou[1], Zengguang Hou[1,2], Liqun Jiao[3], Yan Ma[3], Ye Li[3], Jichang Luo[3], Jia Dong[3], and Bairu Zhang[3]

[1] The State Key Laboratory of Multimodal Artificial Intelligence Systems, Institute of Automation, Chinese Academy of Sciences, Beijing, China
{liushiqi2016,xiaoliang.xie}@ia.ac.cn
[2] School of Artificial Intelligence, University of Chinese Academy of Sciences, Beijing, China
[3] Department of Neurosurgery, Xuanwu Hospital, Capital Medical University, Beijing, China

Abstract. Acquiring accurate anatomy information of intracranial artery from 3D digital subtraction angiography (3D-DSA) is crucial for intracranial artery intervention surgery. However, this task often comes with challenges of large-scale image and memory constraints. In this paper, an effective two-stage framework is proposed for fully automatic morphological analysis of intracranial artery. In the first stage, the proposed Region-Global Fusion Network (RGFNet) achieves accurate and continuous segmentation of intracranial artery. In the second stage, the 3D morphological analysis algorithm obtains the access diameter, the minimum inner diameter and the minimum radius of curvature of intracranial artery. RGFNet achieves state-of-the-art performance (93.36% in Dice, 87.83% in mIoU and 15.64 in HD95) in the 3D-DSA intracranial artery segmentation dataset, and the proposed morphological analysis algorithm also shows effectiveness in obtaining accurate anatomy information. The proposed framework is not only helpful for surgeons to plan the procedures of interventional surgery but also promising to be integrated to robotic navigation systems, enabling robotic-assisted surgery.

Keywords: Intracranial Artery · Segmentation · Morphological Analysis

1 Introduction

Intracranial atherosclerotic stenosis (ICAS) is a leading cause of ischemic stroke worldwide, associated with high risk of recurrent stroke [1]. Image-guide endovascular therapy, such as percutaneous transluminal angioplasty (PTA), has been proved as a necessary treatment for those patients with severe stenosis more than 70% of the diameter of a major intracranial artery [2]. In the procedures of the interventional surgery for ICAS, accurate segmentation and morphological analysis of intracranial artery in 3D-DSA is

---

H. Zhao and T. Wang—Co-first authors.

© The Author(s), under exclusive license to Springer Nature Singapore Pte Ltd. 2024
B. Luo et al. (Eds.): ICONIP 2023, CCIS 1964, pp. 50–61, 2024.
https://doi.org/10.1007/978-981-99-8141-0_4

of great significance for both preoperative planning and intraoperative navigation, as they can provide accurate anatomy information for surgeons. In the preoperative period, the access diameter and minimum diameter of intracranial artery are measured in order to choose suitable intervention instruments. Besides, the surgeons also observe the varied curvature along the artery so that they can plan the procedures of surgery, and the section with maximum curvature will be specifically noted as it will be one of the most difficult parts in surgery. During the intraoperative phase, 3D segmentation of the artery is utilized as supplement of X-ray fluoroscopy to provide 3D spatial information for the navigation of the instruments.

Although 3D-DSA can provide detailed anatomy and spatial information, it also brings some challenges. As the size of 3D-DSA of intracranial artery is very large, it is hard put the entire volume into segmentation models due to the memory constraint. In order to resolve this problem, most researchers applied 3D segmentation model on 3D sub-volumes cropped from original volume [9–11, 15]. This method can encapsulate dependencies in three directions, it only utilized the information of local region for segmentation, and the global semantic of the entire volume is ignored. However, the global information of the entire volume is quite important while segmenting intracranial artery in sub-regions as the artery spans across a wide range in the volume. Besides, the varied diameters along the artery and the existence of bifurcations also bring some obstacles to the morphological analysis.

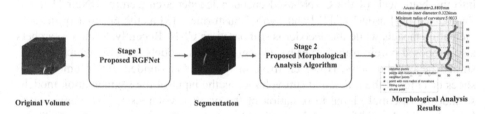

Original Volume   Segmentation   Morphological Analysis Results

Fig. 1. Overview of the framework

In order to tackle these challenges, we propose a two-stage framework for the morphological analysis of intracranial artery in 3D-DSA as shown in Fig. 1. In the first stage, RGFNet is proposed for the segmentation of intracranial artery. RGFNet achieves best performance not only in overlap-based metrics (Dice, mIoU) but also in boundary-based metrics (HD95) for intracranial artery segmentation as it can supplement the cropped-volume segmentation with the context from other regions of the entire volume. In the second stage, an automatic 3D morphological analysis algorithm is developed to obtain access diameter, the minimum inner diameter and the minimum radius of curvature of intracranial artery. The algorithm shows robustness in largely varied artery structures.

Our contributions can be summarized as follows:

- An effective two-stage framework is proposed for the segmentation and morphological analysis of intracranial artery in 3D-DSA. In stage 1, the proposed model obtains more accurate and continuous segmentation results of intracranial artery than other methods, achieving state-of-the-art performance (93.36% in Dice, 87.83% in mIoU and 15.64 in HD95) in the 3D-DSA intracranial artery segmentation dataset. In stage 2, the proposed morphological analysis algorithm is effective in robustly extracting accurate anatomical information from arterial structures that vary largely across subjects.

- In the proposed segmentation model, region-global fuse block is put forward to model the dependencies between the features extracted from the cropped volume and the entire volume, and feature enhance block is put forward to enhance the features of cropped volume with the global information. The experiments demonstrate that these two modules can effectively address the challenges caused by the sparse structure of artery in large-scale volume.
- A 3D-DSA intracranial artery segmentation dataset, named IAS-3D-DSA, is established. It consists of 210 3D-DSA images of intracranial arteries from 205 patients.

2 Related Work

Deep neural network and encoder-decoder architecture with skip connection have been widely adopted in 3D medical image segmentation models since U-Net [3] achieved excellent performance in 2D biomedical image segmentation challenges. Çiçek Ö et al. [4] replaced 2D convolutions in U-Net with 3D convolutions to deal with 3D segmentation tasks. [6, 7] applied residual connections in the encoder of U-structure segmentation models. TransUNet [8] and its 3D version TransBTS [9] integrated transformer blocks into the bottleneck of the CNN-based encoder-decoder architecture. UNetr [10] and SwinUNetr [11] used ViT [12] and Swin Transformer [13] as the encoder of the segmentation models while the decoder is still based on CNN. Recently, Some researchers also proposed purely transformer-based segmentation models [14, 15] which demonstrated better robustness. However, these methods either divided 3D volumes to 2D slices or cropped them to small sub-volumes as the input of the segmentation models, so that the essential global information of the original volumes is lost. Fabian et al. [5] proposed a hybrid framework to refine the coarse segmentation of cropped volume and achieves better performance in peripheral nerve segmentation. Nevertheless, the coarse segmentation of this framework is still lack of global information, which limits its performance.

3 Method

3.1 RGFNet for Segmentation of Intracranial Artery in 3D-DSA

As shown in the Fig. 2, the framework of RGFNet can be divided to four parts: (1) the region segmentation branch including region encoder and region decoder, (2) the global segmentation branch including global encoder and global decoder, (3) the region-global fuse block and (4) the feature enhance block. The original medical image $X \in \mathbb{R}^{1 \times D \times H \times W}$ is divided to small patches $\{X_{1-1}, X_{1-2}, \ldots, X_{1-n}\} \in \mathbb{R}^{1 \times D_1 \times H_1 \times W_1}$ before processed by region segmentation branch, while X is also down sampled to $X_2 \in \mathbb{R}^{1 \times D_2 \times H_2 \times W_2}$ as the input of global segmentation branch.

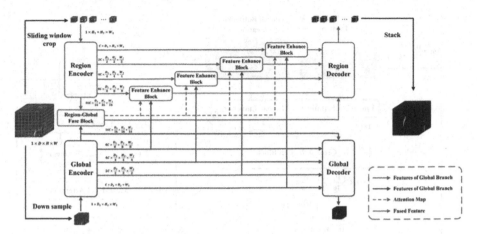

Fig. 2. Overview of RGFNet

Region Segmentation Branch. Region segmentation branch is accomplished on the basis of modified 3D version of UNet, and it is the main branch of RGFNet, responsible for the segmentation of the sub-volumes $\{X_{1-1}, X_{1-2}, \ldots, X_{1-n}\}$ cropped from the original volume X. For each input sub-volume X_1, region encoder will generate hierarchical features $\{F_1^i \in \mathbb{R}^{2^i C \times \frac{D_1}{2^i} \times \frac{H_1}{2^i} \times \frac{W_1}{2^i}}, i = 0, 1, 2, 3, 4\}$. After fused with corresponding global encoder features, those features are fed into the region decoder to produce segmentation result of each sub-volume. These outputs are then stacked to get the final prediction of the original volume as same as most 3D segmentation models do.

Global Segmentation Branch. Global encoder is used to extract hierarchical features $\{F_2^i \in \mathbb{R}^{2^i C \times \frac{D_2}{2^i} \times \frac{H_2}{2^i} \times \frac{W_2}{2^i}}, i = 0, 1, 2, 3, 4\}$, from the down-sampled volume X_2. The global encoder features $\{F_2^0, F_2^1, F_2^2, F_2^3, F_2^4\}$ will be fused into the corresponding region encoder features $\{F_1^0, F_1^1, F_1^2, F_1^3, F_1^4\}$ to get the fused features $\{F_f^0, F_f^1, F_f^2, F_f^3, F_f^4\}$. The output of global segmentation branch is utilized as deep supervision to ensure that the global encoder gets fully trained and it is not required after training process is done, so the global decoder will be removed during inference period.

Region-Global Fuse Block. Region-global fuse block is used to capture the relationship of the bottleneck features and also fuse them. Inspired by self-attention mechanism, we propose cross-attention to fuse the bottleneck features from the two branches as shown in Fig. 3.

As shown in Fig. 3, the bottleneck feature of region encoder $F_1^4 \in \mathbb{R}^{16C \times \frac{D_1}{16} \times \frac{H_1}{16} \times \frac{W_1}{16}}$ and the bottleneck feature of global encoder $F_2^4 \in \mathbb{R}^{16C \times \frac{D_2}{16} \times \frac{H_2}{16} \times \frac{W_2}{16}}$ are firstly reshaped into patches $P_1^4 \in \mathbb{R}^{1 \times \left(16C \times \frac{D_1}{16} \times \frac{H_1}{16} \times \frac{W_1}{16}\right)}$ and $P_2^4 \in \mathbb{R}^{(N_D \times N_H \times N_W) \times \left(16C \times \frac{D_2}{16 N_D} \times \frac{H_2}{16 N_H} \times \frac{W_2}{16 N_W}\right)}$, in which N_D, N_H, N_W are the numbers of the sub-volumes in three directions. Then, P_1^4 and P_2^4 are linear projected to embeddings $z_1^4 \in \mathbb{R}^{1 \times C_{embed}}$ and $z_2^4 \in \mathbb{R}^{N \times C_{embed}}$, where $N = N_D \times N_H \times N_W$ and C_{embed} is the number of

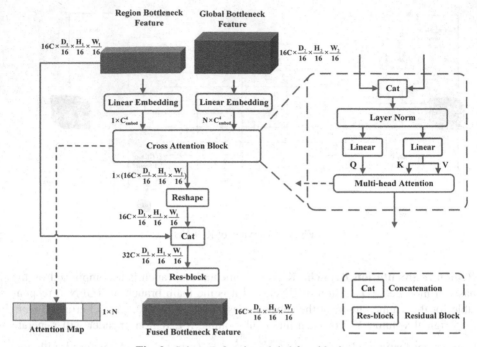

Fig. 3. Scheme of region-global fuse block

embeddings' channels. Subsequently, z_1^4 and z_2^4 are concatenated and layer normed. After that, z_1^4 and z_2^4 are separately linear projected to generate $Q \in \mathbb{R}^{1 \times \left(16C \times \frac{D_1}{16} \times \frac{H_1}{16} \times \frac{W_1}{16} \right)}$ and $\{K, V\} \in \mathbb{R}^{N \times \left(16C \times \frac{D_1}{16} \times \frac{H_1}{16} \times \frac{W_1}{16} \right)}$. Then, Q, K, V are processed according to multi-head attention mechanism to get the output, $Z \in \mathbb{R}^{1 \times \left(16C \times \frac{D_1}{16} \times \frac{H_1}{16} \times \frac{W_1}{16} \right)}$. Besides, the attention map M obtained in multi-head attention block is also output and used later in the feature enhance block of the other feature levels as it models the dependencies between global encoder features and region encoder features. Z is then reshaped to $(16C, \frac{D_1}{16}, \frac{H_1}{16}, \frac{W_1}{16})$ and concatenated with the region bottleneck feature. After that, res-block is used to fuse the concatenated feature and get the final output, $F_{fuse}^4 \in \mathbb{R}^{16C \times \frac{D_1}{16} \times \frac{H_1}{16} \times \frac{W_1}{16}}$.

Feature Enhance Block. The feature enhance block, shown in Fig. 4, is utilized to fuse the features of other levels based on the attention map M obtained from region-global fuse block. Firstly, the global feature $F_2^i \in \mathbb{R}^{2^i C \times \frac{D_2}{2^i} \times \frac{H_2}{2^i} \times \frac{W_2}{2^i}}$ is reshaped and linear projected to embeddings $z_2^i \in \mathbb{R}^{N \times C_{embed}^i}$. Then, z_2^i is layer normed and linear projected to generate $V^i \in \mathbb{R}^{N \times (2^i C \times \frac{D_2}{2^i N_D} \times \frac{H_2}{2^i N_H} \times \frac{W_2}{2^i N_W})}$ and the enhanced information $Z^i \in \mathbb{R}^{1 \times (2^i C \times \frac{D_2}{2^i N_D} \times \frac{H_2}{2^i N_H} \times \frac{W_2}{2^i N_W})}$ is obtained by Eq. (1), where $N = N_D \times N_H \times N_W$.

$$Z^i = M V^i \tag{1}$$

Afterwards, Z^i are reshaped and up sampled to $2^i C \times \frac{D_1}{2^i} \times \frac{H_1}{2^i} \times \frac{W_1}{2^i}$, so that it can be concatenated with the region feature and subsequently put into the res-block to get the fused feature $F^i_{fuse} \in \mathbb{R}^{2^i C \times \frac{D_1}{2^i} \times \frac{H_1}{2^i} \times \frac{W_1}{2^i}}$.

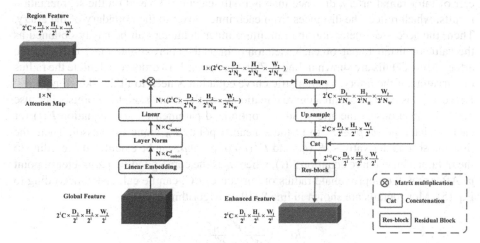

Fig. 4. Scheme of feature enhance block

Loss Function. As shown in Eq. (2), the loss function of RGFNet consists of two parts, L_{region} and L_{global} to ensure region segmentation branch and global segmentation branch are fully trained.

$$L_{total} = \lambda_1 L_{region} + \lambda_2 L_{global} \tag{2}$$

Region segmentation branch is the main branch of RGFNet and used to predict the probability map $\widehat{Y}_{1-i}$ of input sub-volume X_{1-i}. Given the cropped ground truth Y_{1-i}, a weight sum of soft dice loss and cross-entropy loss are used as the loss of region segmentation. λ_1^r, λ_2^r are adjustable hyperparameters.

$$L_{region} = -\lambda_1^r SoftDice\left(\widehat{Y}_{1-i}, Y_{1-i}\right) - \lambda_1^r CrossEntropy(\widehat{Y}_{1-i}, Y_{1-i}) \tag{3}$$

Global segmentation branch is used as a deep supervise branch to train the global encoder. The loss is calculated as same as region segmentation branch. As shown in Eq. (4), $\widehat{Y}_2$ is the output of global decoder and Y_2 is the down sampled ground truth. λ_1^g, λ_2^g are adjustable hyperparameters.

$$L_{global} = -\lambda_1^g SoftDice\left(\widehat{Y}_2, Y_2\right) - \lambda_1^g CrossEntropy(\widehat{Y}_2, Y_2) \tag{4}$$

3.2 Automatic Morphological Analysis Algorithm of Intracranial Artery

Based on the segmented vessel, we develop an automatic method to obtain the access diameter, the minimum inner diameter and the minimum radius of curvature of intracranial artery. This proposed method is demonstrated in details in Algorithm 1.

Considering the varied radius along vessel, a 3D thinning algorithm [16] is firstly applied here to erode the segments to get the skeleton mask to represent the structure of intracranial artery. Skeleton pointset of vessel, $P = \{p_i \in \mathbb{R}^3, i = 1, 2, 3, \ldots M\}$, is then obtained from the skeleton mask. For the access diameter and the minimum inner diameter of intracranial artery, distance map is firstly calculated based on the segmentation results, which reflect the distances from each inner voxel to the boundary of the artery. Then, the access diameter and the minimum inner diameter can be easily obtained as the value of distance map on each skeleton point is the approximate inner radius of the artery. More details are shown in line 4–7 of Algorithm 1. In order to calculate the radius of curvature of the artery, a parametric curve equation is need to fit the skeleton points. However, it is hard to obtain a curve equation to fit the entire skeleton pointset because of the bifurcations of the artery. So, we optimized parametric curve equation $F(t)$ for each skeleton point, $p \in P$, to fit the k nearest points P_n along the vessel. Then, the first and second derivatives, $F'(t_c)$ and $F''(t_c)$ of point p, can be calculated according to the obtained local curve equation $F(t)$, where t_c is the corresponding parameter of point p. After that, the approximate radius of curvature, RC, can be calculated according to Eq. (5). More details are shown in line 8–19 of Algorithm 1.

$$RC = \frac{\|F'(t_c)\|_2^3}{\|F'(t_c) \times F''(t_c)\|_2} \tag{5}$$

Algorithm 1: Morphological Analysis

Input: binarized prediction $\widehat{Y}$
Output: access diameter d_a, the minimum inner diameter d_{min} and the minimum radius of curvature RC_{min}

1   initialize $RC_{min} \leftarrow 100$
2   initialize the number of neighbor points $k \leftarrow 20$
3   extract the skeleton pointset P from $\widehat{Y}$
4   calculate the distance map D from $\widehat{Y}$
5   get the start point p_0 from P
6   $d_a \leftarrow 2 \times D(p_0)$
7   $d_{min} \leftarrow 2 \times \min\{D(p) | p \in P\}$
8   **for** p in P:
9       $P_n, t, t_c \leftarrow$ **LocalPoints**(p, P, k)
10   initialize cubic parametric curve equation $F(t)$
11   optimize $F(t)$ using least squares method according to P_n, t
12   **if** $\text{MSE}(F(t), P_n) < 0.2$
13   calculate the first and second derivatives vector in p, $F'(t_c)$ and $F''(t_c)$
14   calculate the radius of curvature, RC, according to equation (5)
15   **if** $RC < RC_{min}$:
16              $RC_{min} \leftarrow RC$
17   **end if**
18   **end if**
19 **end for**

The function LocalPoints($\cdot$) used in line 9 of Algorithm 1 is applied to get the sorted local points $\boldsymbol{P}_n$, the parameters $\boldsymbol{t}$ corresponding to them and the parameter t_c of point $\boldsymbol{p}$. The details are shown in Algorithm 2. The function MSE($\cdot$) in line 12 of Algorithm 1 is used to calculate the mean square error between curve equation $\boldsymbol{F}(t)$ and local pointset $\boldsymbol{P}_n$, and it is used to remove those curve equations fitting local pointset badly.

Algorithm 2: LocalPoints

Input: center point $\boldsymbol{p}_c$, pointset $\boldsymbol{P}$, number of neighbor points k

Output: local pointset $\boldsymbol{P}_n$, corresponding parameters $\boldsymbol{t}$, parameter of center point t_c

1   initialize pointset $\boldsymbol{P}'_n \leftarrow \{\boldsymbol{p}_c\}$

2   while $|\boldsymbol{P}'_n| \leq k$:

3   find the two points $\{\boldsymbol{p}_{n1}, \boldsymbol{p}_{n2}\}$ of $\boldsymbol{P}$, which has the shortest distance to $\boldsymbol{P}'_n$

4        $\boldsymbol{P}'_n \leftarrow \boldsymbol{P}'_n + \{\boldsymbol{p}_{n1}, \boldsymbol{p}_{n2}\}$

    end while

5     $\boldsymbol{p}_0 \leftarrow \arg\min_{\boldsymbol{p}\in\boldsymbol{P}'_n} \|\boldsymbol{p}\|_2^2$

6   initialize $\boldsymbol{P}_n \leftarrow \{\boldsymbol{p}_0\}$

7   while $|\boldsymbol{P}_n| \leq k$:

8   find the nearest point $\{\boldsymbol{p}_n\}$ of $\boldsymbol{P}$ to $\boldsymbol{P}_n$

9        $\boldsymbol{P}_n \leftarrow \boldsymbol{P}_n + \{\boldsymbol{p}_n\}$

10   if $\boldsymbol{p}_n == \boldsymbol{p}_c$:

11        $t_c \leftarrow |\boldsymbol{P}_n|$

12   end if

13     $\boldsymbol{t} \leftarrow k$ points equidistantly sampled in the interval $[0,10]$

14   end while

15     $t_c \leftarrow t_c \times 10 \,/\, k$

---

4 Experiments

4.1 Datasets

IAS-3D-DSA. This dataset consists of 210 3D-DSA images of intracranial arteries, collected from 205 patients. 96 images are captured with size of $512 \times 512 \times 512$, while 113 images are captured with size of $384 \times 384 \times 384$. The collected images are firstly labelled by 5 clinicians. Then, the results are validated by 3 experienced physicians. The dataset is randomly separated to train dataset with 190 volumes and test dataset with 20 volumes.

4.2 Implementation Details

Hyperparameters of RGFNet In our implementation of the proposed segmentation model, the size of cropped sub-volume $D_1 \times H_1 \times W_1$ was set as $128 \times 128 \times 128$, while the down-sampled volume's size $D_2 \times H_2 \times W_2$ was $64 \times 64 \times 64$. The channels' number of the first feature level C is 16. The adjusted coefficients of the loss function, $\lambda_1 = \lambda_2 = \lambda_1^r = \lambda_2^r = \lambda_1^g = \lambda_2^g = 0.5$.

Training Strategy of RGFNet. Our model was implemented in Pytorch and trained on four 32 GB NVIDIA Tesla V100 GPUs. We train our model using AdamW [17] optimizer with a warm-up cosine scheduler, batch size of 4 per GPU, initial learning rate of 1×10^{-4}, momentum of 0.9 and decay of 1×10^{-5} for 1K iterations.

4.3 Segmentation Results

We compare our method with 6 previous SOTA methods on IAS-3D-DSA dataset. The quantitative results are shown in Table 1. Compared with other methods, our proposed RGFNet achieves best performance. Our method obtains the best score not only in Dice and mIoU but also a dramatic decrease in HD95, which demonstrates our method is able to get accurate segmentation results not only in the overlap area but also in the boundaries of artery. Besides, as shown in Fig. 5, the results of our method show more continuous vascular structures with much less outliers than other baseline methods.

Table 1. Performance on IAS-3D-DSA test dataset

Method	Dice	mIoU	HD95
TransUNet [8]	79.75	67.76	99.93
TransBTS [9]	90.38	82.91	49.62
UNetr [10]	86.65	77.12	96.07
SwinUNetr [11]	88.66	80.22	76.50
Swin UNet [14]	83.15	71.80	134.22
VT-UNet [15]	90.90	83.77	32.58
RGFNet (ours)	**93.36**	**87.83**	**15.64**

4.4 Ablation Study of RGFNet

In order to verify the effectiveness of the components in our proposed method, we perform empirical study on RGFNet with different combination of modules as shown in Table 2. It demonstrates that using region-global fuse block alone to fuse the bottleneck features of region and global segmentation branch only shows slight improvements over the baseline and even a little worse performance in HD95. However, while combining region-global attention and feature enhance blocks, it shows an increase of 0.94% in Dice, 1.39% in mIoU and a significant decrease of 51.38% in HD95. The results demonstrate that the dense fusion of every feature level is need to be done to enhance the features of region segmentation branch.

4.5 Morphological Analysis Results

The morphological analysis results of two cases are demonstrated in Fig. 6. The access diameter, the minimum inner diameter and the minimum radius of curvature are on

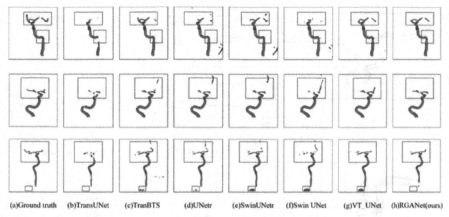

(a)Ground truth (b)TransUNet (c)TranBTS (d)UNetr (e)SwinUNetr (f)Swin UNet (g)VT_UNet (h)RGANet(ours)

Fig. 5. Qualitative visualization of the proposed RGFNet and other baseline methods. Three representative cases are shown.

Table 2. Ablation study on RGFNet

Region Segmentation Branch	Global Segmentation Branch	Region-Global Fuse block	Feature Enhance Block	Dice	mIoU	HD95
√				92.52	86.45	32.17
√	√	√		92.96	87.10	34.35
√	√	√	√	**93.36**	**87.83**	**15.64**

the top of each morphological analysis result. It is difficult to get the exact value of these morphological parameters, so we also visualized the intermediate results in Fig. 6 to verify the effectiveness of our method. (1) For access diameter, the purple point in Fig. 6(b) is located at the entrance of vessel, so its diameter is access diameter. (2) For the minimum inner diameter, the corresponding black boxes in Fig. 6(a) and Fig. 6(b) shows that our method can find the narrowest section of the vessel, which have the minimum inner diameters. (3) For the minimum radius of curvature, it can be shown in Fig. 6(b) that our method can find the point (red) with minimum radius of curvature. And for the accuracy of calculated radius of curvature, because the error of radius of curvature mainly comes from the fitness of the parametric curve equation, so we provide close-up shot of the fitting curve in Fig. 6(c) and the mean square error between optimized curve equation and the local points to verify the accuracy of the radius of curvature obtained by our method.

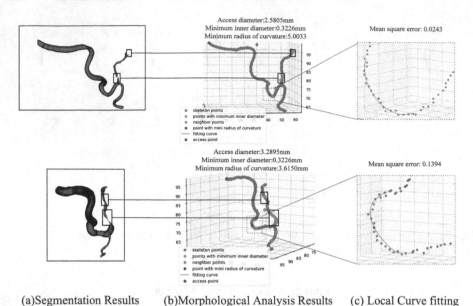

(a)Segmentation Results   (b)Morphological Analysis Results   (c) Local Curve fitting

Fig. 6. Morphological analysis results of two cases. The bule points are vessels' skeleton points extracted from volume data, the purple point is the entrance skeleton point of intracranial artery, the orange points are the skeleton points with minimum inner diameter and the red point is the skeleton point with minimum radius of curvature while the green points are its neighbors. The purple spline is the fitted parametric curve.

5 Conclusion

In this paper, we propose a two-stage framework for segmentation and morphological analysis of the intracranial artery. The SOTA performance in the IAS-3D-DSA dataset shows that the proposed model can effectively enrich the segmentation of local-region with global information. Moreover, the proposed analysis algorithm can obtain morphological information robustly. This framework can be used to help clinicians in preoperative planning and selection of surgical instruments to reduce the operation time and surgery cost.

Acknowledgements. This work was supported in part by the National Key Research and Development Program of China under Grant (2022YFB4700902), the National High Level Hospital Clinical Research Funding (2022-PUMCH-B-125), and the National Natural Science Foundation of China under Grant (62303463, U1913210, 62073325, 62003343, 62222316).

References

1. Nassef, A.M., Awad, E.M., El-bassiouny, A.A., et al.: Endovascular stenting of medically refractory intracranial arterial stenotic (ICAS) disease (clinical and sonographic study). Egypt. J. Neurol. Psychiatry Neurosurg. **56**(1), 1–12 (2020)

2. Gao, P., Wang, T., Wang, D., et al.: Effect of stenting plus medical therapy vs medical therapy alone on risk of stroke and death in patients with symptomatic intracranial stenosis: the CASSISS randomized clinical trial. J. Am. Med. Assoc. **328**(6), 534–542 (2022)
3. Ronneberger, O., Fischer, P., Brox, T.: U-Net: convolutional networks for biomedical image segmentation. In: Navab, N., Hornegger, J., Wells, W., Frangi, A. (eds.) MICCAI 2015, Part III. LNCS, vol. 9351, pp. 234–241. Springer, Cham (2015). https://doi.org/10.1007/978-3-319-24574-4_28
4. Çiçek, Ö., Abdulkadir, A., Lienkamp, S.S., Brox, T., Ronneberger, O.: 3D U-Net: learning dense volumetric segmentation from sparse annotation. In: Ourselin, S., Joskowicz, L., Sabuncu, M., Unal, G., Wells, W. (eds.) MICCAI 2016, Part II. LNCS, vol. 9901, pp. 424–432. Springer, Cham (2016). https://doi.org/10.1007/978-3-319-46723-8_49
5. Balsiger, F., Soom, Y., Scheidegger, O., Reyes, M.: Learning shape representation on sparse point clouds for volumetric image segmentation. In: Shen, D., et al. (eds.) MICCAI 2019, Part II. LNCS, vol. 11765, pp. 273–281. Springer, Cham (2019). https://doi.org/10.1007/978-3-030-32245-8_31
6. Milletari, F., Navab, N., Ahmadi, S.A..: V-Net: fully convolutional neural networks for volumetric medical image segmentation. In: 2016 Fourth International Conference on 3D Vision (3DV), pp. 565–571 (2016)
7. Chen, H., Dou, Q., Yu, L., et al.: VoxResNet: deep voxelwise residual networks for volumetric brain segmentation. arXiv preprint arXiv:1608.05895 (2016)
8. Chen, J., Lu, Y., Yu, Q., et al.: TransUNet: transformers make strong encoders for medical image segmentation. arXiv preprint arXiv:2102.04306 (2021)
9. Wang, W., Chen, C., Ding, M., Yu, H., Zha, S., Li, J.: TransBTS: multimodal brain tumor segmentation using transformer. In: de Bruijne, M., et al. (eds.) MICCAI 2021, Part I. LNCS vol. 12901, pp. 109–119. Springer, Cham (2021). https://doi.org/10.1007/978-3-030-87193-2_11
10. Hatamizadeh, A., Tang, Y., Nath, V., et al.: UNETR: transformers for 3D medical image segmentation. In: Proceedings of the IEEE/CVF Winter Conference on Applications of Computer Vision, pp. 574–584 (2022)
11. Tang, Y., Yang, D., Li, W., et al.: Self-supervised pre-training of swin transformers for 3D medical image analysis. In: Proceedings of the IEEE/CVF Conference on Computer Vision and Pattern Recognition (CVPR), pp. 20730–20740 (2022)
12. Dosovitskiy A., Beyer, L., Kolesnikov, A., et al.: An image is worth 16x16 words: transformers for image recognition at scale. arXiv preprint arXiv:2010.11929 (2020)
13. Liu, Z., Lin, Y., Cao, Y., et al.: Swin transformer: hierarchical vision transformer using shifted windows. In: Proceedings of the IEEE/CVF International Conference on Computer Vision (ICCV), pp. 10012–10022 (2021)
14. Cao, H., et al.: Swin-Unet: Unet-like pure transformer for medical image segmentation. In: Karlinsky, L., Michaeli, T., Nishino, K. (eds.) ECCV 2022, Part III. LNCS vol. 13803, pp. 205–218. Springer, Cham (2023). https://doi.org/10.1007/978-3-031-25066-8_9
15. Peiris, H., Hayat, M., Chen, Z., Egan, G., Harandi, M.: A robust volumetric transformer for accurate 3D tumor segmentation. In: Wang, L., Dou, Q., Fletcher, P.T., Speidel, S., Li, S. (eds.) MICCAI 2022, Part V. LNCS, vol. 13435, pp. 162–172. Springer, Cham (2022). https://doi.org/10.1007/978-3-031-16443-9_16
16. Palágyi, K., et al.: A sequential 3D thinning algorithm and its medical applications. In: Insana, M.F., Leahy, R.M. (eds.) IPMI 2001. LNCS, vol. 2082, pp. 409–415. Springer, Heidelberg (2001). https://doi.org/10.1007/3-540-45729-1_42
17. Loshchilov, I., Hutter, F.: Decoupled weight decay regularization. arXiv preprint arXiv:1711.05101 (2017)

Effective Domain Adaptation for Robust Dysarthric Speech Recognition

Shanhu Wang[1] ⓘ, Jing Zhao[1,2]([✉]) ⓘ, and Shiliang Sun[1,2] ⓘ

[1] School of Computer Science and Technology, East China Normal University,
Shanghai, China
shwang1202@163.com, slsun@cs.ecnu.edu.cn
[2] Key Laboratory of Advanced Theory and Application in Statistics and Data
Science, Ministry of Education, Shanghai, China
jzhao@cs.ecnu.edu.cn

Abstract. By transferring knowledge from abundant normal speech to limited dysarthric speech, dysarthric speech recognition (DSR) has witnessed significant progress. However, existing adaptation techniques mainly focus on the full leverage of normal speech, discarding the sparse nature of dysarthric speech, which poses a great challenge for DSR training in low-resource scenarios. In this paper, we present an effective domain adaptation framework to build robust DSR systems with scarce target data. Joint data preprocessing strategy is employed to alleviate the sparsity of dysarthric speech and close the gap between source and target domains. To enhance the adaptability of dysarthric speakers across different severity levels, the Domain-adapted Transformer model is devised to learn both domain-invariant and domain-specific features. All experimental results demonstrate that the proposed methods achieve impressive performance on both speaker-dependent and speaker-independent DSR tasks. Particularly, even with half of the target training data, our DSR systems still maintain high accuracy on speakers with severe dysarthria.

Keywords: Domain Adaptation · Dysarthric Speech Recognition · Low-Resource Speech

1 Introduction

Benefiting from large-scale speech datasets and advanced machine learning approaches, Automatic Speech Recognition (ASR) has achieved excellent recognition performance in speech-based interaction. However, there still exist huge challenges to boost dysarthric speech recognition (DSR) for helping individuals with dysarthria, who are generally characterized by effortful, slurred, slow or prosodically abnormal speech that is indiscernible to human listeners [22]. The well-trained deep neural network (DNN) based ASR framework on normal speech cannot be suitably applied to DSR task with a small amount of dysarthric speech, in which the high inter- and intra-speaker variability severely hinders the modeling of the dysarthric speech features.

B. Luo et al. (Eds.): ICONIP 2023, CCIS 1964, pp. 62–73, 2024.
https://doi.org/10.1007/978-981-99-8141-0_5

Previous works have made great efforts to tackle these problems, where large amounts of out-of-domain data are leveraged for transfer learning due to the shared lexical knowledge with dysarthric speech. As one prominent form of transfer learning, domain adaptation aims to reduce the mismatch between normal speech (source domain) and dysarthric speech (target domain), which draws the interest of many researchers. Among the existing adaptation methods of DSR, most rely on at least one of the following adaptation mechanisms: simulating dysarthric speech for data augmentation or modeling the latent features of dysarthric speech in the acoustic space.

For the aspect of data inconsistency, data augmentation has been widely exploited to generate additional intermediate domain data to formulate an explicit transformation from typical speech to atypical speech. Vachhani et al. [22] modifies the tempo and speed of healthy speech to simulate dysarthric speech with different severity levels. Similarly, the non-linear speech tempo transformations in [25] are employed to adjust typical speech towards dysarthric speech for personalized DSR training. Soleymanpour et al. [19] synthesize dysarthric speech through multi-speaker Text-To-Speech technology to enrich training data. Additionally, the conventional Data combination scheme [4] makes full use of out-of-domain knowledge to enhance DSR, but it is not always feasible due to the size and structure differences between source and target data. Therefore, the potentially beneficial source domain data is selected by the entropy of posterior probability for more efficient transfer learning towards the target domain [26].

These methods mentioned above attach great importance to the utilization of source data, and another is the model-centric approach [15], which focuses on feature extraction, loss function or model architecture. For example, previous works [3,18] have investigated speaker adaptation technologies like maximum a posteriori [7] or maximum likelihood linear regression [12] to build the speaker-adapted system, which requires a baseline of speaker-independent (SI) system for fine-tuning. In general, the purpose of these methods is to minimize the mismatch between a generic baseline of SI model and the intended target speaker [18], where the performance of SI system is not as good as that of speaker-dependent (SD) system. Besides, DNN weight adaptation [8] is proposed to add new task-specific adaptation layers over the internal layers of DNN trained on a large dataset for transfer learning. Recently, the hybrid adaptation methods combining data augmentation with elaborate paradigms obtain promising DSR performance. The contrastive learning of multiple data augmentations is implemented with an attentive temporal pyramid network to learn powerful speech representation [24]. Multi-task learning [5] based on the DSR task and the reconstruction of simulated dysarthric speech is proposed to explicitly learn robust intermediate representations and better adapt to the target domain. On the contrary, another line of work concentrates on the target-to-source transformation, adopting denoising autoencoder [1] to enhance dysarthric speech to match healthy control speech, which is empirically more difficult than the source-to-target adaptation.

Nowadays, there is a growing need to investigate DSR systems with limited target speech data to reduce specific speaker dependency, which is conducive to the generalization of dysarthric speakers with various severity degrees. Despite data augmentation and data selection on the source domain can make the performance improvement for DSR, in fact, they just change the source data distribution, without modifying the sparsity of target data distribution. Moreover, the scale difference between source and target data is greater in low-resource scenarios. Therefore, in this paper, we present effective hybrid domain adaptation methods to get robust DSR systems, which are more suitable for low-resource scenarios. The data balance strategy is designed to mitigate the data distribution shift between source and target domains, simultaneously the imbalance among different speakers in the target domain is also considered. Besides, our proposed data augmentation is integrated with data balance as the joint strategy of data preprocessing to overcome the sparsity and feature inconsistency of target data. We further explore multiple encoders with domain feature fusion to perceive domain-specific features, enhancing DSR for all levels of dysarthria. Experimental results demonstrate that our proposed domain adaptation framework is conducive to getting advanced and robust DSR performance, especially when half of the target training data is exploited to simulate extremely low-resource scenarios.

2 Proposed Domain Adaptation Methods

In this section, our proposed data preprocessing and model architecture will be described respectively, which aims to make full utilization of limited target data and performs effective domain adaptation for robust DSR systems. Figure 1 depicts the overall structure of the proposed Domain-adapted Transformer (DA Transformer), as well as the initial data preprocessing.

2.1 Data Preprocessing

It is well known that normal speech and dysarthric speech differ significantly in pronunciation characteristics and data size, which leads to obvious inconsistency in the distribution of source and target data. When incorporating ample normal speech for domain adaptation, the previous data augmentation and data selection methods do not change the sparse nature of target data. As a consequence, DSR models tend to exhibit biases towards the source or intermediate domain data during training. This point is often overlooked and hinders models from perceiving the unique features of dysarthric speech. Therefore, we propose data-centric domain adaptation methods that involve data balance and data augmentation to modify the distribution shift in the source and target domains.

Data Balance. To relieve the imbalance between source and target data, as well as the imbalance of dysarthric speakers with different severity levels in the target domain, a weighted random data sampler is designed to sample target utterances with higher weights. The sampling weights of each speaker's utterance from source and target domains are formulated as follows

$$p_{s_i} = (\frac{N}{N_s})^{1-\tau}, \quad p_{t_i} = (\frac{N}{N_t})^{1-\tau}(\frac{N_t}{n_{t_i}})^{\tau}, \tag{1}$$

where N_s and N_t denote the total utterance number of source and target domains respectively, having $N = N_s + N_t$, n_{t_i} indicates the utterance number of the i-th dysarthric speaker, and τ is a balance factor to avoid too high weight when fewer target utterances are trained. Therefore, the target domain data can be sampled preferentially and repeatedly to overcome its sparsity.

Data Augmentation. Previous studies [22,25] have shown the effectiveness of tempo adjustment for DSR task, so we are motivated to construct some disturbances on the time direction of frame sequence in the source domain. The devised augmentation method is called variable time masking (VTM), which further reduces the distribution distance between source and target data. Specifically, we randomly mask variable-number frame blocks with the mean of the whole log mel spectrogram. The starting and ending positions of k-th masked frame blocks are acquired by random function (Rand):

$$
\begin{aligned}
N_{mask} &= L_x/10 * \rho, \\
start_k &= \text{Rand}(0, L_x - T), \\
end_k &= start_k + \text{Rand}(0, T) \quad k = 1, ..., N_{mask}.
\end{aligned}
\tag{2}
$$

N_{mask} is the variable number of frame blocks to mask. L_x denotes the total length of source frame sequence x and ρ is the masking rate. T denotes the maximum length of masked frame block, set to 40.

On the whole of data preprocessing, we argue that only by solving the sparsity of data in the target domain can source data distribution match target data distribution as much as possible with the help of data augmentation. In this way, DSR system can overcome the under-fitting or over-fitting problem [20] and maximize the perception of domain-invariant features, which is an important premise of sensing domain-specific features.

2.2 Domain-Adapted Transformer

In order to improve the adaptability of dysarthric speakers with different severity levels, we expect to enhance DSR systems to learn the unique features of source and target data during training. Motivated by the Mixture-of-Expert [6], we design the multi-encoder DA Transformer to perform a domain feature fusion mechanism, which aims to better model the inter- and intra-speaker variability in dysarthric speech.

As the Fig. 1 shows, we adopt the joint CTC-attention-based Transformer framework [9] as the backbone, whose attention mechanism is able to capture long-term dependence in parallel [14]. After data preprocessing, the acoustic feature x of healthy control speech or dysarthric speech is sent to the Transformer-based shared encoder and two specific encoders to learn different acoustic representations z_s and z_t. The shared encoder is designed to extract shallow speech features, which corresponds to the lower layers of Transformer encoder. Two specific encoders corresponding to the upper layers of Transformer encoder, namely

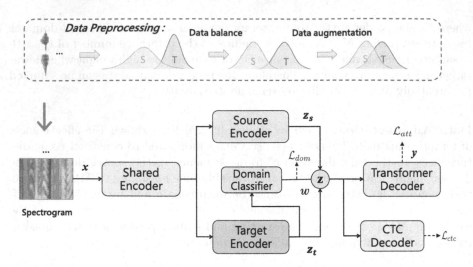

Fig. 1. The whole pipeline of our proposed domain adaptation framework with the data preprocessing and the Domain-adapted Transformer.

source encoder and target encoder, learn the domain-specific features under the driving of the fusion weight.

Domain feature fusion is implemented by integrating z_s and z_t through fusion weight learned from the domain classifier, which takes the acoustic representation z_t as input and outputs the probability p_{dom} to determine how much information of speech representation stems from target domain. The domain classifier is composed of a Bi-LSTM [2], followed by a linear projection and a sigmoid function on the last frame. The domain labels are divided into 0 (source) and 1 (target). We regard the domain probability p_{dom} as a query mapping the output z_t of the target encoder via dot product. Through a linear projection and a sigmoid function, the fusion weight w is learned. The final acoustic representation z is obtained by fusing two specific encoder outputs z_t and z_s as following

$$w = \text{Sigmoid}[\text{Linear}(z_t * p_{dom})],$$
$$z = w * z_t + (1 - w) * z_s. \tag{3}$$

It is precisely the different fusion weights that enable the model to adapt to dysarthric speakers with different severity levels. Then the fused acoustic representation z is sent to the CTC decoder and Transformer decoder for the generation of text token sequence. CTC decoder performs monotonic alignment between acoustic representation and text [23]. The Transformer decoder is to autoregressively generate hypothesis $y_{1:t}$ according to predicted text tokens $y_{1:t-1}$ and global acoustic representation z.

DA Transformer is optimized by the following loss function, which consists of attention loss, domain classification loss and CTC loss, and can be written as the weighted sum of three negative log-likelihood

$$
\begin{aligned}
\mathcal{L} =& (1 - \lambda)(\mathcal{L}_{att} + \mathcal{L}_{dom}) + \lambda\mathcal{L}_{ctc} \\
=& (1 - \lambda)(- \log p_{att}(y_{1:t}|y_{1:t-1}, z) - \log p_{dom}(z|c)) \\
& + \lambda(- \log p_{ctc}(y_{1:t}|y_{1:t-1}, z)),
\end{aligned} \tag{4}
$$

where λ is a coefficient and set to 0.3 empirically. The Transformer benchmark for comparison neglects the domain classification loss.

3 Experiments

3.1 Data Description

We first exploit the TED-LIUM v2 [16] corpus to pre-train the Transformer benchmark [9] for a typical ASR task, which contains around 200-hour clear English TED talks. The fine-tuning experiments are conducted on TORGO [17] corpus, which consists of about 15 h of aligned acoustic and articulatory recordings with 7 healthy control speakers and 8 dysarthric speakers. The healthy control speech is considered as source data, and the dysarthric speech is seen as target data, including 5500 dysarthric speech and 10894 normal speech utterances. The severity degree of the disorder for each of the dysarthric speakers was grouped into four types including severe, severe-moderate, moderate and mild, evaluated by a speech-language pathologist based on the overall clinical intelligibility. We fine-tune and evaluate the speaker-dependent and the speaker-independent DSR with two data settings:

- Speaker-dependent (SD): the healthy control speech and the dysarthric speech are both divided into the training set and test set with a 4:1 split, remaining the proportion of each speaker's utterance number unchanged at the same time.
- Speaker-independent (SI): the training data contains the other 14 speakers except for the particular target dysarthric speaker used in the test stage. Hence, the SI system is personalized for each dysarthric speaker.

The 80-dimensional acoustic features are extracted by filterbank (Fbank) technology as input sequences for all the experiments. The widely used character error rate (CER) and word error rate (WER) metrics for each dysarthric speaker are calculated for the performance estimation of DSR systems.

3.2 Experimental Setup

Model Configurations. All experiments are conducted on the Transformer benchmark [9], which includes a 12-layer Transformer encoder and a 6-layer Transformer decoder with a hidden dimension of 256 and 8 self-attention heads following [10]. It is pre-trained on the TED-LIUM v2 corpus for 50 epochs on 2x NVIDIA RTX 2080Ti GPUs with a minibatch size of 32.

For the speaker-dependent DSR task, we first evaluate the effect of proposed data preprocessing on the SD data setting by fine-tuning the Transformer benchmark with different data preprocessing for 100 epochs with a minibatch size of 16. Then we further make comparisons between our DA Transformer and other models to verify the effect of domain feature fusion. For DA Transformer, we initialize the shared encoder and two specific encoders using the parameters of the lower 6-layer and upper 6-layer Transformer encoder of the pre-trained model respectively, as well as the Transformer decoder and CTC decoder. The domain classifier has a one-layer Bi-LSTM with a hidden dimension of 128, whose parameters are randomly initialized. Additionally, the Transformer benchmark with domain classifier (DC) and Transformer benchmark with domain adversarial training (DAT) [21] are designed to make comparisons on the learned features. DAT is implemented by inserting a gradient reverse layer [21] between Transformer encoder and the following domain classifier to reverse the gradient from domain classifier. On the basis of data preprocessing, we fine-tune these models for 100 epochs with a minibatch size of 16. Finally, we evaluate the hybrid domain adaptation framework with a combination of the proposed data preprocessing and DA Transformer on the speaker-independent DSR task. For all experiments, we make a comparison between 50%-shot and 100%-shot settings to further verify the robustness of our DSR systems.

Parameters Settings. The τ is separately set to 0.8 and 0.4 in 50%-shot and 100%-shot settings. In 50%-shot setting, the source data becomes relatively large when the target data is halved, so the sampling weights of target data are amplified during data balance. To avoid excessive sampling of target data, we reduce the value of $1 - \tau$ by setting τ to 0.8 in 50%-shot setting. Similarly, the masking rate ρ of VTM is set to 10% in 50%-shot setting, and 5% in 100%-shot setting. Because more disturbances in healthy control speech are needed to map dysarthric speech in the low-resource case.

3.3   Evaluation on Speaker-Dependent DSR

We first evaluate our proposed domain adaptation framework on the SD task. The comparisons of different data preprocessing methods are shown in Table 1, and the comparisons of different models are shown in Table 2.

Effect of Data Preprocessing. Table 1 demonstrates that our proposed data preprocessing modifies the distribution shift between source and target data. With the data balance strategy, obvious WER reduction is achieved in 50%-shot setting. In order to analyze the effect of our VTM augmentation, we also make a comparison with TM, which masks a fixed number of frame blocks in [13]. The results in the last row show that our variable-number masked frame blocks are more conducive to bridging the huge gap between source and target data.

Although the Transformer baseline with TM outperforms that with VTM when only considering data augmentation, the VTM augmentation on source data turns the tide when combined with the data balance strategy. This phenomenon reveals that after data balance, VTM can easily transfer source domain features to the intermediate domain closer to the target domain. But before modifying the sparse target data distribution, model pays more attention to the source domain, resulting in excessive interference from VTM on the source data. Therefore, it explains why the data balance and the VTM augmentation should be jointly utilized. Our data preprocessing finally achieves 48% CER reduction and 51% WER reduction in 50%-shot setting, which indicates that our method can overcome the difficulty of extremely imbalanced training data.

Table 1. Illustration of all CER and WER results on data preprocessing for speaker-dependent DSR task, where 'Da_ba' and 'Da_aug' indicate data balance and data augmentation, 'Trans' denotes the Transformer benchmark and 'TM' denotes the time masking augmentation in [13].

Model	Da_ba	Da_aug	100%-shot		50%-shot	
			CER	WER	CER	WER
Trans	–	–	3.1	3.8	7.9	8.8
Trans	–	TM [13]	3.1	3.2	4.9	5.3
Trans	✓	TM [13]	2.8	**3.0**	4.2	4.6
Trans	✓	-	3.2	3.7	5.6	6.0
Trans	–	VTM	3.1	3.8	5.1	5.4
Trans	✓	VTM	**2.6**	**3.0**	**4.1**	**4.3**

Table 2. Illustration of all CER and WER results on different models for speaker-dependent DSR task, 'Trans+DAT' denotes Transformer benchmark with domain adversarial training [21] as a comparative model, 'Trans+DC' denotes Transformer benchmark with domain classifier to ablate domain feature fusion mechanism.

Model	Da_ba	Da_aug	100%-shot		50%-shot	
			CER	WER	CER	WER
Trans	✓	VTM	2.6	3.0	4.1	4.3
Trans+DAT [21]	✓	VTM	2.9	3.1	4.0	4.3
Trans+DC	✓	VTM	2.6	3.0	3.6	3.9
DA_Trans	✓	VTM	**2.4**	**2.8**	**3.4**	**3.6**

Effect of Domain-Adapted Transformer. Table 2 shows the CER and WER results of ablation and comparative models. On the basis of data preprocessing, our **DA_Trans** achieves remarkable CER reduction to 3.4% and WER reduction to 3.6% in 50%-shot setting. To further analyze the role of domain feature fusion in our **DA_Trans**, we list the specific WER results varying from dysarthric speakers for 50%-shot in Table 3.

Trans+DAT model is designed to learn domain-invariant features, where the lowest WER on severe-moderate speaker can be obviously observed. Generally, the mild dysarthric speech is much closer to the source domain, but the severe dysarthric speech is far away from the source domain. Therefore, the domain-invariant feature is limited to model speaker variability of dysarthric speech.

For **Trans+DC** model, it achieves lower WER on both severe and mild dysarthric speakers, which demonstrates that the domain classifier is able to capture the specific features of source and target domains simultaneously. More importantly, we can clearly observe that **DA_Trans** achieves lower WER than **Trans+DC** on the severe (M01, M02) and moderate (F03) dysarthric speakers, which indicates that the domain feature fusion mechanism can better model both the inter- and intra-speaker variability of dysarthric speech. Overall, our domain adaptation framework concentrates more on learning complementary features in the source and target domains.

Table 3. Illustration of WER for the speaker-dependent DSR varying from dysarthric speakers and different models in 50%-shot setting. 'Transformer' is abbreviated as 'Trans'.

Severity	Speaker	Trans	Trans+DAT	Trans+DC	DA_Trans
Severe	F01	**0.0**	1.7	**0.0**	**0.0**
	M01	5.4	4.6	3.8	**3.5**
	M02	2.1	2.8	3.4	**1.8**
	M04	6.6	6.3	**4.6**	5.4
Severe-Mod.	M05	11.2	**10.2**	10.5	10.5
Moderate	F03	**4.5**	5.6	5.4	**4.5**
Mild	F04	3.3	2.5	**1.9**	**1.9**
	M03	0.5	1.5	**0.3**	**0.3**
	Average	4.3	4.3	3.9	**3.6**

3.4   Evaluation on Speaker-Independent DSR

The effectiveness of the data preprocessing and DA Transformer model has been evaluated detail by detail in the above speaker-dependent DSR task. Here, we further apply the hybrid domain adaptation framework to the SI system, whose WER results varying from dysarthric speakers are shown in Fig. 2.

It is seen that the impact of reducing training data is loosely correlated with the severity of speaker's impairments. Ordinarily, data from other dysarthric speakers usually improve the recognition accuracy for the dysarthric speakers with moderate or severe dysarthria [26]. So the recognition performance of speakers with severe and severe-moderate dysarthria is more affected than that of mild

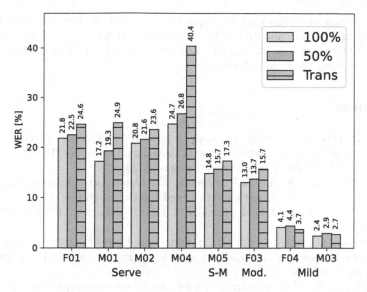

Fig. 2. Comparison of the speaker-independent DSR system in 100%-shot and 50%-shot settings and the strong Transformer baseline [24] marked with 'Trans' in terms of WER.

dysarthric speakers as the reduction of training data. Nevertheless, there is no notable decline in recognition performance in 50%-shot setting for our SI system, which even outperforms the strong Transformer baseline [24] trained in 100%-shot setting on speakers with severe and moderate dysarthria. Therefore, it is concluded that our methods can mitigate the negative impact on learning domain-specific features caused by the deficiency of target data. Moreover, our SI system is only fine-tuned with 50 epochs, which burns off less time and resources than the strong baseline fine-tuned with a wider network for 100 epochs.

4   Conclusions

In this paper, we propose the hybrid domain adaptation framework to improve the performance of DSR in the low-resource case. We investigate the data mismatch between source and target domains and provide feasible solutions from different perspectives. We first present the joint strategy of data balance and data augmentation to relieve the performance degradation caused by the deficiency of dysarthric data. Then the DA Transformer with domain feature fusion mechanism has been demonstrated to successfully capture the domain-specific features while sensing the domain-invariant features, enhancing the generalization of DSR systems over various dysarthric speakers. For future work, we intend to explore ways of continual learning for personalized DSR to better reflect the speech impairment characteristics of individuals, and investigate the unsupervised domain adaptation without target label [11] for the generalization of DSR.

Acknowledgements. This work was supported by the STCSM Project 22ZR1421700, NSFC Projects 62006078 and 62076096, Shanghai Knowledge Service Platform Project ZF1213, the Open Research Fund of KLATASDS-MOE in ECNU and the Fundamental Research Funds for the Central Universities.

References

1. Bhat, C., Das, B., Vachhani, B., Kopparapu, S.K.: Dysarthric speech recognition using time-delay neural network based denoising autoencoder. In: Proceedings of INTERSPEECH 2018, pp. 451–455, September 2018
2. Chan, W., Jaitly, N., Le, Q.V., Vinyals, O.: Listen, attend and spell: a neural network for large vocabulary conversational speech recognition. In: Proceedings of ICASSP 2016, pp. 4960–4964, March 2016. https://doi.org/10.1109/ICASSP.2016. 7472621
3. Christensen, H., et al.: Combining in-domain and out-of-domain speech data for automatic recognition of disordered speech. In: Proceedings of INTERSPEECH 2013, pp. 3642–3645, August 2013
4. Deng, L., Li, X.: Machine learning paradigms for speech recognition: an overview. IEEE Trans. Speech Audio Process. **21**(5), 1060–1089 (2013)
5. Ding, C., Sun, S., Zhao, J.: Multi-task transformer with input feature reconstruction for dysarthric speech recognition. In: Proceedings of ICASSP 2021, pp. 7318–7322, June 2021
6. Gaur, N., et al.: Mixture of informed experts for multilingual speech recognition. In: Proceedings of ICASSP 2021, pp. 6234–6238, June 2021
7. Gauvain, J., Lee, C.: Maximum a posteriori estimation for multivariate gaussian mixture observations of markov chains. IEEE Trans. Speech Audio Process. **2**(2), 291–298 (1994)
8. Ghahremani, P., Manohar, V., Hadian, H., Povey, D., Khudanpur, S.: Investigation of transfer learning for ASR using LF-MMI trained neural networks. In: Proceedings of ASRU 2017, pp. 279–286, December 2017
9. Karita, S., Soplin, N.E.Y., Watanabe, S., Delcroix, M., Ogawa, A., Nakatani, T.: Improving transformer-based end-to-end speech recognition with connectionist temporal classification and language model integration. In: Proceedings of INTERSPEECH 2019, pp. 1408–1412. ISCA, September 2019
10. Karita, S., et al.: A comparative study on transformer vs RNN in speech applications. In: Proceedings of ASRU 2019, pp. 449–456, December 2019. https://doi.org/10.1109/ASRU46091.2019.9003750
11. Kouw, W.M., Loog, M.: A review of domain adaptation without target labels. IEEE Trans. Pattern Anal. Mach. Intell. **43**(3), 766–785 (2021)
12. Leggetter, C.J., Woodland, P.C.: Maximum likelihood linear regression for speaker adaptation of continuous density hidden markov models. Comput. Speech Lang. **9**(2), 171–185 (1995)
13. Park, D.S., et al.: Specaugment: a simple data augmentation method for automatic speech recognition. In: Proceedings of INTERSPEECH 2019, pp. 2613–2617, September 2019
14. Qin, Y., Ding, J., Sun, Y., Ding, X.: A transformer-based model for low-resource event detection. In: Mantoro, T., Lee, M., Ayu, M.A., Wong, K.W., Hidayanto, A.N. (eds.) ICONIP 2021. LNCS, vol. 13111, pp. 452–463. Springer, Cham (2021). https://doi.org/10.1007/978-3-030-92273-3_37

15. Ramponi, A., Plank, B.: Neural unsupervised domain adaptation in NLP - a survey. In: Proceedings of COLING 2020, pp. 6838–6855. International Committee on Computational Linguistics, December 2020

16. Rousseau, A., Deléglise, P., Estève, Y.: Enhancing the TED-LIUM corpus with selected data for language modeling and more TED talks. In: Proceedings of LREC 2014 - Proceedings of the Ninth International Conference on Language Resources and Evaluation, pp. 3935–3939. European Language Resources Association (ELRA), May 2014

17. Rudzicz, F., Namasivayam, A.K., Wolff, T.: The TORGO database of acoustic and articulatory speech from speakers with dysarthria. Lang. Resour. Eval. **46**(4), 523–541 (2012)

18. Sehgal, S., Cunningham, S.P.: Model adaptation and adaptive training for the recognition of dysarthric speech. In: Proceedings of INTERSPEECH 2015, pp. 65–71. Association for Computational Linguistics, September 2015

19. Soleymanpour, M., Johnson, M.T., Soleymanpour, R., Berry, J.: Synthesizing dysarthric speech using multi-speaker tts for dysarthric speech recognition. In: Proceedings of ICASSP 2022, pp. 7382–7386, May 2022. https://doi.org/10.1109/ICASSP43922.2022.9746585

20. Sun, S., Zhao, J.: Pattern Recognition and Machine Learning. Tsinghua University Press, China (2020)

21. Sun, S., Yeh, C., Hwang, M., Ostendorf, M., Xie, L.: Domain adversarial training for accented speech recognition. In: Proceedings of ICASSP 2018, pp. 4854–4858, April 2018

22. Vachhani, B., Bhat, C., Kopparapu, S.K.: Data augmentation using healthy speech for dysarthric speech recognition. In: Proceedings of INTERSPEECH 2018, pp. 471–475, September 2018

23. Watanabe, S., Hori, T., Kim, S., Hershey, J.R., Hayashi, T.: Hybrid CTC/attention architecture for end-to-end speech recognition. IEEE J. Sel. Top. Sig. Process. **11**(8), 1240–1253 (2017)

24. Wu, L., Zong, D., Sun, S., Zhao, J.: A sequential contrastive learning framework for robust dysarthric speech recognition. In: Proceedings of ICASSP 2021, pp. 7303–7307, June 2021

25. Xiong, F., Barker, J., Christensen, H.: Phonetic analysis of dysarthric speech tempo and applications to robust personalised dysarthric speech recognition. In: Proceedings of ICASSP 2019, pp. 5836–5840, May 2019

26. Xiong, F., Barker, J., Yue, Z., Christensen, H.: Source domain data selection for improved transfer learning targeting dysarthric speech recognition. In: Proceedings of ICASSP 2020, pp. 7424–7428, May 2020. https://doi.org/10.1109/ICASSP40776.2020.9054694

TCNet: Texture and Contour-Aware Model for Bone Marrow Smear Region of Interest Selection

Chengliang Wang[1]([✉]), Jian Chen[1], Xing Wu[1]([✉]), Zailin Yang[2], Longrong Ran[2], and Yao Liu[2]

[1] College of Computer Science, Chongqing University, Chongqing 400000, China
{wangcl,wuxing}@cqu.edu.cn
[2] Department of Hematology-Oncology, Chongqing Key Laboratory of Translational Research for Cancer Metastasis and Individualized Treatment, Chongqing University Cancer Hospital, Chongqing 400000, China
{zailinyang,liuyao77}@cqu.edu.cn

Abstract. Bone marrow smear cell morphology is the quantitative analysis of bone marrow cell images. Due to the cell overlap and adhesion in bone marrow smears, it is essential to select uniformly distributed and clear sections as regions of interest (ROIs). However, current ROI selection models have not considered the characteristics of bone marrow smears, resulting in poor performance in practical applications. By comparing bone marrow smear ROIs and non-ROIs, we have identified significant differences in fundamental features, such as texture and contour. Therefore, we propose a texture and contour-aware bone marrow smear ROI selection model (TCNet). Inspired by multi-task learning, this model enhances its feature extraction capabilities for texture and contour by constructing different prediction modules to learn feature representations of texture and contour, and applying multi-level deep supervision with pseudo labels. To validate the effectiveness of the proposed method, we evaluate it on a self-built dataset. Experimental results show that the proposed model achieves a 2.22% improvement in classification accuracy compared to the baseline model. In addition, we verify the proposed module's generalizability by testing it on different backbone networks, and the results demonstrate its strong universality.

Keywords: Deep Learning · Bone Marrow Smear · Region of Interest · Cell Morphology · Deep Supervision

1 Introduction

Cell morphology is the detailed observation and analysis of morphology, size, proportion, and staining characteristics in blood and bone marrow smears to detect and classify blood cells, providing doctors with information on patients' blood function and pathological conditions. Current smear examinations are mainly performed manually through microscopic observation and analysis. With the development of imaging technology, high-resolution digital scans of smears, such as whole slide images (WSIs),

B. Luo et al. (Eds.): ICONIP 2023, CCIS 1964, pp. 74–85, 2024.
https://doi.org/10.1007/978-981-99-8141-0_6

can be obtained using automatic scanners, followed by manual observation and analysis of the scanned images. Both methods heavily rely on manual labor, making them time-consuming and labor-intensive. Consequently, there is a significant demand for intelligent image detection approaches.

A large number of studies have been conducted on artificial intelligence-assisted morphological analysis of peripheral blood smears, and satisfactory results have been achieved [1–4]. Compared to peripheral blood smears, bone marrow smears are more complex cytological specimens with more impurities in the background, more diverse cell types and morphologies, higher cell density, and more severe cell adhesion [5]. These issues lead to fewer regions suitable for cytomorphological analysis in bone marrow smears, necessitating the introduction of ROI selection. In fact, only a small portion of the area in whole slide images (WSIs) obtained from electron microscope scans is suitable for cell morphological analysis. In addition, the dimensions of bone marrow smear whole slide images (WSIs) are substantial, resulting in a large number of smaller patches after splitting. Currently, the majority of the region of interest (ROI) selection processes are carried out manually, resulting in a considerable expenditure of time and effort. Consequently, deep learning-based ROI selection methods have gradually been applied to bone marrow smears to improve the efficiency of ROI selection. However, these methods do not consider the characteristics of bone marrow smears, especially in cases of cell adhesion, resulting in poor practical performance.

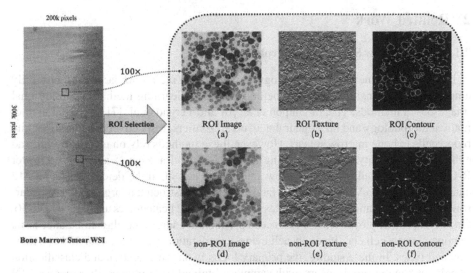

Fig. 1. ROI selection process and comparison between ROI and non-ROI (a) is the ROI image, (b) is the ROI texture image, (c) is the ROI contour image, (d) is the non-ROI image, (e) is the non-ROI texture image, and (f) is the non-ROI contour image.

By comparing the ROI and non-ROI in bone marrow smears, it was found that there were significant differences in texture and contour between the two types of images. The ROI (Fig. 1(a)) have a clear background, while the non-ROI (Fig. 1(d)) contain a large amount of impurities in the background. From the corresponding texture images

(Fig. 1(b) and Fig. 1(e)), these impurities and cells show significant differences in texture. In addition, the cells in the ROI (Fig. 1(a)) are evenly distributed and have clear contours (Fig. 1(c)), while the cells in the non-ROI (Fig. 1(d)) have severe adhesion and blurred contours (Fig. 1(f)). Based on the above analysis, texture and contour are the key features to distinguish between ROI and non-ROI, and the model should enhance the ability to extract texture and contour features.

In this study, we designed a texture and contour-aware ROI classification model for bone marrow smears, named TCNet. This model utilizes texture and contour pseudo labels, conducts texture and contour deep supervision at multiple network levels, and enhances the network's ability to extract texture and contour features from bone marrow smears. To achieve this goal, the following work was carried out:

- Designed a Texture Deep Supervision Module (TDSM) that extracts and predicts textures by calculating feature covariance, using texture pseudo labels for deep supervision to achieve optimized texture extraction.
- Designed a Contour Deep Supervision Module (CDSM) that introduces reverse attention to enhance contour sensitivity and achieve contour prediction, using contour pseudo labels for deep supervision to optimize contour extraction.
- Constructed a bone marrow smear ROI selection dataset to validate the effectiveness of the proposed model TCNet. Simultaneously, TCNet was built on different backbone networks to verify the universality of the proposed modules.

2 Related Work

2.1 Bone Marrow Smear Cell Morphology

In bone marrow smear cell morphology research, two main approaches are primarily employed: traditional image-based methods and deep learning methods. In traditional image-based methods, Theera-Umpon et al. [6] and Pergad et al. [7] employed mathematical morphology and novel neural network classifiers, respectively, to classify white blood cells in bone marrow smears. However, these methods rely on manual design and exhibit poor stability. In deep learning methods, Chandradevan et al. [8] used Faster R-CNN to detect cells in bone marrow smears and classified the detected cells into 12 categories using the VGG network. Matek et al. [9] constructed a large-scale bone marrow smear classification dataset and classified cells into 21 categories using ResNeXt-50. Guo et al. [10] employed a class-balanced approach to address the class imbalance issue in bone marrow cell classification, achieving classification recognition for 15 bone marrow cell types. In these studies, the primary focus is on the detection and classification of cells in bone marrow smears, with samples obtained from images after manual ROI selection.

2.2 Bone Marrow Smear ROI Selection

Currently, in bone marrow smear morphology analysis, ROI selection is primarily done manually. However, with advancements in technology, the ROI selection method based on Convolutional Neural Networks (CNN) has been gradually applied to bone marrow smears. Wang et al. [11] used cascade R-CNN to detect rough regions of interest on the thumbnail of whole slide images (WSIs) of bone marrow smears, and then mapped them to the original image and cropped to obtain ROI slices. This method reduces the computation at high zoom levels by performing preliminary ROI selecting but generates ROI slices containing blurry backgrounds and overlapping cells. Tayebi et al. [12] divided the whole slide image (WSIs) of bone marrow smears into small patches, and then used DenseNet [13] classification network to classify each patch, dividing them into ROI and non-ROI. However, the methods presented in these articles employ general domain models, which do not optimize for the complex background, high cell density, and severe adhesion characteristics of bone marrow smears, resulting in poor performance in actual applications.

3   Method

Based on the above analysis, we thoroughly considered the characteristics of bone marrow smears when designing the model, enhanced texture and contour feature extraction, and designed a texture and contour-aware bone marrow smear ROI selection model (TCNet). In this section, we will introduce the overall structure of the model, followed by the Texture Deep Supervision Module (TDSM) and the Contour Deep Supervision Module (CDSM).

3.1   Overall Model Structure

The overall structure of the model is shown in Fig. 2. The proposed TCNet consists of three basic components: (1) CNN backbone network (ResNet-50 [14]) for extracting image feature representations, (2) TDSM, and (3) CDSM.

As shown in Fig. 2, the input of TCNet is the RGB color image of the bone marrow smear patch, and the output is the image classification result, i.e., ROI or non-ROI. In the training process, the texture pseudo labels and contour pseudo labels used for deep supervision are generated by OpenCV using the LBP (Local Binary Pattern) algorithm and Canny edge detection algorithm, respectively.

For TCNet, given an input $I \in \mathbb{R}^{3 \times H \times W}$, we use a convolutional neural network (ResNet-50) to extract multi-scale features $\{F_i\}_{i=2}^5$. The extracted features (F_2, F_3, F_4, F_5) are the input to the Texture Deep Supervision Module (TDSM) and the Contour Deep Supervision Module (CDSM). For feature $F_2 \in \mathbb{R}^{C_2 \times \frac{H}{4} \times \frac{W}{4}}$, it is sent to both TDSM and CDSM. The TDSM processes the input feature, predicting the texture $T_2 \in \mathbb{R}^{1 \times \frac{H}{4} \times \frac{W}{4}}$ and subsequently upscaling T_2 to match the dimensions of the original image ($H \times W$). The predicted results and texture pseudo labels are used to compute the auxiliary loss, and gradient backpropagation is then performed. Similarly, the CDSM processes the input feature, predicting contour $C_2 \in \mathbb{R}^{1 \times \frac{H}{4} \times \frac{W}{4}}$ and upscaling C_2 to the same size as the original image ($H \times W$). The predicted results and contour pseudo labels are used to compute the auxiliary loss, and gradient backpropagation is then performed. The same operation is performed for F_3, F_4, F_5 to achieve deep supervision of texture and contour at each level. The model finally outputs the classification result.

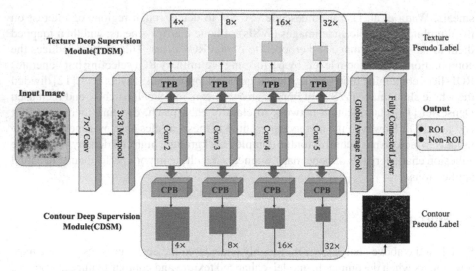

Fig. 2. Schematic diagram of the overall structure of the model (TCNet).

3.2 Texture Deep Supervision Module (TDSM)

In the process of selecting ROI in bone marrow smears, texture features are of great significance for distinguishing between background impurities and cells. Therefore, we specifically designed a Texture Deep Supervision Module (TDSM) to improve the model's texture extraction ability. The module mainly includes a Texture Prediction Block (TPB) and a Texture Auxiliary Loss Calculation Block (TALCB).

The TPB extracts texture and performs texture prediction by calculating feature covariance [15].It takes a feature map with a resolution of $C \times H \times W$ as the input, and multiple calculations using a 1×1 convolution kernel result in a feature map with a resolution of $C_1 \times H \times W$. To improve computational efficiency, C_1 is much smaller than C (in this paper, C_1 is set to 24), and the resulting feature maps are used to learn multiple aspects of texture in subsequent operations. Next, the feature map is fed into the texture extractor, where we calculate the covariance matrix between feature channels at each position to capture the correlation between different responses in convolutional features. The covariance matrix between features measures the co-occurrence of features, describes the combination of features, and is used to represent texture information [16, 17]. As shown in Fig. 3, for each pixel f_n^k in the feature map $f^k \in C_1 \times H \times W$, we compute its covariance matrix as the inner product between f_n^k and $f_n^{k^T}$. Since the covariance matrix ($C_1 \times C_1$) has the property of diagonal symmetry, we use the upper triangular part of the matrix to represent texture features and reshape the result into a feature vector. Perform the same operation on each pixel in the feature map and concatenate the results to obtain $t^k \in \frac{C_1 \times (C_1+1)}{2} \times H \times W$, which contains texture information. Perform a convolution operation with a 1×1 convolution kernel on the obtained feature map to get the predicted texture map.

The TALCB upsamples the texture maps predicted by the TPB at each level to the original image size, calculates auxiliary loss using texture pseudo labels, and performs gradient backpropagation to achieve optimized texture extraction.

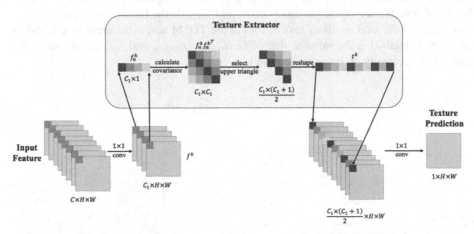

Fig. 3. Schematic diagram of the Texture Prediction Block (TPB).

3.3 Contour Deep Supervision Module (CDSM)

To enhance the network's ability to extract contours, we specifically designed a Contour Deep Supervision Module (CDSM). CDSM includes a Contour Prediction Block (CPB) and a Contour Auxiliary Loss Calculation Block (CALCB).

In the CPB, we introduce reverse attention mechanism [18] to enhance the sensitivity to contour features. As shown in Fig. 4, in the Block, we first perform average pooling and max pooling on the input features and concatenate them, then perform a convolution operation with a 7×7 convolution kernel and Sigmoid operation on the resulting features to obtain attention weights. Next, subtract each element of the attention weights from 1 and perform element-wise multiplication with the input features to achieve feature enhancement. Through the aforementioned processing, the contour features have been enhanced. Finally, use a 3×3 convolution kernel for contour prediction.

The CALCB upsamples the contour maps predicted by the CPB at each level to the original image size, calculates auxiliary loss with contour pseudo labels, and performs gradient backpropagation to achieve optimized contour extraction.

3.4 Loss Function

During the training process, the total loss function is:

$$Loss_{total} = \lambda_{cls}Loss_{cls} + \lambda_{texture}Loss_{texture}$$
$$+ \lambda_{contour}Loss_{contour} \tag{1}$$

In this context, $Loss_{cls}$ represents the class loss, which uses cross-entropy loss. λ_{cls} is the weight of the classification loss, which is set to 1 by default. $Loss_{texture}$ is the total auxiliary texture loss in the TDSM across different levels. The auxiliary texture loss can be defined as $Loss_{texture} = \sum_{j=1}^{J} L_{BCE}^{(j)}$, where J represents the total number of levels of edge features used for supervision, and L_{BCE} is the binary cross-entropy loss. $Loss_{contour}$ is the total auxiliary texture loss in the TDSM across different levels, and its calculation method is the same as that of $Loss_{texture}$. $\lambda_{texture}$ and $\lambda_{contour}$ are set to 0.4 and 0.4, respectively, to balance the total loss.

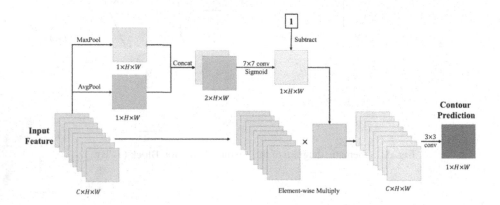

Fig. 4. Schematic diagram of the Contour Prediction Block (CPB).

4 Experiment

4.1 Experimental Setup

Dataset. Since the bone marrow smear ROI dataset constructed in article [12] is not publicly available, we built our own bone marrow smear ROI dataset. The dataset is sourced from different patients and collected from analyzers. We slice the collected bone marrow smear whole slide images (WSIs) with each slice having a size of 1600 × 1600. We manually annotate selected slices, marking slices with uniformly distributed cells and clear contours as ROI, and classifying slices with blurry backgrounds and cell adhesion as non-ROI. A total of 2030 slices were annotated, including 1051 ROI slices and 979 non-ROI slices, which were divided into training and validation sets at an 8:2 ratio. The statistical information of the dataset is shown in Table 1.

Evaluation Metrics. To evaluate the effect of the ROI classification model in predicting ROI and non-ROI, we calculated commonly used performance indicators, such as accuracy, precision, recall, and F1 score. In each of the following tables, the specific values for each metric are presented as percentages without the "%" symbol.

Technical Details. Our TCNet is implemented in PyTorch on a single Nvidia RTX 4090 GPU and trained using the Adam optimizer. All models use pre-trained weights trained

Table 1. Statistical information of the dataset

	ROI	non-ROI	Total
Train	841	784	1625
Validation	210	195	405
Total	1051	979	2030

on ImageNet. During the training process, all models are trained for 50 epochs, with a batch size of 16 and a learning rate of 1e−4. All images are resized to 224 × 224.

4.2 Comparison Experiments

We trained and evaluated TCNet on our self-built dataset, and since our network is based on ResNet-50, we also trained ResNet-50 on the dataset. To compare with the method used in article [12], we trained and evaluated DenseNet121 on our self-built dataset. To more objectively evaluate the performance of the proposed network, we also trained other convolutional neural network-based models on the dataset. Other networks include AlexNet [19] and VGG16 [20].

Table 2. Comparison with other CNN-based methods

Method	Accuracy	Precision	Recall	F1 Score
AlexNet	85.43	85.61	85.29	85.36
VGG16	86.91	87.01	86.81	86.87
ResNet-50	85.43	85.56	85.31	85.37
DenseNet121	86.17	86.28	86.06	86.12
TCNet	**87.65**	**87.88**	**87.51**	**87.59**

The data in Table 2 demonstrates that TCNet significantly outperforms other comparison models in evaluation metrics such as accuracy, precision, recall, and F1 score. Specifically, compared to the baseline network ResNet-50, TCNet achieved improvements of 2.22, 2.32, 2.20, and 2.22% points in terms of accuracy, precision, recall, and F1 score, respectively. This indicates that after optimizing the ResNet-50 structure, TCNet successfully achieved better performance. Additionally, TCNet showed excellent performance compared to DenseNet121, with improvements of 1.48, 1.60, 1.45, and 1.47% points in terms of accuracy, precision, recall, and F1 score, respectively. Despite having a simpler network structure and fewer layers, TCNet outperforms DenseNet121. TCNet's performance improvement was also considerable compared to other models, particularly in terms of precision and F1 score, indicating that it can more accurately make category judgments in the bone marrow smear ROI classification problem and achieve a good balance between precision and recall.

To evaluate the adequacy of our self-built dataset, we divided the training set into 10 parts and gradually expanded the training data scale. Specifically, we first used only one part of the data as the training set, then gradually added training data, adding one part at a time until all 10 parts were used. For each expanded training set, we trained and evaluated the models. Figure 5 shows the trend of model accuracy as the amount of training data increases.

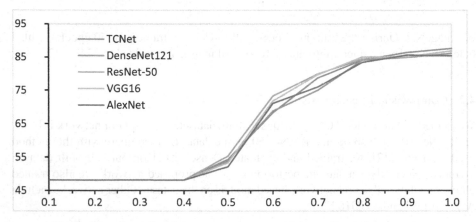

Fig. 5. Line chart of model accuracy changes with increasing training set size.

Figure 5 shows that as the amount of training data increases, the classification accuracy of the model shows an upward trend. However, when the training data volume reaches 80% of the total training set data volume, the accuracy curve tends to stabilize with a small change in magnitude. This indicates that the dataset we built is already sufficient to meet the model's training needs.

4.3 Ablation Experiments

In order to verify the effectiveness of each module, we sequentially added TDSM (Texture Deep Supervision Module) and CDSM (Contour Deep Supervision Module) to the baseline model. The experimental results of each model are shown in Table 2.

Table 3. Ablation studies for different components in our TCNet

TDSM	CDSM	Accuracy	Precision	Recall	F1 Score
		85.43	85.56	85.31	85.37
✓		86.42(0.99 ↑)	87.55(1.99 ↑)	86.26(0.95 ↑)	86.34(0.97 ↑)
	✓	86.91(1.48 ↑)	86.94(1.38 ↑)	86.95(1.64 ↑)	86.88(1.51 ↑)
✓	✓	87.65(2.22 ↑)	87.88(2.32 ↑)	87.51(2.20 ↑)	87.59(2.22 ↑)

Effectiveness of TDSM. From Table 3, it can be seen that after adding TDSM to the baseline model, accuracy, precision, recall, and F1 score increased by 0.99, 1.99, 0.95, and 0.97% points, respectively. The experimental results show that the TDSM can enhance the network's ability to extract texture features, thereby improving the performance of ROI classification.

Effectiveness of CDSM. From Table 3, it can be seen that after adding CDSM to the baseline model, accuracy, precision, recall, and F1 score increased by 1.48, 1.38, 1.64, and 1.51% points, respectively. The experimental results show that the CDSM can enhance the network's ability to extract contour features, thereby improving the performance of ROI classification.

Different Backbone Networks. The proposed network is based on ResNet-50. To verify the universality of the proposed modules, we built TCNet on ResNet-18, ResNet-34, and ResNet-101. The experimental results are shown in Table 4.

Table 4. Performance of TCNet with different backbone networks

Mothod	Accuracy	Precision	Recall	F1 Score
ResNet-18	84.44	84.43	84.41	84.42
TCNet (ResNet-18)	85.43(0.99 ↑)	85.67(1.24 ↑)	85.27(0.86 ↑)	85.35(0.93 ↑)
ResNet-34	85.43	85.41	85.42	85.41
TCNet (ResNet-34)	85.93(0.50 ↑)	86.47(1.06 ↑)	85.70(0.28 ↑)	85.80(0.39 ↑)
ResNet-101	85.68	85.71	85.60	85.64
TCNet (ResNet-101)	87.65(1.97 ↑)	87.95(2.24 ↑)	87.49(1.89 ↑)	87.58(1.94 ↑)

The experimental results show that when TCNet is built on different backbone networks, accuracy, precision, recall, and F1 score all have good improvements, proving that our designed modules have good universality.

Different Weight Coefficient Combinations. When calculating loss, the texture loss weight coefficient and contour loss weight coefficient need to be set. To achieve better results, we conducted experiments with different weight coefficient combinations and selected the experimental results of some weight coefficient combinations, as shown in Table 5.

Table 5. Performance of TCNet with different weight coefficient combination

$\lambda_{texture}$	$\lambda_{contour}$	Accuracy	Precision	Recall	F1 Score
0.0	0.0	85.43	85.56	85.31	85.37
0.2	0.2	86.17	86.52	85.99	86.08
0.4	**0.4**	**87.65**	**87.88**	**87.51**	**87.59**
0.6	0.6	86.17	86.17	86.14	86.15
0.8	0.8	86.67	87.83	86.41	86.53

The experimental results show that when $\lambda_{texture}$ is set to 0.4 and $\lambda_{contour}$ is set to 0.4, the model achieves optimal performance.

5 Conclusion

This paper designs a novel texture and contour-aware deep neural network architecture (TCNet) for bone marrow smear ROI selection. Our key idea is to construct texture and contour prediction modules for multitask learning to perceive texture and contour features and perform multi-level Deep supervision through texture and contour pseud labels. In the model, we designed the Texture Deep Supervision Module (TDSM) and the Contour Deep Supervision Module (CDSM), which supervise each network level through texture and contour pseudo labels, respectively, achieving optimized extraction of texture and contour features. We evaluated our method on our self-built dataset and compared it with the baseline model, showing good improvements in the proposed model's performance. At the same time, we verified the effectiveness of the proposed modules on different backbone networks. In the future, we will explore applying the proposed method for the detection and classification of bone marrow smear cells.

Acknowledgement. This work is supported by the Fundamental Research Funds for the Central Universities (No. 2022CDJYGRH-001) and the Chongqing Technology Innovation & Application Development Key Project (cstc2020jscx-dxwtBX0055; cstb2022tiad-kpx0148).

References

1. Matek, C., Schwarz, S., Spiekermann, K., et al.: Human-level recognition of blast cells in acute myeloid leukaemia with convolutional neural networks. Nat. Mach. Intell. **1**, 538–544 (2019). https://doi.org/10.1038/s42256-019-0101-9
2. Tiwari, P., et al.: Detection of subtype blood cells using deep learning. Cognit. Syst. Res. **52**, 1036–1044 (2018)
3. Hegde, R.B., Prasad, K., Hebbar, H., Singh, B.M.K.: Comparison of traditional image processing and deep learning approaches for classification of white blood cells in peripheral blood smear images. Biocybern. Biomed. Eng. **39**(2), 382–392 (2019)
4. Rastogi, P., Khanna, K., Singh, V.: LeuFeatx: deep learning–based feature extractor for the diagnosis of acute leukemia from microscopic images of peripheral blood smear. Comput. Biol. Med. **142**, 105236 (2022)

5. Lee, S.H., Erber, W.N., Porwit, A., Tomonaga, M., Peterson, L.C., International Councilfor Standardization In Hematology: ICSH guidelines for the standardization of bone marrow specimens and reports. Int. J. Lab. Hematol. **30**(5), 349–364 (2008)
6. Theera-Umpon, N., Dhompongsa, S.: Morphological granulometric features of nucleus in automatic bone marrow white blood cell classification. IEEE Trans. Inf. Technol. Biomed. **11**(3), 353–359 (2007)
7. Pergad, N.D., Hamde, S.T.: Fractional gravitational search-radial basis neural network for bone marrow white blood cell classification. Imaging Sci. J. **66**(2), 106–124 (2018)
8. Chandradevan, R., et al.: Machine-based detection and classification for bone marrow aspirate differential counts: initial development focusing on nonneoplastic cells. Lab. Investig. **100**(1), 98–109 (2020)
9. Matek, C., Krappe, S., Münzenmayer, C., Haferlach, T., Marr, C.: Highly accurate differentiation of bone marrow cell morphologies using deep neural networks on a large image data set. Blood J. Am. Soc. Hematol. **138**(20), 1917–1927 (2021)
10. Guo, L., et al.: A classification method to classify bone marrow cells with class imbalance problem. Biomed. Signal Process. Control **72**, 103296 (2022)
11. Wang, C.W., Huang, S.C., Lee, Y.C., Shen, Y.J., Meng, S.I., Gaol, J.L.: Deep learning for bone marrow cell detection and classification on whole-slide images. Med. Image Anal. **75**, 102270 (2022)
12. Tayebi, R.M., et al.: Automated bone marrow cytology using deep learning to generate a histogram of cell types. Commun. Med. **2**(1), 45 (2022)
13. Huang, G., Liu, Z., Van Der Maaten, L., Weinberger, K.Q.: Densely connected convolutional networks. In: Proceedings of the IEEE Conference on Computer Vision and Pattern Recognition, pp. 4700–4708 (2017)
14. He, K., Zhang, X., Ren, S., Sun, J.: Deep residual learning for image recognition. In: Proceedings of the IEEE Conference on Computer Vision and Pattern Recognition, pp. 770–778 (2016)
15. Ren, J., et al.: Deep texture-aware features for camouflaged object detection. IEEE Trans. Circuits Syst. Video Technol. (2021)
16. Gatys, L.A., Ecker, A.S., Bethge, M.: Image style transfer using convolutional neural networks. In: Proceedings of the IEEE Conference on Computer Vision and Pattern Recognition, pp. 2414–2423 (2016)
17. Karacan, L., Erdem, E., Erdem, A.: Structure-preserving image smoothing via region covariances. ACM Trans. Graph. (TOG) **32**(6), 1–11 (2013)
18. Pei, J., Cheng, T., Fan, D.P., Tang, H., Chen, C., Van Gool, L.: OSFormer: one-stage camouflaged instance segmentation with transformers. In: Avidan, S., Brostow, G., Cissé, M., Farinella, G.M., Hassner, T. (eds.) ECCV 2022, Part XVIII. LNCS, vol. 13678, pp. 19–37. Springer, Cham (2022). https://doi.org/10.1007/978-3-031-19797-0_2
19. Krizhevsky, A., Sutskever, I., Hinton, G.E.: Imagenet classification with deep convolutional neural networks. Commun. ACM **60**(6), 84–90 (2017)
20. Simonyan, K., Zisserman, A.: Very deep convolutional networks for large-scale image recognition. arXiv preprint arXiv:1409.1556 (2014)

Diff-Writer: A Diffusion Model-Based Stylized Online Handwritten Chinese Character Generator

Min-Si Ren[1], Yan-Ming Zhang[1]($\boxtimes$), Qiu-Feng Wang[2], Fei Yin[1], and Cheng-Lin Liu[1]

[1] State Key Laboratory of Multimodal Artificial Intelligence Systems, Institute of Automation, Chinese Academy of Sciences, Beijing, China
ymzhang@nlpr.ia.ac.cn
[2] School of Advanced Technology, Xi'an Jiaotong-Liverpool University, Suzhou, China

Abstract. Online handwritten Chinese character generation is an interesting task which has gained more and more attention in recent years. Most of the previous methods are based on autoregressive models, where the trajectory points of characters are generated sequentially. However, this often makes it difficult to capture the global structure of the handwriting data. In this paper, we propose a novel generative model, named Diff-Writer, which can not only generate the specified Chinese characters in a non-autoregressive manner but also imitate the calligraphy style given a few style reference samples. Specifically, Diff-Writer is based on conditional Denoising Diffusion Probabilistic Models (DDPM) and consists of three modules: character embedding dictionary, style encoder, and an LSTM denoiser. The character embedding dictionary and the style encoder are adopted to model the content information and the style information respectively. The denoiser iteratively generates characters using the content and style codes. Extensive experiments on a popular dataset (CASIA-OLHWDB) show that our model is capable of generating highly realistic and stylized Chinese characters.

Keywords: Generative model · Conditional diffusion model · Online handwriting generation

1 Introduction

Reading and writing are among the most important abilities for human intelligence. In the past decades, great efforts have been devoted to teaching machines how to read in the field of optical character recognition (OCR). Recently, with the rapid development of deep learning-based generation methods such as Variational Autoencoder (VAE) [1] and Generative Adversarial Network (GAN) [2], researchers become more and more interested in teaching machines how to write [3–5]. Handwriting generation is not only an interesting and creative task

B. Luo et al. (Eds.): ICONIP 2023, CCIS 1964, pp. 86–100, 2024.
https://doi.org/10.1007/978-981-99-8141-0_7

but also of great value in practice. For example, the related technique have been proved to be very useful in data augmentation for handwritten text recognition [6,7] and handwritten signature verification [8].

In general, handwriting data has two different representations. The first one is based on raster images which is called offline handwriting data, and the other one is based on sequences of trajectory points which is called online handwriting data. The generation of offline data has been widely studied and obtained considerable progress. But one major shortcoming of this representation is that since images are static, it can not reflect the dynamic process of writing or painting. On the other hand, as the prevalence of touchscreen devices, online handwriting data such as texts and sketches becomes increasingly popular in our daily life [17]. However, compared to the generation of 2D images, the generation of online data is much less explored.

In this work, we consider the task of generating stylized online handwritten Chinese character. More concretely, given a character and a few style reference samples, we try to generate the point sequence of the specified character with the same calligraphy style as the reference samples. Most previous work [9–13] use autoregressive models to generate the character point-by-point, which makes it difficult to capture the global structure of the handwriting data [15].

In this paper, we propose a novel method, named Diff-Writer, that adopts conditional Denoising Diffusion Probabilistic Models (DDPM) [16] as the backbone, which works in non-autoregressive manner. Furthermore, compared with sketches or English letters, the generation of Chinese characters is more challenging because of its huge amount of classes and complex structures. To overcome this problem, inspired by [10], we equip the model with a character embedding dictionary to store the content information for each character. Besides, we adopt a style encoder to extract calligraphy style information from a few reference samples which enable our model to imitate the specified calligraphy style. By quantitative and qualitative experiments on the public dataset CASIA-OLHWDB [25], we show Diff-Writer can generate highly realistic and stylized Chinese characters.

Our contributions can be summarized as follows:

- To the best of our knowledge, we are the first to apply conditional Diffusion Models for stylized online handwritten Chinese character generation successfully.
- We experimentally show that non-autoregressive diffusion models outperform traditional autoregressive models dramatically with the same model architecture and parameter quantity.

It should be emphasized that our method is not in conflict with previous work, but can complement each other. While previous work mainly focuses on better calligraphy feature extraction and network structure design, our work concentrates on utilizing non-autoregressive diffusion process to replace traditional autoregressive methods for modeling the generation process.

2   Related Work

Online handwritten generation has been studied by many researchers using various models: autoregressive models, recurrent neural networks (RNNs), GAN and diffusion models. Some typical works are reviewed in the following.

2.1   Autoregressive Methods for Online Handwriting Generation

Online handwriting data is a kind of sequential data which includes handwritten texts, sketches, flow charts and so on. With the development of natural language processing, autoregressive methods are usually used to process sequential data. Take some typical work for example: As early as 2013, Graves et al. [18] first proposed to use a single directional RNN model to sequentially generate handwritten English texts. More specifically, they use the Gaussian mixture model (GMM) for the pen-direction. The parameters of GMM are obtained from each hidden state of the RNN. In the next few years, most of work using autoregressive methods are based on this work. SketchRNN (2018) [19] adopts the VAE framework for the generation of sequential data. More specifically, they use two RNNs as the encoder and decoder of a VAE respectively. SketchRNN can implement unconditional natural speech generation as well as handwriting generation. But the disadvantage is that one trained model can only generate one class of sketches. CoSE (2020) [20] also focus on the generation of sketches and diagrams. Unlike previous work, it generates whole sketches stroke by stroke. It considers one single stroke as a curve in the differential geometry and uses MLPs to generate one single stroke, which is in non-autoregressive manner. However, to generate the whole sketch, strokes are still generated in autoregressive way.

2.2   Online Chinese Character Generation

In 2015, [9] improved [18] and proposed an LSTM-based model which can generate fake unreadable Chinese characters. Then [10] proposed a conditional RNN-based generative model that can draw readable Chinese characters of specified classes. Since then, the following work starts to consider how to generate characters with specified calligraphy style instead of just content. Utilizing transfer learning strategy, FontRNN [11] can imitate calligraphy style to generate characters. However, it can only generate the same one style as the train set. Recently, DeepImitator [12], WriteLikeYou [13] and SDT [14] can generate characters of any style given a few of reference samples. DeepImitator uses a CNN as the style encoder to extract style information from handwritten character images. It encodes the calligraphy style information into one vector what we called style code. The advantage of doing this is that it saves computation and is easy to perform operations such as interpolation on the style code. The disadvantage is that the style code is global and rough, many local style information may be ignored. [13] uses a lot of attention mechanisms to pay attention to local content and style information and [14] further explores to extract both writer-wise styles and more complicated character-wise styles using a combination of CNNs and Transformers.

2.3   Non-autoregressive Diffusion Model

As mentioned in Sect. 1, diffusion model has been proved to have strong potential to learn the data distribution by learning to predict the noise of disturbed data. Actually, it was first proposed in [21] in 2015 and has a series of improvement work, such as [16,22,23]. They all mainly focus on the generation of images data while the generation of online handwritten data has not been studied as much. DDPM [16] is one of the most typical representative of them and has been widely used. Recently, based on DDPM, [15] first propose to use diffusion model for chirographic data generation and implement many downstream applications. However, it does not consider much about how to generate samples of the specified class.

3   Methods

In this section, we first give a brief introduction to DDPM, and then propose the Diff-Writer method for online Chinese character generation in details.

3.1   Denoising Diffusion Probabilistic Models

As shown in [24], DDPM can be regarded as a variant of Hierarchical Variational AutoEncoder (HVAE). It is composed of a forward diffusion process and a reverse denoising process. Both processes are formulated as discrete-time Markov chains of T timesteps. In the following section, we denote the forward process as q and the reverse process as p. Starting from the original data x_0, each forward step can be regarded as adding Gaussian noise of different degrees to the current data:

$$x_t = \sqrt{\alpha_t}x_{t-1} + \sqrt{1 - \alpha_t}\epsilon, \quad \epsilon \sim \mathcal{N}(0, I)$$

$$\Rightarrow q(x_t|x_{t-1}) = \mathcal{N}(x_t; \sqrt{\alpha_t}x_{t-1}, (1 - \alpha_t)I) \tag{1}$$

where $\{\alpha_t\}_{t=0}^{T}$ are preset hyperparameters. Due to the Markov property of the forward process, the noisy data distribution at timestep t can be directly calculated as:

$$x_t = \sqrt{\overline{\alpha}_t}x_0 + \sqrt{1 - \overline{\alpha}_t}\epsilon, \quad \epsilon \sim \mathcal{N}(0, I) \tag{2}$$

$$\Rightarrow q(x_t|x_0) = \mathcal{N}(x_t; \sqrt{\overline{\alpha}_t}x_0, (1 - \overline{\alpha}_t)I), \quad \overline{\alpha}_t = \prod_{i=1}^{t} \alpha_i \tag{3}$$

In general, the total number of timesteps T should be large enough to ensure that $\overline{\alpha}_T$ is almost equal to 0, so that the data distribution obtained at timestep T is almost identical to the standard Gaussian distribution. Furthermore, given x_t and x_0, the conditional distribution of x_{t-1} can be calculated using Bayes formula in close form:

$$q(x_{t-1}|x_t, x_0) = \frac{q(x_t|x_{t-1}, x_0)q(x_{t-1}|x_0)}{q(x_t|x_0)} = \mathcal{N}(x_{t-1}; \mu_q(x_t, x_0), \sigma_q^2(t)I)$$

$$\mu_q(x_t, x_0) = \frac{\sqrt{\alpha_t}(1 - \overline{\alpha}_{t-1})x_t + \sqrt{\overline{\alpha}_{t-1}}(1 - \alpha_t)x_0}{1 - \overline{\alpha}_t} \qquad (4)$$

$$\sigma_q^2(t) = \frac{(1 - \alpha_t)(1 - \overline{\alpha}_{t-1})}{1 - \overline{\alpha}_t}$$

The reverse denoising process of DDPM tries to restore x_{t-1} from x_t. Specifically, it learns $p_\theta(x_{t-1}|x_t)$ by optimizing the evidence lower bound of $\log p(x_0)$:

$$L(\theta) = E_{t \sim U\{2,T\}} E_{q(x_t|x_0)} D_{KL}(q(x_{t-1}|x_t, x_0) \| p_\theta(x_{t-1}|x_t)) \qquad (5)$$

Ho et al. [15] proposed an equivalent but more effective objective function to train the DDPM model:

$$L_{\text{simple}}(\theta) = E_{t \sim U\{1,T\}, \epsilon \sim N(0,I)} \| \epsilon - \epsilon_\theta(x_t(x_0, \epsilon), t) \|^2 \qquad (6)$$

where $x_t(x_0, \epsilon)$ computes x_t by Eq. (2). Intuitively, the goal of Eq. (6) is to learn a denoiser that predicts the added noise given the corrupted data x_t and timestep t. We adopt this loss function in this work.

3.2   Data Representation

An online handwritten Chinese character is composed of a sequence of trajectory points $[p_1, p_2, p_3, ..., p_n]$, where each point is represented by a three-dimensional vector: $p_i = (\triangle x^i, \triangle y^i, s^i)$. Specifically, $(\triangle x^i, \triangle y^i)$ represents the positional offset from the previous point p_{i-1}, and $s^i \in \{-1, 1\}$ represents the state of the pen: 1) $s^i = 1$ denotes the pen touching state, indicating that p^i is within a stroke. 2) $s^i = -1$ denotes the pen lifting state, indicating that p^i is the end of a stroke. It should be noted that since the diffusion model needs to add zero mean Gaussian noise to the data, $s = -1$ is chosen as the opposite state with $s = 1$ instead of $s = 0$, so that after adding Gaussian noise, we can distinguish the two states by simply adopting dichotomy with threshold 0. To sum up, each character can be represented as a matrix of size $n \times 3$ in which each row vector denotes a trajectory point. Following the tradition of DDPM literature, we denote this original data matrix as x_0. During the diffusion process, we directly add noise using Eq. (2) to sample the noisy data x_t at timestep t.

3.3   Diff-Writer Model

The overall framework of our model is shown in Fig. 1. Compared to traditional DDPM, to generate characters of specified content and calligraphy style, we adopt a character dictionary and a style encoder to capture the content and style information. As a result, Diff-Writer consists of a character embedding dictionary, a style encoder, and an LSTM denoiser.

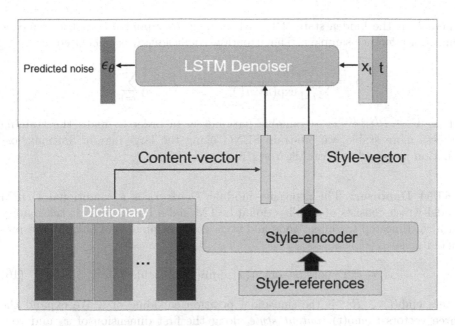

Fig. 1. The framework of our model which consists of three modules: character embedding dictionary, style encoder, and LSTM denoiser.

Character Embedding Dictionary: To encode the content information of characters, inspired by [10], we build a dictionary $E \in R^{d \times N_z}$, where d represents the number of categories and N_Z represents the code dimension. Specifically, we randomly initialize the matrix E and treat it as a learnable parameter of Diff-Writer. After training, the c-th row of the dictionary $E[c] \in R^{N_z}$ is the corresponding content code of the c-th character.

Style Encoder: In order to generate characters with any specified style, we employ a style encoder SENC to extract the style information from a few reference samples written by a given writer:

$$style = \text{SENC}(Ref) = \frac{1}{m} \sum_{i=1}^{m} \text{senc}(r_i) \in R^{N_s}, \qquad (7)$$

where $Ref = \{r_1, r_2, ..., r_m\}$ is m reference samples and N_s represents the dimension of the style codes. More specifically, we adopt a bidirectional LSTM equipped with the attention mechanism as the style encoder:

$$\text{senc}(r) = \text{Attention}(h_1, h_2, ..., h_n) = \sum_{i=1}^{n} \alpha_i h_i, \qquad (8)$$

where h_i is the hidden state of LSTM at trajectory point i, n is the sequence length of reference sample r. The attention coefficient α_i is calculated as:

$$\alpha_i = \frac{\exp(\overline{h}V h_i^T)}{\sum_{j=1}^n \exp(\overline{h}V h_j^T)} \quad \text{with} \quad \overline{h} = \frac{1}{n}\sum_{i=1}^n h_i, \tag{9}$$

where $V \in R^{N_s \times N_s}$ is a trainable parameter. In order to make the training process more stable, we pre-train SENC using the large-margin Softmax loss function proposed in WriteLikeYou [13].

LSTM Denoiser: The denoising module $Deniser_\theta$ is a conditional DDPM model which consists of a bidirectional LSTM and a linear layer. It takes noisy data x_t, timestep t, content code and style code as input, and outputs the predicted noise:

$$\epsilon_\theta = \text{Denoiser}_\theta(\text{concat}(x_t, \text{emb}(t), content, style)), \tag{10}$$

where $\text{emb}(t) \in R^{32}$ is the sinusoidal position encoding of t. We expand the three vectors : $\text{emb}(t), content, style$ along the first dimension of x_t and concatenate them together. Note that the output should have the same dimension as the original data x_0, which is guaranteed by simply adopting the linear layer mentioned before.

3.4   Training Process

The training process of Diff-Writer is very simple and stable. Given a target sample x_0 and a few reference samples written by the same writer as x_0, we firstly query the character embedding dictionary according to the class of x_0 and get its content code. Then the reference samples are input into the style encoder to calculate the style code according to Eq. (7). Next, we randomly sample time step t and a random noise ϵ, and calculate the noisy data x_t by Eq. (2). All the information is input into $Denoiser_\theta$ to obtain the predicted noise ϵ_θ. Finally, our model is trained by minimizing the L_2 distance between ϵ and ϵ_θ defined as Eq. (6).

3.5   Generation Process

The generation process of Diff-Writer is as follows. Given a specified Chinese character and a set of style reference samples, the corresponding content code and style code are extracted by the dictionary and the style encoder, respectively. We randomly sample standard Gaussian noise as $x_T \in R^{n \times 3}$. The character is then generated by performing the reverse denoising process of DDPM. The whole procedure is summarized in Algorithm 1. The visualization of the generation process is shown as Fig. 2.

Algorithm 1. Generation process of Diff-Writer

Input: character c, reference samples $Ref = \{r_i\}, T, \{\alpha_t\}$

1   $content = E[c]$

2   $style = \text{SENC}(Ref)$

3   $x_T \sim \mathcal{N}(0, I)$

4   **for** $t=T, T\text{-}1,...,1$ **do**

5      $\epsilon_\theta = \text{Denoiser}_\theta(\text{concat}(x_t, \text{emb}(t), content, style))$

6      $\mu_\theta(x_t) = \frac{1}{\sqrt{\alpha_t}} x_t - \frac{1-\alpha_t}{\sqrt{1-\overline{\alpha}_t}\sqrt{\alpha_t}} \epsilon_\theta$

7      $\sigma_t^2 = \frac{(1-\alpha_t)(1-\overline{\alpha}_{t-1})}{1-\overline{\alpha}_t}$

8      $x_{t-1} = \mu_\theta(x_t) + \sigma_t \epsilon, \quad \epsilon \sim \mathcal{N}(0, I)$

9   **end**

10   Output x_0

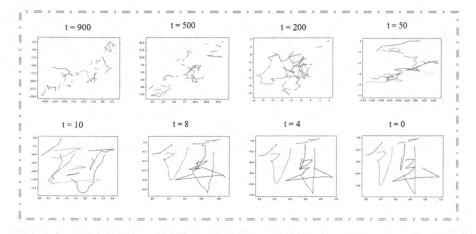

Fig. 2. Visualization of the generation process. Different colors denotes different strokes.

4   Experiment and Results

4.1   Dataset

We use CASIA-OLHWDB1.0-1.1 [25] to train and test our model, which contains 720 different writers. We take 600 writers as the training set and left 120 writers as the test set. We choose 1,000 classes of commonly used Chinese characters in our experiments, which results in 600,000 training samples and 120,000 testing samples. For data preprocessing, we only do normalization for each online handwritten character to ensure that the coordinates of every key point are between 0 and 1.

4.2   Implementation Details

First, we use a 3-layer bidirectional LSTM with the hidden size of 200 as the denoiser model and a 3-layer bidirectional LSTM with the hidden size of 128 as the style encoder. We set $N_z = 128$ and $N_s = 128$ as the dimension of content code and style code respectively. For the hyperparameters of DDPM, we adopt the traditional linear noising schedule proposed by (Nichol et al. 2021): $\{T = 1000; \quad \alpha_0 = 1 - 1e^{-4}; \quad \alpha_T = 1 - 2e^{-2}\}$. During the experiment, we empirically find that in the last 20 or so time steps of the denoising process, setting $\sigma_t = 0$ while $\sigma_t = 0.8 \frac{(1-\alpha_t)(1-\overline{\alpha}_{t-1})}{1-\overline{\alpha}_t}$ otherwise will make the generation process more stable. We implement our model in Pytorch. Adam optimizer with initial learning rate 0.01 is adopted to train our model. We set learning rate decay = 0.9999. We run experiments on NVIDIA TITAN RTX GPUs.

Fig. 3. Visualization of results generated by our diffusion model and autoregressive model.

4.3   Evaluations

In this work, we adopt four evaluation metrics, namely subjective visual Turing test, Dynamic Time Warping(DTW) distance, style score and content score, which will be demonstrated in the following part.

Visualization: To better demonstrate the effectiveness of our method, we first do visualization of the generated results, as shown in Fig. 3. It is easy to see that our model can not only perfectly generate complete character structures, but also effectively imitate the specified calligraphy styles. What is more, the generation of the diffusion model is more stable than that of the autoregressive model, especially for characters with complex structures. That is perhaps due to the better ability of the diffusion model to capture the global structure of handwritten data as we mentioned before.

Subjective Test: Subjective test is an intuitive and effective evaluation method to test the performance of our model. We invited 20 participants to perform two tests. The first one is to test the verisimilitude and structural correctness of generated characters. As shown in Fig. 4, we provide 25 pairs of character samples to each participant and ask them to tell which one is more likely written by real people. The second test is to evaluate the imitating ability of our model. As shown in Fig. 5, we provide 12 sets of character samples to each participant. In each set, the left character is generated by our model while the other 4 characters are written by 4 different people, one of which is the imitating target. We ask the participants to guess which one is the target. The results are reported in Table 1.

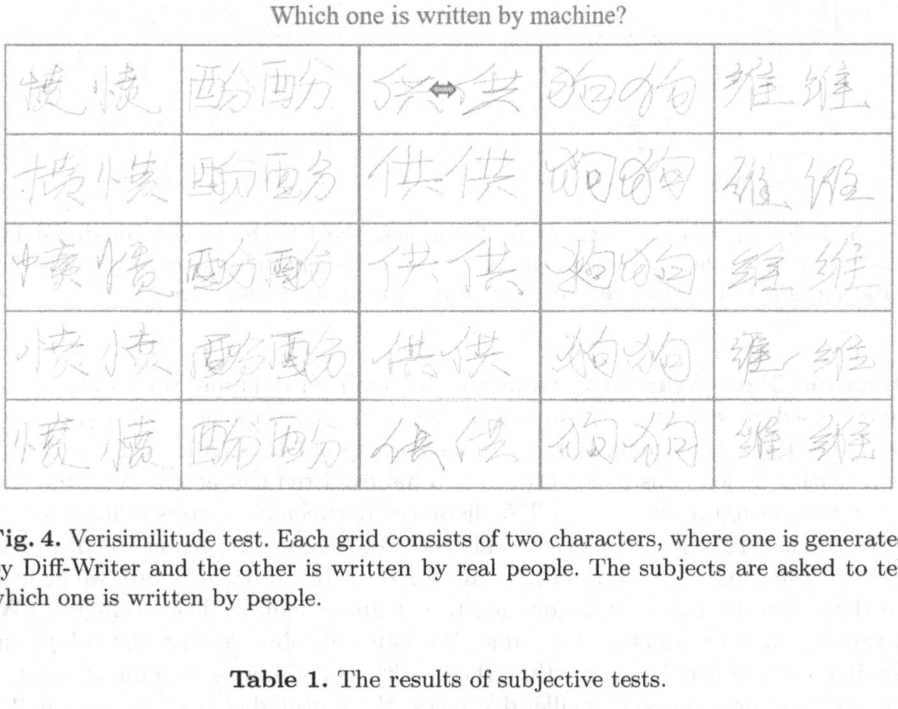

Which one is written by machine?

Fig. 4. Verisimilitude test. Each grid consists of two characters, where one is generated by Diff-Writer and the other is written by real people. The subjects are asked to tell which one is written by people.

Table 1. The results of subjective tests.

Test	Readability test	Imitating test
Accuracy	48.0%	50.8%

As we can see, the accuracy of the first test is close to 50%, which indicates that it is quite difficult for the participants to distinguish the generated characters from the real ones. And in the second test, the accuracy is far greater than 25% which is the accuracy of random guess. It further proves that our model can effectively imitate the specified style to some extent.

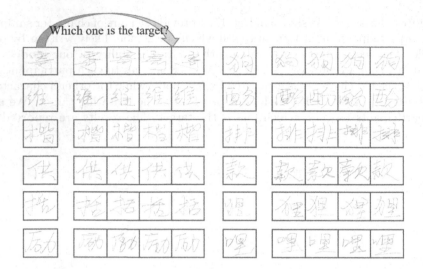

Fig. 5. Imitating test. In each group of samples, the left character is produced by Diff-Writer. The other 4 characters are written by different people, one of which is the imitating target. The subjects are asked to tell which one is the target.

Dynamic Time Warping Distance: Normalized dynamic time warping is widely used to calculate the distance between two sequences with different lengths. In the heat map, darker colors correspond to smaller DTW values, representing a closer distance between two handwritten characters. As shown in Fig. 6, the left figure shows the DTW distances between characters written by 60 different authors in the test set, the values of the diagonal represent the distance between the same character, so they are all zero. In the right figure, we generate three imitation characters for each target and calculate their average DTW distance from the target as the value. We can still observe that the colors on the diagonal are usually darker than those in the same row or column. Recalling the darker colors represent smaller distances, the imitated characters are usually closer to characters written by the target writer than the others, which indicates that our model has the ability to imitate specific styles.

Comparison with Autoregressive Models: As mentioned before, DDPM is a non-autoregressive model which can capture the global structure of handwritten characters more easily compared to autoregressive models. The visualization experiment in Fig. 3 shows that the characters generated by our model is very consistent and usually do not have structural errors.

Besides, we do quantitative experiments to compare our model with autoregressive models. We adopt style score and content score as the evaluation metric. For style score, we train a style classifier on the test set which contains 120 different writers and can achieve over 96% accuracy. Style score is the classification accuracy on generated samples using this classifier. A higher style score indicates

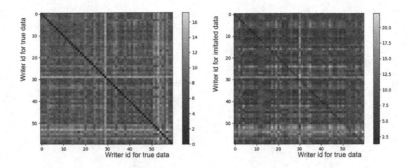

Fig. 6. The Heatmap of DTW distance matrix.

a better style imitation ability of the model. Since the style of different handwritten characters even from the same writer may have great differences, the classifier accepts ten characters written by one person and calculates the mean value of their features, then uses the mean features to complete the classification task. Similarly, the content score is obtained using a Chinese character classifier trained on the training set with an accuracy of 99.2% and 95.0% on the test set. The accuracy of the classifier on the generated samples is used as the content score.

For a fairer comparison, we contrast our model with DeepImitator [12] and WriteLikeU (without ASB) [13] which have similar parameter quantity. Furthermore, we achieved an autoregressive model which has the same architecture as our proposed model. More specifically, it also contains a character embedding dictionary and a style encoder. The only difference is that our proposed model uses a bidirectional LSTM as a DDPM denoiser while the autoregressive model uses a bidirectional LSTM to directly generate the motion of the pen.

As shown in Table 2, our achieved autoregressive model can achieve competitive performance with respective to previous work while diffusion-based model performs better dramatically, especially for the content score. We attribute this to the better ability of diffusion models to capture global features, which tends to generate samples with more accurate and complete structures.

Recalling Eq. (7), we will randomly select m style reference samples for the style encoder. In this experiment, we further use different m (from 5 to 15) to compare the performance of our achieved autoregressive model and diffusion model. The results is shown in Fig. 7.

Table 2. Comparison with autoregressive models.

	DeepImitator	WriteLikeU(without ASB)	Our autoregressive	**Diff-Writer**
Style score	43.2%	42.2%	41.3%	46.6%
Content score	83.4%	-	84.6%	94.8%

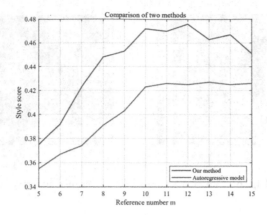

Fig. 7. Style score of two models with different m.

Style Code Interpolation Experiments: Recalling Sect. 3.3, we use a style encoder to extract a style code for given reference samples which is used to imitate the specified calligraphy style. Suppose we have two style codes for two different writers respectively and we adopt simply linear interpolation method to them: $style_{merge} = 0.5 * style_1 + 0.5 * style_2$. We use these style codes to generate characters respectively. Figure 8 shows the results. The first and last columns represent real samples from writer1 and writer2. The second and fourth columns represent the samples generated by using style code 1 and style code 2. The third column represents samples generated using merged style code. It can be seen that the calligraphy style of writer1 is neat, while the calligraphy style of writer2 is cursory with very few strokes. The merged style code generates a style between them. This not only proves the ability of our style encoder to extract style, but also can be used for calligraphy style synthesis.

Training Strategy for Style Encoder: Style encoder is the key module to extract calligraphy style features from reference samples. In our proposed method, we adopt pre-training with finetuneing strategy for the style encoder. To test its effectiveness, we conduct several ablation studies. The results are reported in Table 3. As we expected, the model with both pre-train and fine-tuning has better performance than that with pre-train only. The model only with pre-train can achieve 37.2% style score, which indicates it already has the ability of style imitation. If we remove the style encoder, that is to say we set it to untrainable from the beginning and we get 0.8% style score. Since there are 120 writers in the test set, this result equals to a random guess. Furthermore, we find that pre-training is an essential strategy for our model. We try to initialize the style encoder randomly and train our model from scratch. However, although the model can still generate high quality characters, the corresponding

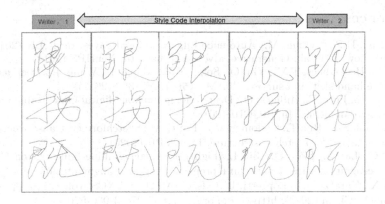

Fig. 8. Style code interpolation results.

style score drops to 0.8% as same as that without a style encoder, which means the model completely loses the ability to imitate style.

Table 3. Effect of different training strategies for style encoder.

	Without style encoder	Random initialization	Pre-train only	Pre-train +finetuning
Style score	0.8%	0.8%	37.2%	46.6%

5   Conclusion and Future Work

In this paper, we proposed Diff-Writer, a diffusion model-based stylized online handwritten Chinese character generator. A character embedding dictionary and a style encoder are integrated to provide content and style information respectively. However, one limitation is that we only use one vector (style code) to represent the complex and varied style information, which may limit the ability of calligraphy style imitating. This can be partially resolved by integrating our approach with other relevant methods which use more complicated calligraphy style extractor and generator. In addition, the diffusion process introduces unique conditional generation methods, such as classifier guidance and classifier-free guidance. Moreover, how to use diffusion-based model to generate strings instead of one single character is an exciting and challenging task. We leave them as our potential future work.

Acknowledgements. This work is supported by the Major Project for New Generation of AI under Grant No. 2018AAA0100400 and the National Natural Science Foundation of China (NSFC) Grant No. 62276258.

References

1. Kingma, D.P., Welling, M.: Auto-encoding variational bayes. In: ICLR (2014)
2. Goodfellow, I., et al.: Generative adversarial nets. In: NeurIPS (2014)
3. Xu, P., Hospedales, T.M., Yin, Q., Song, Y.Z., Xiang, T., Wang, L.: Deep learning for free-hand sketch: a survey. TPAMI **45**(1), 285–312 (2022)
4. Aksan, E., Pece, F., Hilliges, O.: DeepWriting: making digital ink editable via deep generative modeling. In: SIGCHI (2018)
5. Ribeiro, L.S.F., Bui, T., Collomosse, J., Ponti, M.: Sketchformer: transformer-based representation for sketched structure. In: CVPR (2020)
6. Xie, C., Lai, S., Liao, Q., Jin, L.: high performance offline handwritten Chinese text recognition with a new data preprocessing and augmentation pipeline. In: Bai, X., Karatzas, D., Lopresti, D. (eds.) DAS 2020. LNCS, vol. 12116, pp. 45–59. Springer, Cham (2020). https://doi.org/10.1007/978-3-030-57058-3_4
7. Kang, L., Riba, P., Rusinol, M., Fornes, A., Villegas, M.: Content and style aware generation of text-line images for handwriting recognition. In: TPAMI (2022)
8. Lai, S., Jin, L., Zhu, Y., Li, Z., Lin, L.: SynSig2Vec: forgery-free learning of dynamic signature representations by Sigma Lognormal-based synthesis and 1D CNN. In: TPAMI (2021)
9. Ha, D: Recurrent net dreams up fake Chinese characters in vector format with tensorflow. http://blog.otoro.net/2015/12/28/recurrent-netdreams-up-fake-chinese-characters-in-vectorformat-with-tensorflow/
10. Zhang, X.Y., Yin, F., Zhang, Y.M., Liu, C.L., Bengio, Y.: Drawing and recognizing Chinese characters with recurrent neural network. In: TPAMI, Yoshua Bengio (2018)
11. Tang, S., Xia, Z., Lian, Z., Tang, Y. and Xiao, J.: FontRNN: generating large-scale Chinese fonts via recurrent neural network. In: CGF (2019)
12. Zhao, B., Tao, J., Yang, M., Tian, Z., Fan, C., Bai, Y.: Deep imitator: handwriting calligraphy imitation via deep attention networks. In: PR (2020)
13. Tang, S., Lian, Z.: Write like you: synthesizing your cursive online Chinese handwriting via metric-based meta learning. In: CGF (2021)
14. Dai, G., et al.: Disentangling writer and character styles for handwriting generation. In: CVPR, Zhuoman Liu and Shuangping Huang (2023)
15. Das, A., Yang, Y., Hospedales, T., Xiang, T., Song, Y.Z.: ChiroDiff: modelling chirographic data with diffusion models. In: ICLR (2023)
16. Ho, J., Jain, A., Abbeel, P.: Denoising diffusion probabilistic models. In: NeurIPS (2020)
17. Yun, X.L., Zhang, Y.M., Yin, F., and Liu, C.L.: Instance GNN: a learning framework for joint symbol segmentation and recognition in online handwritten diagrams. In: TMM (2021)
18. Graves, A.: Generating sequences with recurrent neural networks. In: ArXiv (2013)
19. Ha, D., Eck, D.: A neural representation of sketch drawings. In: ICLR (2018)
20. Aksan, E., Deselaers, T., Tagliasacchi, A., Hilliges, O.: CoSE: compositional stroke embeddings. In: NeurIPS (2020)
21. Sohl-Dickstein, J., Weiss, E., Maheswaranathan, N., Ganguli, S.: Deep unsupervised learning using nonequilibrium thermodynamics. In: ICML (2015)
22. Dhariwal, P.: Alexander quinn nichol: diffusion models beat GANs on image synthesis. In: NeurIPS (2021)
23. Song, J., Meng, C., Ermon, S.: Denoising diffusion implicit models. In: ICLR (2021)
24. Luo, C.: Understanding diffusion models: a unified perspective. In: arXiv (2022)
25. Liu, C.L., Yin, F., Wang, D.H., Wang, Q.F.: CASIA online and offline Chinese handwriting databases. In: ICDAR, Qiu-Feng Wang (2011)

EEG Epileptic Seizure Classification Using Hybrid Time-Frequency Attention Deep Network

Yunfei Tian[1,2], Chunyu Tan[1,2(✉)], Qiaoyun Wu[1,2], and Yun Zhou[1,2]

[1] Engineering Research Center of Autonomous Unmanned System Technology,
Ministry of Education, Anhui University, Hefei, China
{cytan,wuqiaoyu,zhouy}@ahu.edu.cn
[2] Anhui Provincial Engineering Research Center for Unmanned System and
Intelligent Technology, School of Artificial Intelligence, Anhui University, Hefei, China

Abstract. Epileptic seizure is a complex neurological disorder and is difficult to detect. Observing and analyzing the waveform changes of EEG signals is the main way to monitor epilepsy activity. However, due to the complexity and instability of EEG signals, the effectiveness of identifying epileptic region by previous methods using EEG signals is not very satisfactory. On the one hand, these methods use the initial time series directly, which reflect limited epilepsy related features; On the other hand, they do not fully consider the spatiotemporal dependence of EEG signals. This study proposes a novel epileptic seizure classification method using EEG based on a hybrid time-frequency attention deep network, namely, a time-frequency attention CNN-BiLSTM network (TFACBNet). TFACBNet firstly uses a time-frequency representation attention module to decompose the input EEG signals to obtain multiscale time-frequency features which provides seizure relevant information within the EEG signals. Then, a hybrid deep network combining convolutional neural network (CNN) and bidirectional LSTM (BiLSTM) architecture extracts spatiotemporal dependencies of EEG signals. Experimental studies have been performed on the benchmark database of the Bonn EEG dataset, achieving 98.84% accuracy on the three-category classification task and 92.35% accuracy on the five-category classification task. Our experimental results prove that the proposed TFACBNet achieves a state-of-the-art classification effect on epilepsy EEG signals.

Keywords: Epileptic seizure classification · Electroencephalogram (EEG) · Time-frequency attention · CNN · Bi-LSTM

1 Introduction

The automatic analysis of epileptic seizure classification is a key research tool for brain functional diseases and one of the challenges in the analysis of EEG signals. Many researchers have been identifying epileptic and non-epileptic activity using EEG signals for a long time [1–3]. Most studies detecting epileptic seizure

© The Author(s), under exclusive license to Springer Nature Singapore Pte Ltd. 2024
B. Luo et al. (Eds.): ICONIP 2023, CCIS 1964, pp. 101–113, 2024.
https://doi.org/10.1007/978-981-99-8141-0_8

mainly fall into two methodologies: traditional methods, which combine feature extraction and machine learning methods, and deep learning methods. [4,5].

Traditional methods mainly include time-domain analysis methods, frequency-domain analysis methods, and time-frequency analysis methods. Vijayalakshmi and Appaji [6] proposed a method based on time-domain template matching for EEG peak detection in epilepsy patients. However, due to the complexity of EEG and the diversity of epileptic discharges, it is difficult to directly analyze EEG in the time domain. Frequency domain analysis methods point out another direction, which transforms EEG signals into frequency domain for processing. In [7], Wulandari et al. proposed an EEG visualization method by combining the genetic algorithm and Fourier spectrum analysis. Still, the instability and nonlinearity of EEG signals results in limited physiological information extracted by the frequency-domain analysis methods. The time-frequency analysis method can simultaneously extract features related to both time and frequency domains, making it more suitable for analyzing signals with severe oscillations such as EEG signals. Wavelet transform has its multi-resolution and multi-scale analysis ability to generate time-scale-frequency descriptions of signals. Tang et al. [8] proposed a seizure detection algorithm by combining wavelet transform and detrended fluctuation analysis. Even though time-frequency analysis methods can effectively characterize EEG signals, it is difficult to further obtain discriminative features related to epilepsy.

Deep learning methods, which uses multi-layer nonlinear processing units, can automatically learn and discover nonlinear relationships in data, making them better extract relevant features of EEG signals. Deep leaning methods is generally divided into CNN based methods, RNN based methods and hybird methods. For CNN based methods, Elakkiya et al. developed a 1DCNN model for epileptic seizure detection [9] and Antoniades et al. detected epileptic discharges in intracranial EEG by designing a four layer CNNs [10]. Besides, Shoji et al. proposed a multi-channel CNN networks to detect abnormal patterns in EEG signals [11]. For RNN based methods, the common type is their evolution, long short-term memory network (LSTM) and bidirectional LSTM (BiLSTM). Tuncer et al. introduced a method for classifying epilepsy EEG by BiLSTM [12]. Manohar et al. used short-time Fourier transform to preprocess EEG signals which are input to BiLSTM for epilepsy detection [13]. Recently, hybrid networks are developing rapidly, as their ability to explore different types of data structures allows for better performance in the EEG epileptic seizure detection. A 1DCNN-LSTM network used in [14] has a 5% higher F1-score than CNN-based methods. Xu et al. improved the epilepsy detection performance by combining CNN and RNN instead of using them separately. Qiu et al. proposed a difference attention ResNet-LSTM model (DRARNet) [15], obtaining 98.17% and 90.17% accuracy for three classes and five classes epileptic seizure detection, respectively.

Although hybrid methods can improve the effect of classification for epilepsy EEG signals, the input of hybrid network is the single raw signal, which results in insufficient features. In addition, the multi-scale and multi-structural information between EEG signals is not fully utilized by existing methods, so the implied epilepsy information is limited. Finally, the irrelevant features generated by the deep layer structure will distract the network's attention, making it impossible to further improve classification performance. Accordingly, we propose a time-frequency attention CNN-BiLSTM network (TFACBNet) for EEG epileptic seizure classification to make full use of multi-scale time-frequency features of EEG signals, and to extract effective epilepsy discriminative information. The experimental results show that the proposed TFACBNet yields an impressive classification performance on EEG epileptic seizure classification. The key contributions of this study are summarized as follows:

(1) Combining with a time-frequency attention module, TFACBNet is proposed for EEG epileptic seizure classification. This proposed network structure can be also applicable to other non-stationary signal processing problems.
(2) By using time-frequency representation to describe the time-frequency characteristics of EEG signals and further enhancing epilepsy related features through attention, the proposed method can extract relevant information that is helpful for distinguishing epileptic and non-epileptic activity within EEG signals.
(3) Experimental results based on the Bonn EEG database verifies the superiority of the proposed TFACBNet compared to previous epilepsy detection methods.

The rest of this paper is organized as follows, the construction of TFACBNet is described in Sect. 2. Section 3 provides experimental results and analysis. Finally, conclusion is given in Sect. 4.

2   The Construction of TFACBNet

Aiming to extract time-frequency features of EEG signals and focus on the critical information related to epilepsy EEG signals, a time-frequency attention CNN-BiLSTM network is constructed in this paper and applied to the classification of epilepsy EEG signals. The global flowchart of the proposed method is shown in Fig. 1. As it can be seen from Fig. 1, after the original EEG signal is input into the network, the time-frequency representation is firstly carried out through wavelet transform. Then, multi-channel features is enhanced by the attention model. Furthermore, a deep hybrid network composed of CNN and BiLSTM is used to extract temporal and spatial epilepsy seizure information. Finally, classification results are output through the fully connected layer that performs feature fusion.

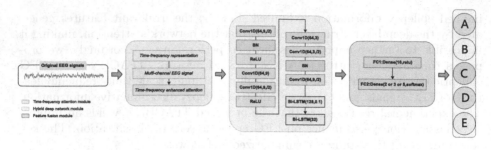

Fig. 1. The global framework of TFACBNet.

2.1 Time-Frequency Attention Module

Time-Frequency Representation. EEG signals are chaotic and highly non-stationary. Directly processing the original EEG signals and inputting them into the constructed deep network, the extracted epilepsy related information is limited, because brain significant activities often involve joint variations of time and frequency [16]. The time-frequency representation is well suited to describe and characterize non-stationary phenomena of the EEG signal, and can obtain time-frequency structures that are more discriminative between epileptic and non-epileptic activities. Wavelet transform is a continuously developing time-frequency representation method with rigorous and complete theoretical support [17]. It is suited for non-stationary time-series data like EEG signal.

The continuous wavelet transform of a signal $F(t)$ is defined as the correlation between the function $F(t)$ with a wavelet $\psi_{a,b}$,

$$(W_\psi F)(a,b) = |a|^{-\frac{1}{2}} \int_{-\infty}^{\infty} F(t)\psi^*(\frac{t-b}{a})dt \tag{1}$$

where $a, b \in \mathbf{R}$, $a \neq 0$, the asterisk denotes complex conjugation, and $\psi_{a,b}(t) = |a|^{-\frac{1}{2}}\psi(\frac{t-b}{a})$. For some specific forms of functions, set the discrete values of $a_k = 2^{-j}$ and $b_l = 2^{-l}k$ to obtain discrete transformations,

$$\psi_{k,l} = 2^{\frac{k}{2}}\psi(2^k t - l) \tag{2}$$

where $k, l \in \mathbf{Z}$. In this way, an EEG signal can be expanded in a series of wavelets. This discrete wavelet transform decomposition method is also known as the multi-resolution analysis according a hierarchical scheme of nested subspaces. For any resolution level $K < 0$, the wavelet expansion of $F(t)$ is

$$F(t) = \sum_{k=-K}^{-1} \sum_{l=-\infty}^{\infty} C_k(l)\psi_{k,l}(t), \tag{3}$$

$C_k(l) = \langle F, \psi_{k,l} \rangle$ are the wavelet coefficients. Through the wavelet expansion, the decomposition component of each scale k contains the information of the signal $F(t)$ corresponding to the frequencies $2^k\pi \leq |\omega| \leq 2^{k+1}\pi$. Therefore, wavelet transform performs a successive decomposition of the signal in different frequency scales. At each step, the corresponding details of signals are separated, providing useful information for detecting and characterizing the short time-frequency phenomena of the EEG epilepsy signal. EEG samples and their corresponding time-frequency representations are shown in Fig. 2. Figure 2(a) shows an ECG time series of healthy people and its corresponding time-frequency diagram, and Fig. 2(b) is an EEG time series of the epileptic patient and its corresponding time-frequency diagram. It can be seen from Fig. 2 that the time domain wave shape of EEG signals oscillates violently and changes quickly, making it difficult for visual observation to distinguish between normal ECG signals and epileptic ones. However, from the corresponding time-frequency diagrams, it can be observed that compared to the normal EEG signal, the time-frequency features of the epileptic EEG signal are more concentrated, making the time-frequency related features more discriminative.

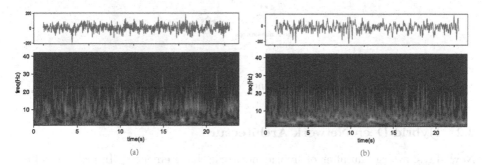

Fig. 2. EEG samples and their corresponding time-frequency representations. (a) an ECG time series of healthy people and its corresponding time-frequency diagram. (b) an EEG time series of the epileptic patient and its corresponding time-frequency diagram.

Time-Frequency Enhanced Attention. The time frequency representation method represents the original EEG signal as multi-channel signals with different time-frequency characteristics, some of which are closely related to the pathological features of epilepsy. However, this inevitably generates some redundant information, especially affected by noise. So, we embed a time-frequency enhanced attention to adaptively focus on relevant time-frequency features, weakening irrelevant information.

The enhanced attention structure is shown in Fig. 3. As seen from Fig. 3, multi-channel EEG signals $\tilde{F} \in \mathbf{R}^{C \times L}$ (C is the number of channels, and L is the number of signal points) output by the time-frequency representation are converted into the channel attention map $M \in \mathbf{R}^C$. Firstly, $\tilde{F}$ is reshaped to

$\mathbf{R}^C$ by a global average pooling processing. Next, the channel attention map $M \in \mathbf{R}^C$ is obtained by the fully connected layer, which is

$$M_i = f_2\left(w_2 f_1\left(w_1\frac{\sum_{j=1}^{L}\tilde{F}_{i,j}}{|L|}\right)\right) \tag{4}$$

where w_1, w_2 are the learnable weights, and f_1, f_2 denote the ReLU function and Sigmoid function, respectively. Then the output is obtained by

$$\tilde{F}' = M \bullet \tilde{F}. \tag{5}$$

Therefore, through the learned channel attention map, channels containing more epilepsy relevant information will be strengthened, while channels containing redundant information will be weakened.

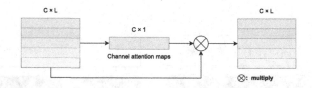

Fig. 3. The time-frequency enhanced attention model

2.2   Hybrid Deep Network Architecture

Nowadays, a large number of hybrid networks have emerged. In order to effectively extract both temporal and spatial information from multi-channel EEG signals simultaneously, and also lighten the network structure, we utilized a deep hybrid network combining CNN and BiLSTM. The detailed structure of the deep hybrid network and its output feature size of each layer are summarized in Fig. 1.

As shown in Fig. 1, the layers of CNN typically consist of convolutional layers, batch normalization (BN) layers and rectifier linear units (ReLU). Convolutional layers apply a convolution operation to the input, transferring the learned features to the next layer. Due to the sequential characteristic of EEG signals, we chose one-dimensional convolution (Conv1D) operation. The spatial correlations of EEG signals are obtained by the constructed CNN. LSTM is a variant of RNN, which alleviates the problem of gradient vanishing in RNN and is very suitable for modeling sequential data. LSTM is composed of LSTM units, each containing input gate, forget gate, and output gate [18]. The input gate is used to retain input information, the forget gate controls the retention of information at the previous moment, and the output gate learns to obtain information for the next unit. By learning the parameters in each gate, LSTM can fully extract the temporal dependency between sequences. BiLSTM is an improved version of LSTM. It is a combination of forward LSTM and backward LSTM. Through the bidirectional structure, BiLSTM can obtain information transmitted from previous data and also learn the influence of subsequent sequences. In the proposed deep network, we use a two-layer BiLSTM network to fully capture the implicit temporal information between EEG signals, which can be seen in Fig. 1.

3 Experiment and Results

In this section, experiments are conducted to evaluate the effectiveness of the proposed TFACBLNet. The experimental results are given below.

3.1 Database

We use EEG signals from the Bonn EEG database [19], which is the most widely used benchmark database for epileptic seizure classification. This database consists of five EEG subsets containing healthy and epilepsy disease which are labelled as A, B, C, D, and E. Each subset includes 100 EEG recordings of 23.6 s duration sampling at 173.61 Hz on the standard 10–20 electrode system. Subsets A and B are collected from healthy people with open and closed eyes, respectively. Subsets C, D and E are recorded EEG signals of epileptic patients, where subset C and D is collected from the epileptic interval, and subset E is collected from the epileptic period. Each EEG recording is divided into 23 signals segments. A total of 11500 signal segments are extracted from the EEG recordings, and each segmentation is 178 points long in this study. All category segments are randomly divided into training set and test set. That is, 90% of each category segments are randomly selected for training set, and the remaining 10% are used for the test set.

To keep pace with consistency with other epileptic seizure detection methods and make fair comparisons, experiments are conducted with two-category classification task to classify normal and epileptic conditions, three category classification task to classify AB of the normal condition, CD of the epileptic condition, and E of the seizure condition, and the five-category classification task to classify epileptic states and four non-epileptic states. The proposed method is implemented based on the Keras architecture using the Windows 10 operating system with CPU of Intel i7-12700 and GPU of Nvidia 4070Ti. The dynamic learning rate is adopted and it is multiplied by 0.1 each 25 epochs with the initial value of 0.001. Batch size is set 128, the cross entropy function is used as the loss function, and the adam optimizer is used as the optimizer. All experimental tasks are trained for 100 epochs, with the best result being the final result.

3.2 Evaluation Metric

We use four typical classification indicators to evaluate the epileptic seizure detection performance. There are accuracy, precision, recall, and F1-score. These indicators are defined as follows:

$$Accuracy = \frac{TP + TN}{TP + TN + FP + FN}, \ Precision = \frac{TP}{TP + FP} \quad (6)$$

$$Recall = \frac{TP}{TP + FN}, \quad F1 - score = 2 \times \frac{Precision \times Recall}{Precision + Recall} \quad (7)$$

where TP, TN represent the number of true positive samples and true negative samples, respectively, and FP, FN denote the number of false positive samples and false negative samples, respectively. A larger value of accuracy, precision, recall, and F1-score means better epileptic seizure classification performance, and F1-score is a comprehensive metric of the classification performance.

3.3 Experimental Results

Ablation Studies. Ablation research is conducted to verify the effect of the time-frequency representation and time-frequency enhanced attention module. BNet, CBNet, and TFCBNet models are constructed as downgraded versions of TFACBNet. Specifically, CBNet is removed the time-frequency attention module, TFCBNet is removed the time-frequency enhanced attention layer, and BNet only uses the BiLSTM. Experimental results are shown in Fig. 4. As it is can be seen in Fig. 4(a), for the three-category classification task, the evaluation indicators of the four models are all above 90%, and results of TFCBNet and TFACBNet reach to 98%. Compared with BNet, CBNet and TFCBNet, the accuracy of TFACBNet has an improvement of 5.2%, 2.34%, and 0.49%. Figure 4(b) presents the experimental results of the five-category classification task. The accuracy, precision, recall, and F1-score of TFACBNet reach 92.35%, 92.13%, 92.17% and 92.14%, respectively, which are higher than the results of other three downgraded models. According to Fig. 4, the F1-score of TFACBNet is 0.36% and 0.43% higher than that of TFCBNet in (a) and (b), respectively, while the F1-score of TFCBNet is increased by 5.49% and 8.88% compared with CBNet, respectively. It indicates that the time-frequency representation of EEG signals provides extra useful information for the model to obtain more discriminative epileptic features, and the attention to some extent enhances distinctiveness. In addition, the F1-score curves of the four models are shown in Fig. 5. Figure 5(a) is the F1-score curves of three-category classification task and Fig. 5(b) is the F1-score cures of the five-category classification task. Obviously, as shown in Fig. 5, TFACBNet is superior to others, which once again proves the effectiveness of the time-frequency attention model in the proposed method.

Comparative Experiment. Comparative experiments are conducted to further evaluate the validity of the proposed TFACBNet. We compare TFACBNet with five epileptic seizure classification methods that are DARLNet, ResNet, Vgg16, AlexNet, and 1DCNN-LSTM. Experimental results of two-category and five-category classification are recorded in Table 1 and Table 2, respectively. As shown in Table 1 and Table 2, the proposed TFACBNet outperforms other compared methods, especially for the multi-classification in Table 2. The F1-score of TFACBNet is 13.07%, 12.11%, 8.1%, 10.58%, and 2.09% higher than that of AlexNet, Vgg16, Resnet, 1DCNN-LSTM [14], and DARLNet [15], which demonstrates the superiority of the TFACBNet. Furthermore, for a more fair comparison, the comparison between the recently proposed methods using the Bonn EEG

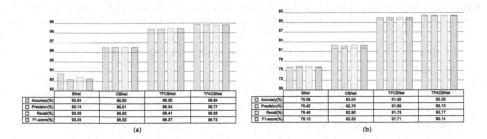

Fig. 4. (a) Ablation experimental results of three-category classification task. (b) Ablation experimental results of five-category classification task.

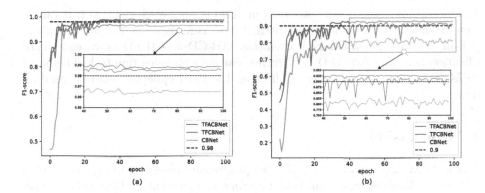

Fig. 5. F1-score curves of CBNet, TFCBNet, and TFACBNet. (a) Three-category classification: AB-CD-E; (b) Five-category classification:A-B-C-D-E.

Table 1. The performance of Alexnet, Vgg16, Resnet, 1DCNN-LSTM, DARLNet and the proposed TFACBNet model on the two-category classification task.

Models	Accuracy (%)	Precision (%)	Recall (%)	F1-score (%)
AlexNet	96.17	97.39	96.28	96.83
Vgg16	96.61	97.68	96.70	97.19
ResNet	97.04	97.99	97.67	97.53
1DCNN-LSTM [14]	99.39	98.39	98.79	98.59
DARLNet [15]	98.87	98.99	99.13	99.06
TFACBNet	**99.78**	**99.78**	**99.77**	**99.78**

Table 2. The performance of Alexnet, Vgg16, Resnet, 1DCNN-LSTM, DARLNet and the proposed TFACBNet on the five-category classification task.

Models	Accuracy (%)	Precision (%)	Recall (%)	F1-score (%)
AlexNet	79.48	79.37	79.24	79.12
Vgg16	80.52	80.41	81.15	80.03
ResNet	84.17	84.08	84.07	84.04
1DCNN-LSTM [14]	82.00	81.78	81.70	81.56
DARLNet [15]	90.17	90.00	90.16	90.05
TFACBNet	**92.35**	**92.13**	**92.17**	**92.14**

database [15, 20–25] and the proposed method is presented in Table 2. It can be seen in Table 2 that the proposed method has made significant improvements in D-E of the two-category classification task, AB-CD-E of the three-category classification task, and A-B-C-D-E of the five-category classification task. Therefore, TFACBNet does have a superior effect on the EEG epileptic seizure classification.

4   Conclusion

A new TFACBNet model for EEG epileptic seizure classification is proposed in this paper. Wavelet transform is first applied to represent EEG signals as time-frequency representations, providing multi-scale time-frequency characteristics to obtain more information related to epilepsy EEG. Then, a time-frequency attention is introduced to enhance epilepsy relevant features of multi-channel signals adaptively. Finally, a hybrid deep network combining CNN and BiLSTM is constructed to extract patio-temporal dependence within EEG signals. The effectiveness of TFACBNet is verified by experiments on the Bonn EEG database. Compared with the previous studies, the proposed methods exhibits a superlative performance for classifying EEG epileptic seizure. Furthermore, the proposed network structure can be also applicable to other similar signal processing problems for disease detection and analysis (Table 3).

Table 3. Performance comparison of the proposed TFACBNet with other state-of-the-art studies

Classification task	Authors	Years	Accuracy (%)	Precision (%)	Recall (%)	F1-score(%)
A-E	The proposed method		100.00	100.00	100.00	100.00
	DARLNet [15]	2023	100.00	100.00	100.00	100.00
	Tajmirriahi et al. [22]	2021	99.00	\	99.00	\
	Ramos-Aguilar et al. [23]	2020	100.00	\	\	\
	Siuly et al. [24]	2019	99.50	\	100.00	\
B-E	The proposed method		99.78	99.78	99.77	99.78
	DARLNet [15]	2023	100.00	100.00	100.00	100.00
	Tajmirriahi et al. [22]	2021	99.50	\	100.00	\
	Ramos-Aguilar et al. [23]	2020	99.80	\	\	\
	Siuly et al. [24]	2019	99.00	\	99.09	\
	Deng et al. [25]	2018	97.01	\	\	\
C-E	The proposed method		98.69	98.48	98.49	98.47
	DARLNet [15]	2023	99.78	100.00	99.56	99.78
	Tajmirriahi et al. [22]	2021	99.60	\	98.00	\
	Ramos-Aguilar et al. [23]	2020	99.25	\	\	\
	Siuly et al. [24]	2019	98.50	\	98.18	\
D-E	The proposed method		**100.00**	**100.00**	100.00	**100.00**
	DARLNet [15]	2023	99.57	99.12	100.00	99.56
	Tajmirriahi et al. [22]	2021	99.20	\	98.80	\
	Ramos-Aguilar et al. [23]	2020	97.75	\	\	\
	Siuly et al. [24]	2019	97.50	\	98.09	\
AB-CD-E	The proposed method		**98.84**	**98.62**	**98.70**	**98.66**
	DARLNet [15]	2023	98.17	97.46	97.50	97.46
	Acharya et al. [20]	2018	88.67	\	95.00	\
A-B-C-D-E	The proposed method		**92.35**	**92.13**	**92.17**	**92.14**
	DARLNet [15]	2023	90.17	90.00	90.16	90.05
	Xu et al. [21]	2020	82.00	81.78	81.70	81.56

Acknowledgment. The authors express gratitude to the anonymous referee for his/her helpful suggestions and the partial supports of the National Natural Science Foundation of China (62206005/62236002/62206001).

Declaration of Competing Interest. No author associated with this paper has disclosed any potential or pertinent conflicts which may be perceived to have impending conflict with this work.

References

1. Fisher, R.S., et al.: Epileptic seizures and epilepsy: definitions proposed by the international league against epilepsy (ILAE) and the international bureau for epilepsy (IBE). Epilepsia **46**(4), 470–472 (2005)
2. MacAllister, W.S., Schaffer, S.G.: Neuropsychological deficits in childhood epilepsy syndromes. Neuropsychol. Rev. **17**, 427–444 (2007)

3. Acharya, U.R., Fujita, H., Sudarshan, V.K., Bhat, S., Koh, J.E.: Application of entropies for automated diagnosis of epilepsy using EEG signals: a review. Knowl. Based Syst. **88**, 85–96 (2015)

4. Zhou, M., et al.: Epileptic seizure detection based on EEG signals and CNN. Front. Neuroinform. **12**, 95 (2018)

5. Sahu, R., Dash, S.R., Cacha, L.A., Poznanski, R.R., Parida, S.: Epileptic seizure detection: a comparative study between deep and traditional machine learning techniques. J. Integr. Neurosci. **19**(1), 1–9 (2020)

6. Vijayalakshmi, K., Abhishek, A.M.: Spike detection in epileptic patients EEG data using template matching technique. Int. J. Comput. Appl. **2**(6), 5–8 (2010)

7. Wulandari, D.P., Suprapto, Y.K., Juniani, A.I., Elyantono, T.F., Purnami, S.W., Islamiyah, W.R.: Visualization of epilepsy patient's brain condition based on spectral analysis of EEG signals using topographic mapping. In: 2018 International Conference on Computer Engineering, Network and Intelligent Multimedia (CENIM), pp. 7–13. IEEE (2018)

8. Tang, L., Zhao, M., Wu, X.: Accurate classification of epilepsy seizure types using wavelet packet decomposition and local detrended fluctuation analysis. Electron. Lett. **56**(17), 861–863 (2020)

9. Elakkiya, R.: Machine learning based intelligent automated neonatal epileptic seizure detection. J. Intell. Fuzzy Syst. **40**(5), 8847–8855 (2021)

10. Antoniades, A., et al.: Detection of interictal discharges with convolutional neural networks using discrete ordered multichannel intracranial EEG. IEEE Trans. Neural Syst. Rehabil. Eng. **25**(12), 2285–2294 (2017)

11. Shoji, T., Yoshida, N., Tanaka, T.: Automated detection of abnormalities from an EEG recording of epilepsy patients with a compact convolutional neural network. Biomed. Sig. Process. Control **70**, 103013 (2021)

12. Tuncer, E., Bolat, E.D.: Classification of epileptic seizures from electroencephalogram (EEG) data using bidirectional short-term memory (bi-LSTM) network architecture. Biomed. Sig. Process. Control **73**, 103462 (2022)

13. Beeraka, S.M., Kumar, A., Sameer, M., Ghosh, S., Gupta, B.: Accuracy enhancement of epileptic seizure detection: a deep learning approach with hardware realization of STFT. Circ. Syst. Sig. Process. **41**, 461–484 (2022)

14. Xu, G., Ren, T., Chen, Y., Che, W.: A one-dimensional CNN-LSTM model for epileptic seizure recognition using EEG signal analysis. Front. Neurosci. **14**, 578126 (2020)

15. Qiu, X., Yan, F., Liu, H.: A difference attention ResNet-LSTM network for epileptic seizure detection using EEG signal. Biomed. Sig. Process. Control **83**, 104652 (2023)

16. Craley, J., Johnson, E., Venkataraman, A.: A spatio-temporal model of seizure propagation in focal epilepsy. IEEE Trans. Med. Imaging **39**(5), 1404–1418 (2019)

17. Daubechies, I.: Orthonormal bases of compactly supported wavelets. Commun. Pure Appl. Math. **41**(7), 909–996 (1988)

18. Greff, K., Srivastava, R.K., Koutník, J., Steunebrink, B.R., Schmidhuber, J.: LSTM: a search space odyssey. IEEE Trans. Neural Netw. Learn. Syst. **28**(10), 2222–2232 (2016)

19. Andrzejak, R.G., Lehnertz, K., Mormann, F., Rieke, C., David, P., Elger, C.E.: Indications of nonlinear deterministic and finite-dimensional structures in time series of brain electrical activity: dependence on recording region and brain state. Phys. Rev. E **64**(6), 061907 (2001)

20. Acharya, U.R., Oh, S.L., Hagiwara, Y., Tan, J.H., Adeli, H.: Deep convolutional neural network for the automated detection and diagnosis of seizure using EEG signals. Comput. Biol. Med. **100**, 270–278 (2018)
21. Chen, X., Ji, J., Ji, T., Li, P.: Cost-sensitive deep active learning for epileptic seizure detection. In: Proceedings of the 2018 ACM International Conference on Bioinformatics, Computational Biology, and Health Informatics, pp. 226–235 (2018)
22. Tajmirriahi, M., Amini, Z.: Modeling of seizure and seizure-free EEG signals based on stochastic differential equations. Chaos, Solitons Fractals **150**, 111104 (2021)
23. Ramos-Aguilar, R., Olvera-López, J.A., Olmos-Pineda, I., Sánchez-Urrieta, S.: Feature extraction from EEG spectrograms for epileptic seizure detection. Pattern Recogn. Lett. **133**, 202–209 (2020)
24. Siuly, S., Alcin, O.F., Bajaj, V., Sengur, A., Zhang, Y.: Exploring Hermite transformation in brain signal analysis for the detection of epileptic seizure. IET Sci. Measur. Technol. **13**(1), 35–41 (2019)
25. Deng, Z., Xu, P., Xie, L., Choi, K.S., Wang, S.: Transductive joint-knowledge-transfer TSK FS for recognition of epileptic EEG signals. IEEE Trans. Neural Syst. Rehabil. Eng. **26**(8), 1481–1494 (2018)

A Feature Pyramid Fusion Network Based on Dynamic Perception Transformer for Retinal OCT Biomarker Image Segmentation

Xiaoming Liu[✉] and Yuanzhe Ding

Wuhan University of Science and Technology, Wuhan, China
lxmspace@gmail.com

Abstract. OCT biomarkers are important for assessing the developmental stages of retinal diseases. However, the biomarkers show diverse and irregular features in OCT images and do not have a fixed location. In addition, irregular and unevenly distributed biomarkers can be accompanied by damage to the retinal layers, which can lead to blurred boundaries, thus making it difficult for physicians to make judgments about biomarkers. Therefore, we propose a dynamic perception Transformer-based feature pyramid fusion network for segmenting retinal OCT biomarker images. Our network consists of two modules: the Feature Pyramid Fusion Module (FPFM) and the Dynamic Scale Transformer Module (DSTM). The FPFM connects features at different scales while incorporates an attention mechanism to emphasize the important features. The DSTM dynamically adjusts the scale of the fused features and captures their long-range dependencies, which enables us to preserve small biomarkers and adapt to complex shapes. In this way, we can cope with the challenge of excessive biomarker scale changes from a variable scale perspective. Our proposed model demonstrates good performance on a local dataset.

Keywords: OCT · Retinal Biomarker Segmentation · Transformer

1 Introduction

Retinal biomarkers can reflect the stages of diseases, predict the progression of the diseases and indicate some prognostic factors, which can help ophthalmologists assess the patients' ophthalmic health more conveniently and provide timely clinical diagnosis and treatment [1, 2]. Treatment success for AMD depends greatly on the stage of disease progression [3]. Studies have shown that early and prompt diagnosis and treatment can slow down the progression of AMD and improve the patients' outcomes [4, 5]. However, early biomarkers of AMD do not change significantly, whereas in mid- to late-stage, the retina is usually severely damaged [6, 7]. Therefore, it is crucial to accurately identify the various biomarkers for early detection of AMD and initiate appropriate therapy [8]. OCT can reveal pathological changes in different ocular structures [9, 10]. To detect some diseases in retinal OCT, early studies used classical algorithms [11–13]. However, these algorithms require pre-processing of OCT images such as denoising, enhancement, etc.

to improve image quality and usability. Figure 1 illustrates the eight retinal biomarkers to be recognized.

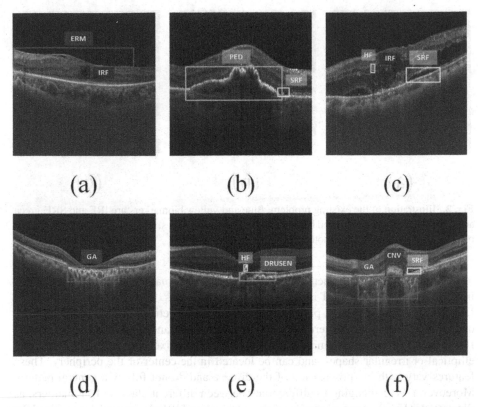

Fig. 1. Examples of eight retinal OCT biomarkers. They are Epiretinal Membrane (ERM), Pigment Epithelial Detachment (PED), Geographic Atrophy (GA), Intraretinal Fluid (IRF), Hyperreflective (HF), Choroidal Neovascularization (CNV), Subretinal Fluid (SRF), and DRUSEN.

As deep learning algorithms are more advanced and powerful, they have been significantly surpassing the effectiveness of traditional methods in retinal OCT biomarker segmentation in recent years [14–16]. To address the challenge of uneven distribution of hyperreflective foci (HF) on retinal OCT images, Xie et al. [17] proposed an effective HF segmentation model. Firstly, the image was pre-processed by denoising and enhancement techniques. Subsequently, dilated convolution was applied in the deepest layer of the network to improve perception. Due to the model's inability to fully consider the small target properties of HF and address the category imbalance problem when segmenting multiple categories, [18] proposed a novel network that can simultaneously segment multiple retinal layers and small foreground macular cystoid edema (MCE). This network effectively utilizes spatial and channel attention to extract features from multiple perspectives. Moreover, the weighted loss approach was used to mitigate category imbalances. Morelle et al. [19] effectively localized drusen by utilizing prior structural information constraints on retinal layers, with dilation convolution and attention gating mechanisms to handle drusen of different sizes.

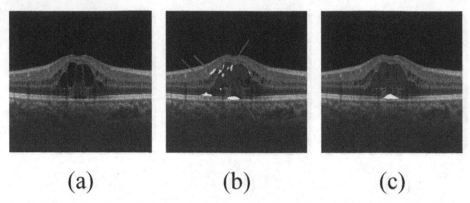

(a)   **(b)**   **(c)**

Fig. 2. Illustration of the existing problem. Blue and yellow biomarkers are IRF and SRF, green and light cyan are mis-segmented PED and HF. (a) original images, (b) U-Net based segmentation results, (c) the ground-truth annotation of biomarkers.

Figure 2 illustrates the performance of U-Net on biomarker segmentation in OCT images. As can be seen, OCT images contain various types of biomarkers, such as SRF, IRF, etc., due to the complex pathological features of the retina. These biomarkers have diverse and uncertain characteristics in terms of size, shape and location. For instance, the IRF in Fig. 2(a) can range from a few pixels to hundreds of pixels in size, can have circular, elliptical or irregular shapes, and can be located in the center or the periphery. These features vary with the progression of the disease and do not follow a general pattern. Moreover, it is challenging to differentiate between different classes of biomarkers, as illustrated in Fig. 2(b), where the arrows indicate cases of IRFs being misclassified as HFs and SRFs. IRFs are typically small and located in the neurosensory layer or the pigmented epithelium, while SRFs are usually irregular and located near neovascularization or pigmented epithelial detachment. Therefore, biomarker segmentation is a difficult and valuable task.

Existing methods for multi-class biomarker segmentation rarely consider the relationship between the biomarker and the image. To explore the information related to the OCT image itself and the biomarker, we propose a dynamic perception Transformer-based feature pyramid fusion network. Our approach to biomarker segmentation is based on spatial pyramid modeling and attention mechanism, which takes into account the scale and positional variations of biomarkers in intra-class and inter-class scenarios. Therefore, our network accepts multi-scale OCT image inputs and uses dilated convolution of different receptive fields to extract multi-scale features for biomarkers of different sizes. Meanwhile, we observe that some small biomarkers have large-scale distribution, so we introduce the attention mechanism to dynamically enhance the long-range dependencies while maintaining performance. Specifically, we introduce an innovative segmentation framework for biomarker segmentation in retinal OCT images. Firstly, we introduce an innovative Feature Pyramid Fusion Module (FPFM) to extract multi-scale information of multi-classes of biomarkers from images and enhance the network's ability to represent the local detailed features of biomarkers with highly variable sizes. Then, we propose a novel Dynamic Scale Transformer Module (DSTM) to address the challenge of low recognition rate of small biomarkers by using variable scale Transformer to complement local and global information while adapting to complex shaped biomarkers.

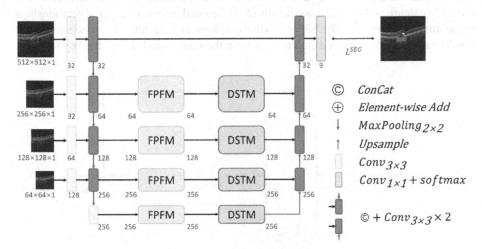

Fig. 3. The overview of the proposed Method.

2   Proposed Method

We introduce a retinal biomarker segmentation network based on an encoder-decoder framework, as depicted in Fig. 3. The primary aim of the network is to categorize each pixel in a retinal OCT image into one of C categories, where $C = 9$, 0 denotes the background, and 1–8 denotes eight different biomarkers. The network consists of four main components: a U-Net encoder, a Feature Pyramid Fusion Module (FPFM), a Dynamic Scale Transformer Module (DSTM), and a U-Net decoder. Our encoder consists of multiple layers, where each layer accepts OCT images that have been adjusted to fit the current encoder size and iteratively combines them with features derived from the previous layer. This way, our encoder can capture both fine-grained and coarse-grained information from the OCT images at different scales. Next, we process each layer of features other than the first one through FPFM and DSTM respectively to improve the expressiveness and long-range dependency of the features, and finally, we input the processed features into the decoder to recover the spatial resolution of the features and output the final segmentation results.

2.1   Feature Pyramid Fusion Module

Multi-class biomarker segmentation is a challenging task, as biomarkers of the same or different classes may vary greatly in their scale and location in the images. Previous models often fail to simultaneously capture the fine-grained details and the global context information of the biomarkers. To address this issue, we introduce the Feature Pyramid Fusion Module (FPFM), which coherently integrates the contextual information and the detailed texture features. It employs a feature pyramid network to amalgamate encoder features at varying resolutions, thereby enabling the capture of both coarse-grained and fine-grained features by using different dilation rates. This fusion process enables the model to capture both the intricate details of the biomarkers and the spatial relationships in the images. Moreover, The FPFM module employs a spatial attention mechanism that considers the importance of different regions in the images and selects the key features.

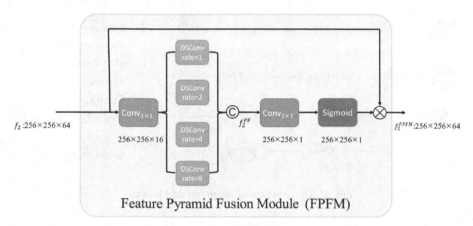

Fig. 4. General structure of the proposed FPFM. The FPFM module processes multi-scale OCT feature inputs. It uses dilated convolution to increase the receptive field and extract richer semantic information. At the same time, it uses spatial pyramid attention mechanism to deal with the complex spatial features of multiple biomarkers.

As depicted in Fig. 3, our proposed FPFM is applied at the skip connections of the 2nd, 3rd, 4th and 5th scales. As shown in Fig. 4, from the second scale onwards, we embed dilated convolution into spatial attention. This way, we can use different sizes of receptive fields to fit biomarkers of different sizes, and highlight the important information in the features. Specifically, we first reduce the channel dimension of the fused feature f_i, and use four parallel depth-wise convolutions with different dilation rates (1, 2, 4, 8) to obtain multi-scale global context information. Then we concatenate the pyramidal features along the channel dimension, and get the merged multi-scale context feature. Next, we squeeze the channel to 1, and use sigmoid to get the weight map of the biomarker. Finally, we multiply the weight with the input feature f_i, resulting in the output feature f_i^{FPFM} of FPFM. The formula for this process is as follows:

$$f_i^{PF} = Concat(DSConv_{rate=1,2,4,8}(Conv_{1\times1}(f_i))), i = 2, 3, 4, 5 \qquad (1)$$

$$f_i^{FPFM} = f_i^{fuse} \otimes \sigma(Conv_{1\times1}(f_i^{PF})), i = 2, 3, 4, 5 \qquad (2)$$

where $DSConv_{rate=1,2,4,8}(\cdot)$ is the deep dilation convolution operation with different dilation rates, $Concat(\cdot)$ is the operation of aggregating parallel dilation convolutions over channels, f_i^{PF} is the feature after fusion of different scale expansion convolutions. $\otimes$ is the feature multiplication operation, $\sigma(\cdot)$ is the sigmoid operation.

2.2 Dynamic Scale Transformer Module

OCT images show diverse biomarker distributions, which makes conventional models prone to fit large targets and fail to detect small targets effectively. Hence, small target detection poses great challenges. We propose a Multi-Head Dynamic Scale Attention (MhDSA) mechanism, which can adaptively adjust the size and location of the attention region based on different levels of the feature map, thereby capturing long-range dependencies in both horizontal and vertical directions and lowering computational complexity.

The brief structure of DSTM, as illustrated in Fig. 5, primarily includes residual connections, layer normalization, MhDSA and feed-forward networks. For the proposed

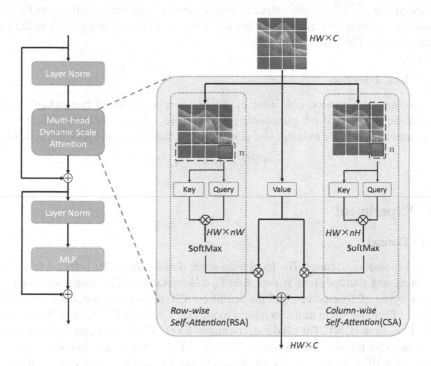

Fig. 5. The overall structure of the proposed DSTM. On the left is a brief composition of the DSTM module. Residual connections and layer normalization can improve the model's stability and convergence rate, and multi-head self-attention and MLP can enhance the model's expressiveness and nonlinearity. On the right is the proposed novel attention mechanism, Multi-Head Dynamic Scale Attention (MhDSA). MhDSA is composed of Row-wise Self-Attention (RSA) and Column-wise Self-Attention (CSA), which capture global dependencies along the horizontal and vertical axes, respectively.

MhDSA, we apply self-attention computation separately in the horizontal and vertical directions to capture the long-range dependencies along both axes. To enhance the model's expressiveness without introducing too much complexity, we design a dynamic attention mechanism. We partition the input features into horizontal or vertical strips of width n in an isometric manner, and dynamically adjust the width based on the network's depth, using smaller n at shallow levels and larger n at deeper levels. The overall formulation of our DSTM is as follows:

$$f_i^{DSTM} = MLP(Norm(f_i^{MhDSA})) + f_i^{MhDSA}, i = 2, 3, 4, 5 \tag{3}$$

$$f_i^{MhDSA} = (RSA(Norm(f_i^{FPFM})) + CSA(Norm(f_i^{FPFM})))$$
$$+ f_i^{FFM}P, i = 2, 3, 4, 5 \tag{4}$$

where f_i^{FPFM} is the output feature of FPFM, $Norm(\cdot)$ is the layer normalization operation, $RSA(\cdot)$ and $CSA(\cdot)$ are row-wise self-attention and column-wise self-attention, respectively, f_i^{MhDSA} is the output feature of multi-head self-attention, $MLP(\cdot)$ is the multilayer perceptual machine of the feedforward network, and f_i^{DSTM} is the output feature of DSTM.

2.3 Loss Function

Since our segmentation task involves a large number of small biomarkers, the class imbalance problem is often encountered. Therefore, our main network combines Focal loss and Dice loss for training. Our main network segmentation loss L^{SEG} is defined as:

$$L^{SEG} = L^{Dice} + L^{Focal} \tag{5}$$

3 Experiment

3.1 Dataset

Local biomarker dataset. The local biomarker dataset, hereafter referred to as the LB dataset, was sourced from Wuhan Aier Eye Hospital. An OCT image contains one or more types of biomarkers, and the specific OCT biomarker categories are shown in Fig. 1. There are eight different retinal biomarkers, namely ERM, PED, SRF, IRF, GA, HF, Drusen and CNV. The LB dataset contains 3000 OCT B-scans, and we set the sample size ratio for training, validation and testing to 6:2:2. The resolution of the LB dataset images is 1024 × 992, and for all images in the LB dataset, we resize them to 512 × 512.

3.2 Comparison Methods and Evaluation Metrics

To examine the segmentation performance of OCT retinal biomarkers, we compared the proposed method with the medical segmentation models U-Net [20] and Att U-Net [21], and the multi-class biomarker segmentation model PASP-Net [22]. In addition, the Dice coefficient was employed as a measure of the segmentation algorithm's accuracy.

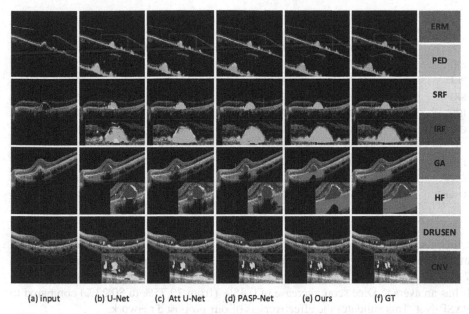

Fig. 6. Segmentation results for each biomarker obtained by comparison with our method.

3.3 Results

Figure 6 (b) shows the segmentation results of U-Net. In the first row, U-Net misclassifies the large Drusen as PED. In the second row, it labels a part of the PED region as SRF, IRF, and CNV. In the third row, it identifies a part of the CNV region as PED, IRF, and Drusen. In the fourth row, it detects the retinal layer as PED and confuses a part of the SRF region with CNV. These results are not satisfactory, and a possible reason is that U-Net's effective receptive field is too small to segment large-scale targets effectively. Figure 6 (c) shows that Att U-Net improves U-Net by introducing the attention gating mechanism, but it still suffers from mis-segmentation. In the first row, Att U-Net also misclassifies Drusen as PED. In the second row, it wrongly labels a part of the PED region as IRF. Figure 6 (d) shows that PASP-Net can process multiple classes of biomarkers with different sizes on a single image better, and it alleviates the problem of mis-segmentation more. In the first row, PASP-Net reduces the region mis-segmented as PED by Drusen. However, it still has serious under-segmentation and over-segmentation problems. Compared with our proposed network, PASP-Net shows region discontinuity in the ERM segmentation in the first row, severe GA under-segmentation in the third row, and HF over-segmentation in the fourth row. Unlike the above methods, our network uses multi-scale pyramid modules and dynamically combines a combination of convolutions and transformers, with minimal mis-segmentation and unrecognition in the segmentation results, especially for small targets.

Table 1 shows the quantitative results. U-Net has an average Dice score of 72.45%. Att U-Net achieves better results than U-Net except for PED and DRUSEN segmentation, with an average Dice score increase of 2.64% (from 72.45% to 75.09%). PASP-Net,

Table 1. The Dice Values of Comparison Methods (unit: %)

	U-Net	Att U-Net	PASP-Net	Ours
ERM	77.81	80.71	82.86	**82.91**
PED	69.16	68.31	76.28	**77.41**
SRF	74.20	78.86	83.66	**83.68**
IRF	70.27	75.19	75.11	**77.95**
GA	74.81	79.70	**84.79**	83.07
HF	63.15	65.80	71.06	**72.27**
DRUSEN	74.03	72.64	78.81	**79.87**
CNV	76.20	79.53	**85.46**	84.69
Overall	72.45	75.09	79.75	**80.23**

which uses a multi-scale pyramid model, outperforms Att U-Net on all biomarker segmentations except IRF, with an average Dice score increase of 4.66% (from 75.09% to 79.75%). Our method obtains the best results for all biomarkers except GA and CNV. It has an average Dice score increase of 0.48% (from 79.75% to 80.23%) compared to PASP-Net. This validates the effectiveness of our proposed network.

3.4 Ablation Experiment

Ablation studies are conducted to affirm the efficacy of the proposed network, with Table 2 illustrating the impact of varying components on the network. Baseline is the model that removes FPFM and DSTM from the proposed network, i.e., U-Net with only encoder and decoder. With the proposed FPFM added to Baseline, all Dice scores are improved, and the average Dice score is increased by 4.10% (from 72.45% to 76.55%). With the proposed DSTM added, there is an enhancement in the Dice score is increased by 3.68% (from 76.55% to 80.23%). Especially for small biomarkers (e.g., IRF, HF, and DRUSEN), significant improvements are obtained. The Dice scores of IRF, HF, and DRUSEN increase by 4.30%, 4.60%, and 2.53%, respectively.

Table 2. The Dice Values of Baselines (unit: %)

	Baseline	Baseline + FPFM	Baseline + FPFM + DSTM (Ours)
ERM	77.81	81.56	**82.91**
PED	69.16	72.94	**77.41**
SRF	74.20	78.52	**83.68**
IRF	70.27	73.65	**77.95**
GA	74.81	82.08	**83.07**
HF	63.15	67.67	**72.27**
DRUSEN	74.03	77.34	**79.87**
CNV	76.20	78.62	**84.69**
Overall	72.45	76.55	**80.23**

4 Conclusion

In this paper, we propose a dynamic perception Transformer-based feature pyramid fusion network to segment multi-class biomarkers in OCT images, which can overcome the challenges faced by existing methods in handling biomarkers of various scales and shapes. Our network consists of two novel modules: the Feature Pyramid Fusion Module (FPFM) and the Dynamic Scale Transformer Module (DSTM). FPFM extracts and fuses biomarker features from different levels and scales, and uses spatial attention to emphasize salient features. DSTM employs a context-based variable scale Transformer to enhance the long range dependency of CNN for fine-grained feature extraction, thus preserving small biomarker features. We performed experiments on the LB dataset and obtained superior results, which show the remarkable effectiveness of our method in segmenting multi-class biomarkers. For future work, we plan to investigate semi-supervised and weakly-supervised methods to address the issue of data scarcity and annotation difficulty.

Acknowledgements. This work was supported in part by the National Natural Science Foundation of China under Grant 62176190.

References

1. Daien, V., et al.: Evolution of treatment paradigms in neovascular age-related macular degeneration: a review of real-world evidence. Br. J. Ophthalmol. **105**, 1475–1479 (2021)
2. Branisteanu, D.C., et al.: Influence of unilateral intravitreal bevacizumab injection on the incidence of symptomatic choroidal neovascularization in the fellow eye in patients with neovascular age-related macular degeneration. Exp. Ther. Med. **20**, 1 (2020)
3. Guymer, R., Wu, Z.: Age-related macular degeneration (AMD): more than meets the eye. The role of multimodal imaging in today's management of AMD—a review. Clin. Exp. Ophthalmol. **48**, 983–995 (2020)
4. Loewenstein, A.: The significance of early detection of age-related macular degeneration: Richard & Hinda Rosenthal Foundation lecture, the Macula Society 29th annual meeting. Retina **27**, 873–878 (2007)
5. Liberski, S., Wichrowska, M., Kocięcki, J.: Aflibercept versus faricimab in the treatment of neovascular age-related macular degeneration and diabetic macular edema: a review. Int. J. Mol. Sci. **23**, 9424 (2022)
6. Kanagasingam, Y., Bhuiyan, A., Abràmoff, M.D., Smith, R.T., Goldschmidt, L., Wong, T.Y.: Progress on retinal image analysis for age related macular degeneration. Prog. Retin. Eye Res. **38**, 20–42 (2014)
7. Weiss, M., et al.: Compliance and adherence of patients with diabetic macular edema to intravitreal anti–vascular endothelial growth factor therapy in daily practice. Retina **38**, 2293–2300 (2018)
8. Schmidt-Erfurth, U., Hasan, T.: Mechanisms of action of photodynamic therapy with verteporfin for the treatment of age-related macular degeneration. Surv. Ophthalmol. **45**, 195–214 (2000)
9. Leitgeb, R., et al.: Enhanced medical diagnosis for dOCTors: a perspective of optical coherence tomography. J. Biomed. Opt. **26**, 100601 (2021)

10. Bussel, I.I., Wollstein, G., Schuman, J.S.: OCT for glaucoma diagnosis, screening and detection of glaucoma progression. Br. J. Ophthalmol. **98**, ii15–ii19 (2014)
11. Wilkins, G.R., Houghton, O.M., Oldenburg, A.L.: Automated segmentation of intraretinal cystoid fluid in optical coherence tomography. IEEE Trans. Biomed. Eng. **59**, 1109–1114 (2012)
12. Wang, T., et al.: Label propagation and higher-order constraint-based segmentation of fluid-associated regions in retinal SD-OCT images. Inf. Sci. **358**, 92–111 (2016)
13. Montuoro, A., Waldstein, S.M., Gerendas, B.S., Schmidt-Erfurth, U., Bogunović, H.: Joint retinal layer and fluid segmentation in OCT scans of eyes with severe macular edema using unsupervised representation and auto-context. Biomed. Opt. Express **8**, 1874–1888 (2017)
14. Liu, X., Wang, S., Zhang, Y., Liu, D., Hu, W.: Automatic fluid segmentation in retinal optical coherence tomography images using attention based deep learning. Neurocomputing **452**, 576–591 (2021)
15. Liu, X., Cao, J., Wang, S., Zhang, Y., Wang, M.: Confidence-guided topology-preserving layer segmentation for optical coherence tomography images with focus-column module. IEEE Trans. Instrum. Meas. **70**, 1–12 (2020)
16. Liu, X., Zhou, K., Yao, J., Wang, M., Zhang, Y.: Contrastive uncertainty based biomarkers detection in retinal optical coherence tomography images. Phys. Med. Biol. **67**, 245012 (2022)
17. Xie, S., Okuwobi, I.P., Li, M., Zhang, Y., Yuan, S., Chen, Q.: Fast and automated hyperreflective foci segmentation based on image enhancement and improved 3D U-Net in SD-OCT volumes with diabetic retinopathy. Transl. Vis. Sci. Technol. **9**, 21 (2020)
18. Cazañas-Gordón, A., da Silva Cruz, L.A.: Multiscale attention gated network (MAGNet) for retinal layer and macular cystoid edema segmentation. IEEE Access **10**, 85905–85917 (2022)
19. Morelle, O., Wintergerst, M.W., Finger, R.P., Schultz, T.: Accurate drusen segmentation in optical coherence tomography via order-constrained regression of retinal layer heights. Sci. Rep. **13**, 8162 (2023)
20. Ronneberger, O., Fischer, P., Brox, T.: U-Net: convolutional networks for biomedical image segmentation. In: Navab, N., Hornegger, J., Wells, W., Frangi, A. (eds.) MICCAI 2015. LNCS, vol. 9351, pp. 234–241. Springer, Cham (2015). https://doi.org/10.1007/978-3-319-24574-4_28
21. Oktay, O., et al.: Attention U-Net: learning where to look for the pancreas. arXiv preprint arXiv:1804.03999 (2018)
22. Hassan, B., Qin, S., Hassan, T., Ahmed, R., Werghi, N.: Joint segmentation and quantification of chorioretinal biomarkers in optical coherence tomography scans: a deep learning approach. IEEE Trans. Instrum. Meas. **70**, 1–17 (2021)

LDW-RS Loss: Label Density-Weighted Loss with Ranking Similarity Regularization for Imbalanced Deep Fetal Brain Age Regression

Yang Liu[1], Siru Wang[1], Wei Xia[2], Aaron Fenster[3], Haitao Gan[1], and Ran Zhou[1](✉)

[1] School of Computer Science, Hubei University of Technology, Wuhan, China
ranzhou@hbut.edu.cn
[2] Wuhan Children's Hospital (Wuhan Maternal and Child Healthcare Hospital), Tongji Medical College, Huazhong University of Science and Technology, Wuhan, China
xiawei@zgwhfe.com
[3] Imaging Research Laboratories, Robarts Research Institute, Western University, London, ON N6A 5B7, Canada
afenster@robarts.ca

Abstract. Estimation of fetal brain age based on sulci by magnetic resonance imaging (MRI) is crucial in determining the normal development of the fetal brain. Deep learning provides a possible way for fetal brain age estimation using MRI. However, real-world MRI datasets often present imbalanced label distribution, resulting in the model tending to show undesirable bias towards the majority of labels. Thus, many methods have been designed for imbalanced regression. Nevertheless, most of them on handling imbalanced data focus on targets with discrete categorical indices, without considering the continuous and ordered nature of target values. To fill the research gap of fetal brain age estimation with imbalanced data, we propose a novel label density-weighted loss with a ranking similarity regularization (LDW-RS) for deep imbalanced regression of the fetal brain age. Label density-weighted loss is designed to capture information about the similarity between neighboring samples in the label space. Ranking similarity regularization is developed to establish a global constraint for calibrating the biased feature representations learned by the network. A total of 1327 MRI images from 157 healthy fetuses between 22 and 34 weeks were used in our experiments for the fetal brain age estimation regression task. In the random experiments, our LDW-RS achieved promising results with an average mean absolute error of 0.760 ± 0.066 weeks and an R-square (R^2) coefficient of 0.914 ± 0.020. Our fetal brain age estimation algorithm might be useful for identifying abnormalities in brain development and reducing the risk of adverse development in clinical practice.

Keywords: Fetal Brain Age Estimation · Deep Imbalanced Regression · Label Density-Weighted Loss · Ranking Similarity Regularization

1 Introduction

Fetal brain age estimation based on sulci development has been widely utilized to characterize the normal development process and variation trend of the fetal brain [7].

B. Luo et al. (Eds.): ICONIP 2023, CCIS 1964, pp. 125–137, 2024.
https://doi.org/10.1007/978-981-99-8141-0_10

Age-related changes in the brain have a significant relationship with the pathogenesis of neurological disorders and the prediction of brain age could be a crucial biomarker for assessing brain health [30, 32]. The difference between the predicted fetal brain age and the gestational age serves as an indicator of deviation from the typical developmental path, which could potentially suggest the presence of neurodevelopmental disorders, such as Ventriculomegaly, Callosal hypoplasia and Dandy-Walker [4, 7]. Fetal brain imaging is an essential cornerstone of prenatal screening and early diagnosis of congenital anomalies [6]. Magnetic resonance imaging (MRI) has unrivaled efficacy in examining the developing fetus in utero, which can clearly visualize the shape, depth, and appearance time of the sulci across different gestational ages [1, 2]. Therefore, an accurate estimation of gestational age from MRI is vital for providing appropriate care throughout pregnancy and identifying complications [3, 5]. Manual identification sulci to estimate fetal brain age is time-consuming and sensitive to the observer's experience. To alleviate this burden, investigators have developed computer-assisted methods for fetal brain age estimation from MRI images. Traditional approaches to fetal brain age estimation involve the use of linear or non-linear regression models that are based on quantitative features such as sulcal depth and curvature [22, 31]. However, these approaches require either manual intervention or complex MRI processing [32].

Recently, several deep learning-based works have shown great potential in predicting fetal brain age. Shen et al. [6] proposed a ResNet [9] architecture with the attention mechanism method to localize the fetal brain region and obtained a mean absolute error (MAE) of 0.971 weeks. Shi et al. [7] employed an attention-based deep residual network, which achieved an overall MAE of 0.767 weeks. Shen et al. [8] designed an attention-guided, multi-view network that analyzed MRI-based features of the normally developing fetal brain to predict gestational age and achieve an MAE of 0.957 weeks. Liao et al. [10] presented label distribution learning to deal with the problem of insufficient training data. They also used a multi-branch CNN with deformable convolutional to aggregate multi-view information and obtained an MAE of 0.751 weeks. Although previous studies have achieved significant progress, all these works rely on balanced datasets for network training. However, in real-world scenarios, the collected fetal datasets are often imbalanced. A highly skewed label distribution would weaken the discriminative power of the regression model.

Most previous works [27–29] handling imbalanced regression problems have primarily focused on synthesizing new samples for minority targets to resample the training set based on targets belonging to different classes. Nevertheless, treating different targets as distinct classes is unlikely to attain the best results because it does not leverage the meaning distance between targets. Moreover, regression labels natively possess attributes of continuity and orderliness, and the target values can even be infinite. As a result, many methods designed for imbalanced classification are untenable. Recently, feature distribution smoothing [12] transfers local feature statistics between adjacent continuous target intervals, relieving potentially biased estimation of the feature distribution for targets with limited samples. However, it overlooks the significance of global feature information among samples. Effective number of samples [13, 23], serving as a novel cost-sensitive learning reweighting scheme in regression, reflects the contribution of each class of effective samples within each interval. Nevertheless, effective samples still exist certain limitations due to regression samples exhibiting unique continuity and similarity distinct from categorical samples.

To address the above challenges, we introduce a novel framework using label density-weighted loss to capture information about the similarity between nearby samples and quickly pay attention to hard continuous samples in the label space, and a ranking similarity regularization to constrain the sorted list of neighbors in the feature space matches the sorted list of neighbors in the label space. To the best of our knowledge, this is the first work that estimates fetal brain age from a deep imbalanced regression perspective. The main contributions are as follows:

- Label density-weighted loss is designed to focus on continuous samples that are hard to distinguish in imbalanced regression, which employs a label distribution smoothing strategy to generate the label density distribution for weighting Focal-MSE loss function.
- RankSim regularization is developed to impose a forceful global inductive bias on the model, ensuring a complete match between the ranking of neighboring samples in the feature space and the ranking of the target values. Mining the similarity and ordinal relationship among the features, effectively calibrating the unjustified feature representations learned by the network.
- Experimental results demonstrate that our method achieves superior predictive performance for fetal brain age estimation on the imbalanced dataset, which will be contributed to precisely evaluating the degree of fetal brain development and detecting developmental abnormalities in clinical practice.

2 Method

Fetal brain MRI datasets often exhibit a highly imbalanced age label distribution. Specifically, most fetal brain age labels are under-represented due to their low frequency, while a minority of age labels dominate the majority of instances. This imbalance can lead to deep learning model learning an unjustified bias towards labels with a higher number of instances while performing poorly on labels with low frequency [12]. In this study, we present a novel framework to revisit the fetal brain age estimation task, mainly including two key techniques: label density-weighted loss (LDW) and ranking similarity regularization (RankSim). Figure 1 is the overview of the proposed imbalanced regression framework. First, LDW captures similar information among adjacent samples with continuous targets by reweighting the Focal-MSE loss function based on effective label density for each target. Second, RankSim encourages the ordinality and similarity relationship of samples in the label space to be consistent in the feature space. Finally, LDW loss and RankSim regularization are combined to form the total loss function.

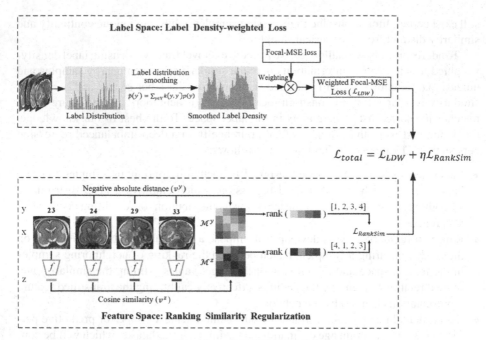

Fig. 1. Overview of the LDW-RS framework

2.1 Label Density-Weighted Loss

Most methods for handling imbalanced data are tailored for classification problems, where hard boundaries exist between different classes. But fetal brain age estimation tasks involve continuous or even infinite target values, and there may be potential missing data in certain target regions. This renders many imbalanced classification methods ineffective. Specifically, the widely used re-sampling and cost-sensitive approaches [24–26] rely on hard boundaries between classes and the frequency of each class, which are eliminated by continuous age labels. How to estimate the effective frequency or distribution of soft labels is still an open question.

Discretization aims to bridge the gap between continuous age labels and hard boundaries. Specifically, let $D = \{(x_i, y_i)\}_{i=1}^{N}$ be a training set, where x_i denotes the input image, y_i is the continuous age label and $\hat{y}_i$ is the predicted label. Discretize the label space $\mathbb{Y}$ into M bins with equal intervals, *i.e.*, $[y_0, y_1), [y_1, y_2), \ldots, [y_{M-1}, y_M)$ and utilize $m \in \Omega$ to represent the bin index of the target value, where $\Omega = \{1, \ldots, M\}$ is the index space. In the fetal brain age estimation task, the defined bins reflect the minimum brain age difference of 1, *i.e.*, $\Delta y = y_{m+1} - y_m = 1$.

Although the class frequencies have been obtained through discretization and can be applied to resampling or reweighting, recent studies have indicated that this empirical label density is not accurate in imbalanced regression [12, 13]. Due to the interdependence among continuous data samples at nearby labels, the coarse empirical label density distribution does not correctly reflect the real density distribution.

Thus, the label distribution smoothing (LDS) strategy is introduced to utilize kernel density estimation for learning an effective imbalance with continuous targets [12]. LDS provides an interesting insight that convolves a symmetric kernel with the empirical label density distribution to obtain an effective label density distribution, which accurately reflects the information about the similarity between adjacent labels. Specifically, the fetal brain image information represented by close age labels is expected to have a certain degree of overlap. The closer the age labels, the more similar the corresponding fetal brain information. The symmetric kernel used in LDS can characterize the similarity between target values y' and any y. In other words, this symmetric kernel function measures the contribution of each sample within its neighborhood. Hence, LDS calculates the effective label density distribution, as follows:

$$\tilde{p}(y') = \sum_{y \in \mathbb{Y}} k(y, y') p(y) \tag{1}$$

where $p(y)$ indicates the number of appearances of label of y in the training data and $\tilde{p}(y')$ denotes the effective density of label y'. The symmetric kernel needs to satisfy: (1) $k(y, y') = k(y', y)$, (2) $\nabla_y k(y, y') + \nabla_{y'} k(y', y) = 0$. Such as the Gaussian kernel $k(y, y') = exp(-\frac{\|y-y'\|^2}{2\sigma^2})$. Note that LDS strategy does not explicitly change the fetal brain MRI dataset (e.g., by creating new samples).

The estimated effective label density for each target serves as a weighting factor to reweight the loss function. Furthermore, we use the Focal-MSE loss to replace the MSE loss, primarily because the MSE may carry the label imbalance into prediction, which leads to the inferior performance on rare labels [12, 18, 19]. The main loss function, called label density-weighted loss ($\mathcal{L}_{LDW}$), is constructed by combining Focal-MSE and LDS. This allows it to not only down-weight the contribution of easy samples and prioritize challenging ones but also accurately learn the real imbalanced distribution by leveraging information about the similarity among neighboring samples with continuous target values and potential missing values:

$$\mathcal{L}_{LDW} = \frac{1}{n} \sum_{i=1}^{n} \frac{1}{\tilde{p}(y_i')} \cdot F(|\beta e_i|)^\gamma (\hat{y}_i - y_i)^2 \tag{2}$$

where $\frac{1}{\tilde{p}(y_i')}$ represents the weight that is the inverse of the estimated effective label density for each sample through LDS, n is the number of each batch, e_i denotes the L_1 error for i th sample, β and γ are hyperparameters. The scaling factor F is a continuous function that can be utilized as a sigmoid function to map the absolute error into the range of $[0, 1]$. $\mathcal{L}_{LDW}$ can reduce the weights of easily distinguishable samples, enabling the model to rapidly focus on training hard continuous samples with rare labels, without being disturbed by frequent easy samples.

2.2 Ranking Similarity Regularization

In regression, the fetal brain age follows a natural ordering (e.g., from 22 weeks to 35 weeks). Intuitively, the orderliness and similarity information of samples in the label space should also be consistent in the feature space. The ordinal pattern of age labels allows us to regularize the feature representations learned by the model. Specifically, the network can be guided in a direction where the feature representation of the fetal brain sulci at 22 weeks old becomes more similar to the representation of a 25-week-old than to that of a 30-week-old. The more accurate the fetal brain sulci feature representation learned by the network, the higher the precision of estimating brain age.

The ranking similarity regularization (RankSim) is proposed to deeply mine the inherent traits of age labels by encoding a global inductive bias, ensuring that the sorted list of neighbors in the feature space matches the sorted list of neighbors in the label space [15]. RankSim is capable of capturing both the ordinal and similarity relationships between samples at both close and distant distances. Especially, it allows fetal brain images with closer age labels to have more similar brain sulci features, while images with noticeably distant age labels tend to have lower feature similarity. For instance, the ages corresponding to the four fetal brain images are 23, 24, 29, and 33, with the corresponding brain sulci features being z^{23}, z^{24}, z^{29}, and z^{33}. The similarity function is denoted as $v(\cdot, \cdot)$, such as the cosine similarity. Based on the distances in the label space (age), RankSim would encourage $v(z^{23}, z^{24}) > v(z^{23}, z^{29}) > v(z^{23}, z^{33})$, with z^{23} serving as the anchor. This operation can be repeated when selecting other brain images as anchors.

More specifically, we let $t \in \mathbb{R}$ represent a vector of H values, Rank($\cdot$) stands for the rank operator and Rank(t) is the permutation of $\{1, \ldots, H\}$ containing the rank (in sorted order) of each element in t. The Rank(t) of the i th element is one plus the number of members in the sequence exceeding its value.

$$\text{Rank}(t)_i = 1 + |\{k : t_k > t_i\}| \tag{3}$$

Let $\mathcal{B} = \{(x_i, y_i)\}_{i=1}^{T}$ is a subset of each batch. In label space, $\mathcal{M}^y \in \mathbb{R}^{T \times T}$ denotes the pairwise similarity matrix computed by a similarity function v^y (negative absolute distance). Therefore, the (i, k) th entry in $\mathcal{M}^y$ can be expressed as:

$$\mathcal{M}_{i,k}^y = v^y(y_i, y_k) \tag{4}$$

In feature space, let $f(\cdot)$ represents feature extractor and $z_i = f(x_i)$ denotes extracted features. $\mathcal{M}^z \in \mathbb{R}^{T \times T}$ stands for the pairwise similarity matrix computed by a similarity function v^z (cosine similarity). Hence, the (i, k) th entry in $\mathcal{M}^z$ can be defined as:

$$\mathcal{M}_{i,k}^z = v^z(z_i, z_k) = v^z(f(x_i), f(x_k)) \tag{5}$$

According to Eq. 3, Eq. 4, and Eq. 5, the RankSim regularization loss is defined as:

$$\mathcal{L}_{RankSim} = \sum_{i=1}^{T} \varphi(\text{Rank}(\mathcal{M}_{[i,:]}^y), \text{Rank}(\mathcal{M}_{[i,:]}^z)) \tag{6}$$

where ranking similarity function φ (cosine distance) penalizes differences between the ranking of neighbors in the label space and the ranking of neighbors in the feature space. Here, $[i, :]$ denotes the i th row in the matrix. The advantage of designing the rank operator Rank($\cdot$) to construct the $\mathcal{L}_{RankSim}$ lies in its ability to not only consider the ordinal relationship between data but also exhibit robustness to anomalous data. When the difference approaches 0, $\mathcal{L}_{RankSim}$ will converge to 0, which indicates that the sorted list of neighbors in the feature space has been successfully matched with the sorted list of neighbors in the label space for a given input sample. This ensures that the orderliness and similarity of the samples in the label space remain consistent in the corresponding feature space. Note that the current batch is sampled so that each age label appears at most once, which can reduce ties and promote the relative representation of infrequent age labels [15].

However, the rank operator Rank$(\cdot)$ is non-differentiable, piecewise constant $\mathcal{L}_{RankSim}$ becomes challenging to optimize as the gradient is zero almost everywhere. Hence, we recast Rank$(\cdot)$ as the minimizer of a linear combinatorial objective [16].

$$\text{Rank}(t) = \arg\min t \cdot \pi \tag{7}$$

where $\pi \in \Pi_s$ and Π_s is the set of all permutations of $\{1, 2, ..., s\}$. Then, we utilize black box combinatorial solvers [17] as an elegant approach to derive meaningful gradient information from the piecewise constant loss. This is achieved through constructing a family of continuous functions parametrized by an interpolated hyperparameter $\lambda > 0$ that trades off the informativeness of the gradient with fidelity to the original function. During the backpropagation process, the calculation of the returned continuous interpolated gradient information is as follows:

$$\frac{\partial \mathcal{L}}{\partial t} = -\frac{1}{\lambda}(\text{Rank}(t) - \text{Rank}(t_\lambda)) \tag{8}$$

$$t_\lambda = t + \lambda \frac{\partial \mathcal{L}}{\partial Rank} \tag{9}$$

where $\frac{\partial \mathcal{L}}{\partial Rank}$ denotes the incoming gradient information and λ represents the interpolation strength. Now, $\mathcal{L}_{RankSim}$ is available and can be directly added to the total loss function to calibrate the detrimental biased feature representations learned by the regression model on imbalanced data. Then, the total loss function is defined as:

$$\mathcal{L}_{total} = \mathcal{L}_{LDW} + \eta \mathcal{L}_{RankSim} \tag{10}$$

where η is the weight to balance the total loss $\mathcal{L}_{total}$.

3 Experiments and Results

3.1 Experimental Settings

Dataset. As shown in Fig. 2, we retrospectively collected 1327 coronal T2-weighted MRI brain images from 157 singleton pregnancy fetuses between 22 and 34 weeks of gestational age. The gestational age at the time of the MRI examination was calculated based on the menstrual period, which was confirmed by crown-rump length during the first trimester ultrasound, and the ultrasound biometry prior to MRI was within the normal range according to gestational age. The clearest slices from MRI stack, examined by professional radiologists, were utilized as training data for the model. Written informed consent was signed by all pregnant women before the examination and this study was approved by the ethics committee of the Wuhan Children's Hospital.

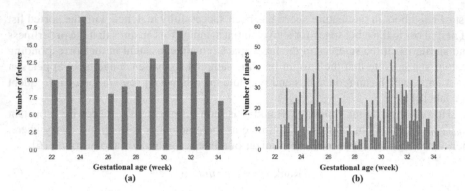

Fig. 2. (a) displays the number of fetuses at each gestational age. (b) shows the number of fetal brain images at each gestational age.

Implementation Details. The fetal brain MRI dataset was randomly divided by patients into a training set (85%) and a test set (15%) without overlap. Ten random experiments were performed for evaluation. Each input image was padded and resized to 96×96 pixels. The model was trained for 230 epochs using Adam optimizer with a learning rate $= 0.001$ and batch size $= 32$. LDW was leveraged with $\beta = 0.2$, $\gamma = 1$, a Gaussian kernel size $= 5$, and a kernel sigma $= 2$. RankSim was employed with λ of 2 and η of 100. The reweighting scheme was sqrt-inverse. All experiments were implemented in Python 3.8 with PyTorch 1.12.1 and NVIDIA GeForce RTX 3090 GPU.

Evaluation Metrics. For the performance evaluation, we followed earlier works [6–8, 10] and employed common metrics for fetal brain age regression, such as the mean absolute error (MAE, lower is better) and coefficient of determination (R^2, higher is better). We further proposed three metrics to enhance age predictive fairness, namely the root mean square error (RMSE, lower is better), mean absolute percentage error (MAPE, lower is better), and error geometric mean (GM, lower is better) [12, 18]. Continuous variables were expressed as mean±standard deviation.

3.2 Comparison with Other State-of-the-Art Methods

Table 1 shows that compared to some popular regression models, such as ResNet, DenseNet, SE-Net, ConvNeXt and Swin Transformer, HS-ResNet achieved the smallest prediction error with MAE of 0.927 weeks and R^2 of 0.874. Hence, we set HS-ResNet as backbone network. By utilizing LDW and RankSim with HS-ResNet, our approach obtained superior performance with an average MAE of 0.760 ± 0.066 weeks and an R^2 of 0.914 ± 0.020, resulting in a significant reduction of 0.167 weeks in MAE. These results indicate that LDW and RankSim can contribute to enhancing the performance of HS-ResNet with an imbalanced training dataset. Additionally, compared with Yang et al. [12], Ding et al. [13] and Gong et al. [15], our imbalanced regression algorithm exhibited excellent performance across all evaluation metrics, with reductions in MAE of 0.424 weeks, 0.396 weeks and 0.102 weeks, respectively. Moreover, compared to the fetal age estimation method proposed by Liao et al. [10], our proposed algorithm LDW-RS obtained better performance with improvements across all metrics, particularly a decrease of 0.134 weeks in MAE. These results indicate that our approach can assist doctors in the earlier detection of disease indicators for fetal brain, thereby effectively reducing the risk of adverse developmental.

Our deep imbalanced regression algorithm required 0.55 h for training and a mean time of 23 ms for testing an image.

Table 1. Performance comparison of fetal brain age estimation methods on imbalanced dataset.

Method	MAE (week)	R^2	RMSE (week)	MAPE (%)	GM (week)
ResNet	0.932 ± 0.075	0.870 ± 0.040	1.229 ± 0.118	3.259 ± 0.214	0.583 ± 0.054
DenseNet	0.938 ± 0.078	0.871 ± 0.029	1.229 ± 0.117	3.269 ± 0.264	0.596 ± 0.056
SE-Net	1.172 ± 0.173	0.794 ± 0.067	1.542 ± 0.250	4.104 ± 0.650	0.776 ± 0.120
ConvNeXt	1.120 ± 0.128	0.824 ± 0.046	1.436 ± 0.169	3.912 ± 0.488	0.759 ± 0.099
HS-ResNet	0.927 ± 0.146	0.874 ± 0.040	1.209 ± 0.164	3.265 ± 0.553	0.594 ± 0.121
Swin Transformer	0.998 ± 0.113	0.855 ± 0.035	1.306 ± 0.133	3.483 ± 0.420	0.647 ± 0.097
Yang et al. [12]	1.184 ± 0.063	0.792 ± 0.052	1.555 ± 0.106	4.147 ± 0.181	0.751 ± 0.046
Ding et al. [13]	1.156 ± 0.079	0.797 ± 0.050	1.535 ± 0.122	4.090 ± 0.300	0.730 ± 0.075
Liao et al. [10]	0.894 ± 0.062	0.877 ± 0.024	1.452 ± 0.231	3.147 ± 0.243	0.555 ± 0.056
Gong et al. [15]	0.862 ± 0.072	0.888 ± 0.028	1.143 ± 0.103	3.024 ± 0.285	0.552 ± 0.053
Ours ($\mathcal{L}_{LDW}$ + RankSim)	$\mathbf{0.760 \pm 0.066}$	$\mathbf{0.914 \pm 0.020}$	$\mathbf{1.006 \pm 0.096}$	$\mathbf{2.684 \pm 0.253}$	$\mathbf{0.489 \pm 0.050}$

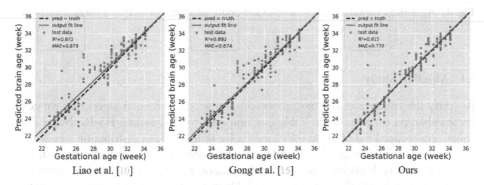

Fig. 3. Correlation of the predicted brain age versus the ground truth from the output fit line.

3.3 Correlation and Agreement Analysis

Figure 3 presents the correlation plot comparing the predicted fetal brain age and the ground truth from the output fit line, resulting from conducting the same random experiments on the test data using different algorithms. Compared to the R^2 of 0.872 and MAE of 0.879 weeks obtained by Liao et al. [10] and R^2 of 0.892 and MAE of 0.874 weeks achieved by Gong et al. [15], our proposed deep imbalanced regression algorithm LDW-RS exhibited a stronger correlation ($R^2 = 0.915$) and lower mean absolute error (MAE = 0.770 weeks), which indicates that LDW-RS can enable the regression model to produce more accurate predictions.

Figure 4 depicts the Bland-Altman plot with 95% limits of agreement (95% LoA) to better visualize the agreement between the predicted fetal brain age and true age, comparing different algorithms in the same random experiments on the test data. Compared to the Mean Difference of 0.36 weeks obtained by Liao et al. [10] and the Mean Difference of 0.20 weeks acquired by Gong et al. [10], our proposed algorithm LDW-RS achieved a smaller bias (Mean Difference = 0.09 weeks) close to zero, signifying a high degree of agreement between the predicted brain age and actual age and most scatter points fall within the 95% LoA (upper and lower limits), indicating a relatively low difference between the prediction and ground truth. These results substantiate that the proposed imbalanced regression algorithm LDW-RS can promote the regression model to generate more stable and reliable predictions.

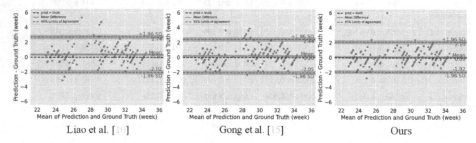

Fig. 4. Agreement between prediction brain age and ground truth with Bland-Altman analysis.

3.4 Ablation Study

To verify the effectiveness of our proposed imbalanced regression algorithm, we conducted ablation experiments in Table 2. Compared to the baseline HS-ResNet, LDW effectively reduced the MAE to 0.876 weeks by utilizing label distribution smoothing strategy to re-weight Focal-MSE based on the estimated effective label density distribution. Additionally, RankSim was leveraged to encourage the sorted list of neighbors in feature space to match the sorted list of neighbors in label space, resulting in a further improvement, with the MAE decreased to 0.843 weeks. By combining LDW loss and RankSim regularization, the orderliness and similarity relationship between samples was effectively constrained in both the label space and the feature space, which led to a significant reduction in MAE, reaching a lower value of 0.760 weeks.

Table 2. Ablation study of the proposed imbalanced regression algorithm using HS-ResNet for fetal brain age estimation ('✓' With; '–' Without).

	LDW	RankSim	MAE (week)	R^2	RMSE (week)	MAPE (%)	GM (week)
(I)	–	–	0.927 ± 0.146	0.874 ± 0.040	1.209 ± 0.164	3.265 ± 0.553	0.594 ± 0.121
(II)	✓	–	$0.876 \pm \mathbf{0.052}$	0.883 ± 0.026	$1.172 \pm \mathbf{0.075}$	$3.080 \pm \mathbf{0.193}$	0.554 ± 0.053
(III)	–	✓	0.843 ± 0.080	0.893 ± 0.023	1.122 ± 0.121	2.975 ± 0.293	0.544 ± 0.062
(IV)	✓	✓	$\mathbf{0.760} \pm 0.066$	$\mathbf{0.914} \pm \mathbf{0.020}$	$\mathbf{1.006} \pm 0.096$	$\mathbf{2.684} \pm 0.253$	$\mathbf{0.489} \pm \mathbf{0.050}$

4 Conclusion

In this paper, we propose a novel imbalanced regression loss (LDW-RS) that combines two key techniques: label density-weighted loss (LDW) and ranking similarity regularization (RankSim). LDW utilizes label distribution smoothing strategy to reweight Focal-MSE, utilizing similar information between nearby samples and focusing on hard continuous samples. RankSim aligns the ranking of samples in the label space with the ranking of samples in the feature space, calibrating the biased feature representations learned by the model. Experiments on the fetal brain dataset demonstrate that the proposed method can effectively address imbalanced regression problems and obtain a lower MAE of 0.760 weeks on the fetal brain age estimation task, which has the potential to be translated to prenatal diagnosis in clinical practice.

The limitation of our method lies in conducting performance evaluation solely on a specific fetal brain dataset, without comparisons to other datasets. We leave this for future extension. Moreover, we plan to further study the intrinsic ordinal pattern of fetal brain to address imbalanced regression more effectively.

Acknowledgments. This work was supported by the National Natural Science Foundation of China under grant No. 62201203, the Natural Science Foundation of Hubei Province, China under grant No. 2021CFB282, the High-level Talents Fund of Hubei University of Technology, China under grant No. GCRC2020016, the Doctoral Scientific Research Foundation of Hubei University of Technology, China under grant No. BSDQ2020064.

References

1. Glenn, O.A., Barkovich, A.J.: Magnetic resonance imaging of the fetal brain and spine: an increasingly important tool in prenatal diagnosis, Part 1. Am. J. Neuroradiol. **27**(8), 1604–1611 (2006)
2. Nie, W., et al.: Deep learning with modified loss function to predict gestational age of the fetal brain. In: International Conference on Signal and Image Processing, pp. 572–575 (2022)
3. Prayer, D., Malinger, G., et al.: ISUOG practice guidelines (updated): performance of fetal magnetic resonance imaging. Int. Soc. Ultrasound Obstet. Gynecol. **61**(2), 278–287 (2023)
4. Hu, D., Wu, Z., et al.: Hierarchical rough-to-fine model for infant age prediction based on cortical features. IEEE J. Biomed. Health Inform. **24**(1), 214–225 (2020)
5. Lee, C., Willis, A., Chen, C., et al.: Development of a machine learning model for sonographic assessment of gestational age. JAMA Netw. Open **6**(1), e2248685 (2023)

6. Shen, L., Shpanskaya, K. S., Lee, E., et al.: Deep learning with attention to predict gestational age of the fetal brain. arXiv preprint arXiv:1812.07102 (2018)
7. Shi, W., Yan, G., Li, Y., et al.: Fetal brain age estimation and anomaly detection using attention-based deep ensembles with uncertainty. Neuroimage **223**, 117316 (2020)
8. Shen, L., Zheng, J., Lee, E.H., et al.: Attention-guided deep learning for gestational age prediction using fetal brain MRI. Sci. Rep. **12**(1), 1408 (2022)
9. He, K., Zhang, X., Ren, S., Sun, J.: Deep residual learning for image recognition. In: IEEE Conference on Computer Vision and Pattern Recognition, pp. 770–778 (2016)
10. Liao, L., Zhang, X., Zhao, F., et al.: Multi-branch deformable convolutional neural network with label distribution learning for fetal brain age prediction. In: IEEE 17th International Symposium on Biomedical Imaging, pp. 424–427 (2020)
11. Yuan, P., Lin, S., Cui, C., et al.: HS-ResNet: hierarchical-split block on convolutional neural network. arXiv preprint arXiv:2010.07621 (2020)
12. Yang, Y., Zha, K., Chen, Y., et al.: Delving into deep imbalanced regression. arXiv preprint arXiv:2102.09554 (2021)
13. Ding, Y., et al.: Deep imbalanced regression using cost-sensitive learning and deep feature transfer for bearing remaining useful life estimation. Appl. Soft Comput. **127**(5) (2022)
14. Sun, B., Feng, J., Saenko, K.: Return of frustratingly easy domain adaptation. arXiv preprint arXiv:1511.05547 (2016)
15. Gong, Y., Mori, G., Tung, F.: RankSim: ranking similarity regularization for deep imbalanced regression. arXiv preprint arXiv:2205.15236 (2022)
16. Rolinek, M., Musil, V., et al.: Optimizing rank-based metrics with blackbox differentiation. In: Conference on Computer Vision and Pattern Recognition, pp. 7617–7627 (2020)
17. MarinVlastelica, P., Paulus, A., Musil, V., et al.: Differentiation of blackbox combinatorial solvers. arXiv preprint arXiv:1912.02175 (2019)
18. Ren, J., Zhang, M., Yu, C., Liu, Z.: Balanced MSE for imbalanced visual regression. In: Conference on Computer Vision and Pattern Recognition, pp. 7916–7925 (2022)
19. Lin, T., Goyal, P., Girshick, R.B., He, K., Dollár, P.: Focal loss for dense object detection. In: IEEE Transactions on Pattern Analysis and Machine Intelligence, pp. 318–327 (2017)
20. Dosovitskiy, A., Beyer, L., et al.: An image is worth 16x16 words: transformers for image recognition at scale. arXiv preprint arXiv: 2010.11929. (2020)
21. Liu, Z., Lin, Y., et al.: Swin transformer: hierarchical vision transformer using shifted windows. In: IEEE/CVF International Conference on Computer Vision, pp. 9992–10002 (2021)
22. Namburete, A.I., Stebbing, R.V., et al.: Learning-based prediction of gestational age from ultrasound images of the fetal brain. Med. Image Anal. **21**, 72–86 (2015)
23. Cui, Y., Jia, M., Lin, T., Song, Y., Belongie, S.J.: Class-balanced loss based on effective number of samples. In: Computer Vision and Pattern Recognition, pp. 9260–9269 (2019)
24. Chawla, N., Bowyer, K., Hall, L.O., Kegelmeyer, W.P.: SMOTE: synthetic minority over-sampling technique. arXiv preprint arXiv: 1106.1813 (2002)
25. Wang, Y., Ramanan, D., Hebert, M.: Learning to model the tail. In: Proceedings of the 31st International Conference on Neural Information Processing Systems, pp. 7032–7042 (2017)
26. Mahajan, D., et al.: Exploring the limits of weakly supervised pretraining. In: Ferrari, V., Hebert, M., Sminchisescu, C., Weiss, Y. (eds.) ECCV 2018. LNCS, vol. 11206, pp. 185–201. Springer, Cham (2018). https://doi.org/10.1007/978-3-030-01216-8_12
27. Chawla, N.V., Bowyer, K.W., Hall, L.O., Kegelmeyer, W.P.: SMOTE: synthetic minority over-sampling technique. J. Artif. Intell. Res. **16**, 321–357 (2002)
28. Torgo, L., Ribeiro, R.P., Pfahringer, B., Branco, P.: SMOTE for regression. In: Correia, L., Reis, L.P., Cascalho, J. (eds.) EPIA 2013. LNCS, vol. 8154, pp. 378–389. Springer, Heidelberg (2013). https://doi.org/10.1007/978-3-642-40669-0_33

29. Branco, P., et al.: SMOGN: a pre-processing approach for imbalanced regression. In: Proceedings of Machine Learning Research, 74, pp. 36–50 (2017)
30. Tanveer, M., Ganaie, M.A., Beheshti, I., et al.: Deep learning for brain age estimation: a systematic review. arXiv preprint arXiv:2212.03868 (2022)
31. Wright, R., Kyriakopoulou, V., et al.: Automatic quantification of normal cortical folding patterns from fetal brain MRI. Neuroimage **91**, 21–32 (2014)
32. Hong, J., Yun, H.J., et al.: Optimal method for fetal brain age prediction using multiplanar slices from structural magnetic resonance imaging. Front. Neurosci. **15** (2021)

Segment Anything Model for Semi-supervised Medical Image Segmentation via Selecting Reliable Pseudo-labels

Ning Li[1], Lianjin Xiong[1], Wei Qiu[1], Yudong Pan[1], Yiqian Luo[1], and Yangsong Zhang[1,2(✉)]

[1] School of Computer Science and Technology, Laboratory for Brain Science and Medical Artificial Intelligence, Southwest University of Science and Technology, Mianyang 621010, Sichuan, China
zhangysacademy@gmail.com

[2] Key Laboratory of Testing Technology for Manufacturing Process, Ministry of Education, Southwest University of Science and Technology, Mianyang 621010, Sichuan, China

Abstract. Semi-supervised learning (SSL) has become a hot topic due to its less dependence on annotated data compared to fully supervised methods. This advantage becomes more evident in the field of medical imaging, where acquiring labeled data is challenging. Generating pseudo-labels for unlabeled images is one of the most classic and intuitive methods in semi-supervised segmentation. However, this method may also produce unreliable pseudo-labels that can provide incorrect guidance to the model and impair its performance. The reliability of pseudo-labels is difficult to evaluate due to the lack of ground truth labels for unlabeled images. In this paper, a SSL framework was presented, in which we proposed a simple but effective strategy to select reliable pseudo-labels by leveraging the Segment Anything Model (SAM) for segmentation. Concretely, the SSL model trained with domain knowledge provides the generated pseudo-labels as prompts to the SAM. Reliable pseudo-labels usually make SAM to conduct predictions consistent with the semi-supervised segmentation model. Based on this result, the reliable pseudo-labels are selected to further boost the existing semi-supervised learning methods. The experimental results show that the proposed strategy effectively improves the performance of different algorithms in the semi-supervised scenarios. On the publicly available ACDC dataset, the proposed method achieves 6.84% and 10.76% improvement over the advanced two baselines respectively on 5% of labeled data. The extended experiments on pseudo-labels verified that the quality of the selected reliable pseudo-labels by the proposed strategy is superior to that of the unreliable pseudo-labels. This study may provide a new avenue for SSL medical image segmentation.

Keywords: semi-supervised learning · medical image segmentation · Segment Anything Model · pseudo labels

B. Luo et al. (Eds.): ICONIP 2023, CCIS 1964, pp. 138–149, 2024.
https://doi.org/10.1007/978-981-99-8141-0_11

1   Introduction

Medical image segmentation is a vital step in medical image diagnosis [16,23]. It enables doctors to locate the region of interest in the image and comprehend the lesion more accurately, as well as to devise a more suitable surgical plan [29]. Various deep learning methods have been developed for this purpose and have shown impressive performance in medical image segmentation [2,8]. A large amount of pixel-level annotations are usually indispensable for deep learning methods. However, annotations at the pixel level for medical images are often difficult to obtain, which is time-consuming and tedious. To alleviate the heavy burden of annotation data, semi-supervised learning is an effective technique that leverages a small amount of labeled data and a large amount of unlabeled data to achieve comparable results to fully supervised training [9].

Semi-supervised learning is typically divided into two categories: Pseudo-label based method [3,11] and consistency regularization [13,22]. In the field of natural images, Lee et al. used the category with the highest confidence in the prediction as the pseudo-label of unlabeled data in the semi-supervised classification task, thereby training the model in a supervised manner [11]. Different from directly using pseudo-labels to constrain unlabeled images, Tarvainen et al. proposed a mean teacher model that consists of two branches, i.e., a teacher model and a student model. The framework enforces consistency constraints between the outputs of the two models for unlabeled images [22]. For unlabeled images, the teacher model outputs more reliable predictions, while the student model takes the teacher model as the learning target. Subsequently, Sohn et al. combined these two approaches. They designed FixMatch, a hybrid framework that applies data augmentations of varying intensity to unlabeled images [21]. This framework imposes consistency constraints on the predictions of unlabeled data under weak and strong data augmentations, and uses the predictions under weak data augmentations as pseudo-labels for the predictions under strong data augmentations. To prevent incorrect pseudo-labels from propagating in the network, Yang et al. proposed a ST++ model that saves different checkpoints in the first training stage of the model, and selects stable pseudo-labels by the prediction consistency of different checkpoints to retrain a more accurate model [27]. Despite employing different strategies to leverage unlabeled data, these methods share the same core idea of providing reliable constraints and learning objectives for unlabeled data.

In the field of medical imaging, these classical methods have been studied and further extended [4,6,14,26]. For instance, Yu et al. developed an uncertainty-aware teacher model based on mean teacher to filter out unreliable pixels from the teacher model [28]. Han et al. improved the quality of pseudo labels by using the features of annotated data and a random patch based on prior locations (RPPL) to filter out unreliable pixels from the teacher model [28]. Luo et al. enforced the consistency of prediction results at different scales and guided unlabeled data to learn from reliable consensus regions with an uncertainty correction module [15]. Similarly, these methods demonstrate that obtaining reliable and accurate learning targets for unlabeled medical images is crucial.

Recently, large language models such as GPT-4 [18] have shown their amazing capabilities to everyone. The large language model in the field of image segmentation, Segment Anything Model (SAM) [10], has attracted great attention with its powerful generalization ability. However, SAM trained on natural images struggles to achieve satisfactory performance in medical image segmentation, due to the low contrast between the target and the background in medical images and the large difference between medical images and natural images [7,20]. Although fine-tuning SAM on medical image datasets can achieve better performance [17,24], a large amount of labeled data and high computational cost still need to be considered. Inspired by ST++ [27], we selected reliable pseudo labels for unlabeled data at the image level in a way that does not require retraining SAM. Unlike Yu et al. [28] and Sohn et al. [21], who seted thresholding to select pixels with higher confidence as pseudo-labels, image-level pseudo-labels provide better context for model training [27]. Moreover, we adopt a different approach from ST++ for selecting reliable pseudo-labels. Rather than relying on the consistency of predictions across different stages of a model, our proposed framework leverages the consistency between predictions from two models: the large language model SAM and the final-stage semi-supervised segmentation model. On one hand, compared with the early stage models, the final stage models often have better feature recognition ability and higher accuracy. On the other hand, although SAM cannot directly provide the final segmentation results, it always tries to find the most suitable segmentation target for the input prompt. Hence, utilizing SAM's characteristics, we employed the final model's output as SAM's input to assess the reliability of the pseudo-labels. We observed that the predictions from SAM are consistent with the reliable pseudo-labels when they are employed as prompts for SAM. However, when the pseudo-labels are unreliable, SAM fails to comprehend the relevant semantic information and produces inconsistent prediction results.

Overall, accurate learning objectives are crucial for unlabeled images. For pseudo-label based methods, unreliable pseudo-labels may cause the model to iterate in the wrong direction in the initial stage, thus hurting the model's final performance [5]. To address this problem, we introduced an intuitive and effective framework to obtain reliable sets from pseudo-labels of unlabeled images. To further improve the performance of existing semi-supervised methods, we retrained them with the reliable sets to obtain a more accurate model.

2   Method

In this paper, the widely-attentioned large segmentation model SAM is introduced into our proposed semi-supervised segmentation framework. This section describes how SAM is employed to generate predictions for unlabeled images and how to determine reliable pseudo-labels in more details. The proposed method is experimented on two models, including the classic mean teacher model [22] and the advanced model SS-Net [25]. In order to obtain a more powerful baseline, strong data augmentation as a powerful means is injected into the mean teacher model. We take the mean teacher with strong data augmentation (termed as SDA-MT) as an example to introduce the complete training process. The overall training process is shown in Fig. 1.

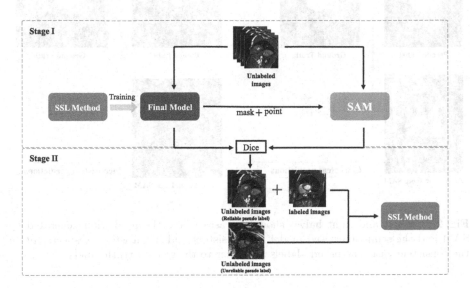

Fig. 1. Visualization of the framework of the proposed method. Final Model is the model saved after the semi-supervised learning method completes the entire training process. The main objective of Stage I is to use the final model to generate pseudo-labels and use SAM to select reliable pseudo-labels. The main objective of Stage II is to train a better-performing model with higher-quality pseudo-labels.

2.1 The Selection of Reliable Pseudo Label (Assisted by SAM)

SAM pre-trained on large-scale data has powerful image encoding and decoding abilities, and can always try to find reasonable segmentation objects for any input prompts. But SAM lacks professional knowledge in the field of medical images, and it is difficult to achieve satisfactory performance by directly relying on its predictions as segmentation results. In contrast, semi-supervised learning models trained on specific datasets can learn the key features of pathological images, but their output predictions still contain unreliable results. Therefore, the proposed method uses the predictions of semi-supervised learning models as prompts for SAM. When the predictions from semi-supervised learning models correspond to plausible lesions or tissue, as well as organ regions, SAM will give consistent results. The judgment process of SAM can be clearly observed in Fig. 2. Evaluate the similarity of the predictions of two models using the Dice index, and set a threshold α to determine whether their predictions are consistent. This process can be formulated as:

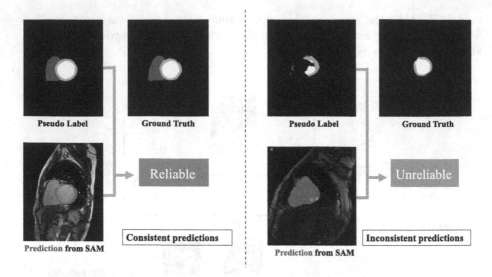

Fig. 2. The left and right halves show the cases where the prediction generated by SAM and the semi-supervised model are consistent and inconsistent, respectively. In the consistent case, the pseudo-labels are closer to the ground truth labels.

$$y = \begin{cases} 1, \ Dice(SAM(x), SSL(x)) > \alpha \\ 0, \ Dice(SAM(x), SSL(x)) \leq \alpha \end{cases}, \tag{1}$$

where x and y are unlabeled images and consistent results, respectively. SSL represents the semi-supervised learning model, and Dice index of two prediction is defined as:

$$Dice = \frac{2 * (SAM(x) \cap SSL(x))}{SAM(x) \cup SSL(x)}. \tag{2}$$

SAM's prompt encoder can accept flexible prompts, such as text, points, boxes, masks. In this experiment, we ultimately adopt the form of points combined with masks as the prompt. Specifically, the segmentation region of the SSL model acts as a mask. In the segmentation probability map, the points with the highest confidence within the foreground region serving as positive prompt points, and the points with the highest confidence within the background region serving as negative prompt points. In order to prevent isolated points in the segmentation result from causing SAM to produce erroneous boundary predictions, the maximum connected graph is adopted to help eliminate isolated points in the segmentation map.

2.2   Re-training the Model with Reliable Pseudo-labels (Taking SDA-MT as an Example)

Denote the labeled and unlabeled datasets as $S_l = (X_l, Y_l)$ and $S_u = (X_u)$, where X_l, Y_l represent the images and the corresponding labels, X_u represents the unlabeled images. The whole dataset could be denoted as $S_l \cup S_u$. Strong data augmentation has been proven to effectively help semi-supervised models to learn additional information and prevent model overfitting to wrong labels [12,21]. We introduced strong data augmentation into the classic mean teacher model to obtain a strong baseline. Specifically, all data are applied with weak data augmentation, such as random flip and random crop. For strong data augmentation, labeled data are further applied with cutout, unlabeled data are applied with RandAdjustContrast and RandGaussianNoise.

The mean teacher method consists of two models with 2D-Unet as the backbone, namely the teacher model (M_t) and the student model (M_s). In the training process, only the student model participates in the gradient backpropagation process, while the teacher model updates its own parameters by exponential moving average of the student model parameters. The overall training loss function of the first stage of the mean teacher can be described as:

$$\mathcal{L} = \mathcal{L}_{sup} + \lambda * \mathcal{L}_{con}. \tag{3}$$

λ is a function that balances the supervised loss and the consistency loss, defined as:

$$\lambda(t) = w_{\max} \cdot e^{\left(-5\left(1-\frac{t}{t_{\max}}\right)^2\right)}, \tag{4}$$

where w_{max} represents the final regularization weight. $\mathcal{L}_{sup}$ represents the supervised loss function, which can be defined as:

$$\mathcal{L}_{sup} = 0.5 * \mathcal{L}_{ce}(M_s(X_l), Y_l) + 0.5 * \mathcal{L}_{dice}(M_s(X_l), Y_l). \tag{5}$$

$\mathcal{L}_{ce}$ and $\mathcal{L}_{dice}$ represent the cross-entropy loss function and the dice loss function respectively. $\mathcal{L}_{con}$ enforces the consistency between the teacher model and the student model, which is formulated as:

$$\mathcal{L}_{con} = \mathcal{L}_{mse}(M_s(X_u), M_t(X_u)). \tag{6}$$

By evaluating the pseudo-labels generated in the first stage with SAM, unlabeled images can be divided into reliable (X_u^r) and unreliable sets (X_u^{un}). In the second stage, the model is retrained with reliable pseudo-labels, where $\mathcal{L}_{sup}$ and $\mathcal{L}_{con}$ can be reformulated as follows:

$$\mathcal{L}_{sup} = 0.5 * \mathcal{L}_{ce}(M_s(X_l + X_u^r), Y_l + Y_u^r) + 0.5 * \mathcal{L}_{dice}(M_s(X_l + X_u^r), Y_l + Y_u^r), \tag{7}$$

$$\mathcal{L}_{con} = \mathcal{L}_{mse}(M_s(X_u^{un}), M_t(X_u^{un})). \tag{8}$$

Y_u^r and X_u^r represent the reliable pseudo-labels and the corresponding unlabeled images, X_u^{un} represents the images corresponding to the unreliable pseudo-labels.

3   Experiment and Results

3.1   Dataset

We evaluated our work with other methods on the publicly available ACDC dataset. The ACDC dataset contains short-axis cardiac MRI scans of 100 patients. Following SS-Net [25], we adopted the same data partitioning and pre-processing methods. Specifically, scans of 70 patients are used for the training set, scans of 10 patients are used for the validation set, and scans of the remaining 20 patients are used for the testing set. To alleviate the large inter-slice spacing in 3D data, we adopted the same method as previous work [1] to use 2D slices as the input to the network. In the inference stage, each predicted 2D slice is restacked into a 3D volume. All methods are compared under three ratios of labeled data, i.e., 5%, 10% and 20%.

Table 1. Quantitative evaluation of all segmentation methods on the ACDC datasets.

Method	# Training set		Metrics		
	Labeled	Unlabeled	Dice (%)↑	95HD (mm)↓	ASD (mm)↓
U-Net [19]	3(5%)	0	47.83	31.16	12.62
U-Net [19]	7(10%)	0	79.41	9.35	2.70
U-Net [19]	14(20%)	0	85.35	7.75	2.22
U-Net [19]	70 (All)	0	91.44	4.30	0.99
MT [22]	3(5%)	67(95%)	55.19	19.92	6.32
UAMT [28]			46.04	20.08	7.75
URPC [15]			55.87	13.60	3.74
SS-Net [25]			65.82	6.67	2.28
SDA-MT			76.85	9.70	2.56
SS-Net++			76.59	6.07	1.68
SDA-MT++			**83.69**	**4.25**	**1.17**
MT [22]	7(10%)	63(90%)	81.70	11.40	3.05
UAMT [28]			81.65	6.88	2.02
URPC [15]			83.10	12.81	4.01
SS-Net [25]			86.78	6.07	1.40
SDA-MT			86.46	7.38	1.53
SS-Net++			**87.93**	3.51	1.03
SDA-MT++			87.63	**2.47**	**0.79**
MT [22]	14(20%)	56(80%)	85.56	4.10	1.32
UAMT [28]			87.80	5.86	1.44
URPC [15]			88.09	4.05	1.19
SS-Net [25]			87.73	4.36	1.24
SDA-MT			88.29	3.75	1.18
SS-Net++			**89.08**	**2.32**	**0.57**
SDA-MT++			88.81	3.06	0.91

3.2   Implementation Details and Evaluation Metrics

In this work, all tasks are performed with a batch size of 4 for 30,000 iterations. The batch size consists of half labeled and half unlabeled data. The threshold α for controlling the similarity between SAM and semi-supervised model predictions is set to 0.9. The learning rate is adjusted by a poly learning rate strategy with an initial value of 0.01 and increases with the model iteration. It can be formulated as: $lr \times \left(1.0 - \frac{t}{t_{\max}}\right)^{0.9}$, where t_{max} and t represent the total and current number of iterations, respectively. Stochastic Gradient Descent optimizer with momentum of 0.9 and weight decay of 0.0001 is used for all methods. In the training stage, we scaled all slices to 256×256 for network input. In the inference stage, we restored all slices to their original size for evaluation. We implemented all tasks using PyTorch 1.11, and training is performed on a NVIDIA 3090TI GPU with 24 G memory. As the three segmentation regions of the ACDC dataset are closely connected, we combined them into a single region to serve as the input of SAM. The final semi-supervised learning model still outputs three segmentation regions. We used three metrics commonly used in segmentation tasks to evaluate all methods, including Dice score (Dice), 95% Hausdorff Distance (95HD) and Average Surface Distance (ASD). Lower 95HD and ASD mean better segmentation performance, while larger Dice indicates better segmentation results.

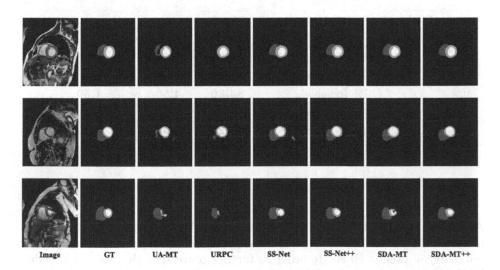

| Image | GT | UA-MT | URPC | SS-Net | SS-Net++ | SDA-MT | SDA-MT++ |

Fig. 3. Visualization of the segmentation results of all methods using 5% of labeled data. SS-Net++ and SDA-MT++ represent the results of the proposed method after improving the baseline. Image and GT represent the images and the corresponding ground truth in the test set, respectively.

3.3 Results

Comparison with Baselines and Existing Works: To verify the effectiveness of our proposed strategy, we used the classic mean teacher model (MT) with strong data augmentation (SDA-MT) and the advanced model SS-Net as baselines, and investigate the performance of the improved variants, i.e., SS-Net++ and SDA-MT++. Table 1 shows that the proposed two methods achieves consistent improvements over the baselines on three evaluation metrics. For SDA-MT++, it achieves 6.84%, 1.17%, and 0.52% improvement on dice metric with 5%, 10%, and 20% labeled data, respectively. For SS-Net++, it achieves 10.76%, 1.34%, and 1.15% improvement, respectively. With increasing labled training data, we could find that as the pseudo-labels become more accurate, the improvement brought by SAM may become more limited. Comparing the results of SDA-MT and MT, it could be found that strong data augmentation as an effective method brings significant improvement for mean teacher. Nevertheless, our strategy still further enhance its performance in segmentation, from 76.85% of SDA-MT to 83.69% of SDA-MT++. The representative segmentation results from all methods are visualized in Fig. 3.

Table 2. The comparison of training reliable and unreliable pseudo-labels in a supervised manner. The pseudo-labels in Stage II are obtained by Stage I using the MT model.

(Stage I) labeled	(Stage II) pseudo label	Metrics		
		Dice (%)↑	95HD (mm)↓	ASD (mm)↓
5%	13% (reliable)	**61.08**	**8.91**	**2.36**
	82% (unreliable)	59.08	18.60	6.03
10%	37% (reliable)	**84.30**	**4.50**	**1.50**
	53% (unreliable)	82.39	6.68	2.00
20%	43% (reliable)	**87.24**	**4.95**	**1.49**
	37% (unreliable)	86.08	6.01	1.79

Ablation Study: Although the comparison between the baseline and our proposed strategy in Table 1 shows improved performance, we further analyzed the reliability of pseudo labels. In order to further prove that the selected pseudo-labels are relatively reliable, related experiments are designed to verify our idea. Specifically, in Stage I, the pseudo-labels generated by the MT model are divided into reliable and unreliable sets by the proposed method. In Stage II, we leverage the reliable pseudo-labels and the unreliable pseudo-labels separately in a supervised manner. In order to intuitively verify their quality, additional technical means (such as strong data augmentation) and real labeled data are not added to Stage II. Table 2 shows that under the conditions with 5%, 10%, and

20% of the labeled data, the reliable pseudo-labels achieved better results than the unreliable pseudo-labels on three evaluation metrics. It is worth noting that on 5% of the labeled data, the reliable pseudo-labels achieved a higher dice with less quantity (13%) than the unreliable pseudo-labels with more quantity (82%). This result implies that compared to the inaccurate information provided by a large number of unreliable pseudo-labels, the accurate information provided by a small number of reliable pseudo-labels is more conducive to boosting the model's performance. To further investigate the impact of different threshold (α) on pseudo-labels, based on 5% labeled data, we compare the performance of SDA-MT++ under various α in Table 3. As shown, when α exceeds 0.99, no pseudo-labels are selected to be included in the training. When α is less than 0.99, the inclusion of pseudo-labels improves the model's performance. As α continues to decrease below 0.9, the model's performance starts to decline. A lower consistency threshold α implies the inclusion of more pseudo-labels into training. However, this also means that more unreliable pseudo-labels might be introduced, potentially harming the model's performance. In our experiments, we set α to 0.9.

Table 3. The influence of different prediction consistency judgment criteria (α).

α	Dice (%)↑	95HD (mm)↓	ASD (mm)↓
1.00	76.85	9.70	2.56
0.99	76.85	9.70	2.56
0.98	81.63	6.61	2.12
0.90	**83.69**	**4.25**	**1.17**
0.80	83.33	4.33	1.23
0.70	81.32	4.72	1.28

4   Conclusion

In this paper, we proposed an effective framework to improve the existing semi-supervised learning methods. We selected reliable pseudo-labels at the image level as the supervised signal for unlabeled images to guide the semi-supervised model. SAM, as a large model in the segmentation field, was first introduced to evaluate the reliability of pseudo-labels. The proposed framework effectively enhances the performance of semi-supervised learning methods on cardiac MR images. Furthermore, the proposed framework can be readily extended to various 2D segmentation tasks and different types of medical images, including CT scans. Since SAM's structure is designed for 2D images, the proposed strategy cannot be directly applied to 3D types of medical image data. In the future work, we will continue to investigate the application of the proposed framework on 3D medical images, contributing more efforts in the semi-supervised learning field to reduce the cost of medical image annotation.

Acknowledgement. This work was supported in part by the National Natural Science Foundation of China under Grant No. 62076209.

References

1. Bai, W., et al.: Semi-supervised learning for network-based cardiac MR image segmentation. In: Descoteaux, M., Maier-Hein, L., Franz, A., Jannin, P., Collins, D.L., Duchesne, S. (eds.) MICCAI 2017, Part II. LNCS, vol. 10434, pp. 253–260. Springer, Cham (2017). https://doi.org/10.1007/978-3-319-66185-8_29
2. Cao, H., et al.: Swin-Unet: Unet-like pure transformer for medical image segmentation. In: Karlinsky, L., Michaeli, T., Nishino, K. (eds.) Computer Vision, ECCV 2022 Workshops, Proceedings, Part III, Tel Aviv, Israel, 23–27 October 2022, vol. 13803, pp. 205–218. Springer, Cham (2023). https://doi.org/10.1007/978-3-031-25066-8_9
3. Chen, X., Yuan, Y., Zeng, G., Wang, J.: Semi-supervised semantic segmentation with cross pseudo supervision. In: Proceedings of the IEEE/CVF Conference on Computer Vision and Pattern Recognition, pp. 2613–2622 (2021)
4. Cui, W., et al.: Semi-supervised brain lesion segmentation with an adapted mean teacher model. In: Chung, A.C.S., Gee, J.C., Yushkevich, P.A., Bao, S. (eds.) IPMI 2019. LNCS, vol. 11492, pp. 554–565. Springer, Cham (2019). https://doi.org/10.1007/978-3-030-20351-1_43
5. Han, K., et al.: An effective semi-supervised approach for liver CT image segmentation. IEEE J. Biomed. Health Inform. **26**(8), 3999–4007 (2022)
6. Hang, W., et al.: Local and global structure-aware entropy regularized mean teacher model for 3D left atrium segmentation. In: Martel, A.L., et al. (eds.) MICCAI 2020. LNCS, vol. 12261, pp. 562–571. Springer, Cham (2020). https://doi.org/10.1007/978-3-030-59710-8_55
7. Hu, C., Li, X.: When SAM meets medical images: an investigation of segment anything model (SAM) on multi-phase liver tumor segmentation. arXiv preprint arXiv:2304.08506 (2023)
8. Isensee, F., Jaeger, P.F., Kohl, S.A., Petersen, J., Maier-Hein, K.H.: nnU-Net: a self-configuring method for deep learning-based biomedical image segmentation. Nat. Meth. **18**(2), 203–211 (2021)
9. Jiao, R., Zhang, Y., Ding, L., Cai, R., Zhang, J.: Learning with limited annotations: a survey on deep semi-supervised learning for medical image segmentation. arXiv preprint arXiv:2207.14191 (2022)
10. Kirillov, A., et al.: Segment anything. arXiv preprint arXiv:2304.02643 (2023)
11. Lee, D.H., et al.: Pseudo-label: the simple and efficient semi-supervised learning method for deep neural networks. In: Workshop on Challenges in Representation Learning, ICML, vol. 3, p. 896 (2013)
12. Liu, Y., Tian, Y., Chen, Y., Liu, F., Belagiannis, V., Carneiro, G.: Perturbed and strict mean teachers for semi-supervised semantic segmentation. In: Proceedings of the IEEE/CVF Conference on Computer Vision and Pattern Recognition, pp. 4258–4267 (2022)
13. Luo, X., Chen, J., Song, T., Wang, G.: Semi-supervised medical image segmentation through dual-task consistency. In: Proceedings of the AAAI Conference on Artificial Intelligence, vol. 35, pp. 8801–8809 (2021)

14. Luo, X., Hu, M., Song, T., Wang, G., Zhang, S.: Semi-supervised medical image segmentation via cross teaching between CNN and transformer. In: International Conference on Medical Imaging with Deep Learning, pp. 820–833. PMLR (2022)

15. Luo, X., et al.: Efficient semi-supervised gross target volume of nasopharyngeal carcinoma segmentation via uncertainty rectified pyramid consistency. In: de Bruijne, M., et al. (eds.) MICCAI 2021. LNCS, vol. 12902, pp. 318–329. Springer, Cham (2021). https://doi.org/10.1007/978-3-030-87196-3_30

16. Luo, X., et al.: MIDeepSeg: minimally interactive segmentation of unseen objects from medical images using deep learning. Med. Image Anal. **72**, 102102 (2021)

17. Ma, J., Wang, B.: Segment anything in medical images. arXiv preprint arXiv:2304.12306 (2023)

18. OpenAI: GPT-4 technical report (2023)

19. Ronneberger, O., Fischer, P., Brox, T.: U-Net: convolutional networks for biomedical image segmentation. In: Navab, N., Hornegger, J., Wells, W.M., Frangi, A.F. (eds.) MICCAI 2015, Part III. LNCS, vol. 9351, pp. 234–241. Springer, Cham (2015). https://doi.org/10.1007/978-3-319-24574-4_28

20. Roy, S., et al.: SAM.MD: zero-shot medical image segmentation capabilities of the segment anything model. arXiv preprint arXiv:2304.05396 (2023)

21. Sohn, K., et al: FixMatch: simplifying semi-supervised learning with consistency and confidence. In: Advances in Neural Information Processing Systems, vol. 33, pp. 596–608 (2020)

22. Tarvainen, A., Valpola, H.: Mean teachers are better role models: weight-averaged consistency targets improve semi-supervised deep learning results. In: Advances in Neural Information Processing Systems, vol. 30 (2017)

23. Wang, G., et al.: DeepIGeoS: a deep interactive geodesic framework for medical image segmentation. IEEE Trans. Pattern Anal. Mach. Intell. **41**(7), 1559–1572 (2018)

24. Wu, J., et al.: Medical SAM adapter: adapting segment anything model for medical image segmentation. arXiv preprint arXiv:2304.12620 (2023)

25. Wu, Y., Wu, Z., Wu, Q., Ge, Z., Cai, J.: Exploring smoothness and class-separation for semi-supervised medical image segmentation. In: Wang, L., Dou, Q., Fletcher, P.T., Speidel, S., Li, S. (eds.) Proceedings of the 25th International Conference on Medical Image Computing and Computer Assisted Intervention, MICCAI 2022, Part V, Singapore, 18–22 September 2022, pp. 34–43. Springer, Cham (2022). https://doi.org/10.1007/978-3-031-16443-9_4

26. Xu, Z., et al.: All-around real label supervision: cyclic prototype consistency learning for semi-supervised medical image segmentation. IEEE J. Biomed. Health Inf. **26**(7), 3174–3184 (2022)

27. Yang, L., Zhuo, W., Qi, L., Shi, Y., Gao, Y.: St++: make self-training work better for semi-supervised semantic segmentation. In: Proceedings of the IEEE/CVF Conference on Computer Vision and Pattern Recognition, pp. 4268–4277 (2022)

28. Yu, L., Wang, S., Li, X., Fu, C.-W., Heng, P.-A.: Uncertainty-aware self-ensembling model for semi-supervised 3D left atrium segmentation. In: Shen, D., et al. (eds.) MICCAI 2019. LNCS, vol. 11765, pp. 605–613. Springer, Cham (2019). https://doi.org/10.1007/978-3-030-32245-8_67

29. Yu, Q., Xie, L., Wang, Y., Zhou, Y., Fishman, E.K., Yuille, A.L.: Recurrent saliency transformation network: incorporating multi-stage visual cues for small organ segmentation. In: Proceedings of the IEEE Conference on Computer Vision and Pattern Recognition, pp. 8280–8289 (2018)

Aided Diagnosis of Autism Spectrum Disorder Based on a Mixed Neural Network Model

Yiqian Luo[1], Ning Li[1], Yudong Pan[1], Wei Qiu[1], Lianjin Xiong[1], and Yangsong Zhang[1,2($\boxtimes$)]

[1] School of Computer Science and Technology, Laboratory for Brain Science and Medical Artificial Intelligence, Southwest University of Science and Technology, Mianyang 621010, China
zhangysacademy@gmail.com
[2] Key Laboratory of Testing Technology for Manufacturing Process, Ministry of Education, Southwest University of Science and Technology, Mianyang 621010, Sichuan, China

Abstract. The diagnosis of autism spectrum disorder (ASD) is a challenging task, especially for children. In order to determine whether a person has ASD or not, the conventional methods are questionnaires and behavioral observation, which may be subjective and cause misdiagnosis. In order to obtain an accurate diagnosis, we could explore the quantitative imaging biomarkers and leverage the machine learning to learn the classification model on these biomarkers for auxiliary ASD diagnosis. At present, many machine learning methods rely on resting-state fMRI data for feature extraction and auxiliary diagnosis. However, due to the heterogeneity of the data, there can be many noisy features that are adverse to diagnosis, and a lot of biometric information may be not fully explored. In this study, we designed a mixed neural network model of convolutional neural network (CNN) and graph neural network (GNN), termed as MCG-Net, to extract discriminative information from the brain functional connectivity based on the resting-state fMRI data. We used the F-score and KNN algorithms to remove the abundant connectivities in the functional connectivity matrix from global and local level. Besides, the brain gradient features were first introduced in the model. A datasets of 848 subjects from 17 sites on ABIDE datasets was adopted to evaluate the methods. The proposed method has achieved better diagnostic performance compared with other existing methods, with 4.56% improvement in accuracy.

Keywords: autism spectrum disorder · deep learning · feature fusion · graph convolutional network · functional gradient

1 Introduction

Autism spectrum disorder (ASD) is a generalized developmental disorder and a developmental disorder with a biological basis [1]. The onset of the disease varies,

B. Luo et al. (Eds.): ICONIP 2023, CCIS 1964, pp. 150–161, 2024.
https://doi.org/10.1007/978-981-99-8141-0_12

with most cases occurring in childhood and continuing throughout life. Its main symptoms are social communication disorder, language communication disorder, emotional defect and repetitive and rigid behavior [2,3]. It leaves patients with huge barriers to communication and learning in their daily lives, causing great damage to their physical and mental health.

Unfortunately, so far, the world has not discovered or invented any specific drug or method that can completely cure ASD. As we all know, school age is an important stage for the development of social interaction and corresponding brain function changes. ASD, as a kind of social dysfunction, has seriously affected various social activities of school-age children with ASD. At present, the best approach is to prevent autism in advance and correct problem behaviors in children with autism through early intervention training [4]. However, there are no reliable biomarkers that can be used for diagnosis [5], and the method through behavioral observation are extremely prone to misdiagnose [6], and not feasible in the early stages of autism.

Functional magnetic resonance imaging (fMRI), as a neuroimaging technique [7], has become an essential tool for investigating neurophysiology [8] and has been widely used to study functional and structural changes in the brain activity of autistic patients [9]. Many studies have demonstrated the feasibility of using resting state fMRI (rs-fMRI) to reveal interactions between regions of interest (ROIs) in the brain of people with psychiatric disorders [8] such as autism [10] and attention deficit hyperactivity disorder (ADHD) [11]. Rs-fMRI signals have shown great potential in identifying diagnostic biomarkers in neuropathology [12]. By sharing and integrating fMRI data samples across different studies and datasets can provide new avenue to seek solutions for clinical applications [13].

In recent years, the brain functional connectivity network (FCN) obtained by rs-fMRI has a great application prospect in the diagnosis of brain diseases. FCN is generally defined as a complex non-Euclidean spatial graph structure [14]. Currently, traditional methods, such as support vector machine and random forest, have not achieved satisfactory results. In recent years, graph network-based methods have emerged rapidly, because they are more in line with the data structure of the brain and can extract more levels of feature information. Graph convolutional network (GCN) is a natural extension of convolutional neural networks in the graph domain [15]. With the continuous development of deep learning, the GCNs achieve good results by using FCN spatial relevant information [16–18]. However, due to the heterogeneity of data and unexplored information in different fMRI datasets, there is still a lot of room for improvement in existing methods.

Functional gradients of the brain represent information about spatial patterns in the embedded space, reflecting the rate of change, or change in relative similarity within the underlying dimensions, i.e., the similarity of functional connectivity patterns in each dimension [20]. It has been confirmed that the functional gradients of patients with mental illness is different from that of normal people [20,21]. In the previous studies, the functional gradient was not adopted for the ASD diagnosis with deep learning (DL) models.

Based on resting-state fMRI data, in this study, we proposed a mixed neural network model, i.e., MCG-Net, with the convolutional neural network (CNN) and graph neural network (GNN) to extract discriminative information from the brain functional connectivity for ASD diagnosis. The functional connectivity matrix was computed with the Pearson correlation, and then the redundant connectivities were removed by the F-score [24] and KNN from the global and local levels.

To the best of our knowledge, this is the first time that functional connectivity features and brain functional gradients are combined for aiding diagnosis. The main contributions of this paper can be summarized as follows.

Firstly, to remove the global noisy features, F-score is utilized for feature selection. For local noisy connections, the KNN algorithm is used to pick the preserved edges for each brain region. This operation removes the noisy connections in the adjacency matrix.

Secondly, the functional gradients was first introduced into deep learning model for assisted diagnosis on ASD. In this study, functional gradients are used as the attention mechanism of brain network.

Thirdly, we proposed a model that combines convolution and graph neural networks. Compared with existing models, the proposed model has better classification accuracy performance.

2 Preliminaries

2.1 Datasets and Preprocessing

We conduct experiments on fMRI images from the ABIDE I datasets [25], which aggregates data from 17 different collection sites [26] and publicly shares rs-fMRI and phenotypic data from 1112 subjects. In our work, we used the Data Processing Assistant for Resting-State fMRI (DPARSF) [27] to pre-process the rs-fMRI data. The list of pipes is as follows: (1) Discard the first 5 volumes. (2) Correction of slice-related delay. (3) Correct head movement. (4) Using EPI template normalization in MNI space, resampling to 3 mm resolution of $3 \times 3 \times 3$. (5) 4 mm full width half maximum Gaussian kernel space smooth. (6) Linear detrending and temporal bandpass filtering of BOLD signal $(0.01-0.10\,\text{Hz})$. (7) Harmful signals that return head movement parameters, white matter, cerebrospinal fluid (CSF), and global signals. A total of 848 subjects, including 396 individuals with ASD and 452 normal controls, were selected for this study.

2.2 The Calculation of Functional Connectivity Matrix

The brain functional connectivity network of the brain is an indicator of the connections between brain regions. After the preprocessing operation on fMRI data, the time series of each brain region can be obtained. To construct FCN, we used the Pearson correlation coefficient (PCC) to calculate the linear relationship between pair-brain regions as shown in Fig. 1. The PCC calculation process is

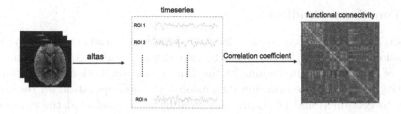

Fig. 1. The computational process of functional connectivity matrix.

shown in formula (1). Here u_t and v_t represent the t-th time point feature values of brain region u and brain region v respectively, $\overline{u}$ and $\overline{v}$ are the average of the time series of brain region u and v respectively, and T represents the total length of time series. In this way, the functional connectivity matrix of $N \times N$ was obtained. N is the number of brain regions.

$$\rho_{uv} = \frac{\sum_{t=1}^{T} (u_t - \overline{u})(v_t - \overline{v})}{\sqrt{\sum_{t=1}^{T} (u_t - \overline{u})^2} \sqrt{\sum_{t=1}^{T} (v_t - \overline{v})^2}} \tag{1}$$

Using the AAL [28] template, the registered fMRI volumes were divided into 116 regions of interest (ROIs). We construct a FC connectivity matrix of size 116×116 for each subject. Then, functional connectivity matrices of 848 subjects were obtained.

2.3 Global Features Selection

In this work, we use the F-score [24] as a global features selection method based on a data-driven approach.

For each subject, the upper triangular part of its functional connectivity matrix was expanded to a vector, if the number of ASD patients is n_p and the number of normal controls is n_n, then the F-score corresponding to i-th feature is calculated by the following formula (2):

$$F(i) = \frac{\left(\overline{x_i^p} - \overline{x_i}\right)^2 + \left(\overline{x_i^n} - \overline{x_i}\right)^2}{\frac{1}{n_p-1} \sum_{k=1}^{n_p} \left(x_{k,i}^p - \overline{x_i^p}\right)^2 + \frac{1}{n_n-1} \sum_{k=1}^{n_n} \left(x_{k,i}^n - \overline{x_i^n}\right)^2} \tag{2}$$

where $\overline{x_i^p}$, $\overline{x_i^n}$ and $\overline{x_i}$ represent the averages of the i-th feature of ASD subjects, control subjects and all subjects, $x_{k,i}^p$ and $x_{k,i}^n$ represent the i-th feature values of the k-th subject in ASD and CN. Through the F-score method, we selected half of the key features for retention.

2.4 Functional Gradient

We used BrainSpace tool to calculate the functional gradient [19]. The main calculation process can be divided into two steps: 1) calculate the cosine similarity of each two brain regions in the functional connection and obtain the similarity matrix; 2) use diffusion dimension reduction operation on the similar matrix to get functional gradients of each subject. We selected the component corresponding to the principal gradient because it has a functional connectome axis that has recently been linked to autism as a possible pathogenic mechanism at the system level [22]. This axis, which represents a large-scale cortical hierarchy stream, specifically demonstrated less differentiation of functional connectivity (FC) in both low-level sensory and high-order transmodal systems in ASD, offering a concise explanation for their altered sensory sensitivity and social impairment [23].

3 Methodology and Materials

3.1 The Architecture of the Proposed Model

The network structure of MCG-Net is shown in Fig. 2. The model is divided into two modules: the first part is the convolutional part of the functional connectivity, the second part is the Graph convolutional network part. The introduction of each part is as follows.

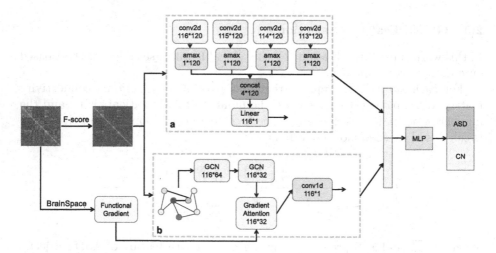

Fig. 2. The digram of the proposed MCG-Net model. (a) Functional connectivity convolution block. (b) GCN block.

Functional Connectivity Convolution Block. In this Block, functional connectivity matrices filtered by F-score are used as inputs. Four convolution layers with different kernel sizes are used for feature extraction, kernel sizes are 1×116, 2×116, 3×116 and 4×116 respectively. Each convolution represents different feature extraction operation. The 1×116 kernel extracts the features of each brain region, and the $s \times 116$ ($s = 2, 3, 4$) kernel extracts the higher-order features of the brain region network. In this way, the model can extract more diagnostic features. After convolution, we use amax function to reduce the dimension of the features, splicing all the features together to get a feature matrix of 4×120, and finally using the linear layer to yield features of size 116×1.

GCN Block. It is well known that graph neural networks can obtain more spatial information. However, in the task of autism diagnosis, the presence of spurious edges and noisy features hinders the model performance. For the processed functional connectivity matrices by the F-score [24] method, we further used the KNN method to reduce the connections for each brain region from local level. We used the Euclidean distance to calculate the distance between brain regions, and then used KNN as a local feature selection method to keep the edges between the k brain regions with the closest distance for each brain region. k was empirically set to be 15 after the preliminary experiment.

For the construction of the graph, we use $G(V, E)$ to represent a graph where V represents the feature matrix and E represents the edge matrix. $v_i(i = 1, 2, \cdots, N)$ represents the embedding of each brain region, N represents the number of brain regions. $E \in R^{N \times N}$ is the adjacency matrix. Each layer of a graph convolutional network can be defined as the following formula (3).

$$H^{l+1} = \sigma(\widetilde{D}^{-\frac{1}{2}} E \widetilde{D}^{\frac{1}{2}} H^l W^l) \tag{3}$$

where E is the adjacency matrix processed by KNN, σ is the activation function, W^l is a layer-specific trainable weight matrix, H^l is the feature matrix of the l-th layer. D is the degree matrix, $\widetilde{D}_i = \sum_{j=1}^{N} E_{ij}$. GCN can be regarded as an operator of node features on a Laplacian smoothing graph structure [29]. A two-layer graph convolution network is used to obtain a feature matrix of size 116×32. Because functional gradients reflect changes in the relative similarity of functional connectivity patterns in each dimension [20]. So we use the component corresponding to the principal gradient (with size 116×1) as the weight of attention mechanism to weight the GCN processed brain networks. Finally, we use a one-dimensional convolution with a kernel size of 1×32 to extract the features of each brain region network and obtain a feature vector of size 116×1.

Feature Fusion for Classification. Through the convolution block and the graph convolution block, two 116×1 feature vectors are obtained. Then, we concatenated the two vectors to get the feature vector of 1×232. Finally, multi-Layer perception (MLP) block was used to achieve classification prediction, in which two fully connected layers were used.

3.2   The Comparison Models

In order to validate the performance of our model, six newly proposed methods are used for comparison. In order to show the fairness of comparison, we used the same processed data in the comparison model. And all the models were tested by the method of 10-fold cross-validation.

ASD-DiagNet [30]: ASD-DiagNet is a collaborative learning strategy that combines single-layer perceptrons (SLPS) with autoencoders, and it leverages data-driven data improvement to boost the quality of extracted features and classification performance. For ASD-DiagNet, we take the same approach as in the original paper, selecting only the largest 25% features and the smallest 25% features of FCN as input features

f-GCN [16]: f-GCN is a graph convolutional neural network with six graph convolution layers to achieve classification. On the basic graph convolution, it uses a hyperparameter threshold to remove redundant connections. We set this hyperparameter to 0.45 for the experiment.

BrainGNN [31]: BrainGNN is an end-to-end graph neural network-based approach for fMRI prediction learns both the downstream whole-brain fMRI categorization and ROIs clustering simultaneously. We employed batch size to 32, and other hyperparameters employed in our comparison are consistent with the setting in the original paper.

Hi-GCN [16]: Hi-GCN uses the features of each subject extracted by f-GCN as a node, and uses the calculation of correlation coefficient to get the edge between subjects to construct the population map. This method uses more available information for classification, but it needs to re-train the model when introducing new test data.

CNN [32]: The CNN method extracts feature information of different levels by using seven convolution layers with different kernel sizes, and splices them together to achieve classification using linear layer.

MVS-GCN [17]: MVS-GCN is a kind of figure convolution network model based on multiple views, using graph structures and multitasking figure embedded learning to brain disease diagnosis and identification of key network.

3.3   Experimental Setting and Evaluation Metrics

The experiment was carried out using 10-fold cross-validation, and the parameter settings in the model are listed in Table 1. The whole experiment was built on the PyTorch environment, on a GeForce RTX 3090 GPU.

The cross entropy was used as the loss function to drive network learning. We assume that the number of training subjects is n, y_i is the true label of the i-th subject, and p_i is the predicted subject label and the formula of cross entropy loss (L_{CE}) is defined as follows:

$$L_{CE} = \frac{1}{n} \sum_{i}^{n} -[y_i \times \log(p_i) + (1 - y_i) \times \log(1 - p_i)] \qquad (4)$$

Table 1. The parameter settings of network training of our method.

Parameter name	Parameters
Optimezer	Adam
Learning rate	0.0001
Dropout rate	0.3
Training epochs	100
Training batch size	32
Activation function	Tanh
Loss	CrossEntropyLoss

Three metrics are used to verify the performance of our model, including accuracy (ACC), sensitivity (SEN), specificity (SPE) and defined as:

$$ACC = \frac{TP + TN}{TP + FN + FP + TN} \tag{5}$$

$$SEN = \frac{TP}{TP + FN} \tag{6}$$

$$SPE = \frac{TN}{TN + FP} \tag{7}$$

where TP, TN, FP and FN denote the numbers of true positive, true negative, false positive and false negative, respectively.

4   Experiment Results and Discussion

The classification results obtained by all the models are shown in Table 2. Compared with other comparison methods, the proposed model improves accuracy by at least 4.56%. Except for the sensitivity on which the Hi-GCN outperformed by 3.25% and MVS-GCN outperformed by 1.29%, the proposed model achieve better results on all the three metrics. Our model achieved the highest specificity value of 73.52% and the highest accuracy value of 70.48%.

Table 2. Performance comparison of multiple models, where ACC, SEN and SPE represent the average after ten fold cross-validation. The best results are marked in bold.

Method	ACC(%)	SEN(%)	SPE(%)
f-GCN	60.03	57.71	62.56
MVS-GCN	62.60	68.37	55.83
BrainGNN	62.28	58.23	65.04
CNN	63.93	58.90	68.39
Hi-GCN	65.33	**70.33**	59.60
ASD-DiagNet	65.92	60.10	70.90
MCG-Net(ours)	**70.48**	67.08	**73.52**

We conducted an ablation experiment based on functional connectivity convolution module, and the experimental results are shown in Table 3. In the table, sub is the short for subtract. We removed the edges selection performed by KNN and used the F-score functional connectivity features as edges (that is, the features of the input GCNs are the same as the adjacency matrix). We also removed the functional gradient attention mechanism and use self-learning attention instead. From the table, we can conclude that after removing each part, the accuracy of classification will decrease, which also confirms the feasibility of functional gradient as the attention mechanism of brain region. In addition, the lowest accuracy rate of MCG-Net in 10-fold is 65.48% (Fig. 3).

Table 3. 10-fold cross-validation results of ablation experiments were measured by acc(%). The best results are bolded.

Method	Fold										
	1	2	3	4	5	6	7	8	9	10	Mean
MCG-Net	**71.43**	75.00	**72.62**	**69.05**	**65.48**	**65.48**	**71.43**	75.00	**67.86**	**71.43**	**70.48**
MCG-Net (sub KNN)	70.24	75.00	69.05	67.86	**65.48**	63.10	70.24	75.00	**67.86**	70.24	69.40
MCG-Net (sub gradient)	67.86	**76.19**	71.43	66.67	**65.48**	64.29	**71.43**	**76.19**	65.48	67.86	69.29

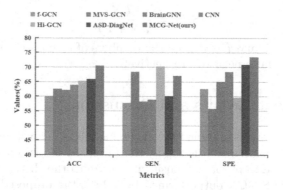

Fig. 3. The comparison between different models.

4.1 Brain Structure Analysis

We used the global average of the component corresponding to the principal gradient of all ASD patients to obtain Gra_{ASD}, performed the same operation for the control subjects to obtain Gra_{CN}. For Gra_{ASD}, the brain regions with the five largest principal gradient values are putamen of the left brain, superior temporal gyrus of the right brain, insula of the right brain, rolandic operculum of the left brain, rolandic operculum of the right brain. For Gra_{CN}, the brain regions with the five largest principal gradient values are superior temporal gyrus of the right brain, superior temporal gyrus of the left brain, rolandic operculum of the right brain, transverse temporal gyrus of the right brain, rolandic operculum of the left brain. We used BrainNet Viewer [33] to map the five brain regions with maximum functional gradients for ASD and CN shown in Fig. 4.

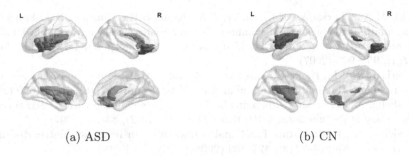

(a) ASD   (b) CN

Fig. 4. Top five brain regions with maximum gradients.

5   Conclusion

It is still challenging to detect the ASD owing to the heterogeneity of data between subjects and excessive spurious connections in the functional connectivity matrices. In this study, we proposed a mixed deep learning model with two branches to extract diversified features conducive to diagnosis. We used the F-score and KNN algorithms to remove the redundant connectivities in the functional connectivity matrix from global and local level, and the brain gradient features to weight the brain regions. Experiments on the public datasets ABIDE demonstrate the effectiveness of our model. Compared with the existing methods, MCG-Net has achieved better performance. The proposed model provides an alternative method for future ASD detection.

References

1. Pandolfi, V., Magyar, C.I., Dill, C.A.: Screening for autism spectrum disorder in children with down syndrome: an evaluation of the pervasive developmental disorder in mental retardation scale. J. Intellect. Dev. Disabil. **43**(1), 61–72 (2018)
2. Bhat, S., Acharya, U.R., Adeli, H., Bairy, G.M., Adeli, A.: Autism: cause factors, early diagnosis and therapies. Rev. Neurosci. **25**(6), 841–850 (2014)
3. Bhat, S., Acharya, U.R., Adeli, H., Bairy, G.M., Adeli, A.: Automated diagnosis of autism: in search of a mathematical marker. Rev. Neurosci. **25**(6), 851–861 (2014)
4. Bradshaw, J., Steiner, A.M., Gengoux, G., Koegel, L.K.: Feasibility and effectiveness of very early intervention for infants at-risk for autism spectrum disorder: a systematic review. J. Autism Dev. Disord. **45**, 778–794 (2015)
5. Lord, C., Elsabbagh, M., Baird, G., Veenstra-Vanderweele, J.: Autism spectrum disorder. The Lancet **392**(10146), 508–520 (2018)
6. Nickel, R.E., Huang-Storms, L.: Early identification of young children with autism spectrum disorder. Ind. J. Pediat. **84**, 53–60 (2017)
7. Huettel, S.A., Song, A.W., McCarthy, G., et al.: Functional Magnetic Resonance Imaging, vol. 1. Sinauer Associates Sunderland (2004)
8. Dvornek, N.C., Ventola, P., Duncan, J.S.: Combining phenotypic and resting-state fMRI data for autism classification with recurrent neural networks. In: 2018 IEEE 15th International Symposium on Biomedical Imaging (ISBI 2018), pp. 725–728. IEEE (2018)

9. Just, M.A., Cherkassky, V.L., Keller, T.A., Kana, R.K., Minshew, N.J.: Functional and anatomical cortical underconnectivity in autism: evidence from an fMRI study of an executive function task and corpus callosum morphometry. Cereb. Cortex **17**(4), 951–961 (2007)

10. Cherkassky, V.L., Kana, R.K., Keller, T.A., Just, M.A.: Functional connectivity in a baseline resting-state network in autism. NeuroReport **17**(16), 1687–1690 (2006)

11. Yu-Feng, Z., et al.: Altered baseline brain activity in children with ADHD revealed by resting-state functional MRI. Brain Develop. **29**(2), 83–91 (2007)

12. Greicius, M.: Resting-state functional connectivity in neuropsychiatric disorders. Curr. Opin. Neurol. **21**(4), 424–430 (2008)

13. Castellanos, F.X., Di Martino, A., Craddock, R.C., Mehta, A.D., Milham, M.P.: Clinical applications of the functional connectome. Neuroimage **80**, 527–540 (2013)

14. Zhang, D., Huang, J., Jie, B., Du, J., Tu, L., Liu, M.: Ordinal pattern: a new descriptor for brain connectivity networks. IEEE Trans. Med. Imaging **37**(7), 1711–1722 (2018)

15. Niepert, M., Ahmed, M., Kutzkov, K.: Learning convolutional neural networks for graphs. In: International Conference on Machine Learning, pp. 2014–2023. PMLR (2016)

16. Jiang, H., Cao, P., Xu, M., Yang, J., Zaiane, O.: HI-GCN: a hierarchical graph convolution network for graph embedding learning of brain network and brain disorders prediction. Comput. Biol. Med. **127**, 104096 (2020)

17. Wen, G., Cao, P., Bao, H., Yang, W., Zheng, T., Zaiane, O.: MVS-GCN: a prior brain structure learning-guided multi-view graph convolution network for autism spectrum disorder diagnosis. Comput. Biol. Med. **142**, 105239 (2022)

18. Fang, Y., Wang, M., Potter, G.G., Liu, M.: Unsupervised cross-domain functional MRI adaptation for automated major depressive disorder identification. Med. Image Anal. **84**, 102707 (2023)

19. Vos de Wael, R., et al.: Brainspace: a toolbox for the analysis of macroscale gradients in neuroimaging and connectomics datasets. Commun. Biol. **3**(1), 103 (2020)

20. Dong, D.: Compression of cerebellar functional gradients in schizophrenia. Schizophr. Bull. **46**(5), 1282–1295 (2020)

21. Guo, S., et al.: Functional gradients in prefrontal regions and somatomotor networks reflect the effect of music training experience on cognitive aging. Cerebral Cortex, bhad056 (2023)

22. Hong, S.J., et al.: Atypical functional connectome hierarchy in autism. Nat. Commun. **10**(1), 1022 (2019)

23. Margulies, D.S., Ghosh, S.S., Goulas, A., Falkiewicz, M., Huntenburg, J.M., Langs, G., Bezgin, G., Eickhoff, S.B., Castellanos, F.X., Petrides, M., et al.: Situating the default-mode network along a principal gradient of macroscale cortical organization. Proc. Natl. Acad. Sci. **113**(44), 12574–12579 (2016)

24. Chen, Y.W., Lin, C.J.: Combining SVMs with various feature selection strategies. In: Feature Extraction: Foundations and Applications, pp. 315–324 (2006)

25. Di Martino, A., et al.: The autism brain imaging data exchange: towards a large-scale evaluation of the intrinsic brain architecture in autism. Mol. Psychiat. **19**(6), 659–667 (2014)

26. Craddock, C., et al.: The neuro bureau preprocessing initiative: open sharing of preprocessed neuroimaging data and derivatives. Front. Neuroinf. **7**, 27 (2013)

27. Yan, C., Zang, Y.: DPARSF: a matlab toolbox for "pipeline" data analysis of resting-state fMRI. Front. Syst. Neurosci. **4**, 1377 (2010)

28. Tzourio-Mazoyer, N., et al.: Automated anatomical labeling of activations in SPM using a macroscopic anatomical parcellation of the MNI MRI single-subject brain. Neuroimage **15**(1), 273–289 (2002)
29. Li, Q., Han, Z., Wu, X.M.: Deeper insights into graph convolutional networks for semi-supervised learning. In: Proceedings of the AAAI Conference on Artificial Intelligence, vol. 32 (2018)
30. Eslami, T., Mirjalili, V., Fong, A., Laird, A.R., Saeed, F.: ASD-DIAGNET: a hybrid learning approach for detection of autism spectrum disorder using fMRI data. Front. Neuroinf. **13**, 70 (2019)
31. Li, X., et al.: BrainGNN: interpretable brain graph neural network for fMRI analysis. Med. Image Anal. **74**, 102233 (2021)
32. Sherkatghanad, Z., et al.: Automated detection of autism spectrum disorder using a convolutional neural network. Front. Neurosci. **13**, 1325 (2020)
33. Xia, M., Wang, J., He, Y.: Brainnet viewer: a network visualization tool for human brain connectomics. PLoS ONE **8**(7), e68910 (2013)

A Supervised Spatio-Temporal Contrastive Learning Framework with Optimal Skeleton Subgraph Topology for Human Action Recognition

Zelin Deng[1], Hao Zhou[1]([✉]), Wei Ouyang[3], Pei He[2], Song Yun[1], Qiang Tang[1], and Li Yu[3]

[1] Changsha University of Science and Technology, Changsha 410114, China
1013552415@qq.com
[2] School of Computer Science and Cyber Engineering, Guangzhou University, Guangzhou 510006, China
[3] Hunan Hkt Technology Co. Ltd., Changsha 410000, China

Abstract. Human action recognition (HAR) is a hotspot in the field of computer vision, the models based on Graph Convolutional Network (GCN) show great advantages in skeleton-based HAR. However,most existing GCN based methods do not consider the diversity of action trajectories, and not highlight the key joints. To address these issues, a supervised spatio-temporal contrastive learning framework with optimal skeleton subgraph topology for HAR (SSTCL-optSST) is proposed. SSTCL-optSST uses the samples with the same lablel as the target action (anchor) to build a positive sample set, each of them represents a trajectory of an action. The sample set is used to design a loss function to guide the model recognize different poses of the action. Furthermore, the subgraphs of an original skeleton graph are used to construct a skeleton subgraph topology space, each subgraph in it is evaluated, and the optimal one is selected to highlight the key joints. Extensive experiments have been conducted on NTU RGB+D 60 and Kinetics datasets, the results show that our model has competitive performance.

Keywords: human action recognition · graph convolutional network · contrastive learning · optimal skeleton subgraph topology

1 Introduction

With the development of computer vision and deep neural networks, human action recognition (HAR) has been more and more widely used, especially in the fields of security surveillance [1], human computer interaction [2] and virtual reality [3]. Among the input types of HAR tasks [4], skeleton data has smaller size and lower computational cost than RGB or optical flow data, and has become an effective model of HAR due to its robustness to body proportion, light changes, dynamic camera views and noise background.

B. Luo et al. (Eds.): ICONIP 2023, CCIS 1964, pp. 162–175, 2024.
https://doi.org/10.1007/978-981-99-8141-0_13

The traditional deep learning methods mainly model the skeleton data as a grid of joint coordinate vectors [5] or a pseudo-image [6]. Convolution neural networks (CNNs) use multiple layer networks to automatically extract feature information, rather than generating hand-craft features by leveraging the local characteristics of the data. Different from the grid data used in CNNs, the graph topology of human skeleton consists of a non-Euclidean space. In order to make full use of the representation ability of graph, more and more methods use graph neural networks (GNNs) to train models. Among them, graph convolutional network (GCN) is the most commonly used in skeleton-based HAR, which can capture non-Euclidean space information to learn the high-level features of graph structure. Yan et al. [7] improved a GCN model with spatio-temporal scheme, which is called spatio-temporal graph convolutional network (ST-GCN). ST-GCN can automatically learn spatial-temporal patterns from data, integrate the information in the context, and make the model have strong expression and generalization capabilities. Duan et al. [8] propoesd a novel framework PoseConv3D that serves as a competitive alternative to GCN-based approaches, which relies on heatmaps of the base representation, enjoys the recent advances in convolutional network architectures and is easier to integrate with other modalities into multi-stream convolutional networks.

Although GCN performs very well on skeleton data, there are also some structural limitations: (i) In a supervised learning environment, most GCN-based methods do not consider the diversity of the action trajectories; (ii) Cross-entropy loss, which is a widely used loss function in GCN-based methods, lacks robustness to noise [9], resulting in insensitivity to different behaviors of the same action; (iii) The original skeleton graph topology is shared by all GCN layers, which makes it difficult for the model to focus on the key joints of the action.

To solve the above problems, a supervised spatio-temporal contrastive learning framework with optimal skeleton subgraph topology for HAR (SSTCL-optSST) is proposed. SSTCL-optSST uses label information to determine the positive samples in each small batch, each of positive sample represents a trajectory of an action. Then a multiple positives noise contrastive estimation loss for supervised learning (MultiPNCE-Sup) is proposed, which uses multiple samples of an action to guide contrastive learning across views, so as to extract the representative features from the multiple trajectories of an action. In addition, an optimal skeleton subgraph topology determination method (optSST) is proposed to obtain the optimal subgraph to highlight the key joints of the action.

Summarily, our main contributions are as follows: (1) A supervised spatio-temporal contrastive learning framework for HAR is proposed, which can enable the model learn different action representations by using trajectories of the action. (2) An optimal skeleton subgraph topology determination method is proposed to search the optimal subgraph of the action to highlight the key joints to further optimize the model. (3) Extensive experiments have been carried out on two large human skeletal action datasets, the results show that the proposed SSTCL-optSST outperforms the existing skeleton-based action recognition methods.

2   Relation Work

2.1   Skeleton-Based Action Recognition

Neural networks can automatically extract features by using relative 3D rotations and spatio-temporal position transformations between joints. Related research on HAR includes RNN, CNN and GCN based methods. (i) RNN-based methods. Zhang et al. [10] proposed VA-LSTM, which can dynamically input the trajectory of samples, and connect the three-dimensional space coordinates of joints in the frame into a time series according to the predefined order. Liu et al. [5] proposed ST-LSTM to take advantage of the contextual dependencies of joints in the spatio-temporal domain and explore the implicit information of action related information in the two domains. (ii) CNN-based methods. Ji et al. [11] used the hand-crafted transformation rules to map the input skeleton into a pseudo image. Skeletal joint nodes and their temporal trajectories are represented by columns and rows in the grid shape structure, respectively. However, neither RNN nor CNN can well represent the structure of human skeleton data, because skeleton data is a natural graph structure. Representing skeleton data in the form of vector sequence or two-dimensional grid will inevitably lose some graph structure information. (3) GCN-based methods. Generally, the GCN-based methods outperform other methods due to its great modeling ability for non-Euclidean data. Shi et al. [12] proposed a two-stream adaptive graph convolutional network (2S-AGCN), which explicitly feeds second-order information of skeleton data into an adaptive graph convolutional network.

Graph Convolutional Network. GCN has efficient performance for processing non-Euclidean data. Graph embedding is calculated by a GCN layer through different aggregation functions to obtain node information from its neighbor nodes. GCNs based methods involve spectral and spatial method. (i) Spectral method. The spectral method performs the graph Fourier transform and transforms the graph convolution to the frequency domain. Most of the work originates from the layer-wise propagation rule [13], which approximates the effect of graph feature representation through Chebyshev expansion. (ii) Spatial method. The works of these methods are to directly design convolution filters to apply to nodes or edges by weighted summation of vertices in spatial domain. The semantics-guided neural network (SGN) proposed by Zhang et al. [14] can capture explicit semantic information by explicitly encoding joint types and frame indexes, so as to retain the spatio-temporal basic information of body structure. Peng et al. [15] introduced a graph pooling method to increase the receptive field and break the structural constraints of high-level features to optimize the graph triplet loss. However, the mentioned networks cannot effectively learn different postures of actions, because they do not pay attention to the diversity of action trajectories. In the traditional GCN layer, an input skeleton sample is a 3D tensor $f_{in} \in R^{C_{in} \times N \times T}$, and C_{in} is number of channels. The spatial sub-module uses a spatial convolution function according to Eq. 1.

$$f_{out} = \sum_{k}^{K_s} W_k(f_{in}A_k) \tag{1}$$

Where K_s is a kernel size on spatial dimension, $A_k = D_k^{-\frac{1}{2}}(\hat{A}_k + I)D_k^{-\frac{1}{2}}$ is the normalized adjacent matrix with $D_{ii} = \sum_{k}^{K_s}(\hat{A}_k^{ij} + I_{ij})$, $\hat{A}_k$ is the adjacent matrix of the undirected graph representing intra-body connections, I is the identity matrix and W_k is a trainable weight matrix, and the final output $f_{out} \in \mathbb{R}^{C_{out} \times N \times T}$.

2.2   Contrastive Learning

Contrastive learning is widely used in self-supervised and unsupervised learning. The contrastive loss is task-related, and the similarity of sample pairs is processed in the representation space. Contrastive learning uses Noise Contrastive Estimation (NCE) [16], and has achieved the most advanced results in self-supervised learning of action recognition. In the discrimination task [17], the contrastive loss related to NCE can draw samples of the same classes closer and push samples of different classes away, which is conducive to improving the recognition accuracy of the model. Khosla et al. [18] proposed a supervised contrastive learning method, which extends the self-supervised batch contrastive to the supervised environment, so that the network can effectively use the label information for image recognition. Chen et al. [19] proposed a method, the SimCLR, which used a larger mini-batch to employ more contrasting negative samples to compute the real-time embedding.

3   Method

3.1   Supervised Spatio-Temporal Contrastive Learning Framework

SSTCL-optSST is a supervised contrastive representation learning method based on graph convolutional network model. For an input skeleton data, SSTCL-optSST applies data preprocessing to obtain two copies of input samples. Both copies are propagated forward through the spatio-temporal feature extraction network to obtain a normalized embedding. In the training, the samples with the same label as the anchor are viewed as positive samples, otherwise, negative ones. The main components of framework are shown in Fig. 1.

Generally, skeletal samples in each frame of the video sequences can be viewed as a skeleton graph topology, which can be represented as G, Skeleton data of a video clip can be represented as a spatio-temporal coordinate vector set $x \in R^{3 \times N \times T}$, N is the number of joints and T is the number of video frames. In each frame t, the x_t denotes the 3D coordinate vector of the nodes. In data preprocessing, the input sample x is reversed to obtain a new sample x'. The two samples are concatenated through $cat(\cdot)$, and then augmented by $aug(\cdot)$, as shown in Eq. (2).

$$\hat{x} = aug(cat(x, x')) \tag{2}$$

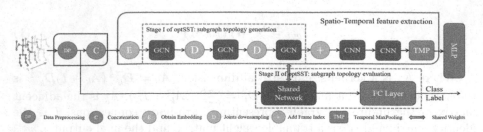

Fig. 1. Supervised spatio-temporal contrastive learning framework with optimal skeleton subgraph topology for HAR.

After data preprocessing, the spatio-temporal feature extraction is used to train the model to extract feature information. There are multiple forward propagation layers, and the preprocessed samples can be learned to obtain a normalized embedding. The encoder embeds $\hat{x}_a$ into a hidden space according to Eq. (3).

$$f_{in} = ReLU(W_2(ReLU(W_1\hat{x} + b_1) + b_2)) \tag{3}$$

Here W_1 and W_2 are the weight matrices of two fully connected layers, b_1 and b_2 are the bias vectors, $ReLU$ denotes a non-linear activation function.

In the feature encoder, normalized embedding f_{in} is fed into graph convolutional networks to explore the correlations for the structural skeleton data, then is fed into other layers to finally obtain spatio-temporal feature representations. The formulas of spatio-temporal feature extraction are shown in Eq. (4) and (5).

$$f_{out} = ReLU(G_s(f_{in}) + G_s(f_{in} \otimes A_k)) \tag{4}$$

$$STF = ReLU(C_s(ReLU(C_s(f_{out} \oplus FI)))) \tag{5}$$

where G_s denotes the graph convolution, C_s denotes the 2D convolution operation. $\otimes$ means the matrix-wise multiplication. $\oplus$ denotes the element-wise addition. FI denotes the frame index obtained by a one-hot vector [14]. STF denotes the spatio-temporal feature representations.

In the feature representation projection stage, the high-dimensional feature representation obtained by the encoder network is reduced to the low dimensional feature representation through the projection network, which is a multi-layer perceptron (MLP) with ReLU activation function. The transformation is done according to Eq. (6).

$$Proj = MLP(MaxPool(STF))) \tag{6}$$

Where MLP is initialized with two linear layers. $MaxPool$ denotes the temporal max pooling. During the evaluation phase, the Shared Network use the same weight of network trained by feature encoder, and the feature representation is passed through a fully connected layer with Softmax to obtain the final action class.

3.2   Supervised Contrastive Loss

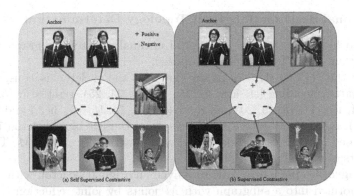

Fig. 2. The difference of contrastive learning between supervised and self-supervised: (a)The traditional self-supervised contrastive learning methods; (b) Our proposed supervised contrastive learning method for skeleton-based HAR.

As shown in Fig. 2(a), the two samples represent the original sample and the reversed sample in the orange and red boxes, respectively. For the self-supervised contrastive learning setting, the augmented skeleton sample is a positive sample, and the samples belonging to waving action are considered as negative samples, which means the model ignores the diversity of action trajectories, and can not recognize different poses of an action. The supervised learning setting is shown in Fig. 2(b). In supervised learning, action labels are known at the training stage, so positive samples can be determined according to whether their labels are the same as anchor. In order to train the model to recognize multiple trajectories of an action, a new loss function MultiPNCE-Sup is proposed, which uses multiple positive samples of action to guide the model learning from different views.

For a group of N samples $\{(x_i, y_i)|i = 1, 2, ..., N\}$, N pairs of augmented samples are obtained through data preprocessing, which constitute the mini-batch for training. Let $\{(x_k, y_k)|k = 1, 2, ..., 2N\}$ be the mini-batch, where x_{2i-1} and x_{2i} are the two augmented samples of x_i and $y_{2i-1} = y_{2i} = y_i$. At each training step, the necessary embeddings are stored in memory bank M. Our proposed MultiPNCE-Sup loss is shown in Eq. (7).

$$\mathcal{L}_{mp} = -\frac{1}{|P|} \sum_{p \in P} log \frac{e^{z \cdot k_p^+ / \tau}}{e^{z \cdot k_p^+ / \tau} + \sum_{i=1}^{M} e^{z \cdot k_i^- / \tau}} \tag{7}$$

Where $\mathcal{L}_{mp}$ denotes the contrastive loss, $P \equiv \{p \in S \setminus M : \hat{y}_p = \hat{y}_z\}$ is set of indices of all positives in the mini-batch, $S \equiv \{1, 2, ..., 2N\}$ indicates a set of all augmented sample, and $|P|$ is the number of set P. τ is a scalar temperature parameter [20], the z is called the anchor, k_p^+ is a positive sample, $M = \{k_i^- | i \in 1, 2, ..., 2(N-1)\}$ is a memory bank that stores negative samples,

k_i^- is the ith negative sample in the memory queue, the · symbol is a dot product of two vectors.

3.3   Optimal Skeleton Subgraph Topology Determination

In order to highlight the key joints, and optimal skeleton subgraph topology determination (optSST) method is proposed.

Taking the NTU60 dataset as an example, the skeleton is consisted of 5 different parts, including left arm (9, 10, 11, 12, 24, 25), right arm (5, 6, 7, 8, 22, 23), left leg (17, 18, 19, 20) and right leg (13, 14, 15, 16), head and body (1, 2, 3, 4, 21), respectively, which can be see from the left blue skeleton in Fig. 3. The proposed optSST can be described as a function $\psi : N \rightarrow M$. The main components of optSST are as follows.

(i) **Subgraph space generation**. Given an original skeleton with N joints, it is transformed into a subgraph with M joints by joint reduction operation. The number of joints should be reduced by $D_1 = \lfloor (N-M)/2 \rfloor$ for the first joint reduction operation and $D_2 = N - M - \lfloor (N-M)/2 \rfloor$ for the second time. Let $N = 20$, $M = 12$ as the example, see Fig. 3, in which the original skeleton graph is regarded as a tree root. The number of joints to be reduced by the two joint reduction operations is $D_1 = D_2 = 4$, and there should be one joint in each part from $\bar{G}_1$ to $\bar{G}_4$ to be deleted by each joint reduction operation. After two joint reduction operations, the action subgraph space is formed.

(ii) **Correlation matrix**. For a selected subgraph topology, a normalized joint correlation function is applied to calculate the affinity between joints in all parts before the GCN layer. Similar to [21], the joint correlation strength function is described as Eq. (8). By calculating the joint affinity of all joint pairs in the same frame, the affinity matrix $g \in R^{N \times N}$ can be obtained.

$$g(\bar{x}_{v_i}, \bar{x}_{v_j}) = \frac{e^{\theta(\bar{x}_{v_i})^T \phi(\bar{x}_{v_j})}}{\sum_{j=1}^{N} e^{\theta(\bar{x}_{v_i})^T \phi(\bar{x}_{v_j})}} \qquad (8)$$

Where $\theta(\cdot)$ and $\phi(\cdot)$ are transformation functions, each of them is implemented by a Linear layer.

Due to the change in skeleton graph topology, the hidden layer feature vector $f_{in} \in R^{C \times N \times T}$ is converted into the new input $\tilde{f}_{in} \in R^{C \times M \times T}$, which corresponds to the current subgraph topology after joint downsampling. The new correlation matrix $\tilde{B}_k$ is updated by $\tilde{g}$ and adjacent matrix $\tilde{A}_k$, which is shown in Eq. (9).

$$\tilde{B}_k = SoftMax(\tilde{g} + \tilde{A}_k) \qquad (9)$$

Here, the correlation matrix $\tilde{B}_k \in R^{M \times M}$ and $SoftMax(\cdot)$ is a normalized exponential function. After joints downsampling, the original sequence sample containing N joints are mapped to a new graph topology of skeleton containing M joints. The model gets a new hidden layer feature input and correlation strength matrix before the GCN layer, and then input them to the GCN layer to obtain the feature representation.

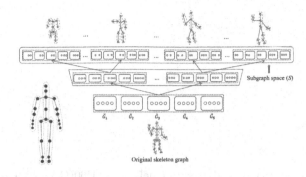

Fig. 3. Illustration of skeleton subgraph segmentation in optSST.

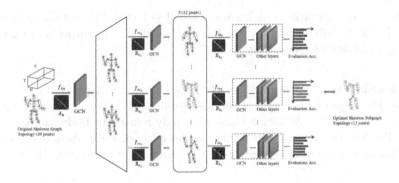

Fig. 4. Illustration of the optimal skeleton subgraph topology determination.

(iii) **Subgraph evaluation and determination.** The input data containing the information of a subgraph topology is sequentially processed to obtain network weight parameters by the encoder network and MLP. To reduce the computational requirements, the learned model shares the encoder weights to the shared network. Then, the shared network with a classifier is trained by a subgraph topology, and evaluated by classifying a small-scale evaluation dataset. Subgraph streams are input to train the model, and the parameters for the skeleton subgraph topology with best performance are retained and the optimal subgraph of the original skeleton is determined. The optSST is shown in Fig. 4

4 Experiments

4.1 Datasets

NTU-RGB+D 60. NTU60 [22] is a large human action dataset, which contains a total of 56880 3D skeleton video samples collected from 60 human action classes of 40 actors. The maximum number of frames for an action sequence is 300, and each frame has 25 joint coordinates for a human skeleton. The original work

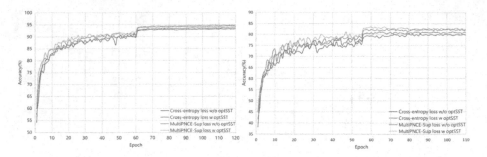

Fig. 5. Accuracy curves of four different models on NTU60(C-View) and KM dataset.

provides two benchmarks, which are Cross-subject (C-Sub) and Cross-view (C-View). In the C-Sub evaluation benchmark, half of the 40 subjects are used for training and the rest for testing.

Kinetics Skeleton. Kinetics [23] is a large-scale noisy dataset, containing about 300000 video clips of 400 classes. Kinetics-Motion (KM) [7] is a subset of Kinetics Skeleton and contains 30 action classes strongly related to body motions. Each video clip does not exceed 300 frames in the released date. Action recognition performance is evaluated by reporting the recognition accuracies of Top-1 and Top-5.

4.2   Experimental Configuration

In order to initialize skeleton data, all datasets were preprocessed same as [24] to speed up the training process before training. The initial learning rate of Adam optimizer is set to 0.001. The hyperparameter τ of MultiPNCE-Sup loss function is set to 0.08. The training of our model is mainly completed by three parts, including data preprocessing module, the encoder network and the projection network. After data preprocessing, augmented samples are fed into the spatio-temporal feature extraction network. In the encoder network, the kernel size of each GCN layer is 1×1. There is a batch normalization (BN) layer and a nonlinear activation function ReLU after each GCN layer. The feature output channels of the three GCN layers are 128, 256 and 256, respectively. Similar to the previous network model [7], the residual network mechanism is used in each GCN layer. Then the feature embedding is sequentially fed into two CNN layers with a kernel size of 3×1 to capture the dynamic information. The number of neurons of the two CNN layers are set to 256 and 512, respectively.

In the evaluation phase, our model is classified by the shared network and full connection (FC) layer. The parameters of the shared network are the same as those of the feature encoder. The final output channel is 60 (the number of action classes). 10% of the training sequences are randomly selected to evaluate the performance of optSST.

During the test, the learning rate is set to 0.1 and is changed at the 60th, 90th and 110th epoch during training. The model stops training at the 120th epoch. We set the batch size as 128. For the Kinetics dataset, we also set the learning rate to 0.1 on the Kinetics Skeleton dataset, and the learning rate decreases sequentially at the 55th, 80th and 100th epoch, and the training process ends at the 110th epoch. The batch size is set to 64. In addition, we set the number of parts of the skeleton in optSST to 5 in all datasets. For the NTU60 datasets, we set the reduction ratio of the number of joints to 20% for each joint downsampling layer, which means that the number of joints should be reduced from 25 to 20, and then reduced from 20 to 15. For Kinetics Skeleton, the reduction ratio is set to 17%, and the 18 joints of original joints is reduced to 12 by the two joint reduction operations.

4.3   Ablation Study

To validate the effectiveness of the components proposed in this study, we conducted several ablation experiments of SSTCL-optSST on the NTU60 dataset and KM dataset. The participating models include "CE w/o optSST", "CE w optSST", "MS w/o optSST" and "MS w optSST". Here "w" and "w/o" denote with and without, respectively. CE stands for Cross-entropy loss and MS is MultiPNCE-Sup loss. The accuracy curves of the four models of using different modules on NTU60(C-View) and KM dataset are shown in Fig. 5. It can be seen that "MS w optSST" model obtained the best performance compared with other models.

As shown in Table 1, "CE w optSST" model is superior to "CE w/o optSST" by 0.5% and 1.1% on the NTU60(C-View) and KM dataset respectively. In addition, "MS w/o optSST" model outperforms "CE w/o optSST" by 1.2% and 2.3% on the NTU60(C-View) and KM dataset respectively. The results show that our proposed MultiPNCE-Sup loss and optSST can improve the recognition accuracy of the model.

Table 1. Accuracy of different modules

Methods	NTU60(C-View)	KM
CE w/o optSST	93.2%	80.1%
CE w optSST	93.7%	81.2%
MS w/o optSST	94.4%	82.4%
MS w optSST	**94.8%**	**83.1%**

4.4   Comparison with Other Methods

To further measure the action recognition performance of our model on skeleton data, we conducted extensive experiments on the NTU60 and Kinetics dataset, and compared with other state-of-the-art methods. To be fair, we compare our model with the ST-GCN baseline and other models on the NTU60 dataset. Taking C-Sub and C-View as the metrics, the Top-1 accuracy of several models is compared. For Kinetics, we use the Top-1 and Top-5 accuracy to evaluate performance of the model. It can be seen from Table 2 that the methods based on deep learning usually outperforms the methods based on hand-crafted feature. As far as the results of both metrics are concerned, the performance of SSTCL-optSST is better than most of the aforementioned methods, which shows that SSTCL-optSST can well distinguish the same action classes with different motions. For the different waving actions, the accuracy of our SSTCL-optSST on the C-View benchmark is improved from 94.5% to 94.8%.

Table 2. Accuracy of different methods on NTU60 dataset.

Methods	C-Sub	C-View
Deep LSTM(2016) [22]	62.9%	70.3%
VA-LSTM (2017) [10]	79.2%	87.7%
ST-GCN (2018) [7]	81.5%	88.3%
ConMLP (2023) [25]	83.9%	93.0%
C-MANs (2021) [26]	83.7%	93.8%
PR-GCN (2021) [27]	85.2%	91.7%
STG-IN (2020) [28]	85.8%	88.7%
AS-GCN (2019) [29]	86.8%	94.2%
GR-GCN (2019) [30]	87.5%	94.3%
KA-AGTN (2022) [31]	88.3%	94.3%
SGN (2020) [14]	**89.0%**	94.5%
Ours (SSTCL-optSST)	88.2%	**94.8%**

Table 3. Accuracy comparison of different methods on Kinetics dataset.

Methods	Top-1	Top-5
Deep LSTM (2016) [22]	16.4%	35.3%
TCN (2017) [6]	20.3%	40.0%
ST-GCN (2018) [7]	30.7%	52.8%
ST-GR (2019) [32]	33.6%	56.1%
PR-GCN (2021) [27]	33.7%	55.8%
PeGCN (2022) [33]	34.8%	57.2%
2S-AGCN (2019) [12]	**35.1%**	57.1%
Ours (SSTCL-optSST)	34.6%	**57.4%**

For the Kinetics dataset as shown in Table 3, we compared our method with seven skeleton-based action recognition methods. On Top-5 accuracy, it

Table 4. Top-1 accuracy of different methods on KM subset of the Kinetics dataset.

Methods	Top-1	
RGB [23]	70.4%	frame
Optical flow [23]	72.8%	frame
ST-GCN [7]	72.4%	skeleton
Ours (SSTCL-optSST)	**83.1%**	skeleton

can be seen that our model achieved the best performance. As for Top-1 accuracy, our model achieved the accuracy of 34.6%, and ranks 2 among of all the compared methods. Table 4 reports the mean class accuracy of skeleton-based and frame-based methods on Kinetics-Motion dataset, and our model achieved better performance than previous frame-based methods and skeleton-based ST-GCN model.

5 Conclusion

In this paper, a supervised spatio-temporal contrastive learning framework with optimal skeleton subgraph topology for HAR is proposed. The MultiPNCE-Sup loss utilizes multiple motion trajectories of skeletal samples to guide contrastive learning across views, and makes the model recognize different postures of an action. Meanwhile, to highlight the key joints of an action, we proposed an optimal skeleton subgraph topology determination method to determine the optimal skeleton subgraph topology of an input skeleton graph topology. By using the subgraph that highlight the key joints of actions, the model can further improve the evaluation performance. The model was evaluated on three large-scale action recognition datasets: NTU60 and Kinetics dataset. The results show that our model can achieve the competitive performance on NTU60(C-View) and Kinetics-Motion.

Acknowledgement. This work was supported in part by the National Natural Science Foundation of China under Grant No. 61977018; Natural Science Foundation of Changsha under Grant No. kq2202215; Practical Innovation and Entrepreneurship Enhancement Program for Professional Degree Postgraduates of Changsha University of Science and Technology (CLSJCX22114).

References

1. Gajjar, V., Gurnani, A., Khandhediya, Y.: Human detection and tracking for video surveillance: a cognitive science approach. In: Proceedings of the IEEE International Conference on Computer Vision Workshops, pp. 2805–2809 (2017)
2. Sahaï, A., Desantis, A., Grynszpan, O., Pacherie, E., Berberian, B.: Action co-representation and the sense of agency during a joint simon task: comparing human and machine co-agents. Conscious. Cogn. **67**, 44–55 (2019)

3. Pilarski, P.M., Butcher, A., Johanson, M., Botvinick, M.M., Bolt, A., Parker, A.S.: Learned human-agent decision-making, communication and joint action in a virtual reality environment. arXiv preprint arXiv:1905.02691 (2019)

4. Wang, H., Schmid, C.: Action recognition with improved trajectories. In: Proceedings of the IEEE International Conference on Computer Vision, pp. 3551–3558 (2013)

5. Liu, J., Shahroudy, A., Xu, D., Wang, G.: Spatio-temporal LSTM with trust gates for 3D human action recognition. In: Leibe, B., Matas, J., Sebe, N., Welling, M. (eds.) ECCV 2016. LNCS, vol. 9907, pp. 816–833. Springer, Cham (2016). https://doi.org/10.1007/978-3-319-46487-9_50

6. Kim, T.S., Reiter, A.: Interpretable 3D human action analysis with temporal convolutional networks. In: Proceedings of the IEEE Conference on Computer Vision and Pattern Recognition Workshops, pp. 20–28 (2017)

7. Yan, S., Xiong, Y., Lin, D.: Spatial temporal graph convolutional networks for skeleton-based action recognition. In: Thirty-Second AAAI Conference on Artificial Intelligence (2018)

8. Duan, H., Zhao, Y., Chen, K., Lin,D., Dai, B.: Revisiting skeleton-based action recognition. In: Proceedings of the IEEE/CVF Conference on Computer Vision and Pattern Recognition, pp. 2969–2978 (2022)

9. Sukhbaatar, S., Bruna, J., Paluri, M., Bourdev, L., Fergus, R.: Training convolutional networks with noisy labels. arXiv preprint arXiv:1406.2080 (2014)

10. Zhang, P., Lan, C., Xing, J., Zeng, W., Xue, J., Zheng, N.: View adaptive recurrent neural networks for high performance human action recognition from skeleton data. In: Proceedings of the IEEE International Conference on Computer Vision, pp. 2117–2126 (2017)

11. Ji, X., Zhao, Q., Cheng, J., Ma, C.: Exploiting spatio-temporal representation for 3D human action recognition from depth map sequences. Knowl.-Based Syst. **227**, 107040 (2021)

12. Shi, L., Zhang, Y., Cheng, J., Lu, H.: Two-stream adaptive graph convolutional networks for skeleton-based action recognition. In: Proceedings of the IEEE/CVF Conference on Computer Vision and Pattern Recognition, pp. 12026–12035 (2019)

13. Kipf, T.N., Welling, M.: Semi-supervised classification with graph convolutional networks. arXiv preprint arXiv:1609.02907 (2016)

14. Zhang, P., Lan, C., Zeng, W., Xing, J., Xue, J., Zheng, N.: Semantics-guided neural networks for efficient skeleton-based human action recognition. In: Proceedings of the IEEE/CVF Conference on Computer Vision and Pattern Recognition, pp. 1112–1121 (2020)

15. Peng, W., Hong, X., Zhao, G.: Tripool: graph triplet pooling for 3d skeleton-based action recognition. Pattern Recogn. **115**, 107921 (2021)

16. Gutmann, M., Hyvärinen, A.: Noise-contrastive estimation: a new estimation principle for unnormalized statistical models. In: Proceedings of the Thirteenth International Conference on Artificial Intelligence and Statistics, pp. 297–304. JMLR Workshop and Conference Proceedings (2010)

17. Wu, Z., Xiong, Y., Yu, S.X., Lin, D.: Unsupervised feature learning via nonparametric instance discrimination. In: Proceedings of the IEEE Conference on Computer Vision and Pattern Recognition, pp. 3733–3742 (2018)

18. Khosla, P.: Supervised contrastive learning. Adv. Neural. Inf. Process. Syst. **33**, 18661–18673 (2020)

19. Chen, T., Kornblith, S., Norouzi, M., Hinton, G.: A simple framework for contrastive learning of visual representations. In: International Conference on Machine Learning, pp. 1597–1607. PMLR (2020)

20. Hinton, G., Vinyals, O., Dean, J., et al.: Distilling the knowledge in a neural network, vol. 2, no. 7. arXiv preprint arXiv:1503.02531 (2015)
21. Wang, X., Girshick, R., Gupta, A., He, K.: Non-local neural networks. In: Proceedings of the IEEE Conference on Computer Vision and Pattern Recognition, pp. 7794–7803 (2018)
22. Shahroudy, A., Liu, J., Ng, T.T., Wang, G.: Ntu rgb+ d: a large scale dataset for 3d human activity analysis. In: Proceedings of the IEEE Conference on Computer Vision and Pattern Recognition, pp. 1010–1019 (2016)
23. Kay, W., et al. The kinetics human action video dataset. arXiv preprint arXiv:1705.06950 (2017)
24. Rao, H., Shihao, X., Xiping, H., Cheng, J., Bin, H.: Augmented skeleton based contrastive action learning with momentum LSTM for unsupervised action recognition. Inf. Sci. **569**, 90–109 (2021)
25. Dai, C., Wei, Y., Xu, Z., Chen, M., Liu, Y., Fan, J.: ConMLP: MLP-based self-supervised contrastive learning for skeleton data analysis and action recognition. Sensors **23**(5), 2452 (2023)
26. Li, C., Xie, C., Zhang, B., Han, J., Zhen, X., Chen, J.: Memory attention networks for skeleton-based action recognition. IEEE Trans. Neural Netw. Learn. Syst. **33**(9), 4800–4814 (2021)
27. Li, S., Yi, J., Farha, Y.A., Gall, J.: Pose refinement graph convolutional network for skeleton-based action recognition. IEEE Rob. Autom. Lett. **6**(2), 1028–1035 (2021)
28. Ding, W., Li, X., Li, G., Wei, Y.: Global relational reasoning with spatial temporal graph interaction networks for skeleton-based action recognition. Signal Process. Image Commun. **83**, 115776 (2020)
29. Li, M., Chen, S., Chen, X., Zhang, Y., Wang, Y., Tian, Q.: Actional-structural graph convolutional networks for skeleton-based action recognition. In: Proceedings of the IEEE/CVF Conference on Computer Vision and Pattern Recognition, pp. 3595–3603 (2019)
30. Gao, X., Hu, W., Tang, J., Liu,J., Guo, Z.: Optimized skeleton-based action recognition via sparsified graph regression. In: Proceedings of the 27th ACM International Conference on Multimedia, pp. 601–610 (2019)
31. Liu, Y., Zhang, H., Dan, X., He, K.: Graph transformer network with temporal kernel attention for skeleton-based action recognition. Knowl.-Based Syst. **240**, 108146 (2022)
32. Li, B., Li, X., Zhang, Z., Fei, W.: Spatio-temporal graph routing for skeleton-based action recognition. In: Proceedings of the AAAI Conference on Artificial Intelligence, vol. 33, pp. 8561–8568 (2019)
33. Yoon, Y., Jongmin, Yu., Jeon, M.: Predictively encoded graph convolutional network for noise-robust skeleton-based action recognition. Appl. Intell. **52**(3), 2317–2331 (2022)

Multi-scale Feature Fusion Neural Network for Accurate Prediction of Drug-Target Interactions

Zhibo Yang[1,2], Binhao Bai[1,2], Jinyu Long[1,2], Ping Wei[3], and Junli Li[1,2(✉)]

[1] School of Computing Science, Sichuan Normal University, Chengdu 610066, China
lijunli@sicnu.edu.cn
[2] Visual Computing and Virtual Reality Key Laboratory of Sichuan, Sichuan Normal University, Chengdu 610068, China
[3] Sichuan Academy of Traditional Chinese Medicine/Sichuan Center of Translational Medicine/Sichuan Key Laboratory of Translational Medicine of Traditional Chinese Medicine, Chengdu 610041, China

Abstract. Identification of drug-target interactions (DTI) is crucial in drug discovery and repositioning. However, identifying DTI is a costly and time-consuming process that involves conducting biological experiments with a vast array of potential compounds. To accelerate this process, computational methods have been developed, and with the growth of available datasets, deep learning methods have been widely applied in this field. Despite the emergence of numerous sequence-based deep learning models for DTI prediction, several limitations endure. These encompass inadequate feature extraction from protein targets using amino acid sequences, a deficiency in effective fusion mechanisms for drug and target features, and a prevalent inclination among many methods to solely treat DTI as a binary classification problem, thereby overlooking the crucial aspect of predicting binding affinity that signifies the strength of drug-target interactions. To address these concerns, we developed a multi-scale feature fusion neural network (MSF-DTI), which leverages the potential semantic information of amino acid sequences at multiple scales, enriches the feature representation of proteins, and fuses drug and target features using a designed feature fusion module for predicting drug-target interactions. According to experimental results, MSF-DTI outperforms other state-of-the-art methods in both DTI classification and binding affinity prediction tasks.

Keywords: Drug-target interaction · Binding affinity · Deep learning

1 Introduction

The drug-target interaction(DTI) prediction is a crucial step in the field of drug development, and largely affects the progress of subsequent research [1]. Although some traditional methods, such as high-throughput screening, are

B. Luo et al. (Eds.): ICONIP 2023, CCIS 1964, pp. 176–188, 2024.
https://doi.org/10.1007/978-981-99-8141-0_14

widely used, they still have high computational costs and time expenditures for large-scale drug-target pair searches [2–4]. Therefore, developing efficient computational methods to explore DTI mechanisms is of great significance.

Understanding the mechanism of interactions between drugs and targets is often a complex process, which poses significant challenges for developing effective computational methods to identify DTI. Currently, there are two main categories of computational methods available: structure-based methods and non-structure-based methods. Structure-based methods, such as molecular docking, can predict DTI, as well as provide valuable biological insights by predicting the binding site of drugs to proteins [5–8]. Nevertheless, their drawbacks are evident, as they overly rely on protein 3D structures with limited available data. Due to this situation, some non-structure-based methods have been proposed that rely on a large amount of known biological data [9–12]. These methods analyze and extract features from the data and then use modeling to achieve the prediction. Based on different modeling methods, they can be classified into three categories: graph-based methods, network-based methods, and machine learning-based methods, among which machine learning-based methods are the most widely used. For instance, Madhukar et al. utilized a Bayesian approach to integrate various types of data in an unbiased manner for predicting DTI [13]. Piazza et al. investigated the potential of detecting kinase or phosphatase inhibitors and membrane protein drugs using a machine learning-based framework, and identified a previously unknown target that could be used for antibacterial purposes [14]. Despite their relative effectiveness, non-structure-based methods still have several limitations. One major drawback of most of these methods is their disregard for the binding affinity values associated with drug-target interactions. Instead, they primarily focus on binary classification problem. Moreover, without utilizing the structural information, the interpretability of these methods remains restricted.

In practical applications, it is not only important to determine whether drugs and targets can bind to each other, but also to know the binding affinity between them, which indicates the strength of their interaction [15]. The binding affinity between drugs and targets is commonly quantified by indicators including dissociation constant (K_d), inhibition constant (K_i), or half-maximal inhibitory concentration (IC_{50}), with lower values indicating stronger binding. Over the last few years, some computational methods have been developed to predict the binding affinity between drugs and targets [16–18]. However, due to the inherent complexity of this problem, it remains a challenging task that continues to receive much attention.

Recently, deep learning has shown remarkable performance in various fields such as image classification and speech recognition [19,20]. In the field of bioinformatics, many researchers have started to apply deep learning to predict drug-target interactions and have achieved outstanding results [21–23]. These deep learning methods usually first encode compounds and proteins to make them processable by computers, then extract potential features of compounds and proteins through various network models and their combinations. Finally, the learned compound features and protein features are combined for the final prediction.

In this research, we propose an end-to-end deep learning model called MSF-DTI, which is based on protein multi-scale feature fusion, for DTI and binding affinity prediction. Proteins, as biological macromolecules, contain rich biological information in their primary sequences, which is not only crucial for protein structure and function but also important for forming interactions with other molecules. However, many studies only extract protein sequence features from a single scale of the entire sequence, which often fails to capture important local information and leads to insufficient protein representation, causing bottlenecks for downstream predictions. To address this issue, we use an improved feature pyramid network to learn multi-scale features of proteins and design an attention-based feature fusion module to integrate the features of drug molecules and the multi-scale features of proteins. Finally, the fused features are used for DTI and binding affinity prediction.

We tested our MSF-DTI model on different datasets for DTI prediction and binding affinity prediction. Comparative analysis reveals superior performance of our proposed method. These findings indicate that our proposed feature pyramid network-based representation module can effectively learn the multi-scale features of protein sequences, greatly enriching the semantic information of proteins. Moreover, the fusion module can effectively combine drug and protein features, which is helpful for modeling interactions. All of these findings indicate that our model is effective in predicting DTI and binding affinity.

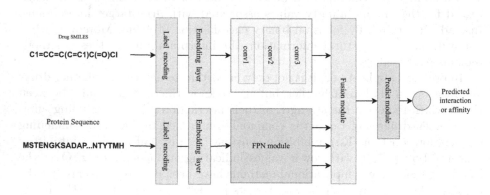

Fig. 1. Architecture of MSF-DTI.

2   Materials and Methods

In this part, we will elaborate on the details of the model. As shown in the Fig. 1, the model consists of four parts: the drug representation learning branch, the protein representation learning branch, the feature fusion module, and the prediction module. The model takes the drug and protein sequence as input, first encoding the drug molecules into character codes and encoding the protein into fixed-length subsequences. Next, the two inputs are processed through their respective branches to extract their latent features. We integrate the features extracted from both branches and combine them with the drug molecule's fingerprint features to make the final prediction of DTI and binding affinity.

2.1 Drug Representation Learning Branch

We used Simplified Molecular Input Line Entry System (SMILES) to represent the input of each compound molecule. SMILES text is composed of 64 unique characters, and we transformed each compound SMILES into a corresponding label sequence by encoding each character. For example, the label encoding of "CC1 = C" is {42 42 35 40 42}. For the convenience of model processing, we transform drugs with different lengths into label sequences of fixed length 100, padding shorter sequences with 0 and truncating longer ones beyond 100.

We employed a convolution-based network to learn the sequence features of the compounds, which consists of an embedding layer and three stacked one-dimensional convolutional layers. Each convolutional layer has different sizes and numbers of kernels, which directly affect the types of features the model learns from the input. It has been reported that increasing the number of filters can lead to better performance in pattern recognition for the model [24]. Furthermore, each convolutional layer is directly followed by a Rectified Linear Unit (ReLU) [25] layer as the activation function, which is used to improve the model's non-linear learning ability.

Furthermore, to enhance the feature representation of compounds, we introduced extended connectivity fingerprints (ECFPs), a commonly used fingerprint type for characterizing molecular substructures [26]. We utilized the RDKit toolkit to obtain the molecular fingerprints for each compound, which were subsequently fed into an artificial neural network to generate their feature representations.

Table 1. Classification of amino acids based on amino acid side chains and dipole volume.

Cluster	Amino acid
a	A,G,V
b	I,L,F,P
c	Y,M,T,S
d	H,N,Q,W
e	R,K
f	D,E
g	C

2.2   Protein Representation Learning Branch

Protein sequences also need to be encoded into a form that the model can process. There are 20 known standard amino acids, which are divided into 7 categories based on the size of the amino acid side chains and dipole volumes [27], as shown in the Table 1. For non-standard amino acids that may appear in the data, we classify them into the eighth category. These 8 categories, grouped into sets of 3 category numbers, can form a dictionary of size 512. We divide the protein sequence into overlapping subsequence fragments of length 3, and map these subsequences to their corresponding codes for their respective categories to capture potential protein sub sequence patterns. For example, for the protein sequence MSHH...SFK, the set of subsequences is {MSH, SHH, ..., SFK}, the set of corresponding categories is {ccd, cdd, ..., cbe}, and the corresponding codes are {148, 156,..., 141}. Therefore, for each protein sequence, we encode it based on its subsequences, and use an embedding layer to generate the protein's initial feature matrix. Since the length of protein sequences varies, we set the maximum length to 1200, fill the insufficient length with 0, and truncate the excess part beyond the maximum length.

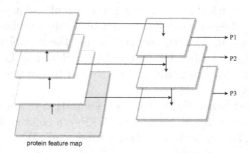

protein feature map

Fig. 2. Architecture of the FPN Module in the MSF-DTI Model.

Protein sequences contain rich semantic information, and effective information extraction is an important step in downstream prediction tasks. In this experiment, we enriched the feature representation of proteins by learning the multi scale representation of amino acid sequences, providing more comprehensive information for drug-target interactions. The feature pyramid network was first used in the computer vision field for object detection and semantic segmentation tasks, mainly to solve the key problem of recognizing objects of different scales, relevant studies have shown that this network has advanced performance [28]. As the name suggests, the feature pyramid network has a hierarchical structure similar to a pyramid, with each level containing different semantic features, and the features of each level are combined through the top-down path and horizontal connections.

In this study, we borrowed the design structure of the feature pyramid network to learn the feature representation of proteins from different scales, as shown in the Fig. 2. Since the protein sequence is one-dimensional, we used a bottleneck structure composed of one-dimensional convolution as the backbone in the model to learn protein features, and set the stride to 2 to decrease the size of the sequence. The whole pyramid is divided into three levels. The initial protein feature matrix is continuously reduced in size through the bottom-up convolutional layer to gradually learn deep semantic features, and the feature matrix output at the topmost level is then upsampled layer by layer through the top-down path, and the upsampled features at each layer are combined with the features at the same layer in the bottom-up path. Each layer in the top-down path outputs a feature map representing the multi-scale features learned by the pyramid network, and these feature maps all have the same number of channels. The final feature output by the protein representation learning module can be represented as {p1, p2, p3}.

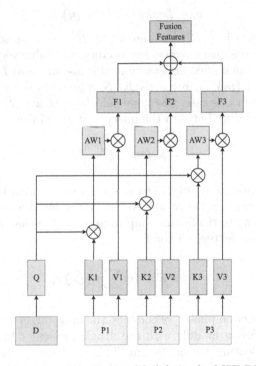

Fig. 3. Architecture of the Fusion Module in the MSF-DTI Model.

2.3   Feature Fusion and Prediction Module

Many deep learning models for predicting drug-target interactions or binding affinity often directly concatenate the learned drug and protein features and feed them into a fully connected network for classification or regression prediction. However, these methods do not explore the interaction relationships between compound molecules and proteins, which may result in poor prediction performance. Hence, we have devised a feature fusion module based on the self-attention mechanism [29], as illustrated in Fig. 3, aimed at extracting potential interaction relationships between compound molecules and proteins. This module seeks to furnish essential insights for predictive tasks, with the ultimate goal of enhancing model performance.

Firstly, we use the feature vector d generated by the compound representation learning module as the query vector, and use the multi-scale features $\{p_i\}_{i=1}^N$ generated by the protein representation learning module as the key and value vectors, and transform them to the same dimensional space through a single-layer neural network.

$$q = LeakyReLU(w_q d) \tag{1}$$

$$k_i = LeakyReLU(w_k^i p_i) \tag{2}$$

$$v_i = LeakyReLU(w_v^i p_i) \tag{3}$$

Here, $w_q \in R^{d \times L_d}$, $w_k^i \in R^{d \times L_{p_i}}$ and $w_v^i \in R^{d \times L_{p_i}}$ represent the learnable weight matrices for the query vector, key vectors, and value vectors in the fusion module, where d is the unified dimension in this module, and L_d and L_{p_i} denote the length of d and $\{p_i\}_{i=1}^N$, respectively. N represents the number of protein features, which is 3 in this case. Then, we calculate the attention scores $\{\alpha_i\}_{i=1}^N$ based on the dot product between the query vector and the key vectors.

$$\alpha_i = softmax(\frac{q \cdot k_i}{\sqrt{d}}) \tag{4}$$

Next, we element-wise multiply the attention scores with the corresponding value vectors, and sum up the results to obtain the fused feature map. Finally, we apply max pooling to the feature map to preserve the channel dimension and obtain the final fused feature vector I.

$$I = MaxPool1d(\sum_{i=1}^N \alpha_i \bigotimes v_i) \tag{5}$$

During the prediction phase, we combine the drug-target interaction features generated by the fusion module with the drug's molecular fingerprint features to form the final feature for prediction. This final feature is then fed into a three-layer fully connected network for prediction.

3 Experiments and Results

3.1 Datasets

For this experiment, we used four datasets to evaluate the predictive performance of DTI and drug-target binding affinity. In the DTI problem, positive samples representing drug-target binding are labeled as 1, while negative samples are

Table 2. The number of distinct compounds, proteins, train and test samples of K_d and K_i datasets

Datasets	No. of compound	No. of proteins	No. of train	No. of test
K_d	5895	812	8778	3811
K_i	93437	1619	101134	43391

labeled as 0. We used human and C.elegans as DTI datasets [30], with positive samples from DrugBank 4.1, Matador, and STITCH 4.0 [31–33]. Human datasets contains 3369 positive samples and C.elegans datasets contains 4000 positive samples. Negative samples in these two datasets were obtained through a systematic screening framework that integrates various compound and protein resources, with high reliability. Human datasets contains 384916 negative samples, while C.elegans datasets contains 88261 negative samples.

In the problem of drug-target binding affinity, the binding affinity values are used as the sample truth, and the ways of representing the binding affinity values include K_d, K_i and IC_{50}. In this experiment, we use two datasets proposed by Karimi et al. [34]. to evaluate the model performance, which measure the binding affinity with K_d and K_i, respectively. As shown in Table 2, The K_d dataset contains 12,589 samples, while the K_i dataset contains 144,525 samples.

3.2 Implementation Details

Our model employs end-to-end learning to represent compound molecules and proteins. We utilize convolution to extract the potential features of compounds. For proteins, we learn multi-scale features based on a feature pyramid network to explore deep semantic information. Additionally, we use an attention-based feature fusion module to model the interaction between compounds and proteins, enriching the downstream feature representation for prediction. Finally, we predict DTI and binding affinity by integrating compound fingerprint features and fusion features. We use cross-entropy loss and mean square error loss as the loss functions for DTI and binding affinity tasks, respectively. Our training objective is to continuously reduce the gap between predicted values and true values, and optimize our learnable parameters through backpropagation.

3.3 DTI Prediction

In line with earlier research, we investigated the efficacy of the MSF-DTI approach in predicting DTI by performing five rounds of 5-fold cross validation [35]. During each round, the dataset is evenly split into 5 groups, with one group used as the test set and the other four groups as the training set. Additionally, 10% of the training set is reserved for validation.

We conducted experiments on two benchmark datasets, human and C.elegans, and used area under the receiver operating characteristic curve (AUC) and area under precision-recall curve (AUPR) as evaluation metrics for our model. It is worth noting that the ratio of positive to negative samples in the human and C.elegans datasets is 1:5, resulting in a highly imbalanced dataset, which is consistent with the real-world scenarios. We compared MSF-DTI with other methods, including CMF [36], BLM-NII [37], NRLMF [38], and a recently developed deep learning method CoaDTI [39], to demonstrate its effectiveness. The results of our experiments are presented in Fig. 4 and Fig. 5, while the results of CMF, BLM-NII, and NRLMF are taken from Li et al.'s research [40]. As for CoaDTI, we used $CoaDTI_{stack}$, which assembles attention units in a stacked manner. We have noted that both the MSF-DTI and $CoaDTI_{stack}$ models surpassed other machine learning models in performance across the two datasets. This underscores the efficacy of end-to-end representation learning methods in extracting latent features from compound molecules and proteins for DTI prediction. Moreover, previous studies have shown that in tasks with imbalanced data, the AUPR metric is more accurate than AUC in reflecting the performance of a model [41]. Our model also had higher AUPR than other methods on both datasets, which demonstrates the strong competitiveness of MSF-DTI in facing imbalanced DTI tasks.

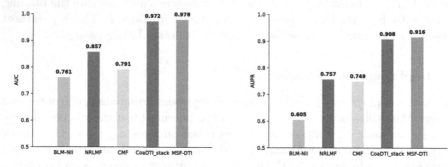

Fig. 4. The AUC and AUPR values of different methods on human datasets for DTI prediction.

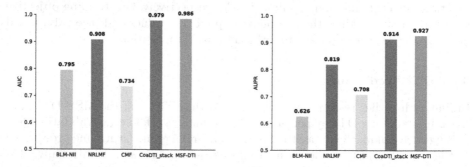

Fig. 5. The AUC and AUPR values of different methods on C.elegans datasets for DTI prediction.

3.4   Binding Affinity Prediction

To assess the efficacy of MSF-DTI in predicting drug-target binding affinity, we conducted experiments on two datasets of different scales that use either K_d or K_i as measures of binding affinity. Each dataset was divided into a training set and a test set, with 10% of the training set reserved for validation. We used the root mean squared error(RMSE) and the Pearson correlation coefficient (PCC) as evaluation metrics to measure model performance, and set up four comparison methods: ridge regression(Ridge), random forest(RF), DeepAffinity, and DeepPurpose [42], of which Ridge and RF belong to traditional machine learning methods, while DeepAffinity and DeepPurpose belong to deep learning methods. The experimental results of Ridge and RF that we used were obtained from Karimi et al.'s study [34], and we evaluated the performance of DeepAffinity and DeepPurpose with their default options. Table 3 and 4 show the RMSE and PCC scores of each model on the K_d and K_i datasets.

Table 3. The RMSE scores of different methods on K_d and K_i datasets

Methods	Datasets	
	K_d	K_i
Ridge	1.24	1.27
RF	1.10	0.97
DeepAffinity	1.17	0.96
DeepPurpose	1.16	0.92
MSF-DTI	1.07	0.88

Table 4. The PCC scores of different methods on K_d and K_i datasets

Methods	Datasets	
	K_d	K_i
Ridge	0.60	0.58
RF	0.70	0.78
DeepAffinity	0.68	0.79
DeepPurpose	0.69	0.81
MSF-DTI	0.74	0.83

As is commonly known, RMSE is a measure frequently used to indicate the deviation between predicted and true values, while PCC can reflect their similarity. These two indicators are often used in drug-target binding affinity prediction. From the results, deep learning methods perform better than traditional machine learning methods in terms of RMSE, indicating that deep learning methods can improve the fitting ability of data by more effectively learning the latent features of data. Based on the outcomes of these two evaluation metrics, our proposed MSF-DTI model excels compared to other methods on both the small-scale K_d dataset and the large-scale K_i dataset, underscoring its effectiveness in addressing tasks related to binding affinity.

4   Conclusion

The discovery of DTI is crucial in drug development, but the experimental identification of such interactions is a time consuming and expensive process, despite the availability of various high-throughput systems biology screenings. The advancement of machine learning and deep learning has brought forth novel insights for tackling such challenges. With a substantially greater volume of known protein sequence data in comparison to available three-dimensional structures, sequence-based deep learning approaches have gained prominence in recent years.

In this paper, we propose an end-to-end deep learning model that only uses the SMILES of compound molecules and the primary sequence of proteins to predict DTI and binding affinity. We use a convolutional neural network with convolutional kernels of different sizes to extract deep features of compounds, a feature pyramid-based network module to extract multiscale features of proteins, and a designed fusion module to explore the potential interactions between compounds and proteins. Experimental results indicate that our model achieves competitive performance in both DTI classification and binding affinity prediction.

It's worth noting that the scope of real-world data is much broader than what's used in experiments. Consequently, enhancing the model's ability to generalize and address real-world problems remains a significant challenge. Moreover, determining whether the model can yield insights into the mechanism of action during inference poses another obstacle to overcome. In our future research endeavors, we will actively explore methods to enhance the model's generalization capabilities and improve its interpretability.

References

1. Paul, S.M., et al.: How to improve R&D productivity: the pharmaceutical industry's grand challenge. Nat. Rev. Drug Disc. **9**(3), 203–214 (2010)
2. Chu, L.-H., Chen, B.-S.: Construction of cancer-perturbed protein–protein interaction network of apoptosis for drug target discovery. In: Choi, S. (ed.) Systems Biology for Signaling Networks. SB, pp. 589–610. Springer, New York (2010). https://doi.org/10.1007/978-1-4419-5797-9_24
3. Ricke, D.O., Wang, S., Cai, R., Cohen, D.: Genomic approaches to drug discovery. Curr. Opin. Chem. Biol. **10**(4), 303–308 (2006)
4. Bakheet, T.M., Doig, A.J.: Properties and identification of human protein drug targets. Bioinformatics **25**(4), 451–457 (2009)

5. Xie, L., Xie, L., Kinnings, S.L., Bourne, P.E.: Novel computational approaches to polypharmacology as a means to define responses to individual drugs. Annu. Rev. Pharmacol. Toxicol. **52**, 361–379 (2012)
6. Śledź, P., Caflisch, A.: Protein structure-based drug design: from docking to molecular dynamics. Curr. Opin. Struct. Biol. **48**, 93–102 (2018)
7. Gschwend, D.A., Good, A.C., Kuntz, I.D.: Molecular docking towards drug discovery. J. Mol. Recogn. Interdiscipl. J. **9**(2), 175–186 (1996)
8. Trott, O., Olson, A.J.: Autodock vina: improving the speed and accuracy of docking with a new scoring function, efficient optimization, and multithreading. J. Comput. Chem. **31**(2), 455–461 (2010)
9. Durrant, J.D., McCammon, J.A.: Nnscore 2.0: a neural-network receptor-ligand scoring function. J. Chem. Inf. Model. **51**(11), 2897–2903 (2011)
10. Ding, H., Takigawa, I., Mamitsuka, H., Zhu, S.: Similarity-based machine learning methods for predicting drug-target interactions: a brief review. Brief. Bioinform. **15**(5), 734–747 (2014)
11. Wan, F., Hong, L., Xiao, A., Jiang, T., Zeng, J.: Neodti: neural integration of neighbor information from a heterogeneous network for discovering new drug-target interactions. Bioinformatics **35**(1), 104–111 (2019)
12. Yuvaraj, N., Srihari, K., Chandragandhi, S., Raja, R.A., Dhiman, G., Kaur, A.: Analysis of protein-ligand interactions of sars-cov-2 against selective drug using deep neural networks. Big Data Min. Anal. **4**(2), 76–83 (2021)
13. Madhukar, N.S., et al.: A Bayesian machine learning approach for drug target identification using diverse data types. Nat. Commun. **10**(1), 5221 (2019)
14. Piazza, I., et al.: A machine learning-based chemoproteomic approach to identify drug targets and binding sites in complex proteomes. Nat. Commun. **11**(1), 4200 (2020)
15. Pahikkala, T., et al.: Toward more realistic drug-target interaction predictions. Brief. Bioinform. **16**(2), 325–337 (2015)
16. Shar, P.A., et al.: Pred-binding: large-scale protein-ligand binding affinity prediction. J. Enzyme Inhib. Med. Chem. **31**(6), 1443–1450 (2016)
17. Gabel, J., Desaphy, J., Rognan, D.: Beware of machine learning-based scoring functions on the danger of developing black boxes. J. Chem. Inf. Model. **54**(10), 2807–2815 (2014)
18. He, T., Heidemeyer, M., Ban, F., Cherkasov, A., Ester, M.: Simboost: a read-across approach for predicting drug-target binding affinities using gradient boosting machines. J. Cheminf. **9**(1), 1–14 (2017)
19. Nassif, A.B., Shahin, I., Attili, I., Azzeh, M., Shaalan, K.: Speech recognition using deep neural networks: a systematic review. IEEE Access **7**, 19143–19165 (2019)
20. Pak, M., Kim, S.: A review of deep learning in image recognition. In: 2017 4th International Conference on Computer Applications and Information Processing Technology (CAIPT), pp. 1–3. IEEE (2017)
21. Öztürk, H., Özgür, A., Ozkirimli, E.: Deepdta: deep drug-target binding affinity prediction. Bioinformatics **34**(17), i821–i829 (2018)
22. Wang, J., Wen, N., Wang, C., Zhao, L., Cheng, L.: Electra-dta: a new compound-protein binding affinity prediction model based on the contextualized sequence encoding. J. Cheminform. **14**(1), 1–14 (2022)
23. Li, F., Zhang, Z., Guan, J., Zhou, S.: Effective drug-target interaction prediction with mutual interaction neural network. Bioinformatics **38**(14), 3582–3589 (2022)
24. Kang, L., Ye, P., Li, Y., Doermann, D.: Convolutional neural networks for no-reference image quality assessment. In: Proceedings of the IEEE Conference on Computer Vision and Pattern Recognition, pp. 1733–1740 (2014)

25. Nair, V., Hinton, G.E.: Rectified linear units improve restricted Boltzmann machines. In: Proceedings of the 27th International Conference on Machine Learning (ICML-10), pp. 807–814 (2010)
26. Rogers, D., Hahn, M.: Extended-connectivity fingerprints. J. Chem. Inf. Model. **50**(5), 742–754 (2010)
27. Shen, J., et al.: Predicting protein-protein interactions based only on sequences information. Proc. Natl. Acad. Sci. **104**(11), 4337–4341 (2007)
28. Lin, T.-Y., Dollár, P., Girshick, R., He, K., Hariharan, B., Belongie, S.: Feature pyramid networks for object detection. In: Proceedings of the IEEE Conference on Computer Vision and Pattern Recognition, pp. 2117–2125 (2017)
29. Vaswani, A., et al.: Attention is all you need. In: Advances in Neural Information Processing Systems, vol. 30 (2017)
30. Liu, H., Sun, J., Guan, J., Zheng, J., Zhou, S.: Improving compound-protein interaction prediction by building up highly credible negative samples. Bioinformatics **31**(12), i221–i229 (2015)
31. Wishart, D.S., et al.: Drugbank: a knowledgebase for drugs, drug actions and drug targets. Nucl. Acids Res. **36**(suppl-1), D901–D906 (2008)
32. Günther, S., et al.: Supertarget and matador: resources for exploring drug-target relationships. Nucl. Acids Res. **36**(suppl-1), D919–D922 (2007)
33. Kuhn, M., et al.: Stitch 4: integration of protein-chemical interactions with user data. Nucl. Acids Res. **42**(D1), D401–D407 (2014)
34. Karimi, M., Di, W., Wang, Z., Shen, Y.: Deepaffinity: interpretable deep learning of compound-protein affinity through unified recurrent and convolutional neural networks. Bioinformatics **35**(18), 3329–3338 (2019)
35. Tsubaki, M., Tomii, K., Sese, J.: Compound-protein interaction prediction with end-to-end learning of neural networks for graphs and sequences. Bioinformatics **35**(2), 309–318 (2019)
36. Zheng, X., Ding, H., Mamitsuka, H., Zhu, S.: Collaborative matrix factorization with multiple similarities for predicting drug-target interactions. In: Proceedings of the 19th ACM SIGKDD International Conference on Knowledge Discovery and Data Mining, pp. 1025–1033 (2013)
37. Mei, J.-P., Kwoh, C.-K., Yang, P., Li, X.-L., Zheng, J.: Drug-target interaction prediction by learning from local information and neighbors. Bioinformatics **29**(2), 238–245 (2013)
38. Liu, Y., Min, W., Miao, C., Zhao, P., Li, X.-L.: Neighborhood regularized logistic matrix factorization for drug-target interaction prediction. PLoS Comput. Biol. **12**(2), e1004760 (2016)
39. Huang, L., et al.: Coadti: multi-modal co-attention based framework for drug-target interaction annotation. Brief. Bioinform. **23**(6), bbac446 (2022)
40. Li, M., Zhangli, L., Yifan, W., Li, Y.: Bacpi: a bi-directional attention neural network for compound-protein interaction and binding affinity prediction. Bioinformatics **38**(7), 1995–2002 (2022)
41. Davis, J., Goadrich, M.: The relationship between precision-recall and ROC curves. In: Proceedings of the 23rd International Conference on Machine Learning, pp. 233–240 (2006)
42. Huang, K., Fu, T., Glass, L.M., Zitnik, M., Xiao, C., Sun, J.: Deeppurpose: a deep learning library for drug-target interaction prediction. Bioinformatics **36**(22–23), 5545–5547 (2020)

GoatPose: A Lightweight and Efficient Network with Attention Mechanism

Yaxuan Sun[1,2], Annan Wang[1], and Shengxi Wu[1,2(✉)]

[1] East China University of Science and Technology, Shanghai, China
[2] School of Information Science and Engineering, ECUST, Shanghai, China
wushengxi@ecust.edu.cn

Abstract. Keypoint detection is an essential part of human pose estimation. However, due to resource constraints, it's still a challenge to deploy complex convolutional networks to edge devices. In this paper, we present GoatPose: a lightweight deep convolutional model for real-time human keypoint detection incorporating attention mechanism. Since the high computational cost is associated with the frequently-use convolution block, we substitute it with our new-designed LiteConv block, which conducts cheap linear operation to generate rich feature maps from the intrinsic features with low cost. This method significantly accelerates the model while inevitably losing a part of spatial information. To compensate for the information loss, we introduce NAM attention mechanism. By applying channel weighting, the model can focus more on the important features and enhance the feature representation. Results on the COCO dataset show the superiority of our model. With the complexity of our model reduced by half and the computational speed doubled, the accuracy of our model is basically the same as that of the backbone model. We further deploy our model on NVIDIA Jetson TX2 to validate its real-time performance, indicating that our model is capable of being deployed and widely adopted in real-world scenarios.

Keywords: Human key point detection · Lightweight network · High-resolution representation · Attention mechanism · Model deployment

1 Introduction

Human pose estimation aims to accurately detect the keypoints of the human body such as the head, shoulders, elbows, wrists, knees, and ankles. The detection task has broad applications, including motion capture, human-computer interaction, pedestrian tracking, etc. Deep learning has been successfully applied in the human body key point detection task, however, it's still a challenge to deploy complex networks to edge devices due to resource constraints.

Lightweight models are needed for human keypoint detection tasks in practical applications. One popular approach to designing lightweight networks is to borrow techniques such as depth-wise separable convolution and channel shuffling from

B. Luo et al. (Eds.): ICONIP 2023, CCIS 1964, pp. 189–202, 2024.
https://doi.org/10.1007/978-981-99-8141-0_15

classification networks [1–3] to reduce computational redundancy. Model compression is also a common approach to generate lightweight models [4–9].

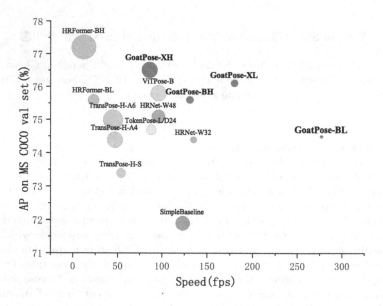

Fig. 1. AP and Speed on MSCOCO val set

In this paper, we present a lightweight high-resolution network GoatPose. Using high-resolution network HRNet [10] as our backbone, GoatPose maintains high-resolution representations through the whole process, therefore capturing strong semantic information. Considering that the frequently-used convolutions excessively consume computational resources, we redesign the convolution block, incorporating cheap linear operation into it, which can generate rich feature maps from the intrinsic features with little cost [11,12]. To compensate for the information loss caused by linear operation, we introduce attention mechanism. By weighting the channels, the model can adaptively find areas that need attention and highlight important features while suppressing irrelevant features [13–15]. Our model demonstrates superior results on the COCO keypoint detection dataset [16], as shown in Fig. 1. We further deploy our model on NVIDIA Jetson TX2 to validate the real-time performance. It turns out that our model has a high AP of 74.5% with only 3.48 GFLOPs, outperforming prior state-of-the-art efficient pose estimation models. We believe our work will push the frontier of real-time pose estimation on edge.

Our main contributions include:

- We propose a lightweight network GoatPose that generates highresolution feature maps with low cost. The key of the model lies in the introduction of cheap operation and attention mechanism.

- We demonstrate the effectiveness of GoatPose on the COCO dataset. Our model outperforms all other models and achieves excellent result, reaching high average precision while maintaining low computational and memory consumption.
- We deploy our model on NVIDIA Jetson TX2 to validate its real-time performance. The results demonstrate that our model is remarkable efficient in practical applications and can be easily generalized to the human keypoint detection task.

2 Related Work

High-Resolution Representation. Tasks that require position-sensitive information rely on high-resolution representations. There are two mainstream approaches. One method is to employ a high-resolution recovery process to enhance the representation resolution. This is achieved by improving the low-resolution output obtained from a classification network [17–22] through sequentially-connected convolutions, typically upsampling. The other way is to replace the downsampling and normal convolution layers with dilated convolutions [23–31]which will significantly increase the computational complexity and number of parameters. HRNet [10,32] is proposed as an efficient way to maintain high-resolution representation throughout the network. HRNet consists of parallel multi-resolution branches. By repeated multi-scale fusion, HRNet can generate high-resolution representation with rich semantics.

Model Lightweighting. Lightweight models are needed for practical real-time application. Separable convolutions and group convolutions, derived from classification networks [2,3,31,33–38] are commonly-used techniques to reduce computational redundancy. MobileNetv3 [3] is built upon depthwise separable convolutions and introduces the inverted residual structure to construct lightweight networks. ShuffleNetv2, on the other hand, incorporates pointwise group convolutions and channel shuffling to maintain model performance. Although these models achieve relatively robust performance with fewer computations, they do not fully explore the redundancy between feature maps to further compress the model. In contrast, the Ghost module [11], proposed by Kai Han et al., can generate additional feature maps from the intrinsic features through cheap linear operations, significantly conserving computational and storage costs.

Attention Mechanisms. Network lightweighting inevitably leads to spatial information loss. Incorporating attention mechanism into convolutional networks allows the model to adaptively allocate weights to different regions, therefore compensate for the loss of information to some extent. Prior studies on attention mechanisms focus on enhancing the performance of neural networks by suppressing insignificant weights [13–15,39] and lack consideration for the contribution factor of the weights, which could further suppress insignificant features. In contrast, NAM [40] uses batch normalization scaling factors to represent the importance of weights and fully exploits the contribution factor of the weights to enhance attention, which not only lowers network complexity and computational costs, but also leads to improved efficiency and model performance.

3 Approach

3.1 Parallel Multi-branch Architecture

The model we proposed in this paper uses HRNet [10] as the backbone. HRNet is a high-resolution convolutional neural network capable of capturing high semantic information while maintaining high-resolution representations. It is characterized by two key features: parallel multi-resolution representations and repeated multi-resolution fusion.

Parallel Multi-resolution Representation. Starting with a high-resolution representation as the first stage, each subsequent stage includes the previous stage's resolution representation and expands a lower-resolution representation, as shown in Fig. 2. The resolutions of all four representations are $1/4$, $1/8$, $1/16$, and $1/32$, respectively. Representations of different resolutions are connected in parallel to avoid information loss. Also, feature extraction is simultaneously performed on multiple scales.

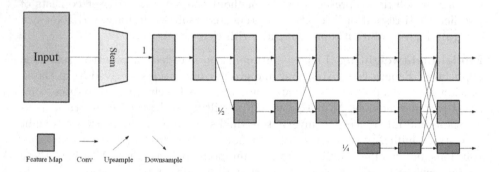

Fig. 2. Parallel Multi-resolution Representation

Repeated Multi-resolution Fusion. The purpose of fusion is to exchange information between representations of different resolutions. Take the fusion of 2-resolution representations for example, the input consists of three representations: $\{R_r^i, r = 1, 2\}$, where r is the resolution index, and the associated output representations are $\{R_r^o, r = 1, 2\}$. Each output representation is the sum of the transformed representations of the two inputs:$R_r^o = f_{1r}(R_1^i) + f_{2r}(R_2^i)$. There will be an additional output from stage 2 to stage 3, which is the expanded lower-resolution representation: $R_3^o = f_{13}(R_1^i) + f_{23}(R_2^i)$. The choice of the transform function $f_{xr}(\cdot)$ is dependent on the input resolution index x and the output resolution index r. If $x = r$, $f_{xr}(R) = R$. If $x < r$, $f_{xr}(R)$ downsamples the input representation R through (rs) stride-2 3×3 convolutions. If $x > r$, $f_{xr}(R)$ upsamples the input representation R through the bilinear upsampling followed by a 1×1 convolution for aligning the number of channels. The functions are depicted in Fig. 3.

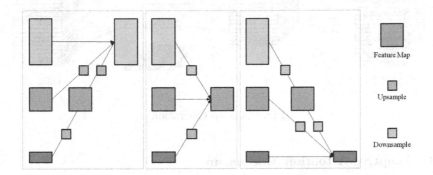

Fig. 3. Repeated Multi-resolution Fusion

3.2 Model Lightweighting

Considering the limited resources of embedded devices, model lightweighting is necessary to deploy convolutional neural networks on edge. By replacing complex convolution operations with cheap linear operations, the computational amount can be greatly reduced [11].

Given the input data $X \in R^{c \times h \times w}$, the operation of an arbitrary convolutional layer for producing n feature maps can be formulated as Eq. 1:

$$Y = X * f + b \tag{1}$$

where $*$ is the convolution operation, b is the bias term, $Y \in R^{h' \times w' \times n}$ is the output feature map, $f \in R^{c \times k \times k \times n}$ is the convolution filters, and the convolution kernel is k. Taking into account the similarity between the generated feature maps, these redundant feature maps are called ghost feature maps. Suppose that those ghost feature maps can be easily transformed from smaller-scale intrinsic feature maps, we can modify the above equation as follows, as Eq. 2 and Eq. 3:

$$Y' = X * f' \tag{2}$$

$$y_{ij} = \Phi_{i,j}(y_i'), \quad \forall i = 1,\ldots,m, \quad j = 1,\ldots,s \tag{3}$$

where $f' \in R^{c \times k \times k \times m}$ is the utilized filters, $m \leq n$, the bias term is omitted for simplicity, y_i' is the i-th intrinsic feature map in Y', and $\Phi_{i,j}$ is the j_{th} linear operation used to generate the j_{th} ghost feature map y_{ij}. The last $\Phi_{i,s}$ is the identity mapping for preserving the intrinsic feature maps. We can obtain $n = m \cdot s$ feature maps as shown in Fig. 4.

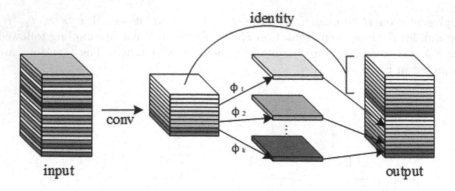

Fig. 4. Cheap Operation

3.3 Adaptive Attention Mechanism

In order to improve the accuracy and robustness of the model without increasing the complexity of the network, we introduce the adaptive attention mechanism NAM.

For the channel attention sub-module, NAM uses a scaling factor from BN to reflect the importance of channels, as shown in Eq. 4.

$$B_{out} = BN(B_{in}) = \gamma \frac{B_{in} - \mu_{\mathcal{B}}}{\sqrt{\sigma_{\mathcal{B}}^2 + \epsilon}} + \beta \tag{4}$$

where $u_{\mathcal{B}}$ and $\sigma_{\mathcal{B}}$ are the mean and standard deviation of the mini-batch $\mathcal{B}$, respectively; γ and β are trainable affine transformation parameters for scaling and translation operations, so that the model can adaptively adjust the activated range and central location. The process of weighting channels is shown in Fig. 5. For each channel, γ_i is the scaling factor, and its weight is $W_i = \gamma_i / \sum_{j=0}^{\gamma_j}$.

In addition, NAM introduces $L1$ regularization as a weight sparsity penalty in the loss function, as shown in Eq. 5, where x represents the input, y represents the output, W represents the network weights, $l(\cdot)$ is the loss function; $g(\cdot)$ is the $L1$ norm penalty function and p is used to balance $g(\gamma)$ and $g(\lambda)$ penalty item. By introducing sparsity constraints into model weights, the model is prompted to focus on more important features while reducing unnecessary calculations.

Fig. 5. Channel Attention Submodule

$$LOSS = \sum_{(x,y)} l(f(x,W),y) + p\sum g(\gamma) + p\sum g(\lambda) \tag{5}$$

3.4 GoatPose

We propose GoatPose as a lightweight high-resolution network. Using HRNet as the backbone, we redesigns the convolution unit and introduces cheap operation to reduce the number of parameters. In order to compensate for the information loss, We incorporates the NAM attention mechanism into the architecture. Goat symbolizes that our model runs as quickly and lightly as a goat. Just as a goat's two horns stand out, certain channels receive more attention due to their higher importance in the task.

Design of Convolution Block. We redesign the convolution block by incorporating cheap linear operation, and name this modified block LiteConv, as shown in Fig. 6. For each LiteConv block, the input X first passes through a normal convolution, followed by BatchNorm and ReLU to upsample the input feature maps. The resulting feature maps are denoted as $X1$. Then, we conduct cheap linear operation on the intrinsic feature map $X1$ to generate Ghost feature maps $X2$, including a sequence of a depthwise separable convolution, BatchNorm and ReLU. $X2$ is concatenated with $X1$ as the final output.

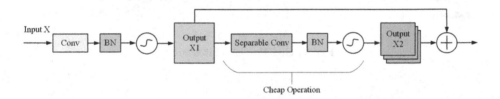

Fig. 6. LiteConv Block

Compensate for the Information Loss. Considering that replacing convolutions with linear operations for feature map generation can inevitably lead to information loss, we introduce the NAM attention mechanism to compensate for this loss. The NAM module is inserted after the stack of Basic Blocks to timely augment the feature representation.

Lightweight High-Resolution Network. GoatPose consists of four stages. Starting from a high-resolution convolutional stream $W \times H \times C$, the first stage includes 4 bottlenecks for extracting image features. Then a 3×3 convolution is applied to generate an additional haif-resolution stream $W/2 \times H/2 \times 2C$. The two convolutional streams are output to the next stage in parallel.

The second, third, and fourth stage consist of 1, 4, and 3 Stage Modules, respectively. Each Stage Module contains parallel multi-resolution convolutional streams. It consists of Basic Block, channel attention module NAM-C, and fusion unit, as shown in Fig. 7. The convolutional streams of different resolutions first pass through four Basic Blocks and then go through the NAM-C module. Finally, they are fused with other convolutional streams through the fusion unit.

The Basic Block contains two LiteConvs followed by BN and ReLU. In the NAM-C module, the weighted feature maps are obtained by element-wise multiplication of the weights with the input, and scaled by sigmoid function as the adjusted feature maps. The fusion unit connects the outputs of different resolutions in a fully connected manner, as illustrated in the previous example. Transition modules are used between stages to expand the convolutional streams, adding an additional parallel convolutional stream with halve resolution using a 3×3 convolution. This stream is then concatenated with the convolutional streams from the previous stage and serves as the input for the next stage.

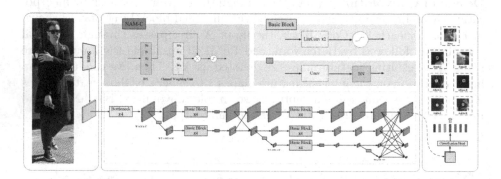

Fig. 7. GoatPose Architecture

4   Experiments

4.1   Dataset

Microsoft COCO. Our model is evaluated on the most popular **MSCOCO** dataset for human pose estimation, which contains an extensive collection of approximately $200,000$ images and $250,000$ person examples. We use the train2017 dataset consisting of $57,000$ images for training, and the val2017 dataset containing $5,000$ images is used for the evaluation process.

Evaluation Metric. In the MS's COCO2017 dataset, the standard evaluation metric is based on Object Keypoint Similarity (OKS), as shown in Eq. 6:

$$\text{OKS} = \frac{\sum_i \exp\left(-\frac{d_i^2}{2\,s^2 k_i^2}\right) \delta\left(v_i > 0\right)}{\sum_i \delta\left(v_i > 0\right)} \tag{6}$$

where d_i is the Euclidean distance between a detected keypoint and its corresponding ground truth position, s is the object scale, k_i is a per-keypoint constant that controls falloff, and v_i denotes the visibility flag of keypoint i.

We report standard average precision and recall scores, as shown in Fig. 8a: AP50 (AP at OKS = 0.50), AP75, AP (the mean of AP scores at OKS = 0.50, 0.55, . . . , 0.90, 0.95), APM for medium objects, APL for large objects, and AR (the mean of recalls at OKS = 0.50, 0.55, . . . , 0.90, 0.95).

4.2 Setting

The network is trained on NVIDIA GeForce RTX 3080 12G, and the installed CUDA Version is Cuda12.0. We train the model for a total of 280 epochs. We adopt Adam optimizer with an initial learning rate of 0.001 and a decay rate of 0.4. The base learning rate dropped to 0.0004 and 0.00016 at the 170th and 230th and 260th epoch respectively, as shown in Fig. 8b.

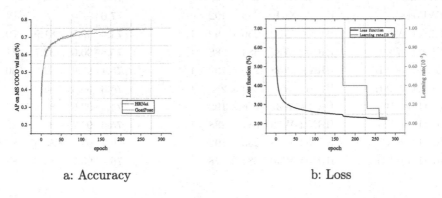

a: Accuracy   b: Loss

Fig. 8. Accuracy and Loss

4.3 Results

Comparisons on COCO Dataset. By adjusting the resolution and number of channels of the input image, we construct four versions of GoatPose, each with the same hyperparameter configuration. Among them, B means that the model input has 32 channels, X means the model input has 48 channels; L means the size of the input is 256 × 192, and H means the size of the input image is 384 × 288. In the following context We mainly analyze the performance improvement of GoatPose-BL.

Table 1 reports the comparison of AP and AR score between our networks and other state-of-the-art methods. Compared to models based on the HRNet-W32 backbone, GoatPose-BL achieves similar average precision to HRNet-W32L and TokenPose-B, and outperforms TransPose-H-S by 1.5%. Meanwhile, GoatPose-BL shows significant improvements in AR score over all aforementioned models, with increases of 3% and 2.9% compared to HRNet-W32 and TokenPose-B, respectively.

Table 1. Comparison on Accuracy.

Model	Backbone	Input Size	Feature Size	AP	AP50	AP75	AR
Simple Baseline [21]	ResNet-152	256 × 192	1/32	72	89.3	79.8	77.8
TokenPose-B [41]	HRNet-W32	256 × 192	1/4	74.7	89.8	81.4	80.0
TokenPose-L/D24	HRNet-W48	256 × 192	1/4	75.8	90.3	82.5	80.9
TransPose-H-S [42]	HRNet-W32	256 × 192	1/4	73.4	91.6	81.1	–
TransPose-H-A6	HRNet-W48	256 × 192	1/4	75	92.2	82.3	80.8
HRFormer-BL [43]	HRFormer-B	256 × 192	1/4	75.6	90.8	82.8	80.8
HRFormer-BH	HRFormer-B	384 × 228	1/4	77.2	91	83.6	82
ViTPose-B [44]	ViT-B	256 × 192	1/16	75.8	90.7	–	81.1
ViTPose-L	ViT-L	256 × 192	1/16	78.3	91.4		83.5
RTMPose-m [45]	CSPNeXt-m	256 × 192	–	75.7	–	–	–
RTMPose-L	CSPNeXt-L	256 × 192	–	76.6	–	–	–
DWPose-m [46]	CSPNeXt-L	256 × 192	–	76.2	–	–	–
DWPose-L	CSPNeXt-L	256 × 192	–	77.0	–	–	–
HRNet-W32L [10]	HRNet-W32	256 × 192	1/4	74.4	90.5	81.9	79.8
HRNet-W32H	HRNet-W32	384 × 288	1/4	75.8	90.6	82.7	81
HRNet-W48L	HRNet-W48	256 × 192	1/4	75.1	90.6	82.2	80.4
HRNet-W48H	HRNet-W48	384 × 288	1/4	76.3	90.8	82.9	81.2
GoatPose-BL	HRNet-W32	256 × 192	1/4	74.5	92.6	82.5	82.3
GoatPose-BH	HRNet-W32	384 × 288	1/4	75.6	92.7	82.3	83.5
GoatPose-XL	HRNet-W48	256 × 192	1/4	76.1	93.5	83.5	83.8
GoatPose-XH	HRNet-W48	384 × 128	1/4	76.5	93.7	83.6	83.9

Table 2 presents the comparison of complexity between our networks and other state-of-the-art methods. GoatPose-BL has fewer parameters compared to the majority of models, meanwhile, its execution speed is significantly faster than all other models. GoatPose-BL has only 15.51M parameters, which is nearly half of the parametes of HRNet-W32L. Additionally, its computational speed is 3.48 GFLOPs, which is approximately twice as fast as HRNet-W32L. Compared to the model TransPose-H-S with the least parameters, GoatPose has slightly more parameters but achieves a four-fold improvement in computation speed.

Table 2. Comparison on Complexity.

Model	Params(M)	GFLOPs	Speed(fps)	AP	AR
Simple Baseline [21]	60	15.7	123	72	77.8
TokenPose-B [41]	14	5.7	–	74.7	80
TokenPose-L/D24	28	11	89	75.8	80.9
TransPose-H-S [42]	8	10.2	54	73.4	–
TransPose-H-A6	18	21.8	46	75	80.8
HRFormer-BL [43]	43	12.2	23	75.6	80.8
HRFormer-BH	43	26.8	12	77.2	82
ViTPose-B [44]	86	17.1	140	75.8	81.1
ViTPose-L	307	70.0+	61	78.3	83.5
RTMPose-m [45]	–	2.2	–	75.7	–
RTMPose-L	–	4.5	–	76.6	–
DWPose-m [46]	–	2.2	–	76.2	–
DWPose-L	–	4.5	–	77.0	–
HRNet-W32L [10]	29	7.1	136	74.4	79.8
HRNet-W32H	29	16	64	75.8	81
HRNet-W48L	64	14.6	96	75.1	80.4
HRNet-w48H	64	32.9	46	76.3	81.2
GoatPose-BL(ours)	**15.51**	**3.48**	**277**	**74.5**	**82.3**
GoatPose-BH(ours)	15.51	7.84	131	75.6	83.5
GoatPose-XL(ours)	34.76	7.81	180	76.1	83.8
GoatPose-XH(ours)	34.76	17.53	86	76.5	83.9

Ablation Study We empirically study how cheap operation and attention mechanism influence the performance, as shown in Table 3. We constructed NAM-HRNet by incorporating the NAM attention mechanism onto the HRNet architecture. On COCO val, NAM-HRNet demonstrates a significant improvement in AP compared to HRNet, with an increase of 1.5%. The improvement confirms that the weight allocation of the attention mechanism can effectively enhance model performance. Compared to NAM-HRNet, GoatPose-BL achieves an AP score of 74.5 while reducing both computational complexity and parameters by half, which validates the efficiency and effectiveness of cheap operation.

Table 3. Ablation about Cheap Operation and Attention mechanism.

Model	Params	GFLOPs	AP
HRNet	29	7.1	74.4
NAM-HRNet(ours)	35	7.8	76.1 (**up 1.5**)
GoatPose(ours)	**15.51**	**3.48**	74.5

4.4 Deployment

In order to verify the real-time performance of our model, We picked NVIDIA TX2 platform as the testing platform for our model.

After preliminary verification, We first install JetPack on the TX2 and configure the necessary environment. Then we debug the code on a server, train the model, and conduct initial tests. Afterwards, the trained model in the .pth format is converted to a compatible format trt with the TensorRT acceleration module. We use TensorRT to accelerate the converted model and perform inference. Our testing result shown in Tabel 4 demonstrates that GoatPose can achieve excellent performance with a high speed even under limited hardware resources.

Table 4. Application on Jetson TX2.

Heading level	Example	Font size and style	speed(fps)	AP(%)
HRNet-W32L	29	7.1	23	74.4
GoatPose-BL(ours)	15.51	3.48	**51**	74.5

5   Conclusion

By incorporating cheap linear operations, we novelly propose a lightweight deep convolutional network GoatPose, which can reduce the amount of computation and parameters by half while maintaining the same or even slightly higher accuracy compared with the backbone HRNet. Specifically, the lightweight modules are combined with NAM attention machanism, greatly improve the effectiveness of GPU computing. Experiments on COCO dataset shows that our model has significantly improved performance. The successful deployment on NVIDIA Jetson TX2 further demonstrates the superiority and generalizability of our model, paving the way to real-time human pose estimation for edge applications.

References

1. Zhang, X., Zhou, X., Lin, M., Sun, J.: ShuffleNet: an extremely efficient convolutional neural network for mobile devices. In: 2018 IEEE/CVF Conference on Computer Vision and Pattern Recognition, pp. 6848–6856 (2017)
2. Howard, A.G., et al.: MobileNets: efficient convolutional neural networks for mobile vision applications (2017). http://arxiv.org/abs/1704.04861, cite arxiv:1704.04861
3. Howard, A., et al.: Searching for mobilenetv3. In: ICCV, pp. 1314–1324. IEEE (2019)
4. He, Y., Zhang, X., Sun, J.: Channel pruning for accelerating very deep neural networks (2017)

5. Liu, Z., Li, J., Shen, Z., Huang, G., Yan, S., Zhang, C.: Learning efficient convolutional networks through network slimming (2017)
6. Han, S., Mao, H., Dally, W.J.: Deep compression: compressing deep neural networks with pruning, trained quantization and Huffman coding (2016)
7. Lin, J., Rao, Y., Lu, J., Zhou, J.: Runtime neural pruning. In: Proceedings of the 31st International Conference on Neural Information Processing Systems, NIPS 2017, pp. 2178–2188, Red Hook, NY, USA. Curran Associates Inc. (2017)
8. Wen, W., Wu, C., Wang, Y., Chen, Y., Li, H.: Learning structured sparsity in deep neural networks (2016)
9. Han, S., Pool, J., Tran, J., Dally, W.J.: Learning both weights and connections for efficient neural networks. In: Proceedings of the 28th International Conference on Neural Information Processing Systems - Volume 1. NIPS 2015, Cambridge, MA, USA. MIT Press (2015)
10. Sun, K., Xiao, B., Liu, D., Wang, J.: Deep high-resolution representation learning for human pose estimation (2019)
11. Han, K., Wang, Y., Tian, Q., Guo, J., Xu, C., Xu, C.: GhostNet: more features from cheap operations (2020)
12. Han, K., et al.: GhostNets on heterogeneous devices via cheap operations. Int. J. Comput. Vis. **130**(4), 1050–1069 (2022). https://doi.org/10.1007/s11263-022-01575-y, https://doi.org/10.10072Fs11263-022-01575-y
13. Hu, J., Shen, L., Albanie, S., Sun, G., Wu, E.: Squeeze-and-excitation networks (2019)
14. Park, J., Woo, S., Lee, J.Y., Kweon, I.S.: BAM: bottleneck attention module (2018)
15. Woo, S., Park, J., Lee, J.Y., Kweon, I.S.: CBAM: convolutional block attention module (2018)
16. Lin, T.Y., et al.: Microsoft COCO: common objects in context (2015)
17. Newell, A., Yang, K., Deng, J.: Stacked hourglass networks for human pose estimation (2016)
18. Badrinarayanan, V., Handa, A., Cipolla, R.: SegNet: a deep convolutional encoder-decoder architecture for robust semantic pixel-wise labelling (2015)
19. Noh, H., Hong, S., Han, B.: Learning deconvolution network for semantic segmentation (2015)
20. Ronneberger, O., Fischer, P., Brox, T.: U-net: convolutional networks for biomedical image segmentation (2015)
21. Xiao, B., Wu, H., Wei, Y.: Simple baselines for human pose estimation and tracking (2018)
22. Peng, X., Feris, R.S., Wang, X., Metaxas, D.N.: A recurrent encoder-decoder network for sequential face alignment (2016)
23. Yu, F., Koltun, V.: Multi-scale context aggregation by dilated convolutions (2016)
24. Chen, L.C., Papandreou, G., Kokkinos, I., Murphy, K., Yuille, A.L.: Semantic image segmentation with deep convolutional nets and fully connected CRFs (2016)
25. Chen, L.C., Papandreou, G., Kokkinos, I., Murphy, K., Yuille, A.L.: DeepLab: semantic image segmentation with deep convolutional nets, Atrous convolution, and fully connected CRFs (2017)
26. Chen, L.C., Papandreou, G., Schroff, F., Adam, H.: Rethinking Atrous convolution for semantic image segmentation (2017)
27. Chen, L.C., et al.: Searching for efficient multi-scale architectures for dense image prediction (2018)
28. Liu, C., et al.: Auto-deepLab: hierarchical neural architecture search for semantic image segmentation (2019)

29. Yang, T.J., et al.: Deeperlab: single-shot image parser (2019)
30. Cheng, B., et al.: Panoptic-deeplab: a simple, strong, and fast baseline for bottom-up panoptic segmentation (2020)
31. Niu, Y., Wang, A., Wang, X., Wu, S.: Convpose: a modern pure convnet for human pose estimation. Neurocomputing **544**, 126301 (2023). https://doi.org/10.1016/j.neucom.2023.126301
32. Wang, J., et al.: Deep high-resolution representation learning for visual recognition (2020)
33. Sandler, M., Howard, A., Zhu, M., Zhmoginov, A., Chen, L.C.: Mobilenetv 2: inverted residuals and linear bottlenecks (2019)
34. Chollet, F.: Xception: deep learning with depthwise separable convolutions (2017)
35. Howard, A.G., et al.: MobileNets: efficient convolutional neural networks for mobile vision applications (2017)
36. Sun, K., Li, M., Liu, D., Wang, J.: Igcv 3: interleaved low-rank group convolutions for efficient deep neural networks (2018)
37. Tan, M., Le, Q.V.: Mixconv: mixed depthwise convolutional kernels (2019)
38. Neff, C., Sheth, A., Furgurson, S., Tabkhi, H.: EfficienthrNet: efficient scaling for lightweight high-resolution multi-person pose estimation (2020)
39. Liu, Z., Wang, L., Wu, W., Qian, C., Lu, T.: Tam: temporal adaptive module for video recognition (2021)
40. Liu, Y., Shao, Z., Teng, Y., Hoffmann, N.: Nam: normalization-based attention module (2021)
41. Li, Y., et al.: Tokenpose: learning keypoint tokens for human pose estimation (2021)
42. Yi, X., Zhou, Y., Xu, F.: Transpose: real-time 3D human translation and pose estimation with six inertial sensors (2021)
43. Yuan, Y., et al.: Hrformer: high-resolution transformer for dense prediction (2021)
44. Xu, Y., Zhang, J., Zhang, Q., Tao, D.: ViTPose: simple vision transformer baselines for human pose estimation (2022)
45. Jiang, T., et al.: RTMPose: real-time multi-person pose estimation based on MMPose (2023)
46. Yang, Z., Zeng, A., Yuan, C., Li, Y.: Effective whole-body pose estimation with two-stages distillation (2023)

Sign Language Recognition for Low Resource Languages Using Few Shot Learning

Kaveesh Charuka(iD), Sandareka Wickramanayake(✉)(iD),
Thanuja D. Ambegoda(iD), Pasan Madhushan(iD), and Dineth Wijesooriya(iD)

Department of Computer Science and Engineering, University of Moratuwa,
Moratuwa, Sri Lanka
{kaveesh.18,sandarekaw,thanuja,pasan.18,dineth.18}@cse.mrt.ac.lk

Abstract. Sign Language Recognition (SLR) with machine learning is challenging due to the scarcity of data for most low-resource sign languages. Therefore, it is crucial to leverage a few-shot learning strategy for SLR. This research proposes a novel skeleton-based sign language recognition method based on the prototypical network [20] called ProtoSign. Furthermore, we contribute to the field by introducing the first publicly accessible dynamic word-level Sinhala Sign Language (SSL) video dataset comprising 1110 videos over 50 classes. To our knowledge, this is the first publicly available SSL dataset. Our method is evaluated using two low-resource language datasets, including our dataset. The experiments show the results in 95% confidence level for both 5-way and 10-way in 1-shot, 2-shot, and 5-shot settings.

Keywords: Sign Language Recognition · Few Shot Learning · Sign Language Recognition(SLR) · Prototypical Network

1 Introduction

Sign languages constitute the principal communication mechanism for approximately 5% of the hearing-impaired global population, as documented by the World Health Organization [1]. Each linguistic community worldwide utilizes a distinct sign language tailored to its cultural and regional context. Recognizing sign languages, especially those considered 'low-resource' like Sinhala Sign Language [26], presents unique challenges. Each sign language varies by region, with distinct gestures and meanings. This diversity, combined with the dynamic nature of sign languages and the limited availability of comprehensive datasets, makes Sign Language Recognition (SLR) a complex task.

In the field of SLR, numerous studies have been conducted, employing both vision-based [4,14] and contact-based approaches [2,10]. While these methods have shown promising results, they also face several limitations. Contact-based methods, for instance, require the user to wear specialized gloves equipped

B. Luo et al. (Eds.): ICONIP 2023, CCIS 1964, pp. 203–214, 2024.
https://doi.org/10.1007/978-981-99-8141-0_16

with sensors, which can be intrusive and cost-prohibitive. On the other hand, vision-based methods often rely on extensive and rich datasets for training, which are not always available for low-resource sign languages. Few-shot learning is designed to build machine learning models that can understand new classes or tasks with minimal examples or training data [25]. Given the scarcity of sign language data, particularly for low-resource sign languages, few-shot learning could play a crucial role in SLR, enabling the development of robust models that can learn from fewer instances.

In this research, we introduce a new dynamic word-level Sinhala Sign Language dataset called SSL50 and propose a new framework called ProtoSign for low-resource SLR. ProtoSign comprises three core components: skeleton location extraction [4,22], a Transformer Encoder (TE) [23] equipped with a novel composite loss function, and the application of ProtoNet [20] for few-shot classification. The new SSL50 dataset comprises 50 classes with over 1000 sign videos, and it is the first publicly available dynamic Sinhala Sign Language dataset. The first step of our proposed ProtoSign is skeleton location extraction, the basis for data preprocessing. Next, we apply a TE, a deep learning model renowned for capturing intricate patterns and dependencies in data. Further, we introduce a composite loss function that combines triplet and classification loss, improving the whole approach's accuracy. Lastly, we employ ProtoNet for the few-shot classification task, which has state-of-the-art results. Experiment results on the newly introduced dataset and two publicly available datasets, LSA64 [15] and GSLL [21] demonstrate the effectiveness of the proposed ProtoSign framework for low-resource SSL.

2   Related Work

Vision-based Sign Language Recognition(SLR) methods use images or videos of hand gestures to recognize the signs. Vision-based SLR has dramatically improved with the advancement of deep learning. For example, Convolutional Neural Networks [9,13], Long Short-Term Memory Networks [6] and Transformers [5,17], have been used for input encoding in SLR.

However, using deep learning in low-resource SLR is challenging because of limited available data. A possible solution is to employ a few-shot learning approach. For instance, Santoro et al. [16] and Vinyals et al. [24] attempted to solve few-shot classification with end-to-end deep neural networks. Metric-based models are commonly used in meta-learning, one of the main types of few-shot learning approaches. Metric learning uses non-parametric techniques to model sample distance distributions, ensuring proximity between similar samples and maintaining distance between dissimilar ones. Core models embodying this principle are Matching Networks [24], Prototypical Networks [20], and Relation Network [19]. In their work on Prototypical Networks [20], Snell et al. expanded the concept from individual samples to a class-based metric. They grouped the descriptors of all samples from a specific class to establish class prototypes. These prototypes are then used for inference. Artem et al. in [7] introduced a meta-learning-based network for American SLR, which acquires the ability to evaluate

(a) SSL50 dataset

(b) LSA64 dataset

(c) GSLL dataset

Fig. 1. Screenshots of sample videos from SSL50, LSA64 and GSLL Sign Language Recognition datasets.

the similarity between pairs of feature vectors. Nevertheless, using metric-based models for low-resource SLR is not well-explored. In this paper, we explore using Prototypical Networks for low-resource SLR.

3   SSL50 Dataset

This paper introduces a diverse Sinhala Sign Language (SSL) dataset called SSL50 to facilitate recognizing dynamic SSL, addressing the lack of dynamic SSL datasets. SSL50 comprises over 1,000 videos, covering 50 classes of commonly used SSL words. We have ensured the SSL50 contains videos representing the most frequently used words by consulting with sign language professionals during the dataset creation and does not have any closely related signs.

Five signers, four female and one male, contributed videos to our dataset. All contributors were right-handed and between the ages of 21–35. Two signers learned SSL from their families, while the other three studied at educational institutions. To ensure a diverse and natural dataset, we conducted orientation sessions with sign language professionals, providing them with an overview of our work. Each participant was requested to produce five sign videos per class using

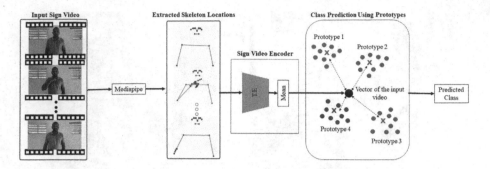

Fig. 2. Overview of the proposed ProtoSign framework

their mobile phones. We encouraged the signers to vary the backgrounds for each video of the same sign, aiming to capture a broader range of real-world scenarios rather than relying on lab-generated datasets. Figure 1a shows screenshots of sample videos in the SSL50 dataset, characterized by its inclusion of natural backgrounds. Conversely, Fig. 1b and 1c showcase the screenshots of sample videos sourced from the LSA64 [15] and GSLL [21] datasets, respectively, created using static backgrounds. Hence, SSL50 better resembles real-world signing practices, incorporating natural variations and promoting inclusivity.

After collecting all the sign videos, we renamed each file in the format of classId_signerId_variantId (e.g., 001_002_001.mp4) to facilitate dataset annotation. Additionally, we converted all the videos into a uniform frame rate of 30fps. The file CSV file contains comprehensive details about the dataset including signer details, gloss details (word, label, word in English), and video details (file name, signer ID, label, duration, fps, video width, video height). The dataset can be downloaded from here

4   Proposed ProtoSign Framework

The overview of the proposed ProtoSign framework is shown in Fig. 2. The Proto-Sign consists of three main steps. First, given the sign video, ProtoSign extracts the skeleton locations of the signer using the MediaPipe model [11]. Second, the extracted skeleton locations are sent to the transformer encoder to obtain a vector representing the input sign video. Finally, following ProtoNet [20], ProtoSign compares the obtained vector with the prototypes of different sign classes to determine the class of the given sign video.

4.1   Skeleton Locations Extraction

In the ProtoSign framework, the first step involves identifying the location of the skeleton in each video frame, including the face, body, and hand landmarks. The process of skeleton extraction is illustrated in Fig. 3. To ensure accuracy, we use the YOLOv4 framework [3] to detect the person in the video. This involves

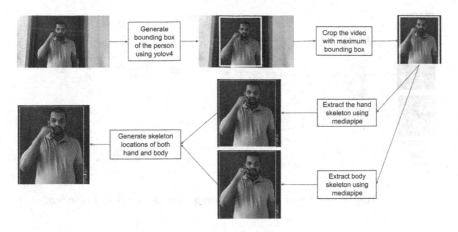

Fig. 3. The process pipeline for extracting and refining skeleton locations.

calculating the coordinates of the bounding box for each frame, which helps us determine the maximum bounding box that includes the person in each frame. Once we have the bounding box, we crop each frame using it. This step is essential in addressing the variability in the distance between the camera and the person. In real-world scenarios, we cannot expect the signer to be at a specific distance from the camera. By isolating the person, we eliminate the effects of distance and potential interference from other objects in the video background.

Next, we use a standard pose estimation algorithm from = MediaPipe [11] to extract skeleton locations. The algorithm utilizes two models: one for the hands and another for the whole body, resulting in a comprehensive extraction of skeleton locations. We then employ a refinement phase to remove any irrelevant locations. This method yields 57 skeleton locations, including 21 per hand, 4 for the body, and 11 for the face. Excluding the face locations, the remaining points represent the body's joints.

4.2 Sign Video Encoder

ProtoSign adapts Prototypical Networks (ProtoNet) [20] to deal with low-resource sign languages. ProtoNet makes the predictions based on prototype representations of each class. To create prototypes of classes, we develop a Sign Video Encoder based on a Transformer Encoder (TE). We use a modified version of the Transformer model [23] with a classification head as the final layer. This TE is first trained using a large sign language dataset, and in our implementation, we use the LSA64 dataset [15]. Next, we finetune the TE following the method used in [20] to create more discriminative prototypes for each sign class. In this section, we first describe how we pre-train the TE using a large sign language model and then detail how we fine-tune it.

Given a sign video v of class y, the input to the transformer is a sequence of normalized skeleton locations x extracted in the previous step, $x \in \mathcal{R}^{n \times d}$

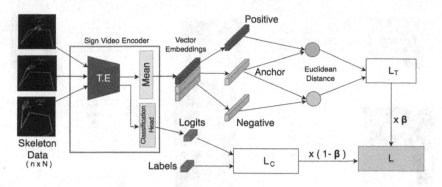

Fig. 4. Training phase of the Transformer Encoder used in ProtoSign

where n is the sequence length and d is the number of features. x is sent through positional encoding, self-attention, and feed-forward layers, mirroring the process in the original TE. Suppose the output of TE is $z \in \mathcal{R}^{n \times \bar{d}}$ where $\bar{d}$ is the output embedding size of the TE. We take the mean of z along the sequence length dimension to get the vector embedding of the input v_x:

$$v_x = \frac{1}{n} \sum_{z_i \in z} z_i \tag{1}$$

Finally, the classification head, another linear layer, is applied to v_x to get the predicted class $\bar{y}$.

We employ a composite objective function of triplet loss and classification loss to train the transformer encoder to generate discriminative vector representations for different sign classes. The classification loss forces the transformer encoder to learn discriminative features for each class. The triplet loss further enhances the discriminativeness of learned feature vectors by forcing higher intra-class and lower inter-class similarities. The training phase of the transformer encoder used in ProtoSign is shown in Fig. 4.

Suppose a positive sample of x, which shares the same label, is denoted by x_p, and a negative sample with a different label is denoted by x_n. Let the output vector embedding of the transformer encoder for x_p and x_n be v_p and v_n, respectively. Then the triplet loss L_T is defined as

$$L_T = \|v_x - v_p\|_2^2 - \|v_x - v_n\|_2^2 + \alpha \tag{2}$$

Here, α is a margin enforced between positive and negative pairs. Through hyperparameter tuning, we set α to 2.0. We adopted an online triplet selection strategy in [18]. Although computationally intensive, the online strategy enhances robustness, expedites convergence, and performs better.

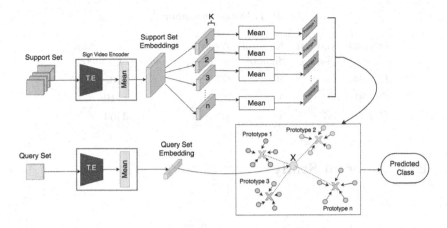

Fig. 5. The Prototypical Network Architecture.

We supplement the Triplet Loss with Classification Loss to provide comprehensive supervision to the TE during training. The Classification Loss aids the model in making accurate classification decisions for individual examples, thus facilitating the distinction between different classes. This, in turn, eases the task of Triplet Loss in refining the relative distances between classes, culminating in enhanced performance. The classification objective of the transformer encoder, L_C, is defined using the multi-class Classification Loss.

$$L_C = -\log P(\bar{y}|v_x) \tag{3}$$

The final objective function of the transformer encoder, L, is

$$L = \beta * L_T + (1 - \beta) * L_C, \tag{4}$$

where β is a hyper-parameter and in our experiment we set *beta* to be 0.9 through parameter tuning.

Next, following [20], we use episodic training to fine-tune the trained TE. Each episode consists of a support set of N samples $S = \{(x_1, y_1), ..., (x_N, y_N)\}$, where x_i is the sample video, $y_i \in \{1, ..., K\}$ is the corresponding label and K denotes the classes randomly selected from the training set comprising C classes ($|C| > |K|$). Suppose S_k is the set of samples belonging to class $k \in K$.

We calculate the prototype for each class $k \in K$, p_k, by taking the average of the embeddings produced TE for S_k. Given a query point ($\bar{x} \in Q$, we obtain its vector representation from TE, $v_{\bar{x}}$ and calculate the Euclidean distance between $v_{\bar{x}}$ and each p_k. To determine the class of $\bar{x}$, we apply softmax over the calculated distances, producing a probability distribution over the classes. Figure 5 shows the fine-tuning phase of ProtoSign following the ProtoNet.

The same approach is employed during the inference, but data is randomly selected from the testing set instead.

Table 1. Datasets Summary

Dataset	Num. of Classes	Num. of Signers	Average Num. of Videos per Class	Total Num. of Videos
LSA64	64	10	50	3200
SSL50	50	5	22	1110
GSLL	347	2	10	3464

5 Experimental Study

5.1 Datasets

In addition to the newly introduced SSL50 dataset, our experiments use LSA64 [15] and GSLL [21] datasets.

- **LSA64:** The LSA64 dataset [15] is a comprehensive database developed for Argentinian Sign Language (LSA). The dataset comprises 3200 videos featuring 64 unique signs performed by ten non-expert, right-handed subjects five times each. The chosen signs, a mix of common verbs and nouns, were recorded in two separate sessions under distinct lighting conditions - outdoors with natural light and indoors with artificial light, providing variety in illumination across the videos.
- **GSLL:** The Greek Sign Language (GSLL) dataset [21] comprises 3,464 videos encapsulating a total of 161,050 frames, with each video representing one of 347 distinct sign classes. Two signers perform these signs, repeating each sign 5–17 times to offer variations.

A summary of the datasets used in our experiments is given in Table 1.

5.2 Implementation Details

We implement ProtoSign using PyTorch framework [12], and it is trained on an NVIDIA Tesla T4 GPU or an NVIDIA GeForce RTX 2040 GPU. We use the ADAM optimizer [8] for training.

We first train the TE in ProtoSign using the LSA64 dataset. We performed a grid search for hyperparameter tuning to optimize the overall performance. We set the number of attention heads to 16, batch size to 32 and learning rate to 0.002 and trained the model for 70 epochs. Further, we experimented with different values for β and set it to 0.9, which gives the best results.

Next, use ProtoSign for few-shot classification of signs by finetuning TE on each low-resource language dataset, SSL50 and GSLL. Here, for each 1-shot, 2-shot and 5-shot scenario for a given low-resource dataset, we create train, validation and test datasets separately. For example, let's consider the SSL50 dataset under the 1-shot scenario. We select two instances from each of the 50 classes for training and the remaining for testing. One sample is assigned as the

Table 2. Few-shot classification accuracies of ProtoSign on SSL and GSLL datasets

	5 way			10 way		
	1 shot	2 shot	5 shot	1 shot	2 shot	5 shot
SSL50	61.2%	81.66%	93.08%	42.75%	69.24%	87.47%
GSLL	73.20%	84.38%	–	62.5%	79.65%	–

Table 3. Few-shot learning accuracies of ProtoSign for different scenarios on SSL50 dataset

Model	5 way			10 way		
	1 shot	2 shot	5 shot	1 shot	2 shot	5 shot
ProtoSign - CL	53.2%	74.3%	87.47%	37.7%	59.8%	84.06%
ProtoSign - TL	57.80%	77.47%	92.47%	41.45%	63.78%	86.73%
Matching Networks	**63.4%**	78.4%	88.45%	**46.87%**	69.09%	83.34%
VE + ProtoNet	41%	43.6%	41.22%	21%	22%	22.25%
ProtoSign	61.21%	**81.66%**	**93.08%**	42.75%	**69.24%**	**87.47%**

support set of two chosen for training, whereas one is assigned as the query set. In each episode, we randomly select 5 (in the 5-way scenario) or 10 (in the 10-way scenario) classes from the training. Each epoch comprises 1000 such episodes. The code is available at https://github.com/ProtoSign

5.3 Experimental Results

Table 2 shows the few-shot classification accuracies of ProtoSign on SSL50 and GSLL datasets under different settings.

To show the effectiveness of our approach, we conducted four main ablation studies on the SSL50 and GSLL datasets. The final model we proposed integrates a combination of classification loss and triplet loss, including an extra classification layer nested within the Transformer Encoder. In the ablation studies, we evaluated the performance of our model by training only using Triplet Loss (TL) or classification loss (CL). Further, we considered replacing the ProtoNet with Matching Networks, allowing for a comparative analysis between these approaches. Furthermore, we assess the performance of ProtoSign when using a Video Encoder (VE) directly instead of skeleton location extractions followed by a Transformer Encoder. In our experiments, we use the r3d-18 video encoder for this evaluation.

Table 3 shows the performance of variants of ProtoSign for various conditions using the SSL50 dataset. We observe that ProtoSign has achieved the best performance among different variants for all the scenarios except the 1-shot scenario, whereas replacing ProtoNet with Matching Networks has yielded the best results in the 1-shot scenario. Table 4 displays the few-shot learning accuracies of variants of ProtoSign for the GSLL dataset. Here, our ProtoSign model

Table 4. N-way k-shot accuracies for different scenarios on GSLL dataset

Model	5 way		10 way	
	1 shot	2 shot	1 shot	2 shot
ProtoSign - CL	66%	74.3%	37.7%	59.8%
ProtoSign - TL	70%	82.02%	60.10%	74.37%
Matching Networks	70.19%	82.78%	61.8%	75.61%
VE + ProtoNet	40.64%	46.45%	36.65%	43.12%
ProtoSign	**73.20%**	**84.38%**	**62.5%**	**79.65%**

surpasses all other variants in all the scenarios. The results of these ablations studies demonstrate the effectiveness of each component of the proposed Proto-Sign framework.

6   Conclusion

This paper presents a new architecture for few-shot learning in low-resource languages called ProtoSign, along with a dynamic dataset of Sinhala Sign Language at the word level called SSL50. The ProtoSign architecture consists of three main steps. Firstly, it extracts the skeleton locations of the signer from the sign video. Secondly, the extracted skeleton locations are sent to the transformer encoder to obtain a vector representing the input sign video. Finally, ProtoSign compares the obtained vector with prototypes of different sign classes to determine the class of the given sign video. The proposed framework's effectiveness is demonstrated through experimental results on two low-resource sign language datasets, the newly introduced SSL50 and GSLL.

References

1. Deafness and hearing loss (2023). https://www.who.int/news-room/fact-sheets/detail/deafness-and-hearing-loss. Accessed 29 May 2023
2. Amin, M.S., Rizvi, S.T.H., Hossain, M.M.: A comparative review on applications of different sensors for sign language recognition. J. Imaging **8**(4), 98 (2022)
3. Bochkovskiy, A., Wang, C.Y., Liao, H.Y.M.: Yolov4: optimal speed and accuracy of object detection. arXiv preprint arXiv:2004.10934 (2020)
4. Boháček, M., Hrúz, M.: Sign pose-based transformer for word-level sign language recognition. In: Proceedings of the IEEE/CVF Winter Conference on Applications of Computer Vision, pp. 182–191 (2022)
5. Camgoz, N.C., Koller, O., Hadfield, S., Bowden, R.: Multi-channel transformers for multi-articulatory sign language translation. In: Bartoli, A., Fusiello, A. (eds.) ECCV 2020. LNCS, vol. 12538, pp. 301–319. Springer, Cham (2020). https://doi.org/10.1007/978-3-030-66823-5_18

6. Cui, R., Liu, H., Zhang, C.: Recurrent convolutional neural networks for continuous sign language recognition by staged optimization. In: Proceedings of the IEEE Conference on Computer Vision and Pattern Recognition, pp. 7361–7369 (2017)
7. Izutov, E.: ASL recognition with metric-learning based lightweight network. arXiv preprint arXiv:2004.05054 (2020)
8. Kingma, D.P., Ba, J.: Adam: a method for stochastic optimization. In: andYann LeCun, Y.B. (ed.) Proceedings of the 3rd International Conference on Learning Representations, ICLR (2015)
9. Koller, O., Zargaran, O., Ney, H., Bowden, R.: Deep sign: hybrid CNN-hmm for continuous sign language recognition. In: Proceedings of the British Machine Vision Conference 2016 (2016)
10. Lee, B.G., Lee, S.M.: Smart wearable hand device for sign language interpretation system with sensors fusion. IEEE Sens. J. **18**(3), 1224–1232 (2017)
11. Lugaresi, C., et al.: MediaPipe: a framework for building perception pipelines. arXiv preprint arXiv:1906.08172 (2019)
12. Paszke, A., et al.: Pytorch: an imperative style, high-performance deep learning library. In: Wallach, H., Larochelle, H., Beygelzimer, A., d' Alché-Buc, F., Fox, E., Garnett, R. (eds.) Proceedings of the Advances in Neural Information Processing Systems, pp. 8024–8035 (2019)
13. Rao, G.A., Syamala, K., Kishore, P., Sastry, A.: Deep convolutional neural networks for sign language recognition. In: 2018 Conference on Signal Processing and Communication Engineering Systems (SPACES), pp. 194–197. IEEE (2018)
14. Rastgoo, R., Kiani, K., Escalera, S.: Hand sign language recognition using multi-view hand skeleton. Expert Syst. Appl. **150**, 113336 (2020)
15. Ronchetti, F., Quiroga, F., Estrebou, C.A., Lanzarini, L.C., Rosete, A.: Lsa64: an argentinian sign language dataset. In: XXII Congreso Argentino de Ciencias de la Computación (CACIC 2016). (2016)
16. Santoro, A., Bartunov, S., Botvinick, M., Wierstra, D., Lillicrap, T.: Meta-learning with memory-augmented neural networks. In: International Conference on Machine Learning, pp. 1842–1850. PMLR (2016)
17. Saunders, B., Camgoz, N.C., Bowden, R.: Continuous 3D multi-channel sign language production via progressive transformers and mixture density networks. Int. J. Comput. Vision **129**(7), 2113–2135 (2021)
18. Schroff, F., Kalenichenko, D., Philbin, J.: FaceNet: a unified embedding for face recognition and clustering. In: Proceedings of the IEEE Conference on Computer Vision and Pattern Recognition, pp. 815–823 (2015)
19. Si, J., et al.: Dual attention matching network for context-aware feature sequence based person re-identification. In: Proceedings of the IEEE Conference on Computer Vision and Pattern Recognition, pp. 5363–5372 (2018)
20. Snell, J., Swersky, K., Zemel, R.: Prototypical networks for few-shot learning. In: Advances in Neural Information Processing Systems, vol. 30 (2017)
21. Theodorakis, S., Pitsikalis, V., Maragos, P.: Dynamic-static unsupervised sequentiality, statistical subunits and lexicon for sign language recognition. Image Vis. Comput. **32**(8), 533–549 (2014)
22. Tunga, A., Nuthalapati, S.V., Wachs, J.: Pose-based sign language recognition using GCN and BERT. In: Proceedings of the IEEE/CVF Winter Conference on Applications of Computer Vision, pp. 31–40 (2021)
23. Vaswani, A., et al.: Attention is all you need. In: Advances in Neural Information Processing Systems, vol. 30 (2017)

24. Vinyals, O., Blundell, C., Lillicrap, T., Wierstra, D., et al.: Matching networks for one shot learning. In: Advances in Neural Information Processing Systems, vol. 29 (2016)
25. Wang, Y., Yao, Q., Kwok, J.T., Ni, L.M.: Generalizing from a few examples: a survey on few-shot learning. ACM Comput. Surv. **53**(3), 1–34 (2020)
26. Weerasooriya, A.A., Ambegoda, T.D.: Sinhala fingerspelling sign language recognition with computer vision. In: 2022 Moratuwa Engineering Research Conference (MERCon), pp. 1–6. IEEE (2022)

T Cell Receptor Protein Sequences and Sparse Coding: A Novel Approach to Cancer Classification

Zahra Tayebi, Sarwan Ali, Prakash Chourasia, Taslim Murad, and Murray Patterson(✉)

Georgia State University, Atlanta, GA 30302, USA
mpatterson30@gsu.edu

Abstract. Cancer is a complex disease marked by uncontrolled cell growth, potentially leading to tumors and metastases. Identifying cancer types is crucial for treatment decisions and patient outcomes. T Cell receptors (TCRs) are vital proteins in adaptive immunity, specifically recognizing antigens and playing a pivotal role in immune responses, including against cancer. TCR diversity makes them promising for targeting cancer cells, aided by advanced sequencing revealing potent anti-cancer TCRs and TCR-based therapies. Effectively analyzing these complex biomolecules necessitates representation and capturing their structural and functional essence. We explore sparse coding for multi-classifying TCR protein sequences with cancer categories as targets. Sparse coding, a machine learning technique, represents data with informative features, capturing intricate amino acid relationships and subtle sequence patterns. We compute TCR sequence k-mers, applying sparse coding to extract key features. Domain knowledge integration improves predictive embeddings, incorporating cancer properties like Human leukocyte antigen (HLA) types, gene mutations, clinical traits, immunological features, and epigenetic changes. Our embedding method, applied to a TCR benchmark dataset, significantly outperforms baselines, achieving 99.8% accuracy. Our study underscores sparse coding's potential in dissecting TCR protein sequences in cancer research.

Keywords: Cancer classification · TCR sequence · embeddings

1 Introduction

T Cell receptors (TCRs) play a crucial role in the immune response by recognizing and binding to antigens presented by major histocompatibility complexes (MHCs) on the surface of infected or cancerous cells, Fig. 1 [44]. The specificity of TCRs for antigens is determined by the sequence of amino acids that make up the receptor, which is generated through a process of genetic recombination

---

Z. Tayebi and S. Ali—Equal Contribution.

and somatic mutation. This enables T Cells to produce a diverse repertoire of receptors capable of recognizing a wide range of antigens [12].

MHC molecules bind to peptide fragments that originate from pathogens and present them on the cell surface. This allows for recognition by the corresponding T Cells [24]. TCR sequencing involves analyzing the DNA or RNA sequences that code for the TCR protein on T Cells, and it can be used to identify changes in the TCR repertoire that occur in response to cancer, as well as to identify specific TCR sequences that are associated with particular types of cancer [27].

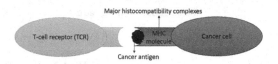

Fig. 1. T Cells mount a targeted immune response against the invading pathogen of cancerous cells.

Embedding-based methods, exemplified by Yang et al. [51], transform protein sequences into low-dimensional vector representations. They find applications in classification and clustering [11]. These embeddings empower classifiers to predict functional classes for unlabeled sequences [23] and identify related sequences with shared features [51].

Embedding-based methods in protein sequence analysis face challenges related to generalizability and complexity [35]. Deep learning techniques, including Convolutional Neural Networks (CNNs), Recurrent Neural Networks (RNNs), and transformer networks, offer promising solutions [37]. These models, trained on extensive and diverse datasets, enhance generalization and enable accurate predictions for new sequences. However, factors like data quality, model architecture, and optimization strategies significantly influence performance [36]. Furthermore, the interpretability of deep learning models remains a challenge, necessitating careful evaluation. While deep learning enhances generalizability, it may not completely resolve the problem.

In our study, we propose a multi-class classification approach for predicting cancer categories from TCR protein sequences. We employ sparse coding [38], a machine-learning technique using a dictionary of sparse basis functions. By representing TCR protein sequences as sparse linear combinations of these basis functions, we capture inherent data structure and variation, improving cancer classification accuracy. Preprocessing involves encoding amino acid sequences as numerical vectors and utilizing a k-mer dictionary for sparse coding.

We assess our approach using multiple metrics and benchmark it against state-of-the-art methods. Our results clearly show its superior classification performance and highlight its key properties: invariance, noise robustness, interpretability, transferability, and flexibility.

In general, our contributions to this paper are the following:

1. We introduce an efficient method using sparse coding and k-mers to create an alignment-free, fixed-length numerical representation for T Cell sequences. Unlike traditional one-hot encoding, we initially extract k-mers and then apply one-hot encoding, preserving context through a sliding window approach.
2. Leveraging domain knowledge, we enhance supervised analysis by combining it with sparse coding-based representations. This fusion enriches the embeddings with domain-specific information, preserving sequence details, facilitating feature selection, promoting interpretability, and potentially boosting classification accuracy.
3. Our proposed embedding method achieves near-perfect predictive performance, significantly surpassing all baselines in classification accuracy. This highlights the effectiveness of our approach in distinguishing T Cell sequences despite their short length.

Considering this, we will proceed as follows: Section 2 covers related work, Sect. 3 presents our approach, Sect. 4 details the dataset and experimental setup, Sect. 5 reports results, and Sect. 6 concludes the paper.

2 Related Work

TCR (T Cell receptor) sequencing data has emerged as a critical asset in biomedical research, revolutionizing our understanding of the immune system and its applications in fields such as immunotherapy and disease response. This technology has enabled the identification of neoantigens, which are crucial targets for immunotherapy in cancer treatment [5]. Moreover, it has played a pivotal role in decoding the adaptive immune response to viral diseases like the formidable SARS-CoV-2 [17]. TCR sequencing has also contributed significantly to the identification of T Cell clones associated with tumor regression in response to immunotherapy [33].

In the realm of biological data analysis, classification methods have proven invaluable for deciphering protein sequences and their implications in cancer research. Techniques like Support Vector Machines (SVM), Random Forest, and Logistic Regression have been instrumental in grouping protein sequences with shared features, aiding in the discovery of subpopulations related to specific cancer types [21]. For instance, SVM was employed to classify TCR sequences for predicting disease-free and overall survival in breast cancer patients [4]. K-means classification has demonstrated promise in distinguishing colorectal cancer patients from healthy individuals based on protein sequence analysis [19].

Deep learning methods have ushered in a new era of protein property prediction, encompassing structure, function, stability, and interactions [49]. Convolutional neural networks (CNNs) have proven effective in forecasting protein structure and function from amino acid sequences [6]. Recurrent neural networks (RNNs) excel at modeling temporal dependencies and long-range interactions within protein sequences [52]. The Universal Representation of Amino Acid Sequences (UniRep) employs an RNN architecture to embed protein sequences, yielding promising results across various biological data analysis tasks [3]. Innovative methods like ProtVec use word embedding techniques to create distributed representations of protein sequences for function prediction [39]. SeqVec introduces an alternative by embedding protein sequences using a hierarchical representation that captures both local and global features [20].

Lastly, ESM (Evolutionary Scale Modeling) employs a transformer-based architecture to encode protein sequences, forming part of the ESM family of protein models [31]. It is trained on extensive data encompassing protein sequences and structures to predict the 3D structure of a given protein sequence [22]. While deep learning enriches the generalizability of embedding-based techniques for protein sequence analysis, robust evaluation and consideration are essential to ensure the reliability of outcomes [31].

3   Proposed Approach

In this section, we delve into the utilization of cancer-related knowledge, introduce our algorithm, and offer an overview through a flowchart.

3.1   Incorporating Domain Knowledge

This section gives some examples of the additional property values for the four cancers we mentioned. Many factors can increase the risk of developing cancer, including Human leukocyte antigen (HLA) types, gene mutations, clinical characteristics, immunological features, and epigenetic modifications.

HLA Types: HLA genes encode cell surface proteins presenting antigens to T Cells. Mutations in these genes due to cancer can hinder antigen presentation, enabling unchecked cancer growth [43].

Gene Mutations: Besides HLA, mutations in genes like TP53 and BRCA1/2 raise cancer risks (breast, ovarian, colorectal) too [40].

Clinical Characteristics: Age, gender, and family history influence cancer risk [29]. Certain cancers are age-specific, some gender-biased, while family history also shapes susceptibility.

Immunological Features: The immune system crucially impacts cancer outcomes. Immune cells recognize and eliminate cancer cells [13]. Yet, cancer cells might evade immune surveillance, growing unchecked. Immune cell presence and activity within tumors influence cancer progression and treatment response [18].

Epigenetic Modifications: Epigenetic changes, like DNA methylation, contribute to cancer development and progression. DNA methylation, adding a methyl group to DNA's cytosine base (CpG sites), leads to gene silencing or altered expression [26].

3.2   Cancer Types and Immunogenetics

Cancer types exhibit intricate connections between immunogenetics, clinical attributes, and genetic factors. Breast cancer, a prevalent malignancy, demonstrates HLA associations such as HLA-A2, HLA-B7, and HLA-DRB1*15:01 alleles, alongside gene mutations (BRCA1, BRCA2, TP53, PIK3CA) that elevate risk [25,30]. Clinical attributes encompass tumor size, grade, hormone receptor presence, and HER2 status. Tumor-infiltrating lymphocytes (TILs) and immune checkpoint molecules (PD-1, PD-L1, CTLA-4) play a substantial role in influencing breast cancer outcomes [32,47]. Similarly, colorectal cancer (CRC), originating in the colon or rectum, shows associations with HLA alleles (HLA-A11, HLA-B44) and gene mutations (APC, KRAS, TP53, BRAF) that impact its development [14,16]. The intricacies of CRC involve tumor attributes, lymph node involvement, TILs, and immune checkpoint molecules, all influencing disease progression [42]. Liver cancer (hepatocellular carcinoma), primarily affecting liver cells, displays connections with HLA alleles (HLA-A2, HLA-B35) and gene mutations (TP53, CTNNB1, AXIN1, ARID1A) [8,34]. Epigenetic DNA methylation abnormalities in CDKN2A, MGMT, and GSTP1 genes contribute to liver cancer's development [53]. Urothelial cancer, affecting bladder, ureters, and renal pelvis cells, is associated with HLA types (HLA-A2, HLA-B7) and gene mutations (FGFR3, TP53, RB1, PIK3CA) [9].

3.3   Algorithmic Pseudocode and Flowchart

The pseudocode to compute sparse coding + k-mers-based embedding is given in Algorithm 1. This algorithm takes a set of sequences $S = \{s_1, s_2, \ldots, N\}$ and k, where N is the number of sequences and k is the length of k-mers. The algorithm computes the sparse embedding by iterating over all sequences and computing the set of k-mers for each sequence; see Fig. 2 (1-b). Then it iterates over all the k-mers, computes the one-hot encoding (OHE) based representation for each amino acid within a k-mer (Fig. 2 (1-c)), and concatenates it with the OHE embeddings of other amino acids within the k-mer (Fig. 2 (1-d)). Finally, the OHE embeddings for all k-mers within a sequence are concatenated to get the final sparse coding-based representation (Fig. 2 (1-e)). To avoid the curse of dimensionality, we used Lasso regression as a dimensionality reduction technique [41], (Fig. 2 (f)). The objective function, which we used in lasso regression, is the following: min(Sum of square residuals + $\alpha \times |slope|$).

In the above equation, the $\alpha \times |slope|$ is referred to as the penalty terms, which reduces the slope of insignificant features to zero. We use $k = 4$ for experiments, which is decided using the standard validation set approach.

The same process repeated while we considered domain knowledge about cancer properties like HLA types, gene mutations, clinical characteristics, immunological features, and epigenetic modifications. For each property, we generate a one-hot encoding-based representation, where the length of the vector equals the total number of possible property values, see Fig. 2 (2-b). All OHE representations of the property values are concatenated in the end to get final representations for all properties we consider from domain knowledge; see Fig. 2 (2-c) to (2-d). The resultant embeddings from Sparse coding (from Algorithm 1) and domain knowledge are concatenated to get the final embedding. These embeddings are used to train machine-learning models for supervised analysis (Fig. 2 (g)).

Algorithm 1. Sparse Coding Algorithm

```
Input: set of sequences S, k-mers length k
Output: SparseEmbedding
1:  totValues = 21                                          ▷ unique amino acid characters
2:  final_sparse_embedding ← []
3:  for i ← 0 to ⌈|S|⌉ do
4:      seq ← S[i]
5:      kmers ← GENERATEKMERS(seq, k)                        ▷ generate set of k-mers
6:      encoded_kmers ← []
7:      for kmer in kmers do
8:          encodedVec ← np.zeros(totValues^k)               ▷ 21³ = 9261 dimensional vector
9:          for i, aa in enumerate(kmer) do
10:             pos ← i × totValues^{k-i-1}
11:             encodedVec[pos:pos+totValues] ← ONEHOTENCODING(aa)
12:         end for
13:         encoded_kmers.append(encodedVec)
14:     end for
15:     final_sparse_embedding.append(np.array(encoded_kmers).flatten())
16: end for
17: SparseEmbedding ← LASSOREGRESSION(final_sparse_embedding)  ▷ dim. reduction
18: return SparseEmbedding
```

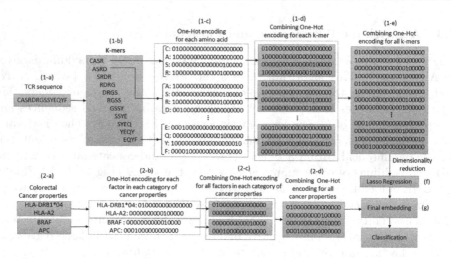

Fig. 2. Flowchart of TCR sequence analysis

4   Experimental Setup

In this section, we describe the details related to the dataset, baseline models, evaluation metrics, and data visualization. We employ a variety of machine learning classifiers, encompassing Support Vector Machine (SVM), Naive Bayes (NB), Multi-Layer Perceptron (MLP), K-Nearest Neighbors (KNN) using K = 3, Random Forest (RF), Logistic Regression (LR), and Decision Tree (DT). We used a 70–30% train-test split based on stratified sampling to do classification. From the training set, we use 10% data as a validation set for hyperparameter tuning. Performance evaluation of the classifiers employed various evaluation metrics. The results are computed 5 times, and the average results are reported.

The experiments were performed on a system with an Intel(R) Core i5 processor @ 2.10 GHz and a Windows 10 64 bit operating system with 32 GB of memory. The model is implemented using Python, which is available online[1].

4.1 Dataset Statistics

We source sequence data from TCRdb, a comprehensive T Cell receptor sequence database with robust search functionality [10]. TCRdb consolidates 130+ projects and 8000+ samples, comprising an extensive sequence repository. Our focus identifies four prevalent cancer types by incidence rates. Extracting 23331 TCR sequences, we maintain the original data proportion via Stratified ShuffleSplit. Table 1a offers dataset statistics. Our embedding approach employs TCR sequences, cancer names, and cancer properties. Table 1b exemplifies TCR sequences, cancer names, and gene mutations specific to each cancer type. This includes 4 cancers with respective gene mutations that impact risk.

Table 1. (a) Dataset Statistics. (b) Examples of sequences for different cancer types, including Breast, Colorectal, Liver, and Urothelial, along with their respective gene mutations.

Cancer Name	Number of Sequences	Sequence Length Statistics		
		Min.	Max.	Average
Breast	4363	8	20	14.2264
Colorectal	10947	7	26	14.5573
Liver	3520	8	20	14.3005
Urothelial	4501	7	24	14.6538
Total	23331	-	-	-

(a)

Sequence	Cancer Name	Gene Mutation
CASSRGQYEQYF	Breast	BRCA1, BRCA2, TP53, PIK3CA
CASSLEAGRAYEQYF	Colorectal	APC, KRAS, TP53, BRAF
CASSLGSGQETQYF	Liver	TP53, CTNNB1, AXIN1
CASSGQGSSNSPLHF	Urothelial	FGFR3,TP53, RB1, PIK3CA

(b)

To assess the effectiveness of our proposed system, we employed a variety of baseline methods, categorized as follows: (i) Feature Engineering methods: (one-hot encoding (OHE) [28], Spike2Vec [1], PWM2Vec [2], and Spaced k-mers [46]), (ii) Kernel method (String kernel [15]), (iii) Neural network (Wasserstein Distance Guided Representation Learning (WDGRL) [45] and AutoEncoder [50]), (iv) Pre-trained Larger Language Model (SeqVec [20]), and (v) Pre-trained Transformer (ProteinBert [7]).

4.2 Data Visualization

To see if there is any natural (hidden) clustering in the data, we use t-distributed stochastic neighbor embedding (t-SNE) [48], which maps input sequences to 2D representation. The t-SNE plots for different embedding methods are shown in Fig. 3 for one-hot encoding (OHE), Spike2Vec, PWM2Vec, Spaced K-mer,

---

[1] https://github.com/zara77/T-Cell-Receptor-Protein-Sequences-and-Sparse-Coding.git.

Autoencoder, and Sparse Coding, respectively. We can generally observe that OHE, Spike2Vec, PWM2Vec, and Autoencoder show smaller groups for different cancer types. However, the Spaced k-mers show a very scattered representation of data. Moreover, the proposed Sparse Coding approach does not show any scattered representation, preserving the overall structure better.

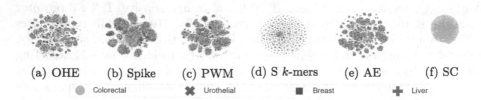

(a) OHE   (b) Spike   (c) PWM   (d) S k-mers   (e) AE   (f) SC

● Colorectal   ✖ Urothelial   ■ Breast   ✚ Liver

Fig. 3. t-SNE plots for different feature embedding methods. The figure best seen in color. Subfigures (a), (b), (c), (d), (e), and (f) are for OHE, Spike2Vec, PWM2Vec, Spaced kmers, Autoencoder (AE), and Sparse Coding (SC), respectively. (Color figure online)

5   Results and Discussion

In this section, we present the classification results obtained by our proposed system and compare them with various machine learning models. The results for different evaluation metrics are summarized in Table 2. This study introduces two novel approaches. First, we employ sparse coding (utilizing k-mers) for T Cell sequence classification. Second, we integrate domain knowledge into the embeddings. To our knowledge, these two methods have not been explored in the context of T Cells in existing literature. Consequently, we exclusively report results for the Sparse Coding method combined with domain knowledge embeddings. Our proposed technique (Sparse Coding) consistently outperforms all feature-engineering-based baseline models (OHE, Spike2Vec, PWM2Vec, Spaced k-mers) across every evaluation metric, except for training runtime. For example, Sparse Coding achieves 48.5% higher accuracy than OHE, 55.8% more than Spike2Vec, 58.4% more than PWM2Vec, and 48.6% more than Spaced k-mers when employing the LR classifier. This suggests that Sparse Coding-based feature embeddings retain biological sequence information more effectively for classification compared to feature-engineering-based baselines. Similarly, our technique outperforms the neural network-based baselines (WDGRL, Autoencoder). For instance, the WDGRL approach exhibits 53.03% lower accuracy than Sparse Coding, while Autoencoder attains 53.83% lower accuracy than Sparse Coding when using the SVM classifier. These results imply that feature vectors generated by neural network-based techniques are less efficient for classification compared to those produced by Sparse Coding. It's worth noting that the SVM consistently delivers the best results among all classifiers, including WDGRL

Table 2. Classification results (averaged over 5 runs) for different evaluation metrics. The best values are shown in bold.

Embeddings	Algo.	Acc. ↑	Prec. ↑	Recall ↑	F1 (Weig.) ↑	F1 (Macro) ↑	ROC AUC ↑	Train Time (sec.) ↓
OHE	SVM	0.5101	0.5224	0.5101	0.4073	0.3152	0.5592	790.5622
	NB	0.2036	0.3533	0.2036	0.0917	0.1213	0.5107	1.0560
	MLP	0.4651	0.4368	0.4651	0.4370	0.3714	0.5764	221.0638
	KNN	0.4464	0.4044	0.4464	0.4100	0.3354	0.5617	7.2748
	RF	0.5156	0.5003	0.5156	0.4521	0.3751	0.5824	18.7857
	LR	0.5143	0.5241	0.5143	0.4327	0.3492	0.5701	61.8512
	DT	0.4199	0.4129	0.4199	0.4160	0.3616	0.5737	1.0607
Spike2Vec	SVM	0.4309	0.4105	0.4309	0.4157	0.3543	0.5711	19241.67
	NB	0.2174	0.3535	0.2174	0.1931	0.2081	0.5221	6.1309
	MLP	0.4191	0.4081	0.4191	0.4128	0.3490	0.5671	227.2146
	KNN	0.4397	0.4105	0.4397	0.4087	0.3400	0.5673	40.4765
	RF	0.5183	0.5078	0.5183	0.4519	0.3738	0.5836	138.6850
	LR	0.4404	0.4189	0.4404	0.4251	0.3626	0.5728	914.7739
	DT	0.4510	0.4307	0.4510	0.4365	0.3745	0.5805	40.3568
PWM2Vec	SVM	0.4041	0.3760	0.4041	0.3836	0.3138	0.5452	10927.61
	NB	0.2287	0.3758	0.2287	0.2109	0.2220	0.5273	23.9892
	MLP	0.4053	0.3820	0.4053	0.3888	0.3226	0.5467	253.6387
	KNN	0.4454	0.3890	0.4454	0.3930	0.3069	0.5475	10.2005
	RF	0.4994	0.4745	0.4994	0.4370	0.3548	0.5716	126.5780
	LR	0.4143	0.3861	0.4143	0.3937	0.3237	0.5484	991.8051
	DT	0.4339	0.4078	0.4339	0.4140	0.3496	0.5636	34.9344
Spaced k-mers	SVM	0.5109	0.5221	0.5109	0.4095	0.3143	0.5582	824.215
	NB	0.2157	0.3713	0.2157	0.1296	0.1510	0.5144	0.1883
	MLP	0.4524	0.4203	0.4524	0.4236	0.3550	0.5663	207.685
	KNN	0.4527	0.4078	0.4527	0.4132	0.3351	0.5607	3.3905
	RF	0.5204	0.5233	0.5204	0.4294	0.3393	0.5679	41.3547
	LR	0.5121	0.5053	0.5121	0.4318	0.3441	0.5674	25.7664
	DT	0.4006	0.4009	0.4006	0.4006	0.3433	0.5629	14.2816
Autoencoder	SVM	0.4597	0.2113	0.4597	0.2896	0.1575	0.5000	14325.09
	NB	0.2601	0.3096	0.2601	0.2682	0.2317	0.5005	0.3491
	MLP	0.3996	0.3132	0.3996	0.3226	0.2252	0.5017	110.76
	KNN	0.3791	0.3245	0.3791	0.3329	0.2445	0.5068	5.9155
	RF	0.4457	0.3116	0.4457	0.3023	0.1792	0.5003	76.2106
	LR	0.4520	0.3170	0.4520	0.2982	0.1712	0.5004	98.5936
	DT	0.3131	0.3119	0.3131	0.3124	0.2525	0.5016	25.1653
WDGRL	SVM	0.4677	0.2188	0.4677	0.2981	0.1593	0.5000	15.34
	NB	0.4469	0.3231	0.4469	0.3397	0.2213	0.5105	0.0120
	MLP	0.4749	0.4659	0.4749	0.3432	0.2184	0.5163	15.43
	KNN	0.3930	0.3415	0.3930	0.3523	0.2626	0.5156	0.698
	RF	0.4666	0.4138	0.4666	0.3668	0.2578	0.5255	5.5194
	LR	0.4676	0.2187	0.4676	0.2980	0.1593	0.4999	0.0799
	DT	0.3604	0.3606	0.3604	0.3605	0.2921	0.5304	0.2610

(continued)

Table 2. (*continued*)

Embeddings	Algo.	Acc. ↑	Prec. ↑	Recall ↑	F1 (Weig.) ↑	F1 (Macro) ↑	ROC AUC ↑	Train Time (sec.) ↓
String Kernel	SVM	0.4597	0.2113	0.4597	0.2896	0.1575	0.5000	2791.61
	NB	0.3093	0.3067	0.3093	0.3079	0.2463	0.4980	0.2892
	MLP	0.3287	0.3045	0.3287	0.3121	0.2402	0.4963	125.66
	KNN	0.3683	0.3106	0.3683	0.3229	0.2330	0.5001	3.1551
	RF	0.4469	0.3251	0.4469	0.3041	0.1813	0.5010	55.3158
	LR	0.4451	0.3080	0.4451	0.3026	0.1787	0.5000	2.0463
	DT	0.3073	0.3082	0.3073	0.3077	0.2502	0.4998	17.0352
Protein Bert	_	0.2624	0.4223	0.2624	0.1947	0.209	0.5434	987.354
SeqVec	SVM	0.432	0.422	0.432	0.415	0.377	0.530	100423.017
	NB	0.192	0.530	0.192	0.086	0.117	0.505	77.271
	MLP	0.428	0.427	0.428	0.427	0.371	0.580	584.382
	KNN	0.432	0.385	0.432	0.391	0.306	0.544	227.877
	RF	0.525	0.524	0.525	0.445	0.360	0.576	350.057
	LR	0.420	0.414	0.420	0.417	0.356	0.571	90626.797
	DT	0.397	0.397	0.397	0.397	0.344	0.562	2626.082
Sparse Coding (ours)	SVM	0.9980	0.9950	0.9980	0.9960	0.9970	0.9950	16965.48
	NB	0.9970	0.9960	0.9970	0.9970	0.9960	0.9980	404.2999
	MLP	0.9950	0.9950	0.9950	0.9950	0.9947	0.9956	5295.811
	KNN	**0.9990**	**0.9990**	**0.9990**	**0.9990**	**0.9989**	**0.9991**	346.7334
	RF	**0.9990**	0.9970	**0.9990**	0.9950	0.9950	0.9960	1805.593
	LR	**0.9990**	0.9940	**0.9990**	0.9980	0.9980	0.9940	**134.8503**
	DT	0.9970	0.9980	0.9970	0.9960	0.9980	0.9970	449.9304

and Autoencoder methods. Our method also surpasses the kernel-based baseline (String Kernel), achieving 53.83% higher accuracy with the SVM model. This highlights the efficiency of Sparse Coding in generating numerical vectors for protein sequences. Additionally, we compare Sparse Coding's performance with pre-trained models (SeqVec and Protein Bert), and Sparse Coding outperforms them across all evaluation metrics, except training runtime. This indicates that pre-trained models struggle to generalize effectively to our dataset, resulting in lower predictive performance. As one of our key contributions is the incorporation of domain knowledge, we do not report results for domain knowledge + baseline methods in this paper. However, our evaluation demonstrates that the proposed method consistently outperforms domain knowledge + baseline methods in terms of predictive performance. Moreover, our proposed method without domain knowledge also outperforms baselines without domain knowledge, thanks to the incorporation of k-mers-based features into Sparse Coding-based embeddings, which enhances their richness.

6   Conclusion

Our study introduced a novel approach for cancer classification by leveraging sparse coding and TCR sequences. We transformed TCR sequences into feature vectors using k-mers and applied sparse coding to generate embeddings, incorporating cancer domain knowledge to enhance performance. Our method achieved a maximum accuracy of 99.9%, along with higher F1 and ROC AUC scores. While our immediate focus is accurate cancer-type classification, our findings have broader implications for personalized treatment and immunotherapy development. Future research can explore sparse coding in other biological data, and optimize it for diverse cancers and TCR sequences.

References

1. Ali, S., Patterson, M.: Spike2Vec: an efficient and scalable embedding approach for covid-19 spike sequences. In: IEEE Big Data, pp. 1533–1540 (2021)
2. Ali, S., Bello, B., et al.: PWM2Vec: an efficient embedding approach for viral host specification from coronavirus spike sequences. MDPI Biol. (2022)
3. Alley, E.C., Khimulya, G., et al.: Unified rational protein engineering with sequence-based deep representation learning. Nat. Methods **16**(12), 1315–1322 (2019)
4. Bai, F., et al.: Use of peripheral lymphocytes and support vector machine for survival prediction in breast cancer patients. Transl. Cancer Res. **7**(4) (2018)
5. van den Berg, J.H., Heemskerk, B., van Rooij, N., et al.: Tumor infiltrating lymphocytes (TIL) therapy in metastatic melanoma: boosting of neoantigen-specific T cell reactivity and long-term follow-up. J. Immunother. Cancer **8**(2) (2020)
6. Bileschi, M.L., et al.: Using deep learning to annotate the protein universe. BioRxiv, p. 626507 (2019)
7. Brandes, N., Ofer, D., et al.: ProteinBERT: a universal deep-learning model of protein sequence and function. Bioinformatics **38**(8), 2102–2110 (2022)
8. Bufe, S., et al.: PD-1/CTLA-4 blockade leads to expansion of CD8+ PD-1int TILs and results in tumor remission in experimental liver cancer. Liver Cancer (2022)
9. Carosella, E.D., Ploussard, G., LeMaoult, J., Desgrandchamps, F.: A systematic review of immunotherapy in urologic cancer: evolving roles for targeting of CTLA-4, PD-1/PD-L1, and HLA-G. Eur. Urol. **68**(2), 267–279 (2015)
10. Chen, S.Y., et al.: TCRdb: a comprehensive database for T-cell receptor sequences with powerful search function. Nucleic Acids Res. **49**(D1), D468–D474 (2021)
11. Chourasia, P., Ali, S., Ciccolella, S., Vedova, G.D., Patterson, M.: Reads2Vec: efficient embedding of raw high-throughput sequencing reads data. J. Comput. Biol. **30**(4), 469–491 (2023)
12. Courtney, A.H., Lo, W.L., Weiss, A.: TCR signaling: mechanisms of initiation and propagation. Trends Biochem. Sci. **43**(2), 108–123 (2018)
13. De Visser, K.E., Eichten, A., Coussens, L.M.: Paradoxical roles of the immune system during cancer development. Nat. Rev. Cancer **6**(1), 24–37 (2006)
14. Dunne, M.R., et al.: Characterising the prognostic potential of HLA-DR during colorectal cancer development. Cancer Immunol. Immunother. **69**, 1577–1588 (2020)
15. Farhan, M., Tariq, J., Zaman, A., Shabbir, M., Khan, I.U.: Efficient approximation algorithms for strings kernel based sequence classification. In: Advances in Neural Information Processing Systems, vol. 30 (2017)
16. Fodde, R.: The APC gene in colorectal cancer. Eur. J. Cancer **38**(7), 867–871 (2002)

17. Gittelman, R.M., Lavezzo, E., Snyder, T.M., Zahid, H.J., Carty, C.L., et al.: Longitudinal analysis of t cell receptor repertoires reveals shared patterns of antigen-specific response to SARS-CoV-2 infection. JCI Insight **7**(10) (2022)

18. Gonzalez, H., et al.: Roles of the immune system in cancer: from tumor initiation to metastatic progression. Genes Dev. **32**(19–20), 1267–1284 (2018)

19. Hee, B.J., Kim, M., et al.: Feature selection for colon cancer detection using k-means clustering and modified harmony search algorithm. Mathematics **9**(5), 570 (2021)

20. Heinzinger, M., Elnaggar, A., et al.: Modeling aspects of the language of life through transfer-learning protein sequences. BMC Bioinformatics **20**(1), 1–17 (2019)

21. Hoadley, K.A., Yau, C., et al.: Cell-of-origin patterns dominate the molecular classification of 10,000 tumors from 33 types of cancer. Cell **173**(2), 291–304 (2018)

22. Hu, M., et al.: Exploring evolution-based & -free protein language models as protein function predictors. arXiv preprint arXiv:2206.06583 (2022)

23. Iqbal, M.J., Faye, I., Samir, B.B., Md Said, A.: Efficient feature selection and classification of protein sequence data in bioinformatics. Sci. World J. **2014** (2014)

24. Janeway, C.A. Jr.: The major histocompatibility complex and its functions. In: Immunobiology: The Immune System in Health and Disease. 5th edn. Garland Science (2001)

25. Johnson, N., et al.: Counting potentially functional variants in BRCA1, BRCA2 and ATM predicts breast cancer susceptibility. Hum. Mol. Genet. **16**(9), 1051–1057 (2007)

26. Kelly, T.K., De Carvalho, D.D., Jones, P.A.: Epigenetic modifications as therapeutic targets. Nat. Biotechnol. **28**(10), 1069–1078 (2010)

27. Kidman, J., et al.: Characteristics of TCR repertoire associated with successful immune checkpoint therapy responses. Frontiers Immunol. **11**, 587014 (2020)

28. Kuzmin, K., et al.: Machine learning methods accurately predict host specificity of coronaviruses based on spike sequences alone. Biochem. Biophys. Res. Commun. **533**(3), 553–558 (2020)

29. Lee, A., et al.: BOADICEA: a comprehensive breast cancer risk prediction model incorporating genetic and nongenetic risk factors. Genet. Med. **21**(8), 1708–1718 (2019)

30. Liang, H., Lu, T., Liu, H., Tan, L.: The relationships between HLA-A and HLA-B genes and the genetic susceptibility to breast cancer in Guangxi. Russ. J. Genet. **57**, 1206–1213 (2021)

31. Lin, Z., Akin, H., Rao, R., et al.: Evolutionary-scale prediction of atomic-level protein structure with a language model. Science **379**(6637), 1123–1130 (2023)

32. Loibl, S., Gianni, L.: HER2-positive breast cancer. Lancet **389**(10087), 2415–2429 (2017)

33. Lu, Y.C., et al.: Single-cell transcriptome analysis reveals gene signatures associated with T-cell persistence following adoptive cell therapygene signatures associated with T-cell persistence. Cancer Immunol. Res. **7**(11), 1824–1836 (2019)

34. Makuuchi, M., Kosuge, T., Takayama, T., et al.: Surgery for small liver cancers. In: Seminars in Surgical Oncology, vol. 9, pp. 298–304. Wiley Online Library (1993)

35. Mikolov, T., Chen, K., Corrado, G., Dean, J.: Efficient estimation of word representations in vector space. arXiv preprint arXiv:1301.3781 (2013)

36. Min, S., Park, S., et al.: Pre-training of deep bidirectional protein sequence representations with structural information. IEEE Access **9**, 123912–123926 (2021)

37. Nambiar, A., Heflin, M., Liu, S., Maslov, S., Hopkins, M., Ritz, A.: Transforming the language of life: transformer neural networks for protein prediction tasks. In: Proceedings of the 11th ACM International Conference on Bioinformatics, Computational Biology and Health Informatics, pp. 1–8 (2020)
38. Olshausen, B.A., Field, D.J.: Sparse coding of sensory inputs. Curr. Opin. Neurobiol. **14**(4), 481–487 (2004)
39. Ostrovsky-Berman, M., et al.: Immune2vec: embedding B/T cell receptor sequences in n using natural language processing. Frontiers Immunol. **12**, 680687 (2021)
40. Peshkin, B.N., Alabek, M.L., Isaacs, C.: BRCA1/2 mutations and triple negative breast cancers. Breast Dis. **32**(1–2), 25–33 (2011)
41. Ranstam, J., Cook, J.: Lasso regression. J. Br. Surgery **105**(10), 1348 (2018)
42. Rotte, A.: Combination of CTLA-4 and PD-1 blockers for treatment of cancer. J. Exp. Clin. Cancer Res. **38**, 1–12 (2019)
43. Schaafsma, E., et al.: Pan-cancer association of HLA gene expression with cancer prognosis and immunotherapy efficacy. Br. J. Cancer **125**(3), 422–432 (2021)
44. Shah, K., Al-Haidari, A., Sun, J., Kazi, J.U.: T cell receptor (TCR) signaling in health and disease. Signal Transduct. Target. Ther. **6**(1), 412 (2021)
45. Shen, J., Qu, Y., Zhang, W., et al.: Wasserstein distance guided representation learning for domain adaptation. In: AAAI Conference on Artificial Intelligence (2018)
46. Singh, R., et al.: GaKCo: a fast gapped k-mer string Kernel using counting. In: Joint European Conference on Machine Learning and Knowledge Discovery in Databases, pp. 356–373 (2017)
47. Stanton, S.E., Disis, M.L.: Clinical significance of tumor-infiltrating lymphocytes in breast cancer. J. Immunother. Cancer **4**, 1–7 (2016)
48. Van, L., Hinton, G.: Visualizing data using t-SNE. J. Mach. Learn. Res. (JMLR) **9**(11) (2008)
49. Wan, F., et al.: DeepCPI: a deep learning-based framework for large-scale in silico drug screening. Genomics Proteomics Bioinform. **17**(5), 478–495 (2019)
50. Xie, J., Girshick, R., Farhadi, A.: Unsupervised deep embedding for clustering analysis. In: International Conference on Machine Learning, pp. 478–487 (2016)
51. Yang, X., Yang, S., Li, Q., Wuchty, S., Zhang, Z.: Prediction of human-virus protein-protein interactions through a sequence embedding-based machine learning method. Comput. Struct. Biotechnol. J. **18**, 153–161 (2020)
52. Zhang, J., et al.: Recurrent neural networks with long term temporal dependencies in machine tool wear diagnosis and prognosis. SN Appl. Sci. **3**, 1–13 (2021)
53. Zhu, J.D.: The altered DNA methylation pattern and its implications in liver cancer. Cell Res. **15**(4), 272–280 (2005)

Weakly Supervised Temporal Action Localization Through Segment Contrastive Learning

Zihao Jiang[1,2](✉) and Yidong Li[1,2](✉)

[1] Key Laboratory of Big Data and Artificial Intelligence in Transportation, Ministry of Education, Beijing, China
[2] School of Computer and Information Technology, Beijing Jiaotong University, Beijing, China
{21120363,ydli}@bjtu.edu.cn

Abstract. Weakly-supervised temporal action localization (WS-TAL) aims to learn localizing action with only video-level labels. In general, traditional methods process each snippet individually, thus ignoring both relationship between different action snippets and the productive temporal contextual relationships which are critical for action localization. In this paper, we propose Bidirectional Exponential Moving Average (BEMA) to utilize contextual information to obtain more stable feature representation. In addition to that, we introduce Inter-Segment Loss to refine the snippet representation in feature space to prevent misidentification of similar actions for accurate action classification, and Intra-Segment Loss to separate action from background in feature space to locate precise temporal boundaries. Substantial analyses show that our strategy is effective and brings comparable performance gain with current state-of-the-art WS-TAL methods on THUMOS'14 and ActivityNet v1.2 datasets. Our code can be found at https://github.com/JiangZhouHao/SCL.

Keywords: Temporal Action Localization · Weakly Supervised Learning · Contrastive Learning

1 Introduction

Temporal action localization (TAL) is one of the challenging tasks of video understanding, which aims to localize and classify action instances in untrimmed videos. Most methods [16,18,35,42] have achieved remarkable performance under the fully supervised manner where requires annotated temporal boundaries of actions. In contrast, weakly-supervised TAL learns how to localize actions only with video-level annotations. Compared with laborious annotations, the video-level labels are easier to collect, thus weakly-supervised TAL has attracted more research interests.

B. Luo et al. (Eds.): ICONIP 2023, CCIS 1964, pp. 228–243, 2024.
https://doi.org/10.1007/978-981-99-8141-0_18

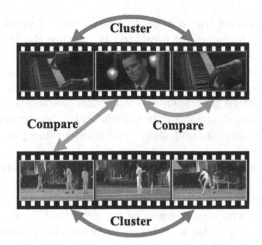

Fig. 1. In the figure, the green boxes represent the action snippets in the segments selected from the individual videos, and the red boxes indicate the background snippets. The action comparing between segments makes them better classified, both the action/background clustering and the action-background comparing within segments make the action background better distinguished. (Color figure online)

Existing approaches [24,30,31,33] for weakly-supervised TAL use video-level annotations and generate a sequence of class-specific scores, called Temporal Class Activation Sequence (T-CAS). As a general rule, a classification loss is used to distinguish the foreground and background to obtain the action regions. And the loss can be generated by two main approaches. One kind of approaches [6,23,24,27,38] introduce attention mechanism to separate foreground with background, then utilize the action classifier to recognize videos. While the other approaches [10,13,21,26,39] formulate this problem as multi-instance learning task and treat the entire video as a bag containing multiple positive and negative instances, i.e. action frames and non-action frames.

As mentioned above, both types of methods aim at learning the effective classification functions to identify action instances from bags of action instances and non-action frames. However, there is still a huge performance gap between the weakly supervised and fully supervised methods. We argue that the reason may lie in the fact that the indistinguishable snippets are easily misclassified and hurt the localization performance. To illustrate it, we take PlayingPiano and Cricket in Fig. 1 as an example, it is challenging to distinguish action instances and their contexts due to that they have similar features resulting in classifier frequently regards these action contexts as action instances.

To realize action precise classification and action contexts suppression with video-level labels, in this paper, we propose segment contrastive learning. Specially, we first generate action proposals using a classification branch, but as mention above this initial CAS is not capable of utilizing ambiguous context information. To tackle this issue, we apply BEMA to initial CAS for obtaining final results by combining the past, present and future. Second, inter- and intra-segment contrastive learning are presented to capture the temporal structures across and within segments at the stage of feature modeling. Specifically, the intra-segment CL utilizes k-means clustering within proposals to model better discriminate action and background snippets. Finally, by optimizing the proposed losses along with a general video-level action classification loss, our method successfully distinguishes action snippets from both other action and background snippets. Moreover, our method shows the considerable performances on the both benchmarks.

The contributions of this paper are three-fold: (1) We introduce BEMA to effectively use context information for video understanding by combining the feature of the previous and the following moments. (2) The proposed segment contrastive learning which fully consider the relationship of intra-segment and inter-segment with only video-level labels according to separating actions from different action classes as well as the background. (3) Comprehensive experiments on THUMOS'14 and ActivityNet v1.2 datasets to validate the effectiveness of our proposed method.

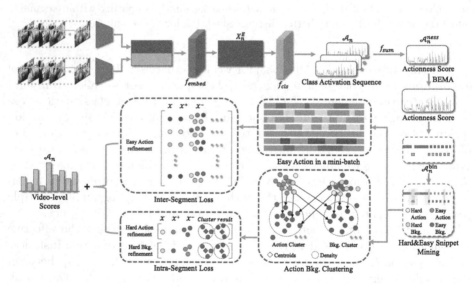

Fig. 2. Overview of the proposed method, which comprises of four parts. (a) The first step involves Feature Extraction and Embedding to obtain the embedded feature X_n^E; (b) Next, Action Proposal is performed to gather class-agnostic action likelihood A_n^{final}; (c) Inter- and Intra-Segment Contrastive Learning to obtain more precise action boundary; (d) Network Training driven by Inter- and Intra-Segment Loss and video-level action Loss.

2   Related Work

Fully-Supervised Action Localization utilizes frame-level labels to classify and find temporal boundaries of action instances from long untrimmed videos. Most existing methods can be categorized into two groups: proposal-based (top-down) and frame-based methods (bottom-up). Proposal-based methods [16,18,35,42] usually generate action proposals with pre-defined regular distributed segments, then classify them to regress temporal boundary. Frame-based methods [2,17] directly predict action category or actionness scores for each snippet and locate action boundary following by some post-processing strategies.

Weakly-Supervised Action Localization only requires video-level labels and has drawn increasing attention. UntrimmedNets [33] is the first method to address this problem by conducting the snippet proposal classification to select relevant segments. STPN [24] improves UntrimmedNets by imposing a sparsity constraint for enforcing the sparsity of the selected segments. Hide-and-seek [31] tries to hide patches randomly in a training image. Zhong et al. [43] train a detector by driving a series of classifiers to achieve similar results. W-TALC [26] utilizes pairwise video similarity constraints via an attention-based mechanism to be complementary with Multiple Instance Learning.

Contrastive Learning uses internal patterns of the data to learn an embedding space in which relevant signals are clustered together and irrelevant signals are distinguished by Noise Contrastive Estimation (NCE). A contrastive learning framework proposed by CMC [32] aims to achieve view-invariant representations by maximizing mutual information between different views of the same scene. SimCLR [5] uses augmented views of other items in a minibatch to select negative samples. To eliminate the batch size limitation and ensure consistent use of negative samples, MoCo [7] uses a memory bank of old negative representations with a momentum update. The experimental results demonstrate that our method refines the representation of both action and background snippets, facilitating action localization.

3   Method

In this section, our method follows the feature extraction and feature embedding (Sect. 3.1), action proposal (Sect. 3.2) and Segment Contrastive Learning (Sect. 3.3) pipeline. The optimization loss terms for training and how the inference is performed are detailed in Sect. 3.4 and Sect. 3.5, respectively.

3.1   Feature Extraction and Embedding

Feature Extraction. For a set of N untrimmed videos $\{V_n\}_{n=1}^N$, which video labels are $\{y_n\}_{n=1}^N$, where $y_n \in \mathbb{R}^C$ is a multi-hot vector, and C is the number of action categories. For an input untrimmed video V_n, we first divide it into multi-frame non-overlapping L_n snippets based on sampling ratio. We keep the fixed number of T snippets as $\{S_n, t\}_{t=1}^T$. Then the RGB features $X_n^R = \{x_t^R\}_{t=1}^T$ and the optical flow features $X_n^O = \{x_t^O\}_{t=1}^T$ are extracted by a pre-trained feature extractor (e.g. , I3D [3]), respectively. Here, $x_t^R \in \mathbb{R}^d$ and $x_t^O \in \mathbb{R}^d$, d is the feature dimension of each snippet. Next, we concatenate all snippets feature to form the video pre-trained feature $X_n \in \mathbb{R}^{T \times 2d}$.

Feature Embedding. In order to project the extracted video features X_n into the task-specific feature space, we apply a function f_{embed} over X_n to map the original video feature $X_n \in \mathbb{R}^{T \times 2d}$ to task-specific video feature $X_n^E \in \mathbb{R}^{T \times 2d}$, f_{embed} is implemented with a single 1-D convolution followed by a ReLU activation function.

3.2   Action Proposal

Action Modeling. Following the previous work [26], we introduce a snippet-level concept Action Proposal referring to the likelihood of action instance for each snippet.

After obtaining the embedded features X_n^E, we apply an action classification branch f_{cls} to generate snippet-level Temporal Class Activation Sequence (T-CAS). Particularly, the branch contains a single 1-D convolution followed by ReLU activation and Dropout function. This can be formulated for a video V_n as following:

$$\mathcal{A}_n = f_{cls}(X_n^E; \phi_{cls}) \tag{1}$$

where ϕ_{cls} represents the trainable parameters. The generated $A_n \in \mathbb{R}^{T \times C}$ represents the action classification results of each snippet.

Afterwards, a general way is to conduct the binary classification on each snippet for action modeling. Since we can already obtain snippet-level classification from the generated T-CAS $A_n \in \mathbb{R}^{T \times C}$ in Eq. 1, then we just sum T-CAS along the class dimension followed by the Sigmoid activation function to aggregate class-agnostic scores and use them to represent the Action Proposal $A_n^{act} \in \mathbb{R}^T$:

$$\mathcal{A}_n^{act} = Sigmoid(f_{sum}(\mathcal{A}^n)) \tag{2}$$

Bidirectional Exponential Moving Average. Exponential Moving Average is an averaging method that gives more weights to recent data. The recent data of the snippets in a video, i.e., the context, is of great significance for video understanding, and the importance of context decays over time. Therefore, we propose a simple but effective method to make full use of the context of Action Proposal named Bidirectional Exponential Moving Average (BEMA).

We obtain its historical and future information only by shifting the tensor of instances, then give different temporal weights to these tensor according to the different moving distance. Finally, we sum these tensor to obtain the final Action Proposal with the following equation:

$$\mathcal{A}_n^{final} = \sum_{m=0}^{M} t^m shift(\mathcal{A}_n^{act}, \pm m) \tag{3}$$

where t represents the time decay. The $A_n^{final} \in \mathbb{R}^T$ represents the final Action Proposal. m represents the moving distance of tensor. When the value of m is positive, this $shift$ function will move the whole tensor backward, on the contrary, the whole tensor will be moved forward.

3.3 Inter- and Intra-segment Contrastive Learning

Pseudo Action/Background Segments Selection. Firstly, we use the same top-k mechanism as Uncertainty Modeling [14] to select easy pseudo action/background segments in the video based on A_n^{final}. Specifically, the top k^{act} segments in terms of the feature magnitude are treated as the easy pseudo action segments $\{s_{n,j} | j \in \mathcal{S}^{EA}\}$, where $\mathcal{S}^{EA}$ indicates the set of pseudo action indices. Meanwhile, the bottom k^{bkg} segments are considered the pseudo background segments $\{s_{n,j} | j \in \mathcal{S}^{EB}\}$, where $\mathcal{S}^{EB}$ denotes the set of indices for easy pseudo background. k^{act} and k^{bkg} represent the number of segments sampled for action and background, respectively. Then, we select hard pseudo action/background segments $\mathcal{S}^{HA}/\mathcal{S}^{HB}$ in the video as a previous study [41] did.

Inter-segment Contrastive Learning. After obtaining the pseudo action segments, we divide them into two parts representing the top-k action segments X_n^a and the remainders X_n^p:

$$X_n^a = \mathcal{S}_n^{EA}[k:]$$
$$X_n^p = \mathcal{S}_n^{EA}[:k] \tag{4}$$

We can construct anchor x_i and positive samples x_i^+, then select the negative samples x_i^- based on the different labels $\hat{y}_n \in \mathbb{R}^C$ of these segments in the same batch:

$$x_i = X_i^a$$

$$x_i^+ = \left\{ \sum_{n \neq i}^N X_n^p \Big| sum(\hat{y}_n \cdot \hat{y}_i) = m, (m > 0) \right\}$$

$$x_i^- = \left\{ \sum_{n \neq i}^N X_n^p \Big| sum(\hat{y}_n \cdot \hat{y}_i) = 0 \right\} \tag{5}$$

Intra-segment Contrastive Learning. In the previous work, CoLA [41] only regards the center of easy actions as positive sample during contrastive learning which is effective for learning discriminative action-background separation. However, it doesn't make the most of intra-class consistency.

Base on the above observations, we argue that making action and background snippets concentrated respectively may achieve better results. Inspired by [15], we propose a loss for clustering action and background snippets respectively. K-means clustering algorithm is used for $\mathcal{S}^{EA}$, $\mathcal{S}^{EB}$, $\mathcal{S}^{HA}$, $\mathcal{S}^{HB}$, and the clustering distribution will be divided into two classes, representing action and background, followed by the clustering center of each class. The concentration estimation ϕ

is obtained from the clustering distribution $\{v'_z\}^Z_{z=1}$ and the clustering center c:

$$\{v'_z\}^Z_{z=1} = Clustering(clus, X, index)$$

$$\phi = \frac{\sum^Z_{z=1} ||v'_z - c||}{Z \log(Z + \theta)} \tag{6}$$

3.4   Training Objectives

Our model is jointly optimized with three losses: 1) video-level classification loss L_{cls} for action classification of videos, 2) Inter-Segment loss L_{intra} for separating the magnitudes of different action feature vectors, and 3) Intra-Segment loss L_{inter} for separating the magnitudes of action and background feature vectors. The overall loss function is as follows:

$$\mathcal{L}_{total} = \mathcal{L}_{cls} + \alpha\mathcal{L}_{Inter} + \beta\mathcal{L}_{Intra} \tag{7}$$

Video-Level Classification Loss. To perform multi-label action classification, we employ the binary cross entropy loss, utilizing normalized video-level labels [33] in the following manner:

$$\mathcal{L}_{cls} = -\frac{1}{N} \sum^N_{n=1} \sum^C_{c=1} y_{n;c} \log p_c(v_n) \tag{8}$$

In this equation, $p_c(v_n)$ denotes the predicted video-level softmax score for the c-th class of the n-th video, while $y_{n;c}$ signifies the normalized video-level label for the c-th class of the n-th video.

Inter-segment Loss. Contrastive learning has traditionally been applied at the image or patch level [1,9], our application employs contrastive learning at the segment level, utilizing embeddings X^E_n. This approach is referred as Inter-Segment Loss, and its objective is to refine the segment-level features of different actions, resulting in a more discriminative feature distribution.

Formally, queries $x \in \mathbb{R}^{1 \times 2d}$, positives $x^+ \in \mathbb{R}^{1 \times 2d}$, and S negatives $x^- \in \mathbb{R}^{1 \times 2d}$ are selected from the segments shown in Eq. 5. As illustrated in Fig. 2, for Inter-Segment Learning, $x \sim x_i$, $x^+ \sim x^+_i$, $x^- \sim x^-_i$. A classification problem of the cross-entropy loss is set to represent the probability of a positive example being chosen over a negative one. Following [7], we use a temperature scale $\tau = 0.07$ to calculate the distance between the querys and the other examples:

$$\mathcal{L}(x, x^+, x^-)$$
$$= -\log \left[\frac{\exp(x^T \cdot x^+ / \tau)}{\exp(x^T \cdot x^+ / \tau) + \sum^S_{s=1} \exp(x^T \cdot x^-_s / \tau)} \right] \tag{9}$$

where x^T is the transpose of x and the proposed Inter-Segment Loss is as follows:

$$\mathcal{L}_{Inter} = \frac{1}{N} \sum^N_{n=1} \mathcal{L}(x_n, x^+_n, x^-_n) \tag{10}$$

where N represents number of different action categories in the same batch.

Intra-segment Loss. Clustering is used to keep the actions and backgrounds closer to their clustering center in the first, and then loss is used to make a better separation between actions and backgrounds within segments to obtain better localization results. The loss is calculated as follows:

$$\mathcal{L}_{Intra} = \sum_{i=1}^{n} \left(\mathcal{L}(x, x^+, x^-) + \mathcal{L}_c \right)$$

$$\mathcal{L}_c = -\frac{1}{M} \sum_{m=1}^{M} \log \frac{\exp(x_i^T \cdot c_s^m / \phi_s^m)}{\sum_{j=0}^{r} \exp(x_i^T \cdot c_j^m / \phi_j^m)}$$

(11)

where n means number of clusters and $\mathcal{L}_c$ represents the clustering loss. During implementation, we adopt a similar strategy to Noise Contrastive Estimation (NCE) and sample r negative prototypes to compute the normalization term. Additionally, we perform clustering on the samples M times with varying numbers of clusters $K = \{k_m\}_{m=1}^{M}$, which results in a more resilient probability estimation of prototypes that encode the hierarchical structure. To maintain the property of local smoothness and aid in bootstrapping clustering, we add the InfoNCE loss.

3.5 Inference

At the test time, for a given video, we first obtain its class activations to form T-CAS and aggregate top-k^{easy} scores described in Sect. 3.4. Then the categories with scores larger than θ_{vid} are selected determining which action classes are to be localized. For each selected category, we threshold its corresponding T-CAS with θ_{seg} to obtain candidate video snippets. Finally, continuous snippets are grouped into proposals and Non-Maximum Suppression (NMS) is applied to remove duplicated proposals.

4   Experiments

4.1   Experimental Setups

Datasets. We evaluate our method on two popular action localization benchmark datasets: THUMOS'14 [11] and ActivityNet v1.2 [8]. THUMOS'14 includes untrimmed videos from 20 categories. The video length varies greatly and multiple action instances may exist in each video. Following convention [13,29,41], we use 200 videos in the validation set for training and 213 videos in the testing set for evaluation. It has 4819 training videos and 2383 validation videos. Following [30,33], we use the training set to train our model and the validation set for evaluation.

Evaluation Metrics. We follow the standard evaluation protocol by reporting mean Average Precision (mAP) values over different intersection over union

236 Z. Jiang and Y. Li

Table 1. Comparisons results on THUMOS'14. The mAP values at different IoU thresholds are reported. AVG is the average mAP under the thresholds [0.1:0.7:0.2]. UNT and I3D denote the use of UntrimmedNets and I3D network as the feature extractor, respectively.

Supervision (Feature)	Method	Publication	mAP@IoU (%)				
			0.1	0.3	0.5	0.7	AVG
Full (-)	R-C3D [35]	ICCV	54.5	44.8	28.9	–	–
	SSN [42]	ICCV	66.0	51.9	29.8	–	–
	TAL-NET [4]	CVPR	59.8	53.2	42.8	20.8	45.1
	P-GCN [40]	ICCV	69.5	63.6	49.1	–	–
	G-TAD [36]	CVPR	–	66.4	51.6	22.9	–
Weak (-)	Hide-and-Seek [31]	ICCV	36.4	19.5	6.8	–	–
	UntrimmedNet [33]	CVPR	44.4	28.2	13.7	–	–
	Zhong et al. [43]	ACMMM	45.8	31.1	15.9	–	–
Weak (UNT)	AutoLoc [30]	ECCV	–	35.8	21.2	5.8	–
	CleanNet [20]	ICCV	-	37.0	23.9	7.1	–
	Bas-Net [13]	AAAI	–	42.8	25.1	9.3	–
Weak (I3D)	STPN [24]	CVPR	52.0	35.5	16.9	4.3	27.0
	Liu et al. [19]	CVPR	57.4	41.2	23.1	7.0	32.4
	Nguyen et al. [25]	ICCV	60.4	46.6	26.8	9.0	36.3
	BaS-Net [13]	AAAI	58.2	44.6	27.0	10.4	35.3
	DGAM [29]	CVPR	60.0	46.8	28.8	11.4	37.0
	ActionBytes [12]	CVPR	-	43.0	29.0	9.5	–
	CoLA [41]	CVPR	66.2	51.5	32.2	13.1	40.9
	Ours		**66.7**	**52.4**	**34.6**	**13.4**	**42.2**

(IoU) thresholds to evaluate our weakly-supervised temporal action localization performance. The evaluation utilize the benchmark code provided by ActivityNet.

Training Details. We employ I3D [3] network pretrained on Kinetics [3] for feature extraction. We apply TVL1 [28] algorithm to extract optical flow from RGB frames in advance. The Adam optimizer is used with the learning rate of 10^{-4}. The number of sampled snippets T for THUMOS'14 and ActivityNet v1.2 is set to 750 and 250, respectively. All hyper-parameters are determined by grid search. We train for total 4.5k epochs with a batch size of 16 for THUMOS'14 and for total 10k epochs with a batch size of 16 for ActivityNet v1.2.

Table 2. Comparison with state-of-the-art methods on ActivityNet v1.2. AVG is the average mAP under the thresholds [0.5:0.95:0.05]. UNT and I3D denote the use of UntrimmedNets and I3D network as the feature extractor, respectively.

Sup.	Method	mAP@IoU (%)			
		0.5	0.75	0.95	AVG
Full	SSN [42]	41.3	27.0	6.1	26.6
Weak (UNT)	AutoLoc [30]	27.3	15.1	3.3	16.0
	UntrimmedNet [33]	7.4	3.2	0.7	3.6
Weak (I3D)	Liu et al. [19]	36.8	22.0	5.6	22.4
	Lee et al. [14]	41.2	25.6	6.0	25.9
	DGAM [29]	41.0	23.5	5.3	24.4
	VQK-Net [34]	44.5	26.4	5.1	23.6
	CoLA [41]	42.7	25.7	5.8	26.1
	WTCL [37]	40.6	24.4	**6.4**	24.9
	Ours	**47.1**	**29.1**	4.8	**28.9**

Testing Details. We set θ_{vid} to 0.2 and 0.1 for THUMOS'14 and ActivityNet v1.2, respectively. For proposal generation, multiple thresholds θ_{seg} is set [0:0.25:0.025] for THUMOS'14 and [0:0.15:0.015] for ActivityNet v1.2, respectively, then we perform non-maxium suppression (NMS) with an IoU threshold of 0.6.

4.2   Comparison with the State-of-the-Arts

We compare our method with the state-of-the-art fully-supervised and weakly-supervised methods under several IoU thresholds. To enhance readability, all results are presented as percentages.

Table 1 demonstrates the results on THUMOS'14. As shown, our method achieves impressive performance on weakly-supervised methods at IoU 0.5 to 0.7 and 42.2% mAP@AVG. Notably, even when provided with a lower level of supervision, our method outperforms several fully-supervised methods, achieving results that closely follow the latest approaches with minimal gap.

The performances on ActivityNet v1.2 are demonstrated in Table 2. Consistent with the results on THUMOS'14, our method outperforms all weakly-supervised approaches, even SSN [42], a fully supervised method, by obvious margins. We can see that our approach outperforms existing weakly supervised methods, including those that use additional information.

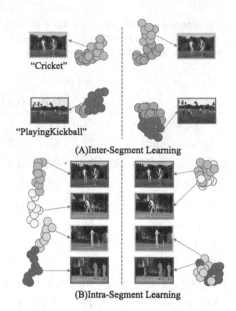

"Cricket"

"PlayingKickball"

(A)Inter-Segment Learning

(B)Intra-Segment Learning

Fig. 3. The UMAP visualizations of the feature embeddings X_E^n are illustrated in Fig. 3. Specifically, the left and right subfigures correspond to the baseline and our proposed method, respectively. As depicted in (A), the green and gray points correspond to the action and non-action embeddings, respectively. Furthermore, (B) depicts the easy and hard action embeddings in the same segment with dark and light green points, respectively, while the dark and light red points correspond to the easy and hard background embeddings in the same segment. Notably, our method yields a more discriminative feature distribution compared to the baseline. (Color figure online)

Table 3. Ablation on the number of time decay and moving distance. AVG is the average mAP under the thresholds [0.1:0.7:0.1]. "0" indicates the base model without corresponding module. For., Bac. and Bid. represents using forward, backward and bidirectional, respectively.

t(m=1)	0	0.05	0.1	0.15
AVG	40.9	41.3	**41.7**	41.4
m(t=0.1)	0	1	2	3
AVG	40.9	41.7	**41.8**	41.6
BEMA	0	For.	Bac.	Bid.
AVG	40.9	41.7	41.6	**41.8**

Table 4. Ablation analysis on each component on THUMOS'14.

Setting	Loss	mAP@0.5
Ours	$\mathcal{L}_{cls} + \mathcal{L}_{Inter} + \mathcal{L}_{Intra}$	**34.6%**
baseline	$\mathcal{L}_{cls}$	24.7%(-9.9%)
BEMA	$\mathcal{L}_{cls} + \mathcal{L}_{Inter}$	31.6%(-3.0%)
BEMA	$\mathcal{L}_{cls} + \mathcal{L}_{Intra}$	32.4%(-2.2%)
w/o BEMA	$\mathcal{L}_{cls} + \mathcal{L}_{Inter} + \mathcal{L}_{Intra}$	33.8%(-0.8%)

4.3 Ablation Study

In this section, we performed several ablation studies to confirm our design intuition. According to the usual practice [24,29], all ablation experiments were conducted on THUMOS'14.

Effects of Bidirectional Exponential Moving Average. In Table 3, we investigate the contribution of each parameter in the BEMA on THUMOS'14. Firstly, we can learn from the table that both the forward and backward exponential averages have some improvement on the results. We have defined two operation degrees (with m and t) for moving distance and time decay in Eq. 3. Here, we tried to evaluate the effect of different parameters. We first vary m from 0 to 3, then we change t from 0.1 to 0.2.

Effects of Loss Components and Score Calculation. To evaluate the effectiveness of our Inter-and Intra-Segment Loss, we perform a comparative analysis by only utilizing the action loss $\mathcal{L}_a$ as the supervisory signal, i.e., the baseline in Table 4. For this purpose, we randomly select two videos from THUMOS'14 testing set and compute the feature embeddings X_n^E for both baseline and our method. Subsequently, these embeddings are projected to a 2-dimensional space using UMAP [22], and the resultant visualizations are presented in Fig. 3.

4.4 Qualitative Results

To demonstrate the effectiveness of our action context modeling mechanism, we present some qualitative results in Fig. 4, two examples from THUMOS'14 are illustrated: sparse and frequent action cases. In both cases, we see that our model detects the action instances more precisely. More specifically, in the red boxes, it can be noticed that baseline splits one action instance into multiple incomplete detection results. On the contrary, our model provides better separation between action and background via uncertainty modeling instead of using a separate class. Consequently, our model successfully localizes the complete action instances without splitting them.

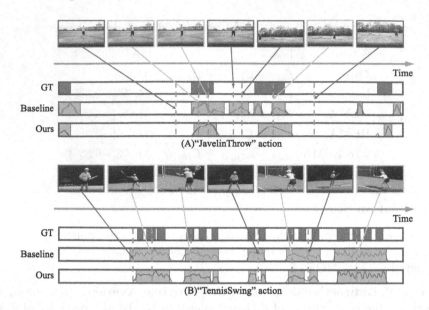

Fig. 4. Qualitative comparison with baseline on THUMOS'14. We provide two different examples: (1) a sparse action case with JavelinThrow, and (2) a frequent action case with TennisSwing. Each example has three plots with sampled frames. The first plot displays the ground truth action intervals. The second plot shows the final scores and the detection results of the corresponding action class from the baseline, while the third plot represents those from our method. The horizontal axis in each plot represents the timesteps of the video, while the vertical axes represent the score values from 0 to 1. The gray dashed lines indicate the action frames that are classified by the baseline and our method. We can observe that some frames are misclassified by both methods, but our method detects more frames correctly.

5   Conclusion

In this paper, we propose a new framework to solve the single snippet cheating problem in weakly supervised action localization. The introduction of the BEMA method integrates the contextual information from past and future frames into each temporal point, resulting in a smoothed estimation of the likelihood of action. Therefore, based on this processed estimation probability, we construct positive and negative samples between segments using differences of video-level labels, and then we apply InfoNCE [7] loss on them, similarly, after constructing the action background snippets within the segments, we use Intra-Segment contrastive learning for these snippets. The feature representation of the mined snippets is refined with the help of simple snippets located in the most discriminative regions. Experiments conducted on two benchmarks, including THUMOS'14 and ActivityNet v1.2, have validated the state-of-the-art performance of our method.

Acknowledgements. This work was supported by the National Natural Science Foundation of China under Grant U1934220 and U2268203.

References

1. Bachman, P., Hjelm, R.D., Buchwalter, W.: Learning representations by maximizing mutual information across views. In: Neural Information Processing Systems (2019)
2. Buch, S., Escorcia, V., Ghanem, B., Fei-Fei, L., Niebles, J.C.: End-to-end, single-stream temporal action detection in untrimmed videos. In: British Machine Vision Conference (2017)
3. Carreira, J., Zisserman, A.: Quo vadis, action recognition? A new model and the kinetics dataset. arXiv Computer Vision and Pattern Recognition (2017)
4. Chao, Y.-W., Vijayanarasimhan, S., Seybold, B., Ross, D.A., Deng, J., Sukthankar, R.: Rethinking the faster R-CNN architecture for temporal action localization. In: Computer Vision and Pattern Recognition (2018)
5. Chen, T., Kornblith, S., Norouzi, M., Hinton, G.E.: A simple framework for contrastive learning of visual representations. In: International Conference on Machine Learning (2020)
6. He, B., Yang, X., Kang, L., Cheng, Z., Zhou, X., Shrivastava, A.: ASM-LOC: action-aware segment modeling for weakly-supervised temporal action localization (2022)
7. He, K., Fan, H., Wu, Y., Xie, S., Girshick, R.: Momentum contrast for unsupervised visual representation learning. arXiv Computer Vision and Pattern Recognition (2019)
8. Heilbron, F.C., Escorcia, V., Ghanem, B., Niebles, J.C.: ActivityNet: a large-scale video benchmark for human activity understanding. In: Computer Vision and Pattern Recognition (2015)
9. Hénaff, O.J., Srinivas, A., Fauw, J.D., Razavi, A., Doersch, C., Eslami, S.M.A.: Data-efficient image recognition with contrastive predictive coding. arXiv Computer Vision and Pattern Recognition (2019)
10. Huang, L., Wang, L., Li, H.: Foreground-action consistency network for weakly supervised temporal action localization. arXiv Computer Vision and Pattern Recognition (2021)
11. Idrees, H., et al.: The THUMOS challenge on action recognition for videos "in the wild". In: Computer Vision and Image Understanding (2017)
12. Jain, M., Ghodrati, A., Snoek, C.G.M.: ActionBytes: learning from trimmed videos to localize actions. In: Computer Vision and Pattern Recognition (2020)
13. Lee, P., Uh, Y., Byun, H.: Background suppression network for weakly-supervised temporal action localization. arXiv Computer Vision and Pattern Recognition (2019)
14. Lee, P., Wang, J., Lu, Y., Byun, H.: Weakly-supervised temporal action localization by uncertainty modeling. arXiv Computer Vision and Pattern Recognition (2020)
15. Li, J., Zhou, P., Xiong, C., Socher, R., Hoi, S.C.H.: Prototypical contrastive learning of unsupervised representations. Learning (2020)
16. Lin, T., Liu, X., Xin, L., Ding, E., Wen, S.: BMN: boundary-matching network for temporal action proposal generation. In: International Conference on Computer Vision (2019)
17. Lin, T., Zhao, X., Shou, Z.: Single shot temporal action detection. In: ACM Multimedia (2017)

18. Lin, T., Zhao, X., Su, H., Wang, C., Yang, M.: BSN: boundary sensitive network for temporal action proposal generation. In: Ferrari, V., Hebert, M., Sminchisescu, C., Weiss, Y. (eds.) ECCV 2018. LNCS, vol. 11208, pp. 3–21. Springer, Cham (2018). https://doi.org/10.1007/978-3-030-01225-0_1

19. Liu, D., Jiang, T., Wang, Y.: Completeness modeling and context separation for weakly supervised temporal action localization (2019)

20. Liu, Z.: Weakly supervised temporal action localization through contrast based evaluation networks. In: International Conference on Computer Vision (2019)

21. Luo, Z.: Weakly-supervised action localization with expectation-maximization multi-instance learning. arXiv Learning (2020)

22. McInnes, L., Healy, J.: UMAP: uniform manifold approximation and projection for dimension reduction, February 2018

23. Min, K., Corso, J.J.: Adversarial background-aware loss for weakly-supervised temporal activity localization. arXiv Computer Vision and Pattern Recognition (2020)

24. Nguyen, P.X., Liu, T., Prasad, G., Han, B.: Weakly supervised action localization by sparse temporal pooling network. arXiv Computer Vision and Pattern Recognition (2017)

25. Nguyen, P.X., Ramanan, D., Fowlkes, C.C.: Weakly-supervised action localization with background modeling. In: International Conference on Computer Vision (2019)

26. Paul, S., Roy, S., Roy-Chowdhury, A.K.: W-TALC: weakly-supervised temporal activity localization and classification (2018)

27. Qu, S., Chen, G., Li, Z., Zhang, L., Lu, F., Knoll, A.: ACM-Net: action context modeling network for weakly-supervised temporal action localization. arXiv Computer Vision and Pattern Recognition (2021)

28. Sánchez, J., Meinhardt-Llopis, E., Facciolo, G.: TV-L1 optical flow estimation. Image Process. On Line (2013)

29. Shi, B., Dai, Q., Mu, Y., Wang, J.: Weakly-supervised action localization by generative attention modeling. arXiv Computer Vision and Pattern Recognition (2020)

30. Shou, Z., Gao, H., Zhang, L., Miyazawa, K., Chang, S.-F.: AutoLoc: weakly-supervised temporal action localization in untrimmed videos. In: Ferrari, V., Hebert, M., Sminchisescu, C., Weiss, Y. (eds.) ECCV 2018. LNCS, vol. 11220, pp. 162–179. Springer, Cham (2018). https://doi.org/10.1007/978-3-030-01270-0_10

31. Singh, K.K., Lee, Y.J.: Hide-and-seek: forcing a network to be meticulous for weakly-supervised object and action localization. In: International Conference on Computer Vision (2017)

32. Tian, Y., Krishnan, D., Isola, P.: Contrastive multiview coding. In: Vedaldi, A., Bischof, H., Brox, T., Frahm, J.-M. (eds.) ECCV 2020. LNCS, vol. 12356, pp. 776–794. Springer, Cham (2020). https://doi.org/10.1007/978-3-030-58621-8_45

33. Wang, L., Xiong, Y., Lin, D., Gool, L.V.: UntrimmedNets for weakly supervised action recognition and detection. In: Computer Vision and Pattern Recognition (2017)

34. Wang, X., Katsaggelos, A.: Video-specific query-key attention modeling for weakly-supervised temporal action localization, May 2023

35. Xu, H., Das, A., Saenko, K.: R-C3D: region convolutional 3D network for temporal activity detection. In: International Conference on Computer Vision (2017)

36. Xu, M., Zhao, C., Rojas, D.S., Thabet, A., Ghanem, B.: G-TAD: sub-graph localization for temporal action detection. In: Computer Vision and Pattern Recognition (2019)

37. Yang, C., Zhang, W.: Weakly supervised temporal action localization through contrastive learning. In: 2022 IEEE 5th International Conference on Multimedia Information Processing and Retrieval (MIPR), August 2022

38. Yu, T., Ren, Z., Li, Y., Yan, E., Xu, N., Yuan, J.: Temporal structure mining for weakly supervised action detection. In: International Conference on Computer Vision (2019)

39. Yuan, Y., Lyu, Y., Shen, X., Tsang, I.W., Yeung, D.-Y.: Marginalized average attentional network for weakly-supervised learning. arXiv Computer Vision and Pattern Recognition (2019)

40. Zeng, R., et al.: Graph convolutional networks for temporal action localization. In: International Conference on Computer Vision (2019)

41. Zhang, C., Cao, M., Yang, D., Chen, J., Zou, Y.: CoLA: weakly-supervised temporal action localization with snippet contrastive learning. In: Computer Vision and Pattern Recognition (2021)

42. Zhao, Y., Xiong, Y., Wang, L., Wu, Z., Tang, X., Lin, D.: Temporal action detection with structured segment networks. In: International Conference on Computer Vision (2017)

43. Zhong, J.-X., Li, N., Kong, W., Zhang, T., Li, T.H., Li, G.: Step-by-step erasion, one-by-one collection: a weakly supervised temporal action detector. In: ACM Multimedia (2018)

S-CGRU: An Efficient Model for Pedestrian Trajectory Prediction

Zhenwei Xu[1,2], Qing Yu[1,3(✉)], Wushouer Slamu[1,3], Yaoyong Zhou[1,2], and Zhida Liu[1,3]

[1] Xinjiang University, Xinjiang Uygur Autonomous Region,
Urumqi 830000, People's Republic of China
{zhenweixu,107552104337,107552103689}@stu.xju.edu.cn, 473461624@qq.com
[2] School of Software, Xinjiang University, Urumqi, People's Republic of China
[3] School of Information Science and Engineering, Xinjiang University, Urumqi,
People's Republic of China

Abstract. In the development of autonomous driving systems, pedestrian trajectory prediction plays a crucial role. Existing models still face some challenges in capturing the accuracy of complex pedestrian actions in different environments and in handling large-scale data and real-time prediction efficiency. To address this, we have designed a novel Complex Gated Recurrent Unit (CGRU) model, cleverly combining the spatial expressiveness of complex numbers with the efficiency of Gated Recurrent Unit networks to establish a lightweight model. Moreover, we have incorporated a social force model to further develop a Social Complex Gated Recurrent Unit (S-CGRU) model specifically for predicting pedestrian trajectories. To improve computational efficiency, we conducted an in-depth study of the pedestrian's attention field of view in different environments to optimize the amount of information processed and increase training efficiency. Experimental verification on six public datasets confirms that S-CGRU model significantly outperforms other baseline models not only in prediction accuracy but also in computational efficiency, validating the practical value of our model in pedestrian trajectory prediction.

Keywords: Pedestrian Trajectory Prediction · Gated Recurrent Unit · Complex number Neural Network · Autonomous Driving

1 Introduction

Trajectory prediction plays a vital role in many critical applications in today's society, such as autonomous driving [2,5], robotic path planning, behavior recognition, security surveillance, and logistics management. Parsing human motion trajectories is a research focus that involves multiple disciplines such as mathematics, physics, computer science, and social science [4]. Current prediction

---

Z. Xu—First author.

B. Luo et al. (Eds.): ICONIP 2023, CCIS 1964, pp. 244–259, 2024.
https://doi.org/10.1007/978-981-99-8141-0_19

methods are typically divided into experience/rule-based methods and deep learning-based methods. Experience or rule-based prediction methods have high interpretability [5–7], but they do not perform well in terms of data fitting ability, often resulting in limited prediction accuracy [11–13]. Deep learning-based methods can provide more accurate prediction results and can automatically adapt to changes in data, demonstrating strong robustness and generalization capabilities [1,8–10]. However, deep learning-based methods also have their limitations. First, such methods typically require a large amount of data to achieve optimal performance. Secondly, the prediction results often lack interpretability. In addition, these methods have high computational resource requirements [14].

In this paper, we propose a brand-new model, the Complex Gated Recurrent Unit (CGRU) model, which aims to combine the advantages of complex numbers and Gated Recurrent Units (GRU) to achieve efficient and accurate trajectory prediction. We leverage the powerful expressive ability of complex neural networks to capture complex patterns in complex space [16–20], thereby enabling the CGRU model to handle intricate trajectory prediction problems while maintaining high computational efficiency. On the other hand, by adopting the efficient simplicity of Gated Recurrent Unit networks [24], our model can effectively handle time-series data and model long-term dependencies. This model also considers the field of view of pedestrians walking in different environments, thus reducing the amount of information that the model needs to process and improving its operational efficiency. In summary, the main contributions of this paper can be encapsulated as follows:

- Innovatively proposed a Complex Gated Recurrent Unit (CGRU) model based on the concept of complex numbers. This deep neural network model combines the advantages of complex numbers in spatial representation and the efficiency of Gated Recurrent Unit networks, thereby constructing a lightweight and efficient Complex Gated Recurrent Unit model (CGRU).
- By integrating the social force model, we designed an S-CGRU model specifically for predicting pedestrian trajectories. This represents a significant extension and enhancement of traditional pedestrian trajectory prediction models.
- We conducted an in-depth study of the field of view of pedestrians walking in different environments, which led to optimization in data processing. This has significantly improved the training efficiency of our model.
- Empirical research results have verified the superiority of our model. Compared to other benchmark models, our model has shown excellent performance on six publicly available datasets, demonstrating its effectiveness and efficiency in pedestrian trajectory prediction.

2   Related Work

2.1   Complex-Valued Neural Network

The use of complex numbers has shown many advantages in various fields, including computation, biology, and signal processing. From a computational perspective, recent research on recurrent neural networks and early theoretical analyses have revealed that complex numbers can not only enhance the expressive power of neural networks but also optimize the memory extraction mechanism for noise suppression [15]. An increasing number of studies are focusing on complex number-based representations because they can improve optimization efficiency [16], enhance generalization performance [17], accelerate learning rates [18–20], and allow for noise-suppression memory mechanisms [19]. Studies in [20] and [18] show that the use of complex numbers can significantly improve the representational capabilities of Recurrent Neural Networks (RNNs).

From a biological perspective, Reichert and Serre [21] constructed a theoretically sound deep neural network model that achieves richer and more universal representations through the use of complex neuron units. This marks the first application of complex representations in deep networks. Conventionally, if the network data is entirely real numbers, we can only describe the specific numerical value of the intermediate output. However, once the data is entirely complex, in addition to being able to express the numerical magnitude of the intermediate output (the modulus of the complex number), we can also introduce the concept of time (the phase of the complex number). In neural network models, especially models like Gated Recurrent Units (GRU), the introduction of complex representations can greatly optimize the ability to process periodic changes and frequency information, as these pieces of information can be encoded through the phase of the complex number. Neurons with similar phases work in synchrony, allowing for constructive superposition, while neurons with different phases will undergo destructive superposition, causing interference with each other. This helps to differentiate and effectively manage the flow of information at different time steps, thereby enhancing the network's efficiency and accuracy in handling time series data [40].

2.2 Recurrent Neural Networks

Recurrent Neural Networks (RNN) and its derivatives, such as Long Short-Term Memory (LSTM) networks [1] and Gated Recurrent Units (GRU) [24], are widely used for sequence prediction tasks. These RNN-based models have achieved remarkable results in fields like machine translation [25] and speech recognition [26]. RNNs are capable of capturing observed sequence patterns and generating predictive sequences based on these patterns.

The Gated Recurrent Unit (GRU) has unique advantages in dealing with these problems. Compared to LSTM, the structure of GRU is simpler, with fewer parameters and more efficient computation. In addition, the gating mechanism of GRU allows it to excel in capturing long-distance dependencies [25]. Therefore, GRU is widely applied in various sequence prediction tasks, including pedestrian trajectory prediction. However, current GRU models still require further research and improvements to better understand and handle interactions between pedestrians.

2.3 Human-Human Interactions

Since the beginning of the research, researchers have recognized that the influence of surrounding neighbors must be fully considered when predicting the future dynamic behavior of agents. As early as the initial stage, some work has first introduced the concept of social forces [30], describing their social interactions by simulating the repulsion and attraction between pedestrians. Subsequent research [31] introduced factors of personal attributes, calculating the impact of stationary crowds on moving pedestrians by classifying pedestrians. Furthermore, S-LSTM [1] successfully integrated the original LSTM network with the time step collection mechanism to simulate the social interaction of pedestrians. Since the social influence between pedestrians mainly depends on their distance from neighbors, research [32] proposed constructing a circular occupancy map to capture the influence of other pedestrians.

In our research, we adopted a data-driven method proposed by [1] to learn the interaction between people more deeply. Then, we expanded on this basis and further explored the field of view that pedestrians pay attention to during their movement. This method can more accurately simulate the behavior of pedestrians in real life because the direction and speed of pedestrian movement are often affected by objects and other pedestrians within their field of view. In addition, by reducing the irrelevant information that the model needs to process, it can improve the prediction accuracy and computational efficiency of the model, making it more practical in actual applications.

3   Method

3.1 Problem Definition

This paper aligns with the works of [1] and [8], assuming that there are N pedestrians during the prediction time period $[1, T_{pred}]$. After preprocessing the pedestrian trajectories in the video, the position of each pedestrian i at each time step t can be defined as a pair of spatial coordinates $P_t^i = (x_i^t, y_i^t)$, where $t \in \{1, 2, 3, ..., t_{pred}\}$ and $i \in \{1, 2, 3, ..., N\}$. Then, the coordinates of each pedestrian in the scene are divided into past trajectories X_i and future trajectories Y_i. As shown in Eq. 1 and Eq. 2:

$$X_i = \left\{ P_t^i \mid t = 1, 2, 3, ..., t_{obs} \right\} \tag{1}$$

$$Y_i = \left\{ P_t^i \mid t = t_{obs} + 1, t_{obs} + 2, t_{obs} + 3, ..., t_{pred} \right\} \tag{2}$$

Finally, this paper takes the past trajectories $\{X_i \mid i = 1, 2, 3, ..., N\}$ of pedestrians in the scene as input, with the objective of generating future trajectories $\left\{ \widehat{Y_i} \mid i = 1, 2, 3, ..., N \right\}$ that closely resemble the actual future trajectories $\{Y_i \mid i = 1, 2, 3, ..., N\}$ of the pedestrians. The generated future trajectories are defined as in Eq. 3:

$$\widehat{Y_i} = \left\{ \widehat{P_t^i} \mid t = t_{obs} + 1, t_{obs} + 2, t_{obs} + 3, ..., t_{pred} \right\} \tag{3}$$

3.2   Architecture Overview

We use an embedding dimension of 64 for the spatial coordinates before using them as input to the LSTM. We set the spatial pooling size N_0 to be 32 and use a 8×8 sum pooling window size without overlaps. We used a fixed hidden state dimension of 128 for all the LSTM models.

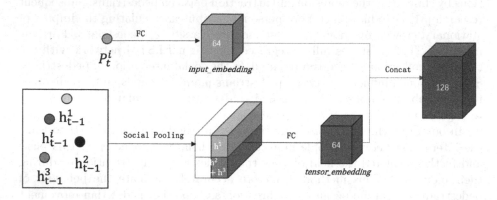

Fig. 1. At time step t, we integrate the neighbor information from the previous step and he current position of pedestrian i, update and obtain the current hidden state of pedestrian i.

Figure 1 is a schematic diagram of the data flow of our model at time t. At time t, the inputs to the S-CGRU include the coordinates of pedestrian i, $P_t^i = (x_i^t, y_i^t)$, as well as the hidden state information at time $t-1$(For example h_{t-1}^i). We use a social pooling layer to acquire the hidden state information of other pedestrians surrounding pedestrian i at time t(For example $h_{t-1}^a, h_{t-1}^b, h_{t-1}^c$). The social pooling layer of the hidden states implicitly infers the behavior of the nearby crowd, thereby adjusting its own path prediction. These nearby pedestrians are also influenced by their surrounding environment, and their behavior may change over time [8]. Next, we stack the hidden tensor at time t with the input data and feed them into the CGRU. After the CGRU processing, we obtain the hidden state at time t and the corresponding prediction results. Our model is built on the foundation of S-LSTM [1] but further optimizes the information processing and prediction mechanism. We particularly focus on how to more effectively extract information from the behavior of surrounding pedestrians, and how to integrate this information into our prediction model. By using the CGRU, we can more effectively utilize historical and environmental information, thereby improving the accuracy of pedestrian trajectory prediction.

3.3   Complex Gated Recurrent Unit

The Gated Recurrent Unit (GRU) is a variant of the Recurrent Neural Network (RNN) and was proposed by Kyunghyun Cho et al. [33]. The GRU solves the problem of gradient vanishing or gradient explosion that traditional RNNs may encounter when dealing with long sequences. The GRU introduces a gating mechanism to regulate the flow of information. Specifically, the GRU has two gates, the update gate and the reset gate:

- The update gate determines to what extent new input information is received. If the value of the update gate is close to 1, then the old memory is mainly retained; if it is close to 0, then the new input is mainly used.
- The reset gate determines how to use the previous hidden state when calculating the new candidate hidden state. If the value of the reset gate is close to 1, then most of the old memory is retained; if it is close to 0, then the old memory is ignored.

The reset gate determines how to use the previous hidden state when calculating the new candidate hidden state. If the value of the reset gate is close to 1, then most of the old memory is retained; if it is close to 0, then the old memory is ignored. Compared with the Long Short-Term Memory network (LSTM), the structure of the GRU is more concise because it only has two gates and does not have a separate cell state [33]. This makes the GRU computationally more efficient and easier to train, while still retaining good performance on many tasks, making it a popular choice in many neural network architectures.

In the Complex Gated Recurrent Unit (CGRU), this paper goes beyond the traditional real-number structure, extending it to the complex domain. In this way, each element not only contains a real part but also an imaginary part. This extension allows the model to perform calculations in the complex space, thus fully leveraging the unique characteristics of complex operations and demonstrating outstanding performance when processing temporal information. As shown in Fig. 2. In CGRU, we pass the input and hidden states separately into the linear transformation layers of the real and imaginary parts, and apply the sigmoid activation function to obtain the real and imaginary parts of the reset gate and update gate. This enables calculations to be performed in the complex space. The characteristics of complex calculations can help the network remember important information in the long term, thereby improving the network's performance in dealing with long sequence problems.

In the reset gate, since the real and imaginary parts of the complex hidden state can encode information independently, the complex reset gate can selectively forget information in higher dimensions. This may allow the CGRU to perform better when dealing with complex, high-dimensional sequence data. The calculation formulas for the real and imaginary parts of the reset gate are shown in Eq. 4 and Eq. 5:

$$r\_r_t^i = \sigma(W_{ir} * P_t^i - W_{hr} * h_{t-1}^i) \tag{4}$$

$$r\_i_t^i = \sigma(W_{ii} * P_t^i - W_{hi} * h_{t-1}^i) \tag{5}$$

W_{ir}, W_{hr}, W_{ii}, and W_{hi} are the weights of the model, P_t^i is the input, and h_{t-1}^i is the previous hidden state. $r\_r_t^i$ and $r\_i_t^i$ are the real and imaginary parts of the reset gate, respectively.

250 Z. Xu et al.

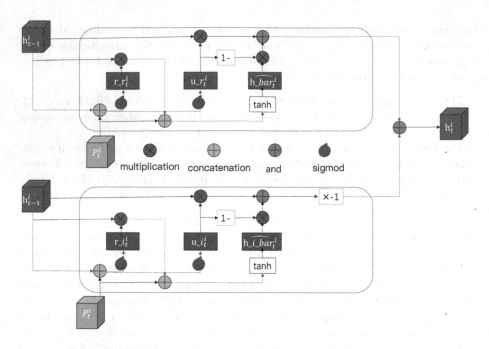

Fig. 2. Complex Gated Recurrent Unit.

The update gate in CGRU functions the same as in GRU, only now it operates on the complex form of the hidden state and the complex form of the input. This could possibly make CGRU perform better when handling complex, high-dimensional sequence data. The formulas for the update gate are shown in Eq. 6 and Eq. 7:

$$u\_r_t^i = \sigma(Wuzr * P_t^i - W_{hr} * h_{t-1}^i) \tag{6}$$

$$u\_i_t^i = \sigma(Wuzi * P_t^i - W_{hi} * h_{t-1}^i) \tag{7}$$

$Wuzr$, W_{hr}, $Wuzi$, and W_{hi} are the weights of the model, P_t^i is the input, and h_{t-1}^i is the previous hidden state. $u\_r_t^i$ and $u\_i_t^i$ are the real and imaginary parts of the update gate, respectively.

Based on the reset gate, we can calculate the real part of the new candidate hidden state $h\_bar_t^i$ (Eq. 8) and the imaginary part $h\_i\_bar_t^i$ (Eq. 9). Equation 10 describes how to combine the update gate, the old hidden state, and the new candidate hidden state in the complex space to obtain the new hidden state h_t^i.

$$h\_\hat{bar}_t^i = \sigma(W_{ir} * P_t^i - r\_r_t^i * W_{hr} * h_{t-1}^i) \tag{8}$$

$$h\_i\_\hat{bar}_t^i = \sigma(W_{ir} * P_t^i - r\_r_t^i * W_{hr} * h_{t-1}^i) \tag{9}$$

$$h_t^i = u\_r_t^i * h_{t-1}^i - u\_i_t^i * h_{t-1}^i + (1 - u\_r_t^i) * h\_\hat{bar}_t^i - (1 - u\_i_t^i) * h\_i\_\hat{bar}_t^i \tag{10}$$

3.4   Exploring Pedestrians' Field of Vision in Different Environments

Understanding and considering the impact of other pedestrians within the field of vision is vital in predicting pedestrian motion patterns. As per studies, pedestrians typically adjust their trajectories based on the positions and movements of other pedestrians within their field of view to avoid collisions. For other pedestrians behind an individual in certain environments, their influence on the trajectory is usually considered negligible. In this paper, we focus on exploring to what extent the movement trajectory of a pedestrian is affected by other pedestrians within a certain angular range. Our aim is to clarify and quantify this influence to accurately simulate this interpersonal interaction behavior when building predictive models. This approach not only helps to simplify the volume of information the model needs to process but also prevents unnecessary or irrelevant information from adversely impacting the accuracy of the prediction model. We hope that this method will enhance the accuracy of pedestrian motion trajectory predictions and provide a useful theoretical foundation and practical reference for related fields.

As shown in Fig. 3, Fig. 3a illustrates the scope of social pooling in [1,8]: based on the hidden state of the LSTM encoder, it uses the method of maximum pooling; then, it merges the relative position coordinates of each neighbor to simulate the interaction between individuals. In models [1,8], taking Fig. 3a as an example, the emphasis is on analyzing the influence of other pedestrians on the trajectory of pedestrian P0 within a circle centered on the coordinates of P0 and with a certain distance as the radius. In Fig. 2a, pedestrians P1, P2, P3, and P4 are all within this range, and they will affect the movement trajectory of P0. Conversely, since P5 is outside this range, we believe that P5 does not have a direct impact on the movement trajectory of P0.

In the process of handling interactions, although pooling all the information of neighboring pedestrians seems to be a solution, we do not recommend this approach. The global information containing all pedestrians' data often contains a large amount of redundant or irrelevant data. For instance, the movement information of pedestrians within a certain angular range behind the target pedestrian may have a negative impact on the accuracy of prediction results [4].

Therefore, we propose a new method that more comprehensively considers various factors that may affect pedestrian interactions. This method pays particular attention to the field of view (FoV) of the pedestrian, only considering those within this FoV as influences on the pedestrian's future movements. Our assumption is based on the fact that people generally pay more attention to the events in their forward direction and ignore those behind. This effectively reduces the volume of unnecessary information and increases the computational efficiency of our model, making it more applicable to real-world scenarios.

In our approach, we set a hypothesis: other pedestrians who influence the trajectory of the target pedestrian should be located within a sector region forming a certain angle in front of the target pedestrian. We also strive to find the optimal angle of this sector region. Taking Fig. 3b as an example, we explore how other

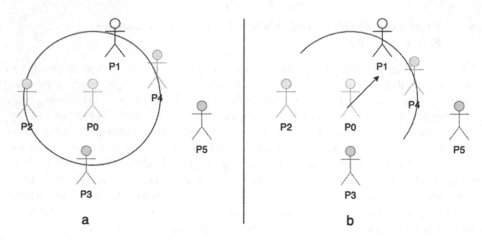

Fig. 3. Figure 3 illustrates the pedestrian movement pooling range. Figure 3a is a demonstration of the social pooling range based on [1,8], while Fig. 3b is a schematic diagram of the social pooling range proposed in this paper.

pedestrians within a certain radius of the sector region (centered on pedestrian P0) influence P0's movement trajectory. Pedestrians in this area, such as P1 and P4, will affect the movement path of P0. However, for pedestrians outside the region, such as P2, P3, and P5, we believe they will not have a direct impact on P0's movement path. We further discussed the method to determine the optimal angle of the sector region in different scenarios. More detailed information can be found in Sect. 4.6.

3.5   Loss Functions

Similar to the method of the loss function in reference [34], the goal of training our model is to minimize the negative log-likelihood loss, as shown in Eq. 11.

$$L = - \sum_{t=t_{obs+1}}^{t_{pred}} \log \left\{ P\left(x_t^i, y_t^i \mid u_t^i, \sigma_t^i, \rho_t^i\right) \right\} \tag{11}$$

4   Experiments

4.1   Dataset and Evaluation Metrics

In the task of human trajectory prediction, this paper adopts six widely used datasets: Stanford Drone Dataset (SDD) [34], ETH datasets: Hotel and ETH [36], UCY datasets: UNIV, Zara1, and Zara2 [37]. The Stanford Drone Dataset (SDD) records videos within a university campus, containing six types of agents and rich interaction scenarios. In this dataset, there are approximately 185,000 interactions between agents and around 40,000 interactions between agents and the environment. The ETH/UCY datasets contain human trajectories from five scenarios, all of which record the movement paths of pedestrians in the world coordinate system. These datasets provide us with a variety of scenarios and rich interactions, which are beneficial to our comprehensive study and understanding of human behavioral trajectories. The use of these datasets offers a broad sample for our research, helping us to understand and predict human behavioral trajectories more deeply.

Table 1. Datasets for human trajectory prediction

Dataset Name	Category
Stanford Drone Dataset	SDD
ETH datasets	Hotel
	ETH
UCY datasets	UNIV
	Zara1
	Zara2

In line with previous studies [1,8], this paper uses two main metrics to evaluate model performance: 1) Average Displacement Error (ADE), 2) Final Displacement Error (FDE). ADE is a method to evaluate the mean square error between the predicted trajectory and the actual trajectory, which is calculated in the real-world coordinate system. It quantifies the average difference between the model's predicted trajectory and the actual trajectory, thereby reflecting the overall prediction accuracy of the model (Eq. 12). On the other hand, FDE focuses on the endpoint of the predicted trajectory. It calculates the L2 distance between the last point of the predicted trajectory and the last point of the actual trajectory in the real-world coordinate system. FDE measures the model's accuracy in predicting the endpoint of the trajectory. These two evaluation metrics can effectively evaluate the model's overall performance in trajectory prediction and the accuracy of endpoint trajectory prediction (Eq. 13).

$$ADE = \frac{\sum_{t=t_{obs}+1}^{t_{pred}} \| p_t^i - \widehat{p}_t^i \|}{t_{pred} - t_{obs}} \tag{12}$$

$$FDE = \| p_{t_{pred}}^i - \widehat{p}_{t_{pred}}^i \| \tag{13}$$

In order to maximize the use of these datasets and optimize our model, we adopt a special form of cross-validation known as the leave-one-out method [38]. Our approach is to train and validate the model on five datasets, then test on the remaining one. This process is iterated on each of the six datasets. Similarly, we also apply this training and testing strategy to the baseline methods used for model comparison.

4.2   Baselines

In order to evaluate the effectiveness of our proposed S-CGRU model, we compared it with some baseline methods for trajectory prediction, specifically as follows:

- Linear Model (Lin.): We use a pre-set Kalman filter, based on the assumption of linear acceleration to extrapolate trajectories.
- LSTM: LSTM is applied to cyclically predict each pedestrian's future location from historical locations.
- S-LSTM [1]: This is a trajectory prediction model that combines LSTM and social pooling layer, capable of integrating the hidden states of neighboring pedestrians.
- S-GRU: This is a novel pedestrian trajectory prediction model based on GRU that we propose. This model serves as a baseline for validating the superior performance of the Complex Gate Recurrent Unit (CGRU) that we designed.

The comparison of the above four methods helps to comprehensively evaluate the performance of our proposed S-CGRU model.

4.3   Evaluations on ADE/FDE Metrics

During the testing phase, we tracked trajectories for 3.2 s and predicted their paths for the next 4.8 s. At a frame rate of 0.4 s, this is equivalent to observing 8 frames and predicting the next 12 frames, which is similar to the setting in reference [8]. In Table 2, the performance of our model is compared with the baseline methods.

Table 2. Results on ETH, UCY and SDD based standard-sampling. 20 samples are used in prediction and the minimal error is reported.

Methods	Lin		LSTM		S-LSTM [1]		S-GRU (ours)		S-CGRU (ours)	
Metrics	ADE	FDE	ADE	FDE	ADE	FDE	ADE	FDE	ADE	FDE
ETH	1.31	1.35	0.96	1.36	0.82	1.11	0.96	1.24	**0.54**	**0.73**
Hotel	1.98	2.20	0.76	1.36	0.56	0.95	0.78	1.10	**0.49**	**0.61**
UNIV	0.71	0.62	0.65	0.68	**0.34**	0.42	0.56	0.83	**0.34**	**0.39**
ZARA1	1.20	1.76	1.09	1.84	0.56	0.95	0.76	1.12	**0.49**	**0.59**
ZARA2	0.72	0.98	0.82	1.18	0.40	0.54	0.65	0.73	**0.39**	**0.44**
SDD	0.83	1.12	0.62	0.93	0.50	0.75	0.60	0.82	**0.48**	**0.63**
AVG	1.13	1.34	0.82	1.23	0.53	0.79	0.72	0.97	**0.46**	**0.57**

From the results in Table 1, we can clearly see that our proposed S-CGRU model significantly outperforms the baseline model S-LSTM on all metrics. This fully validates the effectiveness of our proposed model in pedestrian trajectory prediction tasks. Firstly, compared to S-LSTM, our S-CGRU model has significantly improved in terms of prediction accuracy. This suggests that by utilizing Complex Gate Recurrent Units (CGRU), S-CGRU is able to more accurately capture and predict pedestrian movement trajectories. This can be attributed to CGRU's excellent performance in handling sequential data, especially in capturing long-term dependencies. Secondly, the S-CGRU model offers better generalization performance. Regardless of the scenario or the number of pedestrians, S-CGRU can consistently output high-quality predictions. This indicates that our model can adapt well to various different circumstances, showing good adaptability to various environments. Finally, our S-CGRU model has excellent real-time performance. Compared to S-LSTM, even though both are close in computational complexity, our optimizations to the model structure allow S-CGRU to generate prediction results faster in actual applications, thereby meeting the needs of real-time applications. In summary, our S-CGRU model outperforms S-LSTM in pedestrian trajectory prediction tasks in terms of prediction accuracy, generalization ability, and real-time performance, overcoming the limitations of traditional GRU models. This validates the superiority of our model design and its application potential in pedestrian trajectory prediction tasks.

4.4 Model Parameter Amount and Inference Speed

To accurately assess the inference speed of various models, we specifically conducted an in-depth comparison of our Social-CGRU model and other publicly available models that can serve as comparison benchmarks. This comparison considers two key indicators of the model: parameter scale and inference speed. Test data is derived from time series data, obtained by densely sampling with a time step set to 1, and window size to 20 (including observation period Tobs(8) and prediction period Tpred(12)). When testing inference speed, we calculate the average inference time for all data segments.

Through such testing and comparison, we found that the S-CGRU model showed significant advantages. Since the S-CGRU model only contains two gates, reset and update, the number of its parameters is drastically reduced. Fewer parameters mean fewer variables need to be adjusted during the training process, which can significantly speed up model training. Moreover, fewer parameters make the S-CGRU model more lightweight, making it more suitable for running in resource-constrained environments, thereby improving its practicality. More importantly, the S-CGRU model has excellent performance in terms of inference speed. Fast inference speed not only provides faster feedback in real-time applications but also shows that the model can maintain high efficiency when handling large amounts of data, which helps improve overall work efficiency. This is especially important in application scenarios where quick decisions are needed or large amounts of data need to be processed. In summary, the excellent performance of the S-CGRU model in terms of parameter quantity and inference speed gives it a wide range of potential and practical value in various application scenarios. Detailed results are shown in Table 3.

Table 3. Comparisons of parameter amount and inference speed on ETH, UCY and SDD datasets.

Methods	Parameters (k)	Speed (time/batch)
S-LSTM	264	2.2
S-CGRU(ours)	**231**	**1.6**

4.5　The Range of Visual Focus of Pedestrians While Walking in Different Scenarios

We conducted in-depth research on six datasets: ETH, Hotel, UNIV, ZARA1, ZARA2, and SDD, aiming to reveal the range of vision during pedestrian walking in different scenarios. These datasets cover a variety of environments, such as schools, hotels, streets, intersections, etc., providing us with a wealth of pedestrian behaviors and movement patterns. In our research, we systematically adjusted the angle of the pedestrian's field of view to simulate the degree to which they might be influenced by other pedestrians around them. Then, through the training evaluation system, we detected and recorded the performance of pedestrian trajectory prediction. This method allows us to infer the main field of view that pedestrians pay attention to during walking based on the changes in prediction performance. Detailed results are shown in Table 4.

Table 4. Optimal field of view angles for different scenes on ETH, UCY and SDD datasets.

Scene (Dataset)	Optimal Field of View Angle (Degrees)
Street (ETH)	85
University (UNIV)	135
Hotel (Hotel)	75
Intersection (SDD)	120

We found that in various environments, the range of pedestrian vision is influenced by the surrounding environment. For example, in crowded environments such as hotels and street scenes, the range of pedestrian vision may narrow, mainly focusing on obstacles or other pedestrians directly in front. In contrast, in open environments like intersections, the range of pedestrian vision may expand as they need to consider information from more directions. Additionally, we discovered that the direction of pedestrian movement, speed, and individual characteristics (such as vision conditions) may also affect their field of view.

These findings have significant implications for our model design. Firstly, understanding the range of pedestrian vision can help us better understand and predict their behavior. Secondly, based on understanding the field of view, we can design more accurate and practical pedestrian trajectory prediction models.

By considering the range of pedestrian vision, our model can make more precise predictions about pedestrian behavior in different environments, thereby improving the accuracy of prediction results.

5   Conclusions and Future Works

This research primarily explores the application of Complex Gated Recurrent Units (CGRU) in pedestrian trajectory prediction and proposes a new S-CGRU model, effectively integrating interpersonal interaction information and scene information. We first detailed the basic concepts and theories of complex neural networks and Gated Recurrent Units (GRU) and explored the advantages of using complex parameters in neural network models. Subsequently, we investigated the range of vision that pedestrians pay attention to during walking, providing a new theoretical perspective for understanding pedestrian behavior. Our research results show that the S-CGRU model has a significant advantage over the baseline model S-LSTM in dealing with pedestrian trajectory prediction issues. In the future, we will focus on person-scene interaction modeling (the interaction between pedestrians and their surroundings, and between pedestrians and vehicles). We hope to improve prediction performance by combining person-scene interaction modeling with our S-CGRU model. Moreover, predicting pedestrian trajectories by combining pedestrian movement intentions is also a challenge that we need to address.

Acknowledgements. This work is supported by the National Natural Science Foundation of China (61562082), the Joint Funds of the National Natural Science Foundation of China (U1603262), and the "Intelligent Information R&D Cross-disciplinary Project" (Project Number: 202104140010). We thank all anonymous commenters for their constructive comments.

References

1. Alahi, A., Goel, K., Ramanathan, V., Robicquet, A., Fei-Fei, L., Savarese, S.: Social LSTM: human trajectory prediction in crowded spaces. In: Proceedings of the IEEE Conference on Computer Vision and Pattern Recognition, pp. 961–971 (2016)
2. Deo, N., Trivedi, M.M.: Convolutional social pooling for vehicle trajectory prediction. In: Proceedings of the IEEE Conference on Computer Vision and Pattern Recognition Workshops, pp. 1468–1476 (2018)
3. Rudenko, A., Palmieri, L., Herman, M., Kitani, K.M., Gavrila, D.M., Arras, K.O.: Human motion trajectory prediction: a survey. Int. J. Robot. Res. **39**(8), 895–935 (2020)
4. Yue, J., Manocha, D., Wang, H.: Human trajectory prediction via neural social physics. arXiv preprint arXiv:2207.10435 (2022)
5. Helbing, D., Molnar, P.: Social force model for pedestrian dynamics. Phys. Rev. E **51**(5), 4282 (1995)

6. van den Berg, J., Lin, M., Manocha, D.: Reciprocal velocity obstacles for real-time multi-agent navigation. In: 2008 IEEE International Conference on Robotics and Automation (2008)
7. He, F., Xia, Y., Zhao, X., Wang, H.: Informative scene decomposition for crowd analysis, comparison and simulation guidance. ACM Transaction on Graphics (TOG) 4(39) (2020) 51(5), 4282 (1995)
8. Gupta, A., Johnson, J., Fei-Fei, L., Savarese, S., Alahi, A.: Social GAN: socially acceptable trajectories with generative adversarial networks. In: Proceedings of the IEEE Conference on Computer Vision and Pattern Recognition, pp. 2255–2264 (2018)
9. Sadeghian, A., Kosaraju, V., Sadeghian, A., Hirose, N., Rezatofighi, H., Savarese, S.: SoPhie: an attentive GAN for predicting paths compliant to social and physical constraints. In: Proceedings of the IEEE/CVF Conference on Computer Vision and Pattern Recognition, pp. 1349–1358 (2019)
10. Mangalam, K., An, Y., Girase, H., Malik, J.: From goals, waypoints & paths to long term human trajectory forecasting. In: Proceedings of the IEEE/CVF International Conference on Computer Vision, pp. 15233–15242 (2021)
11. Van Toll, W., Pettr'e, J.: Algorithms for microscopic crowd simulation: advancements in the 2010s. Comput. Graph. Forum 40(2), 731–754 (2021)
12. Wolinski, D., J. Guy, S., Olivier, A.H., Lin, M., Manocha, D., Pettr'e, J.: Parameter estimation and comparative evaluation of crowd simulations. Comput. Graph. Forum 33(2), 303–312 (2014)
13. He, F., Xia, Y., Zhao, X., Wang, H.: Informative scene decomposition for crowd analysis, comparison and simulation guidance. ACM Trans. Graph. (TOG) 39(4), 50:1–50:13 (2020)
14. Korbmacher, R., Tordeux, A.: Review of pedestrian trajectory prediction methods: comparing deep learning and knowledge-based approaches. IEEE Trans. Intell. Transp. Syst. 23(12), 24126–24144 (2022)
15. Bengio, Y., Pal, C.J.: Deep complex networks. In: International Conference on Learning Representations (ICLR) (2018)
16. Nitta, T.: On the critical points of the complex-valued neural network. In: Neural Information Processing (2002)
17. Hirose, A., Yoshida, S.: Generalization characteristics of complex-valued feedforward neural networks in relation to signal coherence. IEEE Trans. Neural Netw. Learn. Syst. 23(4), 541–551 (2012)
18. Arjovsky, M., Shah, A., Bengio, Y.: Unitary evolution recurrent neural networks. arXiv preprint arXiv:1511.06464 (2015)
19. Danihelka, I., Wayne, G., Uria, B., Kalchbrenner, N., Graves, A.: Associative long short-term memory. arXiv preprint arXiv:1602.03032 (2016)
20. Wisdom, S., Powers, T., Hershey, J., Roux, J.L., Atlas, L.: Full-capacity unitary recurrent neural networks. In: Advances in Neural Information Processing Systems, pp. 4880–4888 (2016)
21. Reichert, D.P., Serre, T.: Neuronal synchrony in complex-valued deep networks. arXiv preprint arXiv:1312.6115 (2013)
22. Srivastava, R.K., Greff, K., Schmidhuber, J.: Training very deep net-works. In: Advances in Neural Information Processing Systems, pp. 2377–2385 (2015)
23. Cho, K., Van Merriënboer, B., Bahdanau, D., Bengio, Y.: On the properties of neural machine translation: Encoder-decoder approaches. arXiv pre-print arXiv:1409.1259 (2014)
24. Hochreiter, S., Schmidhuber, J.: Long short-term memory. Neural Comput. 9(8), 1735–1780 (1997)

25. Kingma, D.P., Ba, J.: Adam: a method for stochastic optimization. In: 3rd International Conference on Learning Representations (2015)
26. Kipf, T.N., Welling, M.: Semi-supervised classification with graph convolutional networks. In: International Conference on Learning Representations (2017)
27. Antonini, G., et al.: Discrete choice models of pedestrian walking behavior. Transport. Res. B **40**(8), 667–687 (2006)
28. Bahdanau, D., et al.: Neural machine translation by jointly learning to align and trans-late. In: 3rd International Conference on Learning Representations (2015)
29. Lerner, A., et al.: Crowds by example. Comput. Graphics Forum. **26**, 655–664 (2007)
30. Helbing, D., Molnár, P.: Social force model for pedestrian dynamics. Phys. Rev. E, Stat. Phys. Plasmas Fluids Relat. Interdiscip. Top. **51**(5), 4282 (1995)
31. Yi, S., Li, H., Wang, X.: Understanding pedestrian behaviors from stationary crowd groups. In: Proceedings of IEEE Conference Computer Vision and Pattern Recognition (CVPR), pp. 3488–3496 (2015)
32. Xue, H., Huynh, D.Q., Reynolds, M.: SS-LSTM: a hierarchical LSTM model for pedestrian trajectory prediction. In: Proceedings of IEEE Winter Conference on Applications of Computer Vision (WACV), pp. 1186–1194 (2018)
33. Cho, K., et al.: Learning phrase representations using RNN encoder-decoder for statistical machine translation. In: Proceedings of the 2014 Conference on Empirical Methods in Natural Language Processing (EMNLP) (2014)
34. Mohamed, A., Qian, K., Elhoseiny, M., Claudel, C.: Social-STGCNN: a social spatio-temporal graph convolutional neural network for human trajectory prediction. In: IEEE/CVF Conference on Computer Vision and Pattern Recognition (CVPR) (2020)
35. Robicquet, A., Sadeghian, A., Alahi, A., Savarese, S.: Learning social etiquette: human trajectory understanding in crowded scenes. In: Leibe, B., Matas, J., Sebe, N., Welling, M. (eds.) ECCV 2016. LNCS, vol. 9912, pp. 549–565. Springer, Cham (2016). https://doi.org/10.1007/978-3-319-46484-8_33
36. Pellegrini, S., Ess, A., Van Gool, L.: Improving data association by joint modeling of pedestrian trajectories and groupings. In: Daniilidis, K., Maragos, P., Paragios, N. (eds.) ECCV 2010. LNCS, vol. 6311, pp. 452–465. Springer, Heidelberg (2010). https://doi.org/10.1007/978-3-642-15549-9_33
37. Lerner, A., Chrysanthou, Y., Lischinski, D.: Crowds by example. In: Computer graphics forum. vol. 26, pp. 655–664. Wiley Online Library (2007)
38. Tang, H., Wei, P., Li, J., Zheng, N.: EvoSTGAT: evolving spatio-temporal graph attention networks for pedestrian trajectory prediction. Neurocomputing **491**, 333–342 (2022)
39. Sadeghian, A., Kosaraju, V., Sadeghian, A., Hirose, N., Rezatofighi, H., Savarese, S.: SoPhie: an attentive GAN for predicting paths compliant to social and physical constraints. In: Proceedings of the IEEE/CVF Conference on Computer Vision and Pattern Recognition, pp. 1349–1358 (2019)
40. Danihelka, I., Wayne, G., Uria, B., Kalchbrenner, N., Graves, A.: Associative long short-term memory. In: Proceedings of The 33rd International Conference on Machine Learning (2016)

Prior-Enhanced Network for Image-Based PM2.5 Estimation from Imbalanced Data Distribution

Xueqing Fang[1], Zhan Li[1(✉)], Bin Yuan[2,3], Xinrui Wang[1], Zekai Jiang[1],
Jianliang Zeng[1], and Qingliang Chen[1]

[1] Department of Computer Science, Jinan University, Guangzhou, China
lizhan@jnu.edu.cn
[2] Institute for Environmental and Climate Research, Jinan University, Guangzhou, China
[3] Guangdong-Hongkong-Macau Joint Laboratory of Collaborative Innovation for
Environmental Quality, Guangzhou, China

Abstract. The effective monitoring of PM2.5, a major indicator of air pollution, is crucial to human activities. Compared to traditional physiochemical techniques, image-based methods train PM2.5 estimators by using datasets containing pairs of images and PM2.5 levels, which are efficient, economical, and convenient to deploy. However, existing methods either employ handcrafted features, which can be influenced by the image content, or require additional weather data acquired probably by laborious processes. To estimate the PM2.5 concentration from a single image without requiring extra data, we herein propose a learning-based prior-enhanced (PE) network—comprising a main branch, an auxiliary branch, and a feature fusion attention module—to learn from an input image and its corresponding dark channel (DC) and inverted saturation (IS) maps. In addition, we propose an histogram smoothing (HS) algorithm to solve the problem of imbalanced data distribution, thereby improving the estimation accuracy in cases of heavy air pollution. To the best of our knowledge, this study is the first to address the phenomenon of a data imbalance in image-based PM2.5 estimation. Finally, we construct a new dataset containing multi-angle images and more than 30 types of air data. Extensive experiments on image-based PM2.5 monitoring datasets verify the superior performance of our proposed neural networks and the HS strategy.

Keywords: PM2.5 estimation · Prior information · Imbalanced data distribution · Atmospheric pollution

1 Introduction

As a typical index of outdoor air pollutants, the concentration of PM2.5, or fine particulate matter with aerodynamic diameters of 2.5 μm or less [1], has a significant effect on people's health. PM2.5 encompasses pathogenic microorganisms as well as chemical

---

X. Fang and Z. Li—Both authors contributed equally to this work.

© The Author(s), under exclusive license to Springer Nature Singapore Pte Ltd. 2024
B. Luo et al. (Eds.): ICONIP 2023, CCIS 1964, pp. 260–271, 2024.
https://doi.org/10.1007/978-981-99-8141-0_20

pollutants that cause respiratory obstruction and inflammation. Recent studies have also observed a positive correlation between high PM2.5 concentrations and the fatality rate associated with COVID-19 [17]. Therefore, effective monitoring of PM2.5, especially in cases of high pollution levels, is crucial for people's health.

Physiochemical techniques monitor the PM2.5 concentration by gauging its weight using dedicated devices, which include β-ray absorption, gravimetric analysis, and tapered element oscillating microbalance (TEOM) [2]. Although these sensor-based PM2.5 measurements can produce highly accurate estimations, they inevitably introduce a high maintenance cost and are unavailable in most regions [6]. Therefore, a more effective, efficient, and convenient solution is required to supplement the traditional methods for PM2.5 monitoring.

The concentration of PM2.5 is closely related to the characteristics of atmospheric scattering, which further affects the performance of outdoor imaging systems, as shown in Fig. 1. Therefore, PM2.5 concentrations can be estimated by image analysis [7, 14] in most cases except in extreme weather conditions, such as heavy rain, sandstorm, and sunset. Compared with their physiochemical sensor-based counterparts, image-based methods are economical, convenient to deploy, and available everywhere owing to the widespread use of mobile phones and portable cameras. However, to train image-based PM2.5 estimators, datasets containing pairs of images and PM2.5 levels are required, which is not easy to collect. Moreover, because severe air pollution is unusual, most images are captured in cases of mild or moderate pollution, which could introduce bias into the trained models [21]. Consequently, the estimation accuracy is significantly decreased under heavily polluted conditions with high PM2.5.

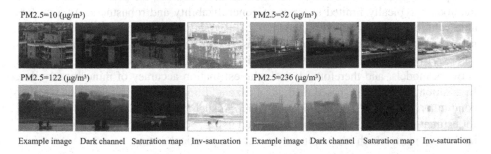

Fig. 1. Examples of images from different scenes under low and high PM2.5 concentrations. Adjacent images are their dark channels, saturation and inverted saturation maps, respectively.

In this paper, we propose a prior-enhanced (PE) network to estimate the PM2.5 concentration from a single image in an end-to-end manner, without handcrafted features or additional weather data. The proposed network comprises two branches: a main branch that aims to extract features from an input natural image, and an auxiliary branch that introduces prior knowledge of the dark channel (DC) and inverted saturation (IS) of the input image. Furthermore, to mitigate the problem of imbalanced data distribution, we propose a histogram smoothing (HS) algorithm, which can improve the estimation accuracy for cases of heavy pollution. This study is the interdisciplinary research of computer vision, deep learning, and atmospheric science, which can be applied as a supplement of the traditional physiochemical techniques. The main contributions of this study can be summarized as follows:

- We propose end-to-end PE neural networks to estimate the PM2.5 concentration from a single image without requiring data from outside sources. The results of extensive experiments verify the superior performance of the proposed networks.
- We introduce prior information in the DC and IS maps of images by designing a lightweight auxiliary branch. The saturation map is inverted to maintain consistency between the two maps and changes in PM2.5 values, thus considerably improving the estimation accuracy.
- To the best of our knowledge, this study is the first to investigate the imbalanced distribution for the task of image-based PM2.5 estimation. As a solution, we propose an HS algorithm to improve the estimation accuracy for cases of heavy pollution.
- We construct a new dataset comprising multi-angle images and some related weather data, as observed in Heshan, Guangdong, China, which can be used to evaluate the performance of the methods for estimating PM2.5 concentration.

2   Related Work

PM2.5 Estimation from Images. Traditional image-based methods focus on mining hand-crafted PM2.5-sensitive features, such as entropy [7,14] and gradient [27] from the input images. These methods are time and resource efficient. However, because the entropy and gradient of an image are sensitive to its content, the performance of these methods is typically limited in terms of generalizability and robustness. As a preferable solution, learning-based methods [6,24] extract image features automatically by neural networks, which learn from the datasets containing pairs of images and PM2.5 values. However, the imbalanced data distribution could introduce bias into the trained network models, and therefore, degrade the estimation accuracy of minority samples. In addition, some image-based methods utilize additional information such as weather conditions [5,14] and field depth [24]. Although such additional information is helpful for estimating PM2.5, it is generally obtained using professional procedures or specialized equipment, which is often inconvenient in a practical setting.

Data-Imbalanced Learning. Imbalanced data are ubiquitous and inherent in the real world. Previous studies on the data imbalance problem mainly focus on classification tasks, which can be divided into data- and model-based solutions. Data-based solutions either oversample the minority classes [3,13] or undersample the majority classes [19], whereas model-based solutions re-weight or adjust their loss functions [25,28] to balance the data distribution. In the classification context, the number of categories is finite, whereas category labels are generally discrete. In contrast, regression tasks such as PM2.5 estimation typically involve continuous and infinite target values with ranges of missing data. Consequently, the issue of imbalanced data in regression tasks has not been sufficiently explored [26]. In the context of PM2.5, the problem of imbalanced data has rarely been discussed. Rijal *et al.* [21] reported that their method failed to accurately estimate high PM2.5 concentrations due to a severely insufficient number of these samples in the training set. However, they did not obtain an effective solution to this problem.

3 Methodology

Given an input image I, PM2.5 concentration estimators aim at outputting a value $\hat{y}$ as an approximation of the actual PM2.5 level y. The mapping function can be formulated as follows:

$$\hat{y} = G_\theta(I), \tag{1}$$

where θ represents learnable parameters of the network G. In order to optimize G_θ, specific loss functions are designed on the training samples. However, due to the diversity of image scenes, in our experiments, we find that networks only fed by an input image itself are hard to train.

We argue that the image degradation priors, *e.g.*, dark channel prior and saturation attenuation, are beneficial for capturing the characteristic of haze in images, and therefore ease the network training. The prior information Ψ can be conveniently represented by the concatenation of different prior maps, wherein we leverage the dark channel I^{dc} and inverted saturation I^{inv-s} derived from the input image,

$$\Psi = P = (I^{dc}, I^{inv-s}). \tag{2}$$

Now, we can reformulate Eq. 1 as:

$$\hat{y} = G_\theta(I|\Psi), \tag{3}$$

where Ψ defines the prior information used as input conditions. In the following section, we introduce the estimation of DC-IS maps, and the proposed PE network architecture.

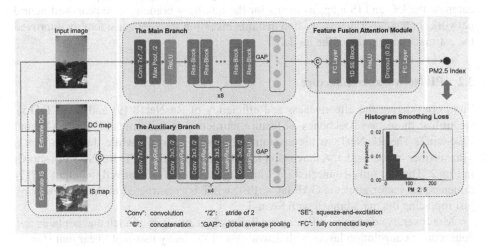

Fig. 2. Architecture of the proposed prior-enhanced neural network (PE-ResNet18).

3.1 Estimation of DC-IS Maps

Following He *et al.* [8], the dark channel of image I is defined as:

$$I^{dc}(x_0) = \min_{c\in\{r,g,b\}} (\min_{x\in\Omega(x_0)} (I^c(x))), \tag{4}$$

where x and x_0 represent the pixel coordinates in horizontal and vertical directions; $\Omega(x_0)$ is a local patch centered at position x_0; r, g, and b represent red, green, and blue, respectively; and I^c is a color component of input image I.

The saturation map [23] of the image I is defined as:

$$I^s(x) = \begin{cases} 0, & \text{if } V(x) = 0 \\ \frac{V(x)-\dot{D}(x)}{V(x)}, & \text{otherwise} \end{cases}, \tag{5a}$$

with

$$V(x) = \max_{c\in\{r,g,b\}} (I^c(x)), \dot{D}(x) = \min_{c\in\{r,g,b\}} (I^c(x)). \tag{5b}$$

Note that $\dot{D}(x)$ can be regarded as a special case of $I^{dc}(x)$ in Eq. (4), with a neighborhood patch $\Omega(x)$ of size 1×1. A higher concentration of PM2.5 increases the risk of haze phenomenon, resulting in grayish-white images with generally higher intensities.

Figure 1 presents several example images under different conditions of PM2.5 concentration, along with their corresponding DC, saturation, and IS maps. As shown in Fig. 1, a high concentration of PM2.5 is associated with high intensities in DC (overall brighter) and low intensities in saturation (overall darker). To keep the changing trends in both maps consistent, we invert the saturation values to obtain an IS map, which is calculated as follows:

$$I^{inv-s}(x) = 255 - I^s(x), \tag{6}$$

where 255 indicates the maximum pixel intensity in an image. Subsequently, we concatenate the DC and IS maps as inputs for the auxiliary branch in the proposed neural network. In our experiments, we find that this inversion strategy considerably improved the estimation accuracy of the model.

3.2 PE Network Architecture

First, considering the effectiveness and efficiency of ResNet18 [9] and Swin-T [15], we utilize these two backbones as feature extractors to construct PE-ResNet18 and PE-SwinT. The ResNet18 is a convolutional neural network, while the Swin-T is a Transformer-based model. As an example, the main branch of PE-ResNet18 is shown in Fig. 2. The last fully-connected (FC) layer of ResNet18 is removed to output a 512-dimensional vector after the GAP layer. PE-SwinT is constructed in the same way by just replacing the main branch with Swin-T.

Moreover, we design an efficient and lightweight auxiliary branch that contains five consecutive convolution layers with activations of the leaky rectified linear unit (Leaky ReLU) [16]. This branch takes the DC-IS maps as inputs, performs downsampling and non-linear transformations, and outputs a 512-dimensional vector corresponding to that generated by the main branch. The process of feature extraction can be expressed as:

$$f_m = G_m(I), f_a = G_a(I^{dc}, I^{inv-s}), \tag{7}$$

where G_m and G_a indicate the main and the auxiliary branch, f_m and f_a indicate the extracted feature vectors.

Finally, the feature fusion attention module is designed to process the concatenation of features output from the main and auxiliary branches, as shown in Fig. 2. In the design of our one-dimensional (1D) squeeze and excitation (SE) block, the GAP layer was removed from the original SE block [11], as the layer is typically used to squeeze multichannel feature maps into vectors. In contrast, all features to be processed in our module are already in the form of one-dimensional vectors. The SE block adaptively learns the weight of each position in the input [11], thereby improving the representational power of the module. The feature fusion stage can be formulated as:

$$f_u = SE(FC(Concatenate(f_m, f_a))), \tag{8}$$

where f_u represents the prior-enhanced features. In addition, a ReLU activation layer and dropout layer are added to introduce nonlinearity and prevent overfitting. At the end of the feature fusion attention module, an FC layer is employed to output the PM2.5 estimation.

3.3   Histogram Smoothing Algorithm

Most datasets used for PM2.5 estimation are highly unbalanced and often exhibit a long tail distribution, which could introduce bias into the trained models, and therefore degrade the estimation accuracy of few-shot samples. To solve this problem, we propose a histogram smoothing (HS) strategy specified in Algorithm 1, which is a cost-sensitive and adaptive re-weighting method for the training of our PE networks.

By the HS algorithm, a histogram of PM2.5 data is created using a predefined bin width and then fitted by a Gaussian kernel to produce a smooth probability density distribution. As increasing proportions of estimation errors for few-shot samples will guide the network to learn more from them, a weight proportional to the reciprocal of the sample number in each interval is assigned. After getting the adaptive weights, we can reformulate a specific loss function, e.g., mean squared error (MSE), as follows:

$$L_w = \frac{1}{n} \sum_{i=1}^{n} w_i (\hat{y}_i - y_i)^2, \tag{9}$$

where $\hat{y}_i$ denotes the estimated PM2.5 value of the i-th sample, y_i denotes the actual PM2.5 value measured by instruments, and w_i denotes the weights computed by the proposed HS algorithm. In our experiment, the Gaussian kernel size and the standard deviation are empirically set to $s = 5$ and $\sigma = 2$, respectively. The selection of bin width is discussed in Sect. 4.3.

Algorithm 1. Histogram smoothing algorithm.

Input:

    Set of PM2.5 indices $\{y_i\}$ from the training data.

    Bin width bw, Gaussian kernel size s, and standard deviation σ.

    Predefined range for clipping sample numbers in each bin with default [5, 500].

Output:

    Weighted loss value L_w.

1: Construct a histogram of the PM2.5 index set $\{y_i\}$ using a bin width of bw, i.e., $b_{j+1} - b_j = bw$, where $j = 1, 2, ..., m$, and $m = \lceil (\max\{y_i\} - \min\{y_i\}) \, / \, bw \rceil$;

2: Clip the number of samples in each bin to mitigate the degree of imbalance;

3: Fit the histogram by a Gaussian kernel with a size of s and standard deviation of σ to smoothen the density distribution of PM2.5 data and obtain fitted sample numbers c_j^b in each bin;

4: Calculate the weight factor of each bin as $w_j^b = k \times (1/c_j^b)$, with $k = m \, / \, \sum_1^m (1/c_j^b)$;

5: Obtain the weight coefficient w_i for each y_i according to its corresponding bin. That is, $w_i = w_j^b$, if $y_i \in [b_j, b_{j+1})$;

6: Calculate the L_w with Eq. (9) using w_i, y_i, and $\hat{y}_i$.

7: **return**

---

4 Experiments

4.1 Experiment Setup

Datasets. We use three datasets to train and test the proposed PE networks: the Beijing, the Heshan, and the various shooting scenes (VSS) datasets. The Beijing dataset [5] comprises 5897 single-scene images, and is randomly split into a training set and a test set at a ratio of 7:3. To evaluate performance for multi-angle scenes, we construct a new Heshan dataset that contains 306 images captured at a fixed location (112.97° E, 22.77° N) from three different angles, and randomly split it at a ratio of 8:2 for training and testing. The VSS dataset was collected from a tour website[1], containing 60 images for the evaluation of performance across various scenes, with corresponding PM2.5 concentrations provided by the Beijing Municipal Ecological and Environmental Monitoring Center. Furthermore, to evaluate the model's performance in heavily polluted conditions (PM2.5 $> 115 \mu g/m^3$), we select a small subset from the Beijing test set as a few-shot set. All images were captured during the day, excluding those that are heavily affected by clouds and sun glows.

Evaluation Metrics. We evaluate the experimental results using three standard metrics: mean absolute error (MAE) [20], root-mean-square error (RMSE) [7], as well as linear correlation coefficient (LCC) [4], which are commonly used for regression tasks [5–7]. Among these criteria, MAE and RMSE compute the average difference between the ground truth and estimation values of PM2.5. The LCC measures the strength of the linear relation of the two vectors. More accurate estimation of PM2.5 can be indicated by smaller values of MAE and RMSE or larger values of LCC.

---

[1] https://www.tour-beijing.com/real_time_weather_photo/.

Implementation Details. The networks were implemented using PyTorch 1.7 [18] on a platform with an RTX 3090Ti GPU. The training datasets were augmented with random horizontal flips, assuming that the sky region was at the top of each image. Input images were scaled to a resolution of 256×256 for normalization. Our network is trained for 150 epochs with a batch size of 16, and an Adam [12] optimizer with default parameters of $\beta_1 = 0.9$ and $\beta_2 = 0.999$ was employed. The learning rate was set to an initial value of 1e-4 and was decremented to zero using the cosine annealing strategy [10].

4.2 Comparison with the SOTA Methods

We compare the performance of our method quantitatively with that of several state-of-the-art (SOTA) image-based approaches for PM2.5 estimation, including PPPC [7], PM-MLP [5], IAWD [6], and MIFF [24]. Among these methods, only the PM-MLP requires additional data related to weather conditions. For a fair comparison, we retrain these models using the Beijing and Heshan training sets, separately, and evaluate them on four test sets. Table 1 shows a quantitative comparison of these methods.

Table 1. Quantitative comparison on the Beijing, Heshan, Few-shot, and VSS test sets. The 1st and 2nd winners of each metric are displayed in **bold** and underline, respectively. "↓" indicates the smaller the better, and "↑" indicates the larger the better.

Method	Beijing			Heshan			Few-shot			VSS		
	RMSE↓	MAE↓	LCC↑	RMSE↓	MAE↓	LCC↑	RMSE↓	MAE↓	LCC↑	RMSE↓	MAE↓	LCC↑
PPPC [7]	126.70	111.48	31.79%	174.34	173.04	11.33%	61.08	52.30	11.18%	85.41	61.67	<u>39.76%</u>
PM-MLP [5]	32.76	22.24	52.06%	<u>15.54</u>	<u>12.08</u>	<u>68.65%</u>	112.60	106.73	34.26%	–	–	–[a]
IAWD [6]	33.65	22.68	43.32%	20.70	15.89	33.92%	120.90	114.70	5.81%	27.82	20.47	18.65%
MIFF [24]	36.83	23.51	32.37%	19.62	14.07	40.15%	139.12	133.73	-0.27%	<u>24.65</u>	<u>19.32</u>	35.28%
PE-ResNet18	<u>9.49</u>	<u>5.37</u>	<u>96.70%</u>	**15.37**	**10.99**	**72.26%**	<u>23.08</u>	<u>14.60</u>	<u>84.82%</u>	**15.09**	**12.98**	**81.56%**
PE-SwinT	**8.92**	**5.31**	**97.10%**	17.30	13.17	58.17%	**20.65**	**12.64**	**88.48%**	30.97	24.19	35.72%

[a] The results of PM-MLP in VSS dataset are unavailable, because PM-MLP requires external whether data, which is unavailable in this dataset.

As shown in Table 1, our PE networks exhibit the best or second best performance for most metrics. Especially, PE-ResNet18 and PE-SwinT achieve a significant performance improvement for the Beijing and Few-shot dataset compared with all previous methods. In addition, our PE-ResNet18 is significantly superior to the others for the VSS dataset, indicating its advantage in generalizability. Further, although better than most other models, the PE-SwinT do not perform as well as PE-ResNet18 on Heshan and VSS datasets, because it tends to be overfitted, due to the model complexity of Swin transformer [15] relative to the volume of these two datasets. Therefore, we design the auxiliary branch and feature fusion module of our PE networks in a lightweight way, which is compatible with replaceable backbones.

4.3 Ablation Studies

We evaluate the proposed PE networks with their backbones on the Heshan test set. The estimation accuracies measured by RMSE and MAE are listed in Table 2, indicating that both PE networks outperformed their baselines at the cost of a small amount of extra computational overhead.

Table 2. Comparison across single- and dual-branch networks.

Models	RMSE	MAE	GFLOPs	Parameters (M)
ResNet18	16.57	12.09	2.38	11.18
PE-ResNet18	**15.37**	**10.99**	2.74	13.34
SwinT	17.55	13.44	3.89	18.85
PE-SwinT	**17.30**	**13.17**	4.24	21.15

To further explore the function of the proposed PE architecture, we compare the training loss of these models on the Beijing dataset, which are shown in Fig. 3. As can be seen, PE networks converge faster than their baseline counterparts, with smoother loss curves. These results verify that the proposed PE framework accelerate the process of the optimization by providing effective auxiliary prior information.

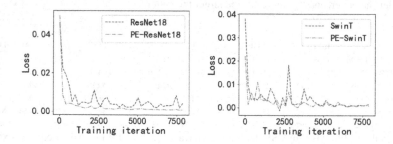

Fig. 3. Training error on the Beijing dataset.

Figure 4 shows the effect of bin width used by the HS algorithm, in which "None" indicates that the HS algorithm was not used. We use the proposed PE-ResNet18 as a baseline model and subsequently applied the HS strategy with varying settings of bin width. Indicated by orange bars of most positive RMSE and MAE reductions, as well as blue bars varying less than 0.5 in Fig. 4, the HS strategy considerably improves the performance of few-shot samples while maintaining the performance on the entire set stable. Therefore, the bin width is selected as 10 to achieve the best accuracy on the few-shot subset and comparable accuracy on the complete dataset.

4.4 Examples on Various Shooting Scenes

To compare the generalizability of existing models, we directly test them on a new VSS dataset containing various scenes without retraining. Figure 5 shows several example images with PM2.5 values listed on the top. PM-MLP is excluded in this study, because

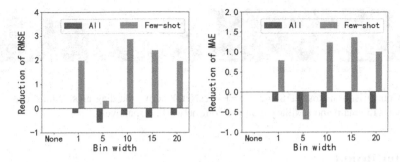

Fig. 4. Reductions of RMSE (left) and MAE (right) using different bin widths on the Beijing test set (all samples) and its subset of heavily polluted conditions (few-shot samples).

additional weather data beyond the input image are required, which is unavailable in these cases. Our PE network yields results with satisfactory accuracy in most cases and outperforms the other competing models. These examples suggest the potential of applying the proposed method to various scenarios.

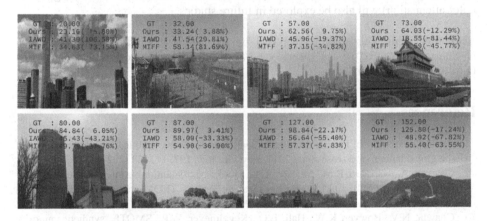

Fig. 5. Example images of various shooting scenes. The observed GT, estimated PM2.5 values, and relative errors are shown on the top of each image.

To visualize the region attention of our network, we compare gradient-weighted class activation mapping (Grad-CAM) diagrams [22] generated by the ResNet18 and our PE-ResNet18 in Fig. 6. As the figure shows, attention regions (in red color) of a ResNet18 model are randomly distributed in the entire input image (the distant or the nearby). In contrast, the main and the auxiliary branch of our network play two complementary roles: the main branch focuses on the sky and distant regions, while the auxiliary branch is interested more in nearby objects. These results demonstrate that our network can distinguish different areas of the input image, and thereby has better generalizability.

Fig. 6. Comparison of ResNet18 and our PE-ResNet18 activation diagrams. 'Main' and 'Auxiliary' means the main and auxiliary branch of our network, respectively.

5   Conclusion

In this paper, we propose end-to-end prior-enhanced (PE) networks for estimating PM2.5 concentrations from a single image, without requiring additional weather data. Furthermore, we propose an adaptive histogram smoothing (HS) algorithm to improve the estimation accuracy for few-shot samples related to heavily polluted cases, which is, to the best of our knowledge, the first solution to the data imbalance problem in image-based PM2.5 estimation. Extensive experiments demonstrate that our method achieves state-of-the-art performance across a variety of scenes with different air pollution levels. Finally, in addition to the DC-IS maps, other information derived from images (e.g. color attenuation) will also be explored in future studies.

Acknowledgment. This work is supported by the National Natural Science Foundation of China (No. 62071201), and Guangdong Basic and Applied Basic Research Foundation (No.2022A1515010119).

References

1. Bu, X., et al.: Global PM2.5-attributable health burden from 1990 to 2017: Estimates from the global burden of disease study 2017. Environ. Res. **197**, 111123 (2021)
2. Charron, A., Harrison, R.M., Moorcroft, S., Booker, J.: Quantitative interpretation of divergence between PM10 and PM2.5 mass measurement by TEOM and gravimetric (partisol) instruments. Atmos. Environ. **38**(3), 415–423 (2004)
3. Chawla, N.V., Bowyer, K.W., Hall, L.O., Kegelmeyer, W.P.: SMOTE: synthetic minority over-sampling technique. J. Artif. Intell. Res. **16**, 321–357 (2002)
4. Chok, N.S.: Pearson's versus Spearman's and Kendall's correlation coefficients for continuous data, Ph. D. thesis, University of Pittsburgh (2010)
5. Feng, L., Yang, T., Wang, Z.: Performance evaluation of photographic measurement in the machine-learning prediction of ground PM2.5 concentration. Atmos. Environ. **262**, 118623 (2021)
6. Gu, K., Liu, H., Xia, Z., Qiao, J., Lin, W., Thalmann, D.: PM2.5 monitoring: use information abundance measurement and wide and deep learning. IEEE Trans. Neural Networks Learn. Syst. **32**(10), 4278–4290 (2021)
7. Gu, K., Qiao, J., Li, X.: Highly efficient picture-based prediction of PM2.5 concentration. IEEE Trans. Ind. Electron. **66**(4), 3176–3184 (2018)
8. He, K., Sun, J., Tang, X.: Single image haze removal using dark channel prior. IEEE Trans. Pattern Anal. Mach. Intell. **33**(12), 2341–2353 (2010)

9. He, K., Zhang, X., Ren, S., Sun, J.: Deep residual learning for image recognition. In: Proceedings of the IEEE/CVF Conference on Computer Vision and Pattern Recognition, pp. 770–778 (2016)

10. He, T., Zhang, Z., Zhang, H., Zhang, Z., Xie, J., Li, M.: Bag of tricks for image classification with convolutional neural networks. In: Proceedings of the IEEE/CVF Conference on Computer Vision and Pattern Recognition, pp. 558–567 (2019)

11. Hu, J., Shen, L., Sun, G.: Squeeze-and-excitation networks. In: Proceedings of the IEEE/CVF Conference on Computer Vision and Pattern Recognition, pp. 7132–7141 (2018)

12. Kingma, D.P., Ba, J.: Adam: A method for stochastic optimization. arXiv preprint arXiv:1412.6980 (2014)

13. Kumari, R., Singh, J., Gosain, A.: SmS: smote-stacked hybrid model for diagnosis of polycystic ovary syndrome using feature selection method. Expert Syst. Appl. **225**, 120102 (2023)

14. Liu, C., Tsow, F., Zou, Y., Tao, N.: Particle pollution estimation based on image analysis. PLoS ONE **11**(2), e0145955 (2016)

15. Liu, Z., et al.: Swin transformer: hierarchical vision transformer using shifted windows. In: Proceedings of the IEEE/CVF International Conference on Computer Vision, pp. 10012–10022 (2021)

16. Maas, A.L., Hannun, A.Y., Ng, A.Y., et al.: Rectifier nonlinearities improve neural network acoustic models. In: Proceedings of the International Conference on Machine Learning, p. 3. Citeseer (2013)

17. Marquès, M., Domingo, J.L.: Positive association between outdoor air pollution and the incidence and severity of COVID-19. a review of the recent scientific evidences. Environ. Res. **203**, 111930 (2022)

18. Paszke, A., et al.: Pytorch: an imperative style, high-performance deep learning library. In: Advances in Neural Information Processing Systems, vol. 32 (2019)

19. Peng, M., et al.: Trainable undersampling for class-imbalance learning. In: Proceedings of the AAAI Conference on Artificial Intelligence, pp. 4707–4714 (2019)

20. Qiao, J., He, Z., Du, S.: Prediction of PM2.5 concentration based on weighted bagging and image contrast-sensitive features. Stochast. Environ. Res. Risk Assess. **34**(3), 561–573 (2020)

21. Rijal, N., Gutta, R.T., Cao, T., Lin, J., Bo, Q., Zhang, J.: Ensemble of deep neural networks for estimating particulate matter from images. In: 2018 IEEE 3rd International Conference on Image, Vision and Computing, pp. 733–738. IEEE (2018)

22. Selvaraju, R.R., Cogswell, M., Das, A., Vedantam, R., Parikh, D., Batra, D.: Grad-cam: visual explanations from deep networks via gradient-based localization. In: Proceedings of the IEEE International Conference on Computer Vision, pp. 618–626 (2017)

23. Shapiro, L.G., Stockman, G.C., et al.: Computer Vision, vol. 3. Prentice Hall New Jersey (2001)

24. Wang, G., Shi, Q., Wang, H., Sun, K., Lu, Y., Di, K.: Multi-modal image feature fusion-based PM2.5 concentration estimation. Atmos. Pollut. Res. **13**(3), 101345 (2022)

25. Wang, T., et al.: C2AM loss: chasing a better decision boundary for long-tail object detection. In: Proceedings of the IEEE/CVF Conference on Computer Vision and Pattern Recognition, pp. 6980–6989 (2022)

26. Yang, Y., Zha, K., Chen, Y., Wang, H., Katabi, D.: Delving into deep imbalanced regression. In: International Conference on Machine Learning, pp. 11842–11851. PMLR (2021)

27. Yue, G., Gu, K., Qiao, J.: Effective and efficient photo-based PM2.5 concentration estimation. IEEE Trans. Instrum. Measur. **68**(10), 3962–3971 (2019)

28. Zhang, Y., Kang, B., Hooi, B., Yan, S., Feng, J.: Deep long-tailed learning: a survey. IEEE Trans. Pattern Anal. Mach. Intell. **45**(9), 10795–10816 (2023). IEEE

Dynamic Data Augmentation via Monte-Carlo Tree Search for Prostate MRI Segmentation

Xinyue Xu[1], Yuhan Hsi[2], Haonan Wang[3], and Xiaomeng Li[1(✉)]

[1] The Hong Kong University of Science and Technology, Hong Kong, China
`xxucb@connect.ust.hk, eexmli@ust.hk`
[2] The Pennsylvania State University, State College, PA, USA
`ybh5084@psu.edu`
[3] The University of Hong Kong, Hong Kong, China
`haonanw@connect.hku.hk`

Abstract. Medical image data are often limited due to the expensive acquisition and annotation process. Hence, training a deep-learning model with only raw data can easily lead to overfitting. One solution to this problem is to augment the raw data with various transformations, improving the model's ability to generalize to new data. However, manually configuring a generic augmentation combination and parameters for different datasets is non-trivial due to inconsistent acquisition approaches and data distributions. Therefore, automatic data augmentation is proposed to learn favorable augmentation strategies for different datasets while incurring large GPU overhead. To this end, we present a novel method, called Dynamic Data Augmentation (**DDAug**), which is efficient and has negligible computation cost. Our DDAug develops a hierarchical tree structure to represent various augmentations and utilizes an efficient Monte-Carlo tree searching algorithm to update, prune, and sample the tree. As a result, the augmentation pipeline can be optimized for each dataset automatically. Experiments on multiple Prostate MRI datasets show that our method outperforms the current state-of-the-art data augmentation strategies.

Keywords: Prostate MRI Segmentation · Data Augmentation · Auto ML

1 Introduction

The prostate is an important reproductive organ for men. The three most prevalent forms of prostate disease are inflammation, benign prostate enlargement, and prostate cancer. A person may experience one or more of these symptoms. Accurate MRI segmentation is crucial for the pathological diagnosis and prognosis of prostate diseases [17]. Manual prostate segmentation is a time-consuming

---

X. Xu and Y. Hsi—Both authors contributed equally to this work.

© The Author(s), under exclusive license to Springer Nature Singapore Pte Ltd. 2024
B. Luo et al. (Eds.): ICONIP 2023, CCIS 1964, pp. 272–282, 2024.
https://doi.org/10.1007/978-981-99-8141-0_21

task that is subject to inter and intra-observer variability [6]. The development of deep learning has led to significant advancements in many fields, including computer-assisted intervention. With the advancement of technology, clinical applications of deep learning techniques have increased. There are multiple deep learning networks [10,18,28] designed to enhance the accuracy of automatic prostatic segmentation. Different neural network structures, such as Vnet [18], U-Net [21] and its variant nnUNet [9], can all be utilized for prostate segmentation. These methods are all from the perspective of modifying the model structure to improve the segmentation accuracy. However, in medical image segmentation tasks, carefully designed network structure is prone to overfitting due to limited data. To alleviate the data shortage, data augmentation is an effective means of enhancing segmentation performance and model generalizability simultaneously on small datasets.

Data augmentation aims to generate more data from the raw samples via pre-defined transformations, which helps to diversify the original dataset [22]. Typical data augmentation techniques include affine transformations (*e.g.*, rotation, flipping, and scaling), pixel-level transformation (*e.g.*, gaussian nosie and contrast adjustment), and elastic distortion. For prostate MRI data, affine transformation or GAN-based methods are frequently used [5]. However, the augment selection and combination process that utilizes these aforementioned transformations are predominantly hand-crafted. It is difficult to identify which operations are actually useful for the prostate segmentation task, thus often resulting in sub-optimal combinations, or even degrading network performance. Automatic data augmentation, with its ability to design different combinations, its flexibility to remove useless augment operations, and its utilization of quantifiable metrics, is a crucial technology that can solve this problem. Approaches to automatic data augmentation need to strike a balance between simplicity, cost, and performance [19]. In the field of natural image, early automatic augmentation techniques [2,8,13,25] were GPU-intensive. Subsequently, RandAugment [3], UniformAugment [14], and TrivialAugment [19] substantially decreased the search cost while maintaining performance. Due to the variation between medical (spatial context information and different morphologies of lesions and tissues) and natural images, directly applying these approaches is either ineffective or unsuitable. The earliest work [27] utilized reinforcement learning (RL) for automatic medical data augmentation, but it required a significant amount of computing resources. The state-of-the-art automatic data augmentation framework (ASNG) algorithm [26] formulated the automatic data augmentation problem as bi-level optimization and applied the approximation algorithm to solve it. Although it is more efficient than reinforcement learning, the time required to find a reasonable strategy can still be highly demanding. Furthermore, using only rudimentary spatial transforms can limit performance, and some state-of-the-art methods involve searching the probability of operation, which can make the training process inefficient.

To this end, we propose a novel automatic augmentation strategy *Dynamic Data Augmentation (DDAug)* for MRI segmentation. Automatic data augmentation problem is formulated into the Monte-Carlo tree search problem for the first time. The augmentation pipeline is represented as a tree structure, which is iteratively refined through updating, pruning, and sampling. In contrast to the previous method, our approach expands the search space by including more spatial augmentations and allows the tree structure to determine the optimal sequence of augmentations while removing redundant ones. Moreover, our method's flexibility in selecting operations without having to search for the probability significantly enhances its search efficiency. Our method adopts a novel approach by using only a few augmentation operations at a time, yet achieving an effect similar to that of manually combining multiple operations. Our DDAug method achieves an optimal balance between simplicity, cost, and performance when compared to previous approaches. Code and documentation are available at https://github.com/xmed-lab/DDAug.

2 Methodology

Automatic augmentation search spaces and Monte-Carlo tree search constitute our method. We meticulously selected a number of dependable operations on medical images to compose our search space for the tree's construction. The search space consists of pixel-level and spatial-level transformations, as well as the left and right ranges of their respective magnitudes. After the tree is constructed, a path is chosen for each training epoch, which is updated by the validation loss, and nodes and their children in the chosen path that degrade model performance are pruned. Finally, the path of the subsequent epoch is chosen by random or Upper Confidence Bounds applied to Trees (UCT) [1] sampling for different layers, and the cycle continues until training is complete. Figure 1 illustrates the procedure of the complete tree and the whole training process can be summarized in Algorithm 1. We will elaborate on each section below.

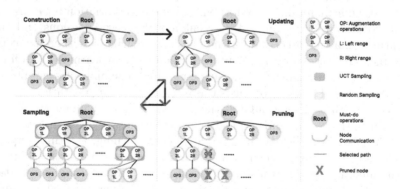

Fig. 1. Four Stages of Monte-Carlo Tree Search.

2.1 Search Space

The design of our search space is crucial to the performance of the final network. To compensate for the absence of spatial-level augmentation operations, optical distortion, elastic transformation, and grid distortion are added. Table 1 displays the complete search space. To better assess the efficacy of various operations in relation to their magnitude, we divide the magnitude into left-range and right-range whenever possible. Operations like brightness transform exhibit left-range (decrease brightness) or right-range (increase brightness). The random crop operation will pad the data by first padding it with a stochastically selected percentage, then randomly crop the padded data to its original size. Unlike brightness transform, random crop only has one magnitude range. This division allows for more precise control over the magnitude range without significantly increasing tree size. The operations of the root type pertain to the subsequent construct tree. These are necessary augmentation operations for each path and are not involved in the search. We forego the search for augmentation probability for two reasons. First, it would significantly increase the size of the tree, making search inefficient. Second, if a particular operation and magnitude range combination increases validation performance, it will be sampled more frequently. And if the combination is prone to degrade network performance, it will be swiftly removed.

Table 1. Augmentation Search Space. The root type refers to must-do operations at the beginning of path selection. '-' denotes range not applicable.

Operations	LR	RR	Type
Mirror	–	–	root
Random Crop	(0%, 33%)	–	root
Contrast Adjustment	(0.5, 1)	(1, 1.5)	pixel-level
Gamma Transform	(0.5, 1)	(1, 1.5)	pixel-level
Brightness Transform	(0.5, 1)	(1, 1.5)	pixel-level
Gaussian Noise	(0, 0.1)	–	pixel-level
Gaussian Blur	(0.5, 1)	(1, 1.5)	pixel-level
Simulate low-res image	(0.5, 1)	–	pixel-level
Scale	(0.5, 1)	(1, 1.5)	spatial-level
Optical Distortion	(0, 0.05)	–	spatial-level
Elastic Transform	(0, 50)	–	spatial-level
Grid Distortion	(0, 0.3)	–	spatial-level

2.2 Tree Construction

Similar to tree nodes, different augmentation operations can be connected to create various augmentation combinations. Additionally, we must consider their order when using a sequence of augmentations. This resembles the structure of a tree very closely. For efficient search purposes, we encode the automatic data augmentation as a tree structure. To construct our tree using the specified search space, we begin by generating a root node with mirror and random crop operations. These operations serve as a set of foundational augmentations and are always applied prior to the augmentation path chosen by tree search for each epoch. The remaining augmentation operations will participate in the search, and we use a three-layer tree structure to load the search space. Each node in the first layer is an augmentation operation, and their child nodes are the augmentation operations that do not include their parent. There are no duplicate operations on a single path, and no two paths have the same order of operations. The first augment path is initialized as the leftmost path of the tree.

Algorithm 1. Training Process of DDAug

1: Initialize the augmentation tree.
2: Set the leftmost path as the first augment path.
3: **for** each epoch **do**
4:     Train the model and calculate the validation loss.
5:     **for** each node in previously selected path **do**
6:         Update node Q-value (Eq. 2) using moving average loss.
7:         Record validation loss change L_{node}.
8:         **if** Past $\sum_{n=1}^{5} L_{node} > 0$ **then**
9:             Delete the current node and subtree of this node.
10:            Break;
11:        **end if**
12:    **end for**
13:    **while** not at leaf node **do**
14:        **if** mean visited times $< k_{uct}$ **then**
15:            Sample node using Random sampling.
16:        **else**
17:            Sample node using UCT sampling (Eq. 4).
18:        **end if**
19:    **end while**
20:    Finish sampling path for next epoch.
21: **end for**
22: Inference and report testing data performance.

2.3   Tree Updating and Pruning

With the initialized path during tree construction, we train the model for one epoch per path and calculate the validation loss L_{val}. The validation loss is computed utilizing the original nnUNet CE + DICE loss. The validation loss is then employed to update the tree by calculating the moving average loss L_{ma} using the following formula:

$$L_{ma} = \beta \cdot L_{val}^{t-1} + (1 - \beta) \cdot L_{val}^t \tag{1}$$

where $\beta \in [0,1]$ controls the ratio of current validation loss. L_{val}^{t-1} is the validation loss of the previous epoch, while L_{val}^t represents the validation loss of the current epoch. We then update the Q-value of all nodes in the previously selected path with:

$$Q = \frac{L_{ma}}{L_{val}^t} \tag{2}$$

A record of validation loss change $L_{node} = L_{val}^t - L_{val}^{t-1}$ is kept for all nodes to evaluate their overall impact on network performance. As we traverse the path to update the Q-value, if the sum of the previous five L_{node} scores is greater than 0, the node is deemed to have a negative impact on the network's performance and is pruned from the tree.

2.4   Tree Sampling

After the pruning process, a new path needs to be chosen for the subsequent epoch. Because the nodes have not yet been visited, we use random sampling rather than Monte-Carlo UCT sampling at the beginning of the network training process. We compare the k_{uct} threshold to the average visited times of the current layer to determine when to switch to UCT sampling. The value of k_{uct} is set to 3, 1, and 1 times, for the first, second, and third layers of the tree, respectively. The number of tree layers is expandable, but increasing the number of layers will lead to the exponential growth of search space, which is not conducive to search efficiency. At the same time, if the tree has less than three layers, the amount of nodes and paths is extremely limited, thus decreasing the diversity introduced via data augmentation.

Inspired by [24], we introduce a node communication term S to better evaluate the current node's efficacy using nodes' Q-value of nodes from the same layer that has the same augment operation and magnitude range as the current node.

$$S(v_i^l) = (1 - \lambda) \cdot Q(v_i^l) + \lambda \cdot \sum_{j=0}^{n} \frac{Q(v_j^l)}{n} \tag{3}$$

where v_i^l is the i-th child node in the l-th layer, v_j^l is the other nodes that have the same operation and magnitude range as v_i in the l-th layer, and n denotes the total number of v_j^l. λ controls the effect of the node communication term.

When the averaged visited times of all children of the current node exceeds k_{uct}, we employ the following equation to calculate the UCT [11,24] score for all children of the current node:

$$UCT(v_i^l) = \frac{Q(v_i^l)}{n_i^l} + C_1 \sqrt{\frac{log(n_p^{(l-1)})}{n_i^l}} + C_2 \cdot S(v_i^l) \tag{4}$$

where n_i^l is the number of visited times of v_i^l, and $n_p^{(l-1)}$ is the visited times of its parent node in the $(l-1)$-th layer.

A temperature term τ [20] is utilized to promote greater discrimination between candidates by amplifying score differences, thus the sampling probability can be calculated as

$$P(v_i^l) = \frac{\exp(\frac{UCT(v_i^l)}{\tau})}{\sum_j^n \exp(\frac{UCT(v_j^l)}{\tau})} \tag{5}$$

where v_i^l are children of the current node. We sample a node from the current group of children using the probabilities calculated, then continue the sampling process until a leaf node is reached. Reaching a leaf node signifies the termination of the sampling process for the current epoch, and the selected path will be adopted in the next epoch. This cycle repeats at the end of every epoch until maximum the training epochs are reached.

3 Implementation and Experiments

3.1 Datasets and Implementation Details

Datasets. We conduct our experiments on several 3d Prostate MRI datasets: subset 1 and 2 are from NCI-ISBI 2013 challenge [4], subset 3 is from I2CVB benchmarking [12], subset 4, 5, 6 are from PROMISE12 [15], and subset 7 is the Task 005 prostate dataset from Medical Segmentation Decathlon [23]. Subsets 1 through 6 are acquired from and have been resized by [16]. All datasets are then resampled and normalized to zero mean and unit variance as described in nnUNet [9].

Implementation Details. For a fair comparison, we base our implementation on the original nnUNet repository. We only inserted additional code for the implementation of DDAug while keeping model architecture and self-configuration process intact. To conduct 5-fold cross-validation, we utilized stochastic gradient descent with Nesterov momentum and set the learning rate of 0.01. Each fold trains for 200 epochs, and each epoch has 250 batches with a batch size of 2. The runtime comparison can be found in Table 2. The utilization of Reinforcement Learning and ASNG method demand substantial GPU resources. In contrast, our approach performs at an equivalent efficiency to the original nnUnet data augmentation.

Table 2. Comparison of GPU costs with different augmentation methods.

Method	RL [27]	ASNG [7,26]	DDAug (Ours)	nnUNet [9]
Cost (h)	768	100	40	40

Compared Methods. Since ASNG [26] requires ten days of GPU processing time and our objective is to create an effective and efficient automated search method, we only compare our implementations on nnUNet that has the same GPU runtime requirements. Limited by the size of each subset, we conduct all of our experiments using 5-fold cross-validation and report the mean validation DICE score inferred with weights from the last epoch. Our baselines are established via training nnUNet using no augmentations (NoDA) and using default augmentations (moreDA). The 'moreDA' is a set of sequential augmentations including scaling, rotating, adding gaussian noise, adding gaussian blur, transforming with multiplicative brightness, transforming with contrast augmentation, simulating low resolution, performing gamma transform, and mirroring axis.

Ablation Study. In our ablation study, we start by replacing the sequential augment operations in moreDA with the uniform sampling mechanism described TrivialAugment [19]. This allows us to assess the viability of using the natural image SOTA approach on medical image. To evaluate the effectiveness of the proposed search space, we extend moreDA's operation search space with additional spatial augmentations (Spatial SS). Finally, we replace moreDA with DDAug to examine the advantage of using the expanded search space inconjunction with Monte-Carlo Tree Search (MCTS).

3.2 Experimental Results

Table 3. Augmentation performance of different Prostate datasets on Dice (%). Subsets represent different prostate datasets and the backbone is nnUNet. NoDA: No augmentation; moreDA: sequential augmentation; Spatial SS: our designed search space; TrivialAugment: natural image SOTA method; DDAug: MCTS + our search space (proposed method). red, blue denote the highest and second highest score.

Method	Subset 1	Subset 2	Subset 3	Subset 4	Subset 5	Subset 6	Subset 7	Average
NoDA	79.12	80.82	84.57	82.02	78.10	82.77	72.36	79.97
moreDA	79.64	81.66	87.60	81.38	83.74	87.12	71.23	81.77
TrivialAugment	80.39	82.21	88.42	82.60	86.36	86.60	72.58	82.74
Spatial SS (**Ours**)	79.96	82.18	87.68	83.74	85.69	86.99	72.90	82.73
DDAug (**Ours**)	80.27	82.72	87.46	88.59	86.40	87.17	73.20	83.69

The five-fold average Dice similarity coefficients of different methods are shown in Table 3. As we can see, in general, adding augmentation to the prostate MRI dataset is better than no augmentation. moreDA demonstrates some improvement from NoDA on most of the datasets, and additional performance increase are observed when expanding the search space by adding spatial-level augmentations. When comparing the performance of Spatial SS and

TrivialAugment, the improvement prove to be inconclusive, as three out of seven dataset exhibits degradation. This is likely due to the fact that TrivialAugment uses uniform sampling over the search space, and unlike our DDAug, does not consider the efficacy of different operations. We are able to further improve the results by utilizing DDAug's full search space and its tree search method. It is important to note that moreDA contains 9 sequential augmentations while DDAug only uses 5. This indicates that simply piling on more augmentations sequentially is not the optimal solution. Though using only a few operations per epoch, DDAug still achieves the highest average DICE when looking at all 7 subsets with near-zero computing consumption.

The performance difference can translate to visual discrepancy between different methods. When inspecting validation segmentation results, we noticed that DDAug is significantly more robust when segmenting validation cases as shown in Fig. 2. DDAug demonstrates enhanced generalizability against its counterparts. Augmenting data sequentially, on the other hand, was not able to handle difficult validation cases.

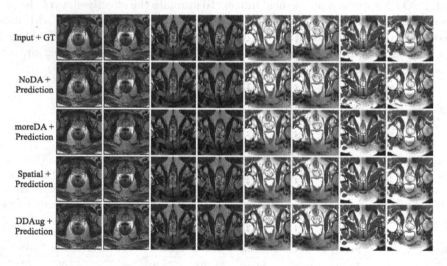

Fig. 2. Comparison of inference results using different augmentation techniques during training. The top row is validation images and their corresponding ground truth. The subsequent rows are inference results using models trained with no augmentation, moreDA augmentation, our designed search space, and DDAug, respectively.

4   Conclusion

We propose an efficient and zero GPU overhead automatic data augmentation algorithm for prostate MRI segmentation. Comparing previous approaches, we include additional spatial transformations into the search space, and adopt a

Monte-Carlo tree structure to store various augmentation operations. An optimal augmentation strategy can be obtained by updating, pruning, and sampling the tree. Our method outperforms the state-of-the-art manual and natural image automatic augmentation methods on several prostate datasets. We show the feasibility of utilizing automatic data augmentation without increasing GPU consumption. In future work, we will further investigate the generalizability of tree search on other medical segmentation datasets, e.g., liver cancer segmentation, brain tumor segmentation and abdominal multi-organ segmentation.

Acknowledgement. This work was supported by the Hong Kong Innovation and Technology Fund under Project ITS/030/21, as well as by Foshan HKUST Projects under Grants FSUST21-HKUST10E and FSUST21- HKUST11E.

Author contributions. Xinyue Xu,Yuhan Hsi :Both authors contributed equally to this work.

References

1. Auer, P., Cesa-Bianchi, N., Fischer, P.: Finite-time analysis of the multiarmed bandit problem. Mach. Learn. **47**, 235–256 (2002)
2. Cubuk, E.D., Zoph, B., Mane, D., Vasudevan, V., Le, Q.V.: Autoaugment: learning augmentation strategies from data. In: Proceedings of the IEEE/CVF Conference on Computer Vision and Pattern Recognition, pp. 113–123 (2019)
3. Cubuk, E.D., Zoph, B., Shlens, J., Le, Q.V.: Randaugment: practical automated data augmentation with a reduced search space. In: Proceedings of the IEEE/CVF Conference on Computer Vision and Pattern Recognition Workshops, pp. 702–703 (2020)
4. Farahani K., KirbyJ., M.A., Huisman H., E.A.: Nci-isbi 2013 challenge - automated segmentation of prostate structures (2015)
5. Garcea, F., Serra, A., Lamberti, F., Morra, L.: Data augmentation for medical imaging: a systematic literature review. Comput. Biol. Med., 106391 (2022)
6. Gardner, S.J., et al.: Contouring variability of human-and deformable-generated contours in radiotherapy for prostate cancer. Phys. Med. Biol. **60**(11), 4429 (2015)
7. He, W., Liu, M., Tang, Y., Liu, Q., Wang, Y.: Differentiable automatic data augmentation by proximal update for medical image segmentation. IEEE/CAA J. Autom. Sinica 9(7), 1315–1318 (2022)
8. Ho, D., Liang, E., Chen, X., Stoica, I., Abbeel, P.: Population based augmentation: efficient learning of augmentation policy schedules. In: International Conference on Machine Learning, pp. 2731–2741. PMLR (2019)
9. Isensee, F., Jaeger, P.F., Kohl, S.A., Petersen, J., Maier-Hein, K.H.: nnu-net: a self-configuring method for deep learning-based biomedical image segmentation. Nat. Methods **18**(2), 203–211 (2021)
10. Jia, H., Song, Y., Huang, H., Cai, W., Xia, Y.: HD-Net: hybrid discriminative network for prostate segmentation in MR images. In: Shen, D., et al. (eds.) MICCAI 2019. LNCS, vol. 11765, pp. 110–118. Springer, Cham (2019). https://doi.org/10. 1007/978-3-030-32245-8_13
11. Kocsis, L., Szepesvári, C.: Bandit based monte-carlo planning. In: Fürnkranz, J., Scheffer, T., Spiliopoulou, M. (eds.) ECML 2006. LNCS (LNAI), vol. 4212, pp. 282–293. Springer, Heidelberg (2006). https://doi.org/10.1007/11871842_29

12. Lemaître, G., Martí, R., Freixenet, J., Vilanova, J.C., Walker, P.M., Meriaudeau, F.: Computer-aided detection and diagnosis for prostate cancer based on mono and multi-parametric mri: a review. Comput. Biol. Med. **60**, 8–31 (2015)

13. Lin, C., et al.: Online hyper-parameter learning for auto-augmentation strategy. In: Proceedings of the IEEE/CVF International Conference on Computer Vision, pp. 6579–6588 (2019)

14. LingChen, T.C., Khonsari, A., Lashkari, A., Nazari, M.R., Sambee, J.S., Nascimento, M.A.: Uniformaugment: a search-free probabilistic data augmentation approach. arXiv preprint arXiv:2003.14348 (2020)

15. Litjens, G., Toth, R.V.W., Hoeks, C., Ginneken, B.K.S.: Miccai grand challenge: prostate mr image segmentation 2012 (2012)

16. Liu, Q., Dou, Q., Yu, L., Heng, P.A.: Ms-net: multi-site network for improving prostate segmentation with heterogeneous mri data. IEEE Trans. Med. Imaging (2020)

17. Mahapatra, D., Buhmann, J.M.: Prostate mri segmentation using learned semantic knowledge and graph cuts. IEEE Trans. Biomed. Eng. **61**(3), 756–764 (2013)

18. Milletari, F., Navab, N., Ahmadi, S.A.: V-net: fully convolutional neural networks for volumetric medical image segmentation. In: 2016 Fourth International Conference on 3D Vision (3DV), pp. 565–571. IEEE (2016)

19. Müller, S.G., Hutter, F.: Trivialaugment: tuning-free yet state-of-the-art data augmentation. In: Proceedings of the IEEE/CVF International Conference on Computer Vision, pp. 774–782 (2021)

20. Neumann, L., Zisserman, A., Vedaldi, A.: Relaxed softmax: efficient confidence auto-calibration for safe pedestrian detection (2018)

21. Ronneberger, O., Fischer, P., Brox, T.: U-Net: convolutional networks for biomedical image segmentation. In: Navab, N., Hornegger, J., Wells, W.M., Frangi, A.F. (eds.) MICCAI 2015. LNCS, vol. 9351, pp. 234–241. Springer, Cham (2015). https://doi.org/10.1007/978-3-319-24574-4_28

22. Shorten, C., Khoshgoftaar, T.M.: A survey on image data augmentation for deep learning. J. Big Data **6**(1), 1–48 (2019)

23. Simpson, A.L., et al.: A large annotated medical image dataset for the development and evaluation of segmentation algorithms. arXiv preprint arXiv:1902.09063 (2019)

24. Su, X., et al.: Prioritized architecture sampling with monto-carlo tree search. In: Proceedings of the IEEE/CVF Conference on Computer Vision and Pattern Recognition, pp. 10968–10977 (2021)

25. Tian, K., Lin, C., Sun, M., Zhou, L., Yan, J., Ouyang, W.: Improving auto-augment via augmentation-wise weight sharing. Adv. Neural. Inf. Process. Syst. **33**, 19088–19098 (2020)

26. Xu, J., Li, M., Zhu, Z.: Automatic data augmentation for 3D medical image segmentaion. In: Martel, A.L., et al. (eds.) MICCAI 2020. LNCS, vol. 12261, pp. 378–387. Springer, Cham (2020). https://doi.org/10.1007/978-3-030-59710-8_37

27. Yang, D., Roth, H., Xu, Z., Milletari, F., Zhang, L., Xu, D.: Searching learning strategy with reinforcement learning for 3D medical image segmentation. In: Shen, D., et al. (eds.) MICCAI 2019. LNCS, vol. 11765, pp. 3–11. Springer, Cham (2019). https://doi.org/10.1007/978-3-030-32245-8_1

28. Yu, L., Yang, X., Chen, H., Qin, J., Heng, P.A.: Volumetric convnets with mixed residual connections for automated prostate segmentation from 3d mr images. In: Proceedings of the AAAI Conference on Artificial Intelligence, vol. 31 (2017)

Language Guided Graph Transformer for Skeleton Action Recognition

Libo Weng, Weidong Lou, and Fei Gao$^{(\boxtimes)}$

College of Computer and Science, Zhejiang University of Technology, Hangzhou, China
gfei_jack@163.com

Abstract. The Transformer model is a novel neural network architecture based on a self-attention mechanism, primarily used in the field of natural language processing and is currently being introduced to the computer vision domain. However, the Transformer model has not been widely applied in the task of human action recognition. Action recognition is typically described as a single classification task, and the existing recognition algorithms do not fully leverage the semantic relationships within actions. In this paper, a new method named Language Guided Graph Transformer (LGGT) for Skeleton Action Recognition is proposed. The LGGT method combines textual information and Graph Transformer to incorporate semantic guidance in skeleton-based action recognition. Specifically, it employs Graph Transformer as the encoder for skeleton data to extract feature representations and effectively captures long-distance dependencies between joints. Additionally, LGGT utilizes a large-scale language model as a knowledge engine to generate textual descriptions specific to different actions, capturing the semantic relationships between actions and improving the model's understanding and accurate recognition and classification of different actions. We extensively evaluate the performance of using the proposed method for action recognition on the Smoking dataset, Kinetics-Skeleton dataset, and NTU RGB+D action dataset. The experimental results demonstrate significant performance improvements of our method on these datasets, and the ablation study shows that the introduction of semantic guidance can further enhance the model's performance.

Keywords: Action Recognition · Transformer · CLIP · Skeleton-based data

1 Introduction

Due to the widespread applications of action recognition in areas such as video surveillance, motion analysis, health monitoring, and human-computer interaction, it has become a popular research topic. In recent years, with the emergence of depth sensors such as Kinect [1] and RealSense [2], acquiring human body joint

© The Author(s), under exclusive license to Springer Nature Singapore Pte Ltd. 2024
B. Luo et al. (Eds.): ICONIP 2023, CCIS 1964, pp. 283–299, 2024.
https://doi.org/10.1007/978-981-99-8141-0_22

information has become easier. Skeleton-based action recognition involves classifying actions based on the two-dimensional or three-dimensional coordinates of
the acquired joint positions. Compared to RGB images or optical flow features
used for action recognition, skeleton data not only preserves the natural topology of the human body but is also less affected by environmental factors such
as lighting, viewpoint, or color, providing the model with stronger robustness.

For skeleton-based action recognition, early methods employed models based
on Recurrent Neural Networks (RNN) or Convolutional Neural Networks (CNN)
to analyze skeleton sequences. For example, Du et al. [3] proposed an end-to-end
hierarchical RNN that divides the human skeleton into five body parts and feeds
them into multiple bidirectional RNNs to capture high-level features of actions.
Li et al. [4] designed a simple yet effective CNN architecture for action classification from trimmed skeleton sequences. In recent years, embedding skeleton
data into a graph has been shown to effectively extract structural features of the
skeleton. Yan et al. [5] introduced the Spatial Temporal Graph Convolutional
Network (ST-GCN) for skeleton-based action recognition. ST-GCN constructs a
spatial graph by considering the natural connections between human joints and
builds temporal edges between corresponding joints across consecutive frames.

Recently, most works on action recognition have utilized Graph Convolutional Networks (GCN) to capture the structured features of spatial-temporal
graphs. Numerous studies have demonstrated that GCN outperforms other network representation learning methods, such as RNN or CNN, in tasks like action
recognition [5–8]. In fact, GCN is a natural extension of convolutional operations in the graph domain. By aggregating features from neighboring regions
using the convolutional operator, GCN can effectively learn spatial-temporal
features of the graph. However, these GCN-based recognition methods still face
two challenging problems: (1) Due to the sparsity of the graph, features of unre-
lated nodes can interfere with each other during the graph convolution process.
Therefore, some existing works also focus on using depth maps, as they not only
provide 3D geometric information but also preserve shape information [9]. (2)
Previous research [10] has shown that fusing features from different modalities
can significantly improve the accuracy of action recognition. For example, textual information contains rich prior knowledge, and its fine-grained semantic
relationships are beneficial for skeleton-based action recognition. However, how
to fully utilize textual information to better explore action features remains a
key issue in action recognition.

In this paper, we propose a novel approach that combines textual information with Graph Transformer to introduce semantic guidance in skeleton-based
action recognition. This network can handle both skeleton data and textual data
simultaneously, extracting their respective features, and improving the accuracy
of action recognition by fusing the features of both types of data. Additionally,
LGGT improves the feature representation of skeleton data to accommodate
sparse skeleton graphs. The main contributions of this paper are as follows:

- We design a Graph Transformer module to extract three-dimensional features from skeletons, accurately capturing the spatial-temporal relationships

between joints. This module maps 2D skeleton data to the 3D space, capturing three-dimensional motion information. Compared to traditional feature extractors, the Graph Transformer mitigates the over-smooth problem and captures long-distance dependencies more effectively.
- We employ a large-scale language model as a knowledge engine, capable of generating meaningful textual descriptions to guide the skeleton action recognition task. We automatically generate detailed descriptions for different actions using textual prompts. This approach enhances the model's understanding of human behavior, enabling more accurate recognition and classification of different actions.
- We design an action recognition network model that integrates textual and skeleton features. Experimental results demonstrate that incorporating detailed textual features significantly improves action recognition performance.

2   Related Works

Skeleton-based Action Recognition. Skeleton-based action recognition is a method that utilizes the rich human motion information contained in skeleton data. The skeleton and joint trajectories of the human body are robust to lighting variations and scene changes, and they are easily obtained due to highly accurate depth sensors or pose estimation algorithms [11]. As a result, there is a wide range of skeleton-based action recognition methods available. These methods can be classified into feature-based handcrafted approaches and deep learning methods. Feature-based handcrafted approaches capture the dynamics of joint movements by designing several handcrafted features. These features include the covariance matrix of joint trajectories [12], relative positions of joints [13], or rotations and translations between body parts [14]. More recently, through exploring the inherent connectivity of the human body, GCN, particularly ST-GCN [5], have achieved significant success in this task, demonstrating the effectiveness of skeleton features in action recognition. The ST-GCN model consists of spatial and temporal convolutional modules, where the graph convolution method employed is based on traditional graph filtering. The graph adjacency matrix encodes the connections between skeleton joints and extracts higher-level spatial representations from the skeleton action sequences. In the temporal dimension, one-dimensional convolutions are beneficial for extracting dynamic information of actions.

Many subsequent works have improved upon ST-GCN to enhance its recognition performance. Li et al. [15] proposed AS-GCN, which extends the connectivity of human skeletons by capturing the latent dependency relationships encoded in the adjacency matrix. Additionally, it directly captures temporal links from actions, enabling the capture of richer dependency relationships. Lei et al. [16] introduced Directed Graph Neural Networks (DGNNs), which merge joint and bone information into a directed acyclic graph to represent skeleton data. Liu et al. [17] proposed a unified spatial-temporal Graph Convolution module (G3D) to aggregate information across spatial and temporal dimensions for effective feature learning. Some studies focus on addressing the computational complexity of GCN-based methods. Cheng et al. [18] introduced Shift-GCN, which utilizes shift convolution to reduce computational complexity. Song et al. [19] proposed a multi-branch Residual GCN (ResGCN) that integrates spatial-temporal features from multiple branches and achieves competitive performance with fewer parameters.

Transformers and Self-Attention Mechanism. Vaswani [20] first introduced the Transformer module for machine translation, which has become the state-of-the-art approach in various NLP tasks. For instance, GPT [21] and BERT [22] are currently the most effective language models based on the Transformer. The core component of the Transformer module is the self-attention mechanism, which learns relationships between each element in a sequence. Unlike recurrent networks that process sequences recursively and are limited to short-term context, the Transformer module supports modeling long-term dependencies in a sequential manner. Additionally, the multi-head self-attention operation is easily parallelizable. In recent years, Transformer-based models have gained significant attention in the field of computer vision. Compared to traditional computer vision models based on convolutional operations, the Transformer brings some unique advantages. Convolutional operations capture short-term dependencies within a fixed-size window, which is also applicable to GCN. However, graph convolution operations cannot capture long-range relationships between joints in terms of spatial and temporal dimensions. Vision Transformer (ViT) [23] was the first successful application of the Transformer model to handle image data. Wang et al. [24] proposed the IIP-Transformer for action recognition, utilizing partial skeleton data encoding. Specifically, the IIP-Transformer considers the body joints as five parts and captures dependencies within and between the parts. To predict molecular structures more accurately, Vijay et al. [25] extended the Transformer to zinc molecular graphs and introduced a Graph Transformer operator. The Graph Transformer operation aggregates features of adjacent vertices and updates the vertices themselves. Therefore, this operator is more suitable for sparse graphs such as our skeleton graph.

Clip. Multi-modal representation learning methods such as CLIP [26] and ALIGN [27] have demonstrated that visual-language co-training can yield powerful representations for downstream tasks, such as zero-shot learning, image captioning, and text-image retrieval. These methods have shown that natural language can be utilized for reasoning and achieve zero-shot transfer learning for downstream tasks. However, these approaches require large-scale image-text paired datasets for training. ActionCLIP [28] follows the training scheme of CLIP for video action recognition. It utilizes a pre-trained CLIP model and adds transformation layers for temporal modeling of video data, achieving promising results in video action recognition.

3   Method

In this section, we will provide a detailed overview of the LGGT framework and its main components. LGGT is a skeleton-based action recognition model based on GCN, and its primary components include the skeleton feature extractor and the text feature extractor. The overview of the LGGT is shown in Fig. 1. For a given input skeleton sequence, we start by transmitting the skeleton data to the spatial transformer. Subsequently, with the incorporation of temporal positional embeddings, the data is fed into the temporal transformer module, thereby obtaining the spatial-temporal feature representation of the skeleton. Following this, the acquired spatial-temporal features are simultaneously fed into the GCN module and the Text Feature Fusion module. Guided by the supervisory role of text features, the final classification outcome is achieved. In the following sections, we will provide a detailed description of the Graph Transformer, the text encoder, and other key components of LGGT.

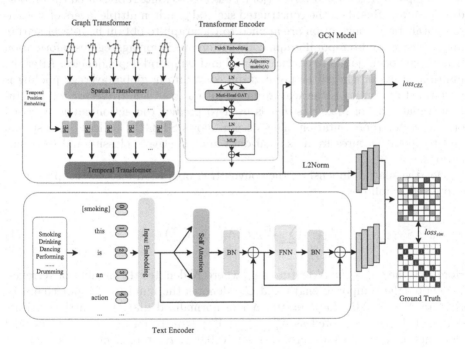

Fig. 1. Architecture of LGGT.

3.1   Skeleton Data Preprocessing

Skeleton data typically consists of a series of consecutive skeleton frames, where each frame contains a set of joint coordinates. Given the skeleton data, we can construct a natural human connectivity graph $G = (V, E)$, $V = \{v_{ti} \mid (t = 1, 2, \cdots, T, i = 1, 2, \cdots, N)\}$ represents all the joint nodes, where N denotes the number of joints and T represents the total number of frames. The set E consists of two parts: spatial edges and temporal edges. The spatial edges, denoted as $E_s = \{< v_{ti}, v_{tj} >\mid (i, j)\}$, capture the spatial connections between joints based on the human body's natural structure, where H represents the set of joint connections. The temporal edges, denoted as $E_t = \{< v_{ti}, v_{(t+1)i} >\}$, represent the connections between the same joint in consecutive frames. Additionally, we can incorporate features such as inter-joint distances, inter-frame joint displacement, and joint angles to enrich the graph representation.

3.2   GCN

In ST-GCN, the skeletal graph is constructed by treating joints as vertices and bones as edges. For consecutive frames, the corresponding joints are connected with temporal edges. Each vertex (joint) has an attribute represented by its coordinate vector. Based on the constructed skeletal graph, multiple layers of spatial-temporal graph convolution are applied to the graph to obtain high-level feature maps. The spatial-temporal graph convolution operation aggregates information from neighboring joints to capture spatial and temporal dependencies. After the multiple layers of spatial-temporal graph convolution, global average pooling is performed to compress the feature maps into a fixed-length vector representation. Finally, a SoftMax classifier is used to predict the action category based on the vector representation. This design allows ST-GCN to learn rich spatial and temporal features from skeletal data and effectively classify and recognize actions.

For the spatial dimension, the convolution operation for each graph can be expressed as follows:

$$f_{out} = \sum_i^{K_v} \overline{W}_i (f_{in} \Lambda_i^{-\frac{1}{2}} A_i \Lambda_i^{-\frac{1}{2}}) \otimes M_i \tag{1}$$

Here, f is the $C_{in} \times T \times N$ feature map where N denotes the number of vertexes, T denotes the temporal length and C_{in} denotes the number of input channels. A is an $N \times N$ adjacency matrix, Λ is a normalized diagonal matrix, with its diagonal elements denoted as $\Lambda_j^{ii} = \sum_k (A_j^{ki}) + \alpha$. To ensure that the diagonal elements of Λ are all non-zero, α is set to 0.001. K_v represents the kernel size in the spatial dimension. According to the aforementioned partitioning strategy, K_v is chosen as 3. $A_0 = I$ represents the self-connections of the vertices, A_1 represents the connections within the centripetal subset, and A_2 represents the connections within the centrifugal subset. $\overline{W}_i$ is a 1×1 convolution operation with a weight vector of size $C_{out} \times C_{in} \times 1 \times 1$. M is an $N \times N$ attention map that represents the importance of each vertex. $\otimes$ represents element-wise matrix multiplication, which means it only affects the vertices connected to the current target.

3.3   Graph Transformer

Although GCN has strong capability in capturing spatial features, the dependencies between joints, especially the implicit long-range dependencies, can still be easily underestimated. Additionally, due to the nature of graph convolution operators as a form of Laplacian smoothing, it is prone to over-smoothing during the stacking of layers, leading to a "deadlock" in feature representation. Therefore, this work designs a skeleton Graph Transformer on top of GCN, specifically tailored for action recognition, to enhance the dependencies between locally adjacent joints using graph self-attention. The effectiveness of this approach is confirmed through ablation studies. This section provides a detailed description of how the Graph Transformer operates on skeleton data.

Graph Self-attention Layer. The attention mechanism used in this paper is a single-layer feed-forward neural network. To incorporate the structural information of the graph, only neighboring nodes are allowed to participate in the attention mechanism of the current node. To ensure that the attention coefficients between different nodes are easily comparable, a softmax function is used to normalize the attention coefficients of all nodes. The fully unfolded formula of the attention mechanism is as follows:

$$\alpha_{ij} = \frac{\exp(\text{LeakyReLU}(\vec{a}^T\left[W\vec{h}_i \| W\vec{h}_j\right]))}{\sum_{k \in N_i}\exp(\text{LeakyReLU}(\vec{a}^T\left[W\vec{h}_i \| W\vec{h}_k\right]))} \tag{2}$$

where α_{ij} represents the importance of the features of node j to node i. $W \in R^{F' \times F}$ is a shared linear transformation that converts the node input features into a high-dimensional feature to achieve sufficient expressive power, $\vec{h}_i$ corresponds to the feature vector of node i. LeakyReLU is a non-linear activation function. $\vec{a} \in R^{2F'}$ is a weight vector that maps the high-dimensional features of length $2F$ to a scalar. $\|$ denotes the concatenation of node features of length F.

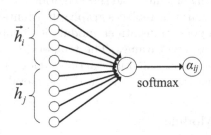

Fig. 2. The multi-head graph self-attention mechanism.

To stabilize the learning process of Graph self-attention, this paper employs a multi-head attention mechanism as shown in Fig. 2. It leverages multiple heads to jointly model information from different positions in different representation subspaces. The formula representation is as follows:

$$\text{GSA}(\overrightarrow{h}_i') = \text{Concat}(\alpha_{i1}, \alpha_{i2}, \cdots, \alpha_{ij})\text{W}\overrightarrow{h}_i \qquad (3)$$

where $\text{Concat}(\cdot)$ represents the concatenation of the features of adjacent nodes.

Spatial Transformer. As shown in Fig. 1, high-dimensional feature embeddings are extracted from a single frame. Given a two-dimensional data with N joints, each joint (i.e., two-dimensional coordinate) is treated as a patch, and feature extraction is performed on all patches using a standard Vision Transformer pipeline. Firstly, trainable linear projections are used to map the coordinates of each joint to a high-dimensional space, known as spatial patch embeddings. These embeddings are then added with learnable spatial position embeddings $E_{SPos} \in R^{J \times C}$. Consequently, the input $x_i \in R^{1 \times (J \cdot 2)}$ of the frame n is transformed into $Z_0^i \in R^{J \times C}$, where 2 represents the two-dimensional coordinates per frame, and C denotes the spatial embedding dimension [23]. The resulting joint sequence of features is fed into a spatial transformer encoder, which employs self-attention mechanisms to integrate information from all joints. The structure of the spatial transformer Encoder with L layers can be represented as follows:

$$Z_l' = \text{GSA}(\text{LN}(Z_{l-1})), l = 1, 2, \cdots, L \qquad (4)$$

$$Z_l = \text{MLP}(\text{LN}(Z_l')), l = 1, 2, \cdots, L \qquad (5)$$

where $Z \in R^{J \times C}$ represents the given embedding features, $\text{LN}(\cdot)$ refers to the normalization layer, and $\text{MLP}(\cdot)$ denotes the multi-layer perceptron.

Temporal Transformer. Since the spatial transformer module encodes high-dimensional features for each individual frame, the goal of the temporal transformer module is to model the dependencies across the frame sequence. For the temporal transformer encoder, we adopt the same structure as the spatial transformer encoder, consisting of multi-head self-attention blocks and MLP blocks. It is worth noting that this module replaces graph self-attention with self-attention. The output of the temporal transformer module is denoted as $Y \in R^{f \times (J \cdot C)}$, where f is the number of input frames. The temporal transformer module captures the temporal dependencies and interactions among the features across the frame sequence, allowing the model to learn the temporal dynamics in the data.

3.4 Text Feature Module

The text feature module consists of three components: Text description template, Text feature extraction, and Text feature fusion.

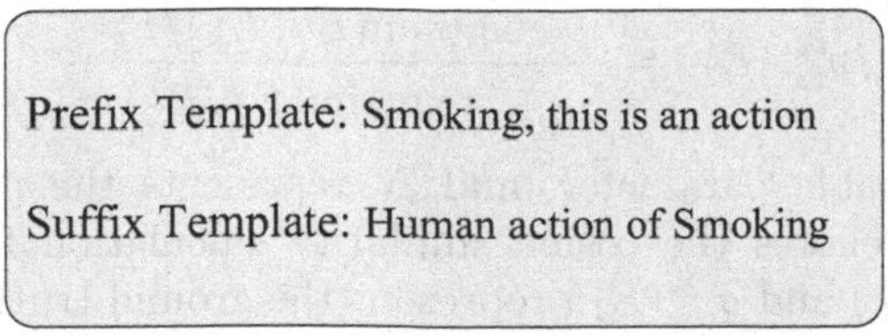

Prefix Template: Smoking, this is an action

Suffix Template: Human action of Smoking

Fig. 3. The multi-head graph self-attention mechanism.

Text Description Template. Appropriate action descriptions play a crucial role in obtaining textual features. In this study, we follow the text description generation template proposed in [28] to generate action descriptions directly using label names. The action labels are inputted into text templates with prefixes and suffixes to generate text descriptions. The prefix template is "[label], this is a human action" and the suffix template is "This is a human action of [label]", where [label] represents the human action label, as shown in Fig. 3.

Text Feature Extraction. We utilize a text encoder to extract textual features in this study. Given the recent success of Transformer models in the field of NLP, we adopt the design proposed by Vaswani [20] as our text encoder. Each layer of the text encoder consists of a self-attention layer, a normalization layer, and a multi-layer perceptron. In addition to the self-attention layer, the text encoder structure is similar to the spatial transformer Encoder. The formula for self-attention is as follows:

$$\text{SA}(Q_l, K_l, V_l) = \text{softmax}(\frac{Q_l K_l^T}{\sqrt{d}} V_l) \tag{6}$$

where Q_l, K_l, V_l represent the query matrix, key matrix, and value matrix of the word embeddings, respectively. They are computed by linearly mapping the word embeddings Z_{l-1} with weight matrices W_Q, W_K and W_V. The scale factor d is a proportionality factor that is equal to the dimension of the query and key matrices. It is used to normalize the dot product between the query and key vectors during the self-attention calculation.

Text Feature Fusion. To obtain the correlation features between the skeleton and text, we employ a similarity calculation module, referred to as the text encoding module in Fig.1. We compute the cosine similarity between the text encoding features E_T obtained by passing the entire label set through the text encoder and the skeleton encoding features E_S obtained by the Graph Transformer. The formulas for computing the skeleton-to-text and text-to-skeleton similarities, after applying softmax normalization, are as follows:

$$p_i^{s2t}(E_S^i) = \frac{\exp(\text{sim}(E_S^i, E_T^i)/\tau)}{\sum_{j=1}^{N} \exp(\text{sim}(E_S^i, E_T^j)/\tau)} \tag{7}$$

$$p_i^{t2s}(E_T^i) = \frac{\exp(\text{sim}(E_T^i, E_S^i)/\tau)}{\sum_{j=1}^N \exp(\text{sim}(E_T^i, E_S^j)/\tau)} \qquad (8)$$

where τ is a learnable parameter, and N represents the number of training instances, $\text{sim}(\cdot)$ denotes the cosine similarity calculation formula for encoding vectors. $q^{s2t}(E_S)$ and $q^{t2s}(E_T)$ represent the ground truth similarity scores, where a score of 0 indicates inconsistency between the labels and the skeleton, and a score of 1 indicates consistency between the labels and the skeleton. We use the Kullback-Leibler (KL) divergence to define the skeleton-text pair loss function for optimizing our model:

$$loss_{\text{sim}} = \frac{1}{2}\text{E}[\text{KL}(p^{s2t}(E_S), q^{s2t}(E_S)) + \text{KL}(p^{t2s}(E_T), q^{t2s}(E_T))] \qquad (9)$$

Combined with the Loss function of action classification, we can get:

$$loss = loss_{\text{CEL}} + \lambda \cdot loss_{\text{sim}} \qquad (10)$$

where $loss_{\text{CEL}}$ represents the cross-entropy loss function for action classification, and λ is the balancing parameter.

4   Experiments

We conduct extensive experiments on our self-collected smoking dataset, as well as the Kinetics-Skeleton [29] and NTU-RGB+D [30] large-scale skeleton action recognition datasets to validate the effectiveness of the proposed method.

4.1   Dataset

Smoking Dataset: Currently, there are many publicly available datasets in the field of action recognition, such as Kinetics-Skeleton [29] and NTU RGB+D [30]. To validate the effectiveness of our algorithm, we created a custom smoking dataset. This dataset was created using OpenPose on Kinetics video data and includes two action categories: smoking and non-smoking. The smoking actions include smoking, holding a cigarette, lighting a cigarette, exhaling smoke, and putting down the cigarette after smoking. The non-smoking actions include drinking water, eating food, riding a bicycle, and walking. The dataset consists of 3,025 samples in the training set and 1,190 samples in the testing set.

Kinetics-Skeleton Dataset. [29]: The Kinetics-Skeleton dataset is a large-scale skeleton-based video action recognition dataset. It is derived from the Kinetics video dataset, which consists of 400 action categories and utilizes Open-Pose skeleton data. The Kinetics-Skeleton dataset contains skeleton sequences composed of 2D joint coordinates and confidences, where each sequence corresponds to a video action. It includes 18 major body joints, such as the head, neck, shoulders, elbows, wrists, hips, knees, and ankles. The dataset consists of 240,000 samples for training and 20,000 samples for testing, making it widely used for research and evaluation of skeleton-based video action recognition algorithms.

NTU-RGB+D [30]: NTU-RGB+D is a commonly used large-scale dataset for evaluating models in skeleton-based action recognition tasks. It consists of 56,880 samples from 40 different subjects, totaling over 4 million frames. The data was captured using the Microsoft Kinect V2 camera from 80 different viewpoints, including RGB videos, depth sequences, skeleton data, and infrared data. The skeleton data consists of the 3D coordinates of 25 joints in the human body and corresponding vectors for the skeleton connections. The NTU-RGB+D dataset provides two different benchmark settings for evaluation: Cross-View and Cross-Subject. In the Cross-View setting, three cameras are positioned at the same height vertically but different angles horizontally (-45°, 0°, 45°), resulting in different views. It includes two frontal views, one left-side view, one right-side view, one left 45° view, and one right 45° view. Camera 1 always captures the 45° view, while cameras 2 and 3 capture the front and side views. Therefore, data captured by cameras 2 and 3 is used for training, and data captured by camera 1 is used for testing. The training set and testing set consist of 37,920 and 18,960 samples, respectively. In the Cross-Subject setting, there are 60 action classes, including 40 daily actions, 11 mutual actions, and 9 health-related actions. The training set and testing set consist of 40,320 and 16,560 samples, respectively.

4.2 Implementation Details

Because the number of frames in each sample may vary in the same dataset, the number of frames in the NTU dataset is standardized to 300 frames, while the Kinetics-Skeleton dataset and the Smoking dataset have 150 frames. Samples longer than the standard length are cropped, and samples shorter than the standard length are padded with copies of the last frame. Additionally, to enhance the robustness of the model, random translation and horizontal flipping are applied to randomly selected 100 frames. Our model is implemented using the PyTorch deep learning framework and trained on four NVIDIA 3090 24GB GPUs. First, stochastic gradient descent (SGD) optimizer is used to speed up the convergence. Second, L2 regularization and dropout (dropout rate of 0.5) strategies are employed to mitigate overfitting, and the weight decay factor is set to 0.0001. The cross-entropy loss function is chosen. For NTU-RGB+D, the batch sizes for training, testing, and the number of training epochs are determined to be (128, 64, 100). For Kinetics-Skeleton, the batch sizes for training, testing, and the number of training epochs are set to (128, 256, 120). For the Smoking dataset, the batch sizes for training, testing, and the number of training epochs are set to (128, 256, 100). Furthermore, the initial learning rate is set to 0.1 and then reduced by a factor of 10 at the 40 and 60 epochs.

4.3 Ablation Study

In this section, we evaluate the impact of skeleton data preprocessing, Graph Transformer, and text encoder through experiments.

Table 1 demonstrates the impact of data preprocessing on model performance. It can be observed that different data processing methods improve the

294 L. Weng et al.

Table 1. Data preprocessing.

Data processing method	Accuracy(%)
None	83.21
Joint point distance	83.95
Joint movement distance between frames	86.75
Joint angle	84.23
All	87.94

accuracy of action classification, with improvements of 0.74%, 3.54%, and 1.02% respectively. Among them, the addition of inter-frame distance data shows the most significant improvement in classification accuracy, indicating that effectively capturing temporal features can significantly enhance the performance of the network. Joint distance can well describe the shape and posture of human movements and is often used to capture local relationships in actions. Joint angle features are used to describe the motion and variation between joints, making them more suitable for extracting global features. Therefore, both methods can effectively improve recognition accuracy.

Table 2. Ablation study on Graph Transformer and Text Encoder in LGGT.

Graph Transformer	Text Encoder	Accuracy(%)
		77.96
	✓	79.39
✓		86.90
✓	✓	87.94

Table 2 presents the impact of the Graph Transformer and the Text Encoder on the model's performance. The results reveal that compared to the model without the Graph Transformer, there is an improvement of 8.55% points in accuracy. Additionally, when the Text Encoder is not used, the accuracy is 86.90%, which increases to 87.94% when the Text Encoder is employed. Furthermore, when neither of them is used, the accuracy is 77.96%. These findings indicate that both the Graph Transformer and the Text Encoder have a positive effect on enhancing the model's performance. Specifically, the Graph Transformer leverages graph attention mechanisms to capture relationships between different parts of the skeleton, while the Text Encoder assists in extracting valuable semantic information from textual descriptions, resulting in better recognition of different actions.

4.4 Comparison with State-of-the-Art Methods

In this section, we evaluated the performance of our method on multiple action recognition datasets: the Smoking dataset, Kinetics-Skeleton [29], and NTU-RGB+D [30]. We noticed that most existing methods adopt an integration strategy that combines joint features, joint motion features, skeleton features, and skeleton motion features. To ensure a fair comparison, we also employed this integration strategy. As shown in Table 3, 4 and 5, our proposed method outperformed the majority of previous methods, demonstrating its superiority in action recognition tasks.

Table 3. Comparison with previous work on Smoking dataset.

Methods	Accuracy(%)
ST-GCN [5]	75.38
2S-AGCN [6]	81.41
Shift-GCN [18]	82.91
2S-AGC-LSTM [31]	81.98
AS-GCN [15]	83.45
DC-GCN [32]	84.10
ST-TR-GCN [33]	80.90
PR-GCN [34]	84.92
CTR-GCN [7]	87.73
Info-GCN [8]	90.45
Ours	87.94

Table 4. Comparison with previous work on Kinetics-Skeleton.

Methods	Top-1 Acc(%)	Top-5 Acc(%)
ST-GCN [5]	30.7	52.8
2S-AGCN [6]	36.1	58.7
AS-GCN [15]	34.8	56.5
ST-TR-GCN [33]	37.4	59.7
PR-GCN [34]	38.1	58.2
SKP [35]	39.2	62.8
Ours	38.6	61.7

Table 5. Comparison with previous work on NTU-RGB+D.

Methods	Accuracy (%)	
	X-Sub	X-View
ST-GCN [5]	81.5	88.3
2S-AGCN [6]	88.5	95.1
Shift-GCN [18]	90.7	96.5
2S-AGC-LSTM [6]	89.2	95.0
AS-GCN [15]	86.8	94.2
SGN [36]	89.0	94.5
ST-TR-GCN [33]	89.9	96.1
DC-GCN [32]	90.8	96.6
cre PR-GCN [34]	85.2	91.7
MST-GCN [37]	91.5	96.6
CTR-GCN [7]	92.4	96.8
Info-GCN [8]	92.7	96.9
Ours	91.6	96.8

Table 3 presents the classification performance of various methods for action recognition on the Smoking dataset. The best-performing method is Info-GCN, achieving an accuracy of 90.45%. Our proposed method achieved an accuracy of 87.94%, surpassing baseline methods such as ST-GCN, 2S-AGCN, and Shift-GCN. This indicates that our method performs well in smoking recognition. Table 4 displays the classification performance of different methods, including ours, on the Kinetics-Skeleton dataset. Our method achieved Top-1 Acc and Top-5 Acc of 38.6% and 61.7%, respectively, outperforming most methods but still lower than the performance of Structured Keypoint Pooling. Structured Keypoint Pooling introduces a point cloud deep learning paradigm for action recognition, leveraging point cloud features to enhance recognition performance, which explains its superior performance compared to our method. Table 5 lists the classification performance of various methods, including ours, on the NTU-RGB+D dataset. Our method achieved accuracies of 91.6% and 96.8% in X-Sub and X-View, respectively, comparable to MST-GCN, CTR-GCN, and Info-GCN. CTR-GCN performs slightly better due to its utilization of dynamic adjacency matrices, which enhances the model's capacity. On the other hand, Info-GCN improves generalization performance by introducing an information bottleneck-based MMD loss and comparing action representations through principal component analysis (PCA). Furthermore, Kinetics-Skeleton exhibits inferior classification performance compared to NTU-RGB+D. This is attributed to the fact that, as shown in Fig. 4(a), most instances in the Kinetics-Skeleton dataset lack complete skeletal structures. In contrast, as depicted in Fig. 4(b), NTU-RGB+D captures rich skeletal information for action recognition, ensuring higher accuracy.

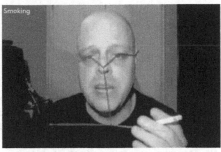 

(a). 'Smoking' in Kinetics-Skeleton   (b). 'Shake hands' in NTU-RGB+D

Fig. 4. Representative frames from Kinetics Skeleton and NTU RGB+D datasets.

5   Conclusion

In this paper, we propose a novel method called LGGT for extracting spatial-temporal features in skeleton action recognition tasks. LGGT utilizes a module called Graph Transformer to extract 3D information from 2D skeleton data. Compared to traditional feature extractors, it better captures long-range dependencies and alleviates the over-smooth issue. Additionally, a large-scale language model is applied as a knowledge engine to guide the action recognition task by generating meaningful textual descriptions, thereby enhancing the model's understanding and accuracy in human behavior. Finally, we propose an action recognition network model that integrates textual features and skeleton features, and experimental results demonstrate that the use of detailed textual features effectively improves action recognition performance.

Acknowledgements. This work is being supported by the Zhejiang Provincial Natural Science Foundation of China under Grant No. LQ22F020008, the National Key Research and Development Project of China under Grant No. 2020AAA0104001 and the "Pioneer" and "Leading Goose" R&D Program of Zhejiang under Grant No. 2022C01120.

References

1. Zhang, Z.: Microsoft kinect sensor and its effect. IEEE Multimedia **19**(2), 4–10 (2012)
2. Keselman, L., Iselin Woodfill, J., Grunnet-Jepsen, A., Bhowmik, A.: Intel realsense stereoscopic depth cameras. In: Proceedings of the IEEE Conference on Computer Vision and Pattern Recognition Workshops, pp. 1–10 (2017)
3. Du, Y., Wang, W., Wang, L.: Hierarchical recurrent neural network for skeleton based action recognition. In: Proceedings of the IEEE Conference on Computer Vision and Pattern Recognition, pp. 1110–1118 (2015)
4. Li, C., Zhong, Q., Xie, D., Pu, S.: Skeleton-based action recognition with convolutional neural networks. In: 2017 IEEE International Conference on Multimedia & Expo Workshops (ICMEW), pp. 597–600. IEEE (2017)

5. Yan, S., Xiong, Y., Lin, D.: Spatial temporal graph convolutional networks for skeleton-based action recognition. In: Proceedings of the AAAI Conference on Artificial Intelligence, vol. 32 (2018)
6. Shi, L., Zhang, Y., Cheng, J., Lu, H.: Two-stream adaptive graph convolutional networks for skeleton-based action recognition. In: Proceedings of the IEEE/CVF Conference on Computer Vision and Pattern Recognition, pp. 12026–12035 (2019)
7. Chen, Y., Zhang, Z., Yuan, C., Li, B., Deng, Y., Hu, W.: Channel-wise topology refinement graph convolution for skeleton-based action recognition. In: Proceedings of the IEEE/CVF International Conference on Computer Vision, pp. 13359–13368 (2021)
8. Chi, H.g., Ha, M.H., Chi, S., Lee, S.W., Huang, Q., Ramani, K.: Infogcn: representation learning for human skeleton-based action recognition. In: Proceedings of the IEEE/CVF Conference on Computer Vision and Pattern Recognition, pp. 20186–20196 (2022)
9. Sun, Z., Ke, Q., Rahmani, H., Bennamoun, M., Wang, G., Liu, J.: Human action recognition from various data modalities: a review. IEEE Trans. Pattern Anal. Mach. Intell. (2022)
10. De Boissiere, A.M., Noumeir, R.: Infrared and 3d skeleton feature fusion for rgb-d action recognition. IEEE Access **8**, 168297–168308 (2020)
11. Cao, Z., Simon, T., Wei, S.E., Sheikh, Y.: Realtime multi-person 2d pose estimation using part affinity fields. In: Proceedings of the IEEE Conference on Computer Vision and Pattern Recognition, pp. 7291–7299 (2017)
12. Hussein, M.E., Torki, M., Gowayyed, M.A., El-Saban, M.: Human action recognition using a temporal hierarchy of covariance descriptors on 3d joint locations. In: Twenty-Third International Joint Conference on Artificial Intelligence (2013)
13. Wang, J., Liu, Z., Wu, Y., Yuan, J.: Mining actionlet ensemble for action recognition with depth cameras. In: 2012 IEEE Conference on Computer Vision and Pattern Recognition, pp. 1290–1297. IEEE (2012)
14. Vemulapalli, R., Arrate, F., Chellappa, R.: Human action recognition by representing 3d skeletons as points in a lie group. In: Proceedings of the IEEE Conference on Computer Vision and Pattern Recognition, pp. 588–595 (2014)
15. Li, M., Chen, S., Chen, X., Zhang, Y., Wang, Y., Tian, Q.: Actional-structural graph convolutional networks for skeleton-based action recognition. In: Proceedings of the IEEE/CVF Conference on Computer Vision and Pattern Recognition, pp. 3595–3603 (2019)
16. Shi, L., Zhang, Y., Cheng, J., Lu, H.: Skeleton-based action recognition with directed graph neural networks. In: Proceedings of the IEEE/CVF Conference on Computer Vision and Pattern Recognition, pp. 7912–7921 (2019)
17. Liu, Z., Zhang, H., Chen, Z., Wang, Z., Ouyang, W.: Disentangling and unifying graph convolutions for skeleton-based action recognition. In: Proceedings of the IEEE/CVF Conference on Computer Vision and Pattern Recognition, pp. 143–152 (2020)
18. Cheng, K., Zhang, Y., He, X., Chen, W., Cheng, J., Lu, H.: Skeleton-based action recognition with shift graph convolutional network. In: Proceedings of the IEEE/CVF Conference on Computer Vision and Pattern Recognition, pp. 183–192 (2020)
19. Song, Y.F., Zhang, Z., Shan, C., Wang, L.: Stronger, faster and more explainable: a graph convolutional baseline for skeleton-based action recognition. In: proceedings of the 28th ACM International Conference on Multimedia, pp. 1625–1633 (2020)
20. Vaswani, A., et al.: Attention is all you need. In: Advances In Neural Information Processing Systems 30 (2017)

21. Radford, A., Narasimhan, K., Salimans, T., Sutskever, I., et al.: Improving language understanding by generative pre-training (2018)
22. Kenton, J.D.M.W.C., Toutanova, L.K.: Bert: pre-training of deep bidirectional transformers for language understanding. In: Proceedings of NAACL-HLT, pp. 4171–4186 (2019)
23. Dosovitskiy, A., et al.: An image is worth 16x16 words: transformers for image recognition at scale. In: International Conference on Learning Representations (2021)
24. Wang, Q., Peng, J., Shi, S., Liu, T., He, J., Weng, R.: Iip-transformer: intra-inter-part transformer for skeleton-based action recognition. arXiv preprint arXiv:2110.13385 (2021)
25. Dwivedi, V.P., Bresson, X.: A generalization of transformer networks to graphs. arXiv preprint arXiv:2012.09699 (2020)
26. Radford, A., et al.: Learning transferable visual models from natural language supervision. In: International Conference on Machine Learning, pp. 8748–8763. PMLR (2021)
27. Jia, C., et al.: Scaling up visual and vision-language representation learning with noisy text supervision. In: International Conference on Machine Learning, pp. 4904–4916. PMLR (2021)
28. Wang, M., Xing, J., Liu, Y.: Actionclip: a new paradigm for video action recognition. arXiv preprint arXiv:2109.08472 (2021)
29. Carreira, J., Zisserman, A.: Quo vadis, action recognition? a new model and the kinetics dataset. In: 2017 IEEE Conference on Computer Vision and Pattern Recognition (CVPR), pp. 4724–4733 (2017). https://doi.org/10.1109/CVPR.2017.502
30. Shahroudy, A., Liu, J., Ng, T.T., Wang, G.: Ntu rgb+ d: a large scale dataset for 3d human activity analysis. In: Proceedings of the IEEE Conference on Computer Vision and Pattern Recognition, pp. 1010–1019 (2016)
31. Si, C., Chen, W., Wang, W., Wang, L., Tan, T.: An attention enhanced graph convolutional lstm network for skeleton-based action recognition. In: proceedings of the IEEE/CVF Conference on Computer Vision and Pattern Recognition, pp. 1227–1236 (2019)
32. Cheng, K., Zhang, Y., Cao, C., Shi, L., Cheng, J., Lu, H.: Decoupling GCN with DropGraph module for skeleton-based action recognition. In: Vedaldi, A., Bischof, H., Brox, T., Frahm, J.-M. (eds.) ECCV 2020. LNCS, vol. 12369, pp. 536–553. Springer, Cham (2020). https://doi.org/10.1007/978-3-030-58586-0_32
33. Plizzari, C., Cannici, M., Matteucci, M.: Skeleton-based action recognition via spatial and temporal transformer networks. Comput. Vis. Image Underst. **208**, 103219 (2021)
34. Li, S., Yi, J., Farha, Y.A., Gall, J.: Pose refinement graph convolutional network for skeleton-based action recognition. IEEE Robot. Autom. Lett. **6**(2), 1028–1035 (2021)
35. Hachiuma, R., Sato, F., Sekii, T.: Unified keypoint-based action recognition framework via structured keypoint pooling. In: Proceedings of the IEEE/CVF Conference on Computer Vision and Pattern Recognition, pp. 22962–22971 (2023)
36. Zhang, P., Lan, C., Zeng, W., Xing, J., Xue, J., Zheng, N.: Semantics-guided neural networks for efficient skeleton-based human action recognition. In: proceedings of the IEEE/CVF Conference on Computer Vision and Pattern Recognition, pp. 1112–1121 (2020)
37. Chen, Z., Li, S., Yang, B., Li, Q., Liu, H.: Multi-scale spatial temporal graph convolutional network for skeleton-based action recognition. In: Proceedings of the AAAI Conference on Artificial Intelligence, vol. 35, pp. 1113–1122 (2021)

A Federated Multi-stage Light-Weight Vision Transformer for Respiratory Disease Detection

Pranab Sahoo[1]([✉])(iD), Saksham Kumar Sharma[2], Sriparna Saha[1], and Samrat Mondal[1]

[1] Indian Institute of Technology Patna, Patna, India
pranab_2021cs25@iitp.ac.in
[2] Maharaja Surajmal Institute of Technology, Delhi, India

Abstract. Artificial Intelligence-based computer-aided diagnosis (CAD) has been widely applied to assist medical professionals in several medical applications. Although there are many studies on respiratory disease detection using Deep Learning (DL) approaches from radiographic images, the limited availability of public datasets limits their interpretation and generalization capacity. However, radiography images are available through different organizations in various countries. This condition is suited for Federated Learning (FL) training, which can collaborate with different institutes to use private data and train a global model. In FL, the local model on the client's end is critical because there must be a balance between the model's accuracy, communication cost, and client-side memory usage. The current DL or Vision Transformer (ViT)-based models have large parameters, making the client-side memory and communication costs a significant bottleneck when applied to FL training. The existing state-of-the-art (SOTA) FL techniques on respiratory disease detection either use small CNNs with insufficient accuracy or assume clients have sufficient processing capacity to train large models, which remains a significant challenge in practical applications. In this study, we tried to find one question: Is it possible to maintain higher accuracy while lowering the model parameters, leading to lower memory requirements and communication costs? To address this problem, we propose a federated multi-stage, light-weight ViT framework that combines the strengths of CNNs and ViTs to build an efficient FL framework. We conduct extensive experiments and show that the proposed framework outperforms a set of current SOTA models in FL training with higher accuracy while lowering communication costs and memory requirements. We adapted Grad-CAM for the infection localization and compared it with an experienced radiologist's findings.

Keywords: Respiratory Disease · Federated Learning · Vision Transformer · Chest X-ray

B. Luo et al. (Eds.): ICONIP 2023, CCIS 1964, pp. 300–311, 2024.
https://doi.org/10.1007/978-981-99-8141-0_23

1   Introduction

Respiratory diseases, such as pneumonia [21], tuberculosis (TB) [13], and COVID-19 [23] are the most common and pose a severe threat to human life. Early screening can successfully stop the disease's progression and improve the chances of effective treatment. Chest X-ray (CXR) plays a crucial role in the preliminary investigation of these diseases, which radiologists interpret to look for infectious regions. However, the overlapping findings between these diseases make it laborious, time-consuming, and error-prone. According to WHO [24], less than one doctor is available for every 1000 people in more than 45% of countries worldwide. Moreover, the COVID-19 pandemic in December 2019 affected millions of individuals and put the patient-doctor ratio out of proportion. Therefore, creating a computer-aided diagnosis (CAD) that can distinguish the overlapping patterns from the CXR images in a specific way is essential and helpful for medical professionals. Over the past several years, the development of Deep Learning (DL)-based models has been crucial to several applications [10–12,14,15]. However, obtaining large data sets in the medical field is challenging. Most of the current research used a small number of radiography datasets for training the DL-based models [16,17,20,23], which limits their applicability in the real applications. Although radiography images are generated regularly, they are still restricted for model training since sharing patient information is challenging due to international laws like the General Data Protection Regulation (GDPR) and the Health Insurance Portability and Accountability Act (HIPAA). To overcome this challenge, some authors have utilized the Federated Learning (FL) approach to collaborate with different hospitals [7]. Recently, the authors in [3] used two pre-trained convolutional neural networks (CNN) as the backbone in client-side and proposed an FL model for lung abnormality detection. In a similar work, the authors [5] used four prominent pre-trained CNNs and performed two types of experiments, with FL and without FL, and attained 96.15% accuracy. In another work, authors [15] proposed an FL-based framework with a Vision Transformer (ViT) as the base model for client-side model training. They have utilized ViT-based models to remove the disadvantages of CNN-based models and reported a performance improvement over CNN-based models. Although, they have not addressed the complexity of training ViT models in resource-limited client settings. Despite its significant success, the current work on respiratory disease detection using FL must address the fundamental challenges. For instance, clients face significant resource limitations regarding memory and communication costs, substantially restricting their capacity to train big models locally. All the above works utilized large models to improve accuracy without addressing the memory and communication costs. Another challenge is that most client-side models are utilized CNN as the base model [3,5]. Although CNNs have demonstrated excellent performance in many vision-related tasks, they are not ideal for complex disease identification, where global characteristics such as multiplicity, distribution, and patterns must be considered. This is due to the convolution operation's intrinsic locality of pixel dependencies. To overcome the limitations of CNNs in vision-related tasks that require the integration of global relationships between pixels, the ViT equipped with the Transformer [22] was proposed. ViT can model long-range dependency among pixels through the self-attention mechanism and has shown SOTA performance [2,16]. These performance enhancements, meanwhile, come at the expense of model size (network parameters). The need for more parameters in ViT-based (ViT-B/16 has 86 million parameters) models are likely because they lack image-specific inductive bias, which is inherent in CNNs [25]. Hybrid approaches that combine convolutions and transformers are gaining interest in recent studies [1,8]. To address the above problems, we have adopted MobileViT, a new architecture combining CNN and ViT with fewer parameters without compromising the performance [8]. In summary, the main contributions are:

1. This study proposed an FL-based architecture using a collaborative technique to benefit from the diversified private data while keeping the data private.
2. The proposed model employs a hybrid lightweight architecture by combining Convolution blocks and Transformer.
3. We have performed an in-depth analysis of centralized, independent, and identically distributed (IID) and non-IID data distributions and reported a detailed analysis.
4. We showed that our framework outperformed state-of-the-art CNN-based and Transformer-based models in FL experiments while lowering the memory and communication cost.
5. Finally, we have compared the model's prediction with the radiologist's for infection localization.

2   Proposed Framework

In the proposed architecture, we have combined the strengths of a light-weight ViT with FL and performed experiments. In the first section, we depict the details of the FL architecture, followed by the light-weight ViT architecture for lung CXR image classification at the client side. The goal of the proposed approach is to maximize the overall performance of the architecture while lowering the number of parameters, memory requirements, and communication costs.

2.1   Federated Learning

In this work, we study a client-server FL framework (Fig. 1) that uses FedAvg [7] algorithms to train a light-weight ViT model for respiratory disease classification collaboratively. In this setup, a central server maintains a global model and shares it with clients. Clients use private datasets and coordinate to build a robust model. We have used a light-weight ViT model on the client side to deal with local model training of local CXR images. The training phase comprises several synchronous communication rounds between the clients and the central server. The ViT model is initialized with random weights (W_0) before the training rounds begin. We consider there are K available clients, with each n_k private CXR image stored locally.

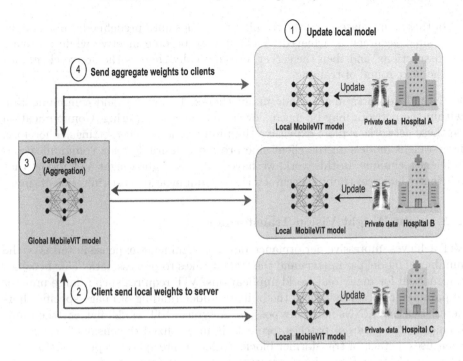

Fig. 1. Architecture details of the Federated Learning

The four steps in each communication round between the client and server are as follows:

1. The server initially maintains a global model (g) with initial pre-trained weights w, which also keeps track of the willing participants (S_k) for collaborative training.
2. Each client $k \in S_k$, requests the central server for the global weights w. Clients load the global weights into the local models and train locally on a mini-batch b of its local data. Each client executes for a pre-defined epoch, E, using mini-batch SGD with a learning rate η_{local} to minimize the local objective F_k.
3. Once the training is done for the pre-defined epochs E, clients send their updated local parameters w_k^t, to the server.
4. After receiving updates from all participating clients, the server updates the global model by aggregating all local updates using the FedAvg method.

$$w^t \leftarrow \sum_{k=1}^{K} \frac{n_k}{n} w_k^t \qquad (1)$$

Here, w_k^t is the parameters sent by client k at round t, n_k is the private dataset stored on client k, and the total dataset size in the collaborative training is n. w^t is the updated parameter at round t. One FL round consists of these four phases, and the training continues until predefined rounds or the desired accuracy is achieved. A client may become inactive while receiving instructions, and then the server automatically chooses the active clients for the next round of training.

Federated networks may include many clients (Hospitals), and communication within the network may be much slower than local computing. Communication on these networks is more expensive than in traditional data settings. Therefore, light-weight models on the client side are most desirable for a communication-efficient technique. In this work, we have utilized a light-weight ViT model that efficiently balances training accuracy and communication cost (in terms of time).

2.2 Light-Weight Vision Transformer

ViT achieves impressive performance but at a significant expense in terms of the number of model parameters and the time it takes to process data. This becomes impractical for many real-world applications. ViT requires such a large number of parameters because it lacks the built-in understanding of image-specific characteristics that CNNs naturally possess. A typical ViT model flattens the input images into a series of patches, projects it into a fixed dimensional space, and then uses a stack of transformer blocks to learn inter-patch representations [2]. These models need additional parameters to learn visual representations since they do not account for the spatial inductive bias found in CNNs. We have adapted MobileViT, a lightweight ViT model in the proposed framework. The architecture details are shown in Fig. 2. The main idea is to learn global representations with lightweight transformers. The MobileViT architecture starts with a 3 × 3 convolution, followed by multiple MV2 blocks and MobileViT blocks. The MV2 blocks are Inverted Residual blocks from MobileNetv2 [18]. The whole architecture takes advantage of the Swish activation function. Typically, feature maps' spatial dimensions (height and width) are chosen to be multiples of 2. Consequently, a 2 × 2 patch size is employed consistently across all levels. The primary purpose of the MV2 blocks is to downsample the feature maps by using the strided convolution. The Inverted Residual block begins with a 1 × 1 point-wise convolution, followed by a depthwise convolution layer, and concludes with a final 1 × 1 point-wise convolution. This block also incorporates a residual connection, which adds the block's input to its output. Next, the MobileViT block further enriches an input tensor with fewer parameters by combining local and global information.

The following sequence of operations is performed on the input tensor:

- The input tensor undergoes an n×n convolution, succeeded by a 1 × 1 convolution, projecting the features into a high-dimensional space.
- The resulting feature maps are then converted into N non-overlapping flattened patches.
- These flattened patches are fed into transformers to learn global representations, benefitting from spatial inductive bias.

- The output from the transformers is transformed back, followed by a 1×1 convolution, and concatenated with the original input tensor.
- Finally, another n × n convolution is applied to merge these concatenated features.

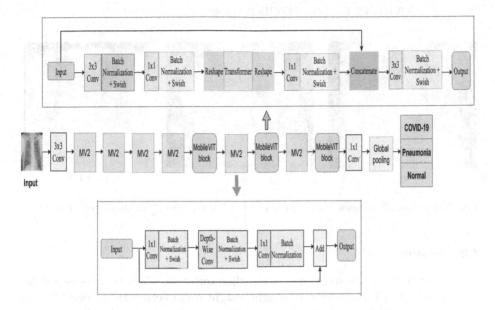

Fig. 2. Architecture details of MobileViT

3   Experimental Dataset and Results

This section describes the datasets used for classification. In addition, the implementation details, evaluation metrics, and experimental results have been discussed.

3.1   Classification Dataset

We created a custom dataset for classification by combining two popular datasets. Pneumonia and normal images were collected from one public repository [9]. COVID-19 CXR images are collected from [26]. We manually curated 4966 COVID-19 images with the help of one experienced radiologist, as some images were not correctly captured. We divided the dataset into a training, validation, and testing set in a 70:10:20 ratio. Image augmentation is a well-known strategy for regularizing the network in supervised learning. We have used data augmentation techniques such as rotation_range=40, shear_range=0.2, horizontal_flip, and zoom_range=0.2. The final dataset and source details are shown in Table 1 and the sample images are shown in Fig. 3.

Table 1. Summary of classification dataset.

Datasets	Classes	Initial Images	Final Images
CXR Images [9]	Normal	1583	9498
CXR Images [9]	Pneumonia	4273	8546
COVIDx CXR-3 [26]	COVID-19	4966	9930

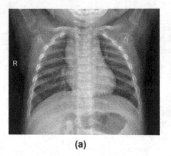 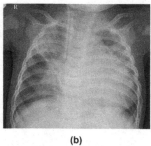 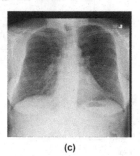

(a)   (b)   (c)

Fig. 3. The sample images from the dataset (a) Normal (b) Pneumonia (c) COVID-19.

3.2 Results

In this section, we try to answer our earlier question: Can we achieve similar performance in FL training with a light-weight architecture than deep CNN or ViT-based models? For this reason, we have experimented with different CNN and ViT-based models in centralized and federated data settings. We considered three clients (hospitals) for the FL training and experimented with IID and non-IID data distribution. We evaluate the model's performance using Precision (Prec), Sensitivity (Sens), F1-score, Accuracy (Acc), number of parameters (Params), and per round time [4].

3.3 Results on Centralized Data

We kept all the data on a single client for the centralized training. The experimental results are shown in Table 2. We have experimented with the original ViT model [2], Swin-Transformer [6] and three different versions of MobileViT models [8]. For the MobileViT, we used a categorical cross-entropy loss function with the Adam optimizer with 0.0001 as the initial learning rate and utilized the Swish activation function. We used the sparse categorical cross-entropy loss function with the 'AdamW' optimizer for ViT and swin-transformer training with an initial learning rate of 0.001. The accuracy of the MobileVit-XS is the highest among other models with a low parameter. Original ViT and swing-Transformer have also achieved a higher accuracy with the cost of large parameters which limits their applicability in the FL setting as it will increase the communication cost in the resource-limited framework.

Table 2. Results on Centralized data settings

Method	Prec. (%)	Sens. (%)	F1-score (%)	Acc. (%)	# Params(Million)
Original ViT	95.31	95.25	95.21	95.29	85.5 M
Swin-Transformer	95.17	94.46	94.90	96.54	1.2 M
MobileViT-S	95.28	95.32	95.30	95.38	1.4 M
MobileViT-XS	**96.67**	**95.92**	**96.28**	**97.46**	**0.6 M**
MobileViT-XXS	94.56	94.70	94.60	94.72	0.3 M

3.4 Results on Federated Learning (IID and Non-IID Data Distribution)

In FL experiments, IID and non-IID data distributions are the most common. We divided the dataset equally among the 3 clients for the IID data distribution. We randomly selected 1583 images from all classes and split them into 527 COVID-19 images, 527 normal images, and 527 pneumonia images to ensure an even data distribution. For the non-IID training, we employ a skewed class distribution and partition the dataset so that each client receives a different amount of samples. We randomly select 50% of COVID-19 and 20% of normal and 20% of pneumonia data for Client-1. For Client-2, we chose 50% of normal images, with 30% of COVID-19 and 30% of pneumonia data. For client-3, we select 50% of pneumonia data along with 20% of Covid-19 and 30% of normal data. The experimental results of the MobileViT-XS model and the original ViT model in FL settings are represented in Table 3. On the new IID dataset, MobileViT reaches 96.32% in 60 communication rounds, whereas the original ViT model achieves 96.08 in 122 rounds, stabilizing after that. For the IID data partition, we can conclude that both models got the same accuracy, but ViT takes more communication rounds and increases the overall communication cost. For non-IID data partition, both models had slightly lower accuracy. This slight degradation in accuracy is because each client has a lot of data from one class and little data from the other. Additionally, the test accuracy for the non-IID data stabilizes as the number of rounds increases, but it continues to converge for the IID data.

Table 3. Results on non-IID and IID data partition

Method	Prec.(%)	Sens.(%)	F1(%)	Acc.(%)	Model Size	Per round time
Original ViT (non-IID)	93.71	94.18	93.94	94.17	870 MB	15 mins
MobileViT-XS (non-IID)	**94.49**	**94.41**	**94.44**	**94.42**	**32.1 MB**	**4 mins**
Original ViT (IID)	96.00	96.33	96.16	96.08	870 MB	15 mins
MobileViT-XS (IID)	**96.43**	**95.83**	**96.12**	**96.32**	**32.1 MB**	**4 mins**

3.5 Communication Time and Client Memory Requirement

Most of the previous approaches [3,5,15] assume that client-server communication time and client-side memory requirements are negligible. This may be true in lab settings, but it is not true in a real-world collaborative application. We have only used 3 clients for the experiment and a small dataset, keeping our system constraints in mind. The experimental results are surprising. Both models got almost the same accuracy but massive differences in memory and communication cost (refer to Table 3). The size of the original ViT is 870 MB, and it took 15 min for one round of communication. In contrast, MobileViT took 4 min for one communication round, with a model size of 32.1 MB. It will have a big impact on a large client setting with a big dataset.

3.6 Infection Localization and Error Analysis

To visualize the infection regions we have utilized Gradient-weighted Class Activation Mapping (Grad-CAM) [19]. It localizes the infected regions on which the model focuses while making decisions. Figure 4 shows the classification model's focused areas. Next, with the help of an experienced radiologist, we annotated infected regions in CXR images. According to the localization maps, our network emphasizes the appropriate lung areas while classifying each disease. From Fig. 4, it is evident that the proposed model performs accurately and can localize the infected regions correctly. Our model also sometimes focuses on more portions of the lungs where there are no markings by the radiologist. Although this can not be a disadvantage as a radiologist will evaluate the outcome while making final decisions. Our model sometimes misclassifies some COVID-19 as pneumonia and vice versa. After a critical analysis with our radiologist, we observed that some characteristics of the COVID-19 are comparable with those of pneumonia. With the availability of more datasets in the future, we will train our model to find the nuanced differences between COVID-19 and pneumonia more accurately. The proposed architecture can be helpful in real clinical systems and assist medical professionals since annotating lungs and localizing the infection is challenging and time-consuming.

Radiologist (A)   Model Prediction (A)   Radiologist (B)   Model Prediction (B)

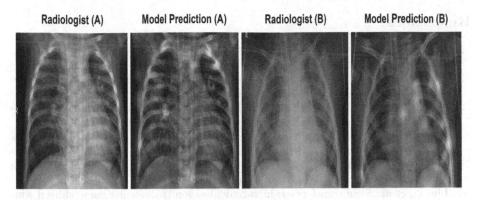

Fig. 4. Radiologist annotated images with the corresponding model prediction.

4   Conclusion and Future Work

This study introduces a multi-stage light-weight ViT model with an FL framework for classifying respiratory disease from CXR images without sharing private data. We have adopted a light-weight ViT model at the client side, incorporating convolution and Transformer blocks and lowering the training parameters. A comparative analysis of traditional CNN and ViT-based models on centralized data has been presented. According to the experiments, MobileViT-XS can accurately identify images while utilizing less memory and per-round communication time. Our findings also imply that the FL strategy, which avoids disclosing private and sensitive data, could produce outcomes comparable to centralized training. Despite the non-IID data distribution, the suggested architecture is reliable and functions similarly to the centralized model. FL-based architecture may connect all research institutions and hospitals, authorizing them to collaborate while maintaining privacy. This collaborative approach will improve the robustness and trustworthiness of the trained model. One drawback of our approach is that we have considered only 3 clients and a small dataset in training, as more clients will increase computational complexity with our limited available resources. By increasing the number of clients, we may get better accuracy. In the future, we will look into replacing the central server with blockchain technology to address single-point failure and security attacks. We have utilized only three lung diseases in this experiment, but it can be extended to other respiratory diseases.

Acknowledgement. Dr. Sriparna Saha gratefully acknowledges the Young Faculty Research Fellowship (YFRF) Award, supported by Visvesvaraya Ph.D. Scheme for Electronics and IT, Ministry of Electronics and Information Technology (MeitY), Government of India, being implemented by Digital India Corporation (formerly Media Lab Asia) for carrying out this research.

References

1. Dai, Z., Liu, H., Le, Q.V., Tan, M.: Coatnet: marrying convolution and attention for all data sizes. Adv. Neural. Inf. Process. Syst. **34**, 3965–3977 (2021)
2. Dosovitskiy, A., et al.: An image is worth 16x16 words: Transformers for image recognition at scale. arXiv preprint arXiv:2010.11929 (2020)
3. Feki, I., Ammar, S., Kessentini, Y., Muhammad, K.: Federated learning for Covid-19 screening from chest x-ray images. Appl. Soft Comput. **106**, 107330 (2021)
4. Japkowicz, N., Shah, M.: Evaluating learning algorithms: a classification perspective. Cambridge University Press (2011)
5. Liu, B., Yan, B., Zhou, Y., Yang, Y., Zhang, Y.: Experiments of federated learning for covid-19 chest x-ray images. arXiv preprint arXiv:2007.05592 (2020)
6. Liu, Z., et al.: Swin transformer: hierarchical vision transformer using shifted windows. In: Proceedings of the IEEE/CVF International Conference on Computer Vision, pp. 10012–10022 (2021)
7. McMahan, B., Moore, E., Ramage, D., Hampson, S., y Arcas, B.A.: Communication-efficient learning of deep networks from decentralized data. In: Artificial intelligence and statistics, pp. 1273–1282. PMLR (2017)
8. Mehta, S., Rastegari, M.: Mobilevit: light-weight, general-purpose, and mobile-friendly vision transformer. arXiv preprint arXiv:2110.02178 (2021)
9. Mooney, P.: Chest x-ray images (pneumonia). kaggle, Marzo (2018)
10. Palmal, S., Arya, N., Saha, S., Tripathy, S.: A multi-modal graph convolutional network for predicting human breast cancer prognosis. In: International Conference on Neural Information Processing, pp. 187–198. Springer (2022). https://doi.org/10.1007/978-981-99-1648-1_16
11. Palmal, S., Saha, S., Tripathy, S.: HIV-1 protease cleavage site prediction using stacked autoencoder with ensemble of classifiers. In: 2022 International Joint Conference on Neural Networks (IJCNN), pp. 1–8. IEEE (2022)
12. Palmal, S., Saha, S., Tripathy, S.: Multi-objective optimization with majority voting ensemble of classifiers for prediction of HIV-1 protease cleavage site. Soft Comput. **27**, 12211–1221 (2023). https://doi.org/10.1007/s00500-023-08431-2
13. Rahman, T., et al.: Reliable tuberculosis detection using chest x-ray with deep learning, segmentation and visualization. IEEE Access **8**, 191586–191601 (2020)
14. Sahoo, P., Saha, S., Mondal, S., Chowdhury, S., Gowda, S.: Computer-aided Covid-19 screening from chest CT-scan using a fuzzy ensemble-based technique. In: 2022 International Joint Conference on Neural Networks (IJCNN), pp. 1–8. IEEE (2022)
15. Sahoo, P., Saha, S., Mondal, S., Chowdhury, S., Gowda, S.: Vision transformer-based federated learning for covid-19 detection using chest x-ray. In: Neural Information Processing: 29th International Conference, ICONIP 2022, Virtual Event, November 22–26, 2022, Proceedings, Part VII, pp. 77–88. Springer (2023). https://doi.org/10.1007/978-981-99-1648-1_7
16. Sahoo, P., Saha, S., Mondal, S., Gowda, S.: Vision transformer based Covid-19 detection using chest CT-scan images. In: 2022 IEEE-EMBS International Conference on Biomedical and Health Informatics (BHI), pp. 01–04. IEEE (2022)
17. Sahoo, P., Saha, S., Mondal, S., Sharma, N.: Covid-19 detection from lung ultrasound images using a fuzzy ensemble-based transfer learning technique. In: 2022 26th International Conference on Pattern Recognition (ICPR), pp. 5170–5176. IEEE (2022)
18. Sandler, M., Howard, A., Zhu, M., Zhmoginov, A., Chen, L.C.: Mobilenetv 2: inverted residuals and linear bottlenecks. In: Proceedings of the IEEE Conference on Computer Vision and Pattern Recognition, pp. 4510–4520 (2018)

19. Selvaraju, R.R., Cogswell, M., Das, A., Vedantam, R., Parikh, D., Batra, D.: Grad-cam: visual explanations from deep networks via gradient-based localization. In: Proceedings of the IEEE International Conference on Computer Vision, pp. 618–626 (2017)
20. Toraman, S., Alakus, T.B., Turkoglu, I.: Convolutional capsnet: a novel artificial neural network approach to detect Covid-19 disease from x-ray images using capsule networks. Chaos, Solitons Fractals **140**, 110122 (2020)
21. Varshni, D., Thakral, K., Agarwal, L., Nijhawan, R., Mittal, A.: Pneumonia detection using CNN based feature extraction. In: 2019 IEEE International Conference on Electrical, Computer and Communication Technologies (ICECCT), pp. 1–7. IEEE (2019)
22. Vaswani, A., et al.: Attention is all you need. In: Advances in Neural Information Processing Systems 30 (2017)
23. Wang, L., Lin, Z.Q., Wong, A.: Covid-net: a tailored deep convolutional neural network design for detection of Covid-19 cases from chest x-ray images. Sci. Rep. **10**(1), 1–12 (2020)
24. WHO: who coronavirus (Covid-19) dashboard. https://covid19.who.int/ (2022)
25. Xiao, T., Singh, M., Mintun, E., Darrell, T., Dollár, P., Girshick, R.: Early convolutions help transformers see better. Adv. Neural. Inf. Process. Syst. **34**, 30392–30400 (2021)
26. Zhao, A., Aboutalebi, H., Wong, A., Gunraj, H., Terhljan, N., et al.: Covidx cxr-2: chest x-ray images for the detection of Covid-19 (2021)

Curiosity Enhanced Bayesian Personalized Ranking for Recommender Systems

Yaoming Deng[1,2], Qiqi Ding[1,2], Xin Wu[1,2], and Yi Cai[1,2(✉)]

[1] School of Software Engineering, South China University of Technology, Guangzhou, China
ycai@scut.edu.cn
[2] The Key Laboratory of Big Data and Intelligent Robot (South China University of Technology), Ministry of Education, Guangzhou, China

Abstract. Curiosity affects the users' selections of items, motivating them to explore the items regardless of their preferences. However, the existing social-based recommendation methods neglect the users' curiosity in the social networks, and it may cause the accuracy decrease in the recommendation. Moreover, only focusing on simulating the users' preferences can lead to users' information cocoons. To tackle the problems above, we propose a Curiosity Enhanced Bayesian Personalized Ranking (CBPR) model for the recommender systems. The experimental results on two public datasets demonstrate the advantages of our CBPR model over the existing models.

Keywords: Recommender Systems · Social Networks · Curiosity

1 Introduction

Recommendation systems are an information filtering technique to predict the "preference" a user would give to an item [31]. Traditional recommender systems often suffer from the data sparsity problem, Therefore, modern recommender systems generally employ side external information in recommendations. To address this issue, the social-based recommendation methods which exploit users' social relationships on social media as supplementary information to enhance the recommender systems has been gaining increasing attention [14,15,18,24].

Existing social-based recommendation methods are often based on the assumption that users tend to share similar preferences with their online friends and then recommend the items that friends preferred to the users [2]. Zhao et al. [33] extend the Bayesian Personalized Ranking(BPR) [20] model based on an observation that users tend to assign higher ranks to items that their friends prefer. However, these methods often ignore the psychological characteristics of individual users. As well established in psychological studies [21], humans' interest is closely associated with curiosity, which is the intrinsic motivation for exploration, learning, and creativity [28]. Users are tend to be curious about new

B. Luo et al. (Eds.): ICONIP 2023, CCIS 1964, pp. 312–324, 2024.
https://doi.org/10.1007/978-981-99-8141-0_24

information about how other people behave, feel, and think, especially in online social media scenarios [27].

Items that users are curious about may not be the same items as their preferred items. In psychology, curiosity is often defined as the desire to acquire new information about the actions, thoughts, and feelings of others as a driving force that can trigger and facilitate exploratory behavior [1]. Social curiosity refers to the curiosity of users on social networks [27], who are always curious about their friends' new behaviors. According to the information gap theory [16], when people are aware of some gap in their knowledge, they become curious and engage in information-seeking behavior to complete their knowledge and resolve the uncertainty [16]. Therefore, without considering the users' curiosity during recommendation, it may lead to a decrease in the accuracy of the recommender systems. Thus, in social networks, when recommending an item to users, it is necessary to take into account both the users' preference and their curiosity.

Furthermore, Information Diffusion in social networks is one of the important contents of social network-related research. Users in social networks influence each other, and it is the influence among users that promotes the diffusion of information in social networks. In social networks, both the establishment of social relations between network nodes (users) and the diffusion of information in the network is often driven by the interaction between users. Thus, it is necessary to consider the impact of information diffusion in social networks.

To solve the above problems, we extend the BPR model and propose a Curiosity Enhanced Bayesian Personalized Ranking (CBPR) model by incorporating the users' curiosity into the BPR model. Our CBPR model is based on two basic hypotheses: (H1) The curiosity of a given user may influence his selections about which item to choose; (H2) It is possible to use the data available on social networks to measure the users' curiosity about an item. The key contributions of this work are summarized below:

- We adopt a psychological perspective to explore the effect of the users' curiosity and its influence on the recommendation problems and propose a method to quantify the curiosity of a user for an item by measuring the uncertainty stimulus. We also make the first attempt to simultaneously consider the influence of direct and indirect friends on the curiosity of a user for an item.
- We build a curiosity-based model for item ranking, called CBPR, which integrates the above methods to measure curiosity into the existing BPR model to produce final recommendations. As far as we know, this is the first work to consider the users' curiosity into the pair-wise item ranking methods.
- We conduct comprehensive experiments on two benchmark datasets. The experimental results demonstrate significant performance improvements over existing approaches, highlighting the superiority of our framework in recommendation.

2   Related Work

Social-Based Recommendation. Most of the existing social-based recommendation methods are based on the assumption that the users and their friends tend to have similar preferences. Neural-based SSVD++ [15] extends the classical SVD++ model with both social information and the powerful neural network framework. Rafailidis et al. [18] propose a joint collaborative ranking model based on the users' social relationships. However, the preferences of friends may not exactly match the users' preferences. To address this, Wang et al. [24] propose a method for an interactive social recommendation, which not only simultaneously explores the users' preferences, but also adaptively learns different weights for different friends. Lin et al. [14] present a characterized social regularization model by designing a universal regularization term for modeling variable social influence. Rafailidis et al. [19] propose a deep learning approach to learn both about users' preferences and the social influence of friends when generating recommendations. Ma et al. [17] propose a matrix factorization framework with social regularization in which friends with dissimilar tastes are treated differently to capture the taste diversity of each friend of users.

Psychological-Based Recommendation. There only exist several studies from psychologically inspired views to explore the effect of the users' curiosity and its impacts on the recommendation system. Wu et al. [27] quantify the surprise stimulus by considering user friends' unexpected behaviors, and then rank items by a weighted sum of preference rating score and curiosity score. In another work [26] they quantify the uncertainty stimulus base on Shannon entropy [5] and Damster-Shafter theory [12], and then rank items by weight borda count [22]. Xu et al. [30] design a novel stimulus-evoked curiosity mechanism SeCM which provides not only a way of measuring novelty stimulus but also a way of measuring conflict stimulus. Then a general recommendation framework CdRF is proposed to rank a list of potential items. Zhao et al. [32] develop a computational model called PCM to model the users' curiosity based on the Wundt Curve, where the input to the model is the novelty stimulus score of the item.

3   Notations and Problem Definition

We denote U and I as the set of users and items, respectively, where $|U| = m$, $|I| = n$. Symbols (u, v) and (i, j, k) are separately preserved for an individual user and item. R is the user-item matrix. Social network $G = (U, E)$, where $(u, v) \in E$ suggests u and v are friends. For each user, total items I can be split into three sets: one with Positive feedback, one with Curiosity feedback, and one with Negative feedback, which are defined as follows:

Positive Feedback. P_u are the set of items that the user u has behaviors. These behaviors could be the user $u's$ rating behavior (not related to ratings), click behavior, buying behavior, and so on.

Curiosity Feedback. C_u are the set of items that the user u has no behaviors but may attract the users' curious online, which represents the users' curiosity caused by inconsistent behaviors (e.g., inconsistent ratings) of the same item in user u's social networks. Thus these items are a subset of those that u's friends have behaviors and more details about how to obtain the curiosity feedback will be shown in Sect. 4.2.

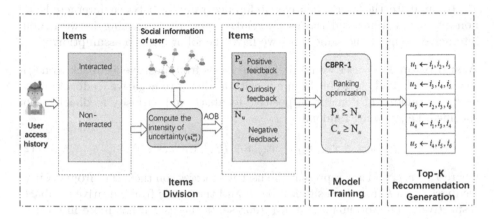

Fig. 1. The architecture of the proposed recommender system framework(CBPR-1).

Negative Feedback. N_u are the set of items that the user u has no behaviors and may not be curious about it. Here *negative* only means no explicit feedback can be observed from the users and does not represent users' dislike of the items.

where $P_u \cap C_u \cap N_u = \emptyset$ and $P_u \cup C_u \cup N_u = I$. Our proposed recommender framework as shown in Fig. 1.

4   Curiosity Enhanced Bayesian Personalized Ranking

4.1   Model Assumption

If a user interacts with an item, it can indicate that the user is interested in the item. Similar to the BPR model [20], this paper assumes that users are more interested in items that have been interacted than items that have not been interacted. This preference relationship of users is expressed by the following formula:

$$x_{ui} \geq x_{uj}, i \in P_u, j \in I \setminus P_u \tag{1}$$

where, x_{ui}, x_{uj} represent the interest values of user u for item i and item j respectively. Item i comes from the items that users have interacted, which is called the positive feedback of user u. Item j comes from the items that users have not interacted, which includes the curiosity feedback and the negative feedback of user u. $x_{ui} \geq x_{uj}$ indicates that user $u's$ interest in item i is higher than user $u's$ interest in item j.

However, the user has no behavior towards the item doesn't mean the user is not interested in it. Due to the existence of users' curiosity, items that users have not yet interacted in behaviors often contain part of items that may arouse users' curiosity, which becomes user $u's$ curiosity feedback items, the rest are user $u's$ negative feedback items.

To simulate the user's relative preference ranking between interacted and non-interacted items and the user's relative preference ranking between non-interacted items at the same time, we have the following two assumptions:

Assumption 1. A user would prefer the items from positive feedback to negative feedback, and the items from curiosity feedback to negative feedback. However, the ranking of items from positive feedback and curiosity feedback is not specified, i.e., ranking randomly.

$$x_{ui} \geq x_{uj}, x_{uk} \geq x_{uj}, i \in P_u, k \in C_u, j \in N_u \tag{2}$$

where x_{ui}, x_{uk}, and x_{uj} represent the user u's interest on the item i from positive feedback, item k from curiosity feedback, and the item j from negative feedback, respectively. The formula $x_{ui} \geq x_{uk}$ represents the user u has more interest in the item i than k.

Assumption 2. A user would prefer the items from positive feedback to negative feedback, and the items from positive feedback to curiosity feedback.

$$x_{ui} \geq x_{uj}, x_{ui} \geq x_{uk}, i \in P_u, k \in C_u, j \in N_u \tag{3}$$

4.2 Modeling of Curiosity

Uncertainty stimulus is one of the key factors that can lead to users' curiosity [1]. Information entropy theory has been used to quantify the intensity of such uncertain stimulus faced by users since it can represent the degree of chaos and uncertainty [1], which involves the number of types of feedback that users are exposed to, and the intensity of competition for each type of feedback. The user u's uncertainty for item i using information entropy [1] as follow:

$$si_{u,i}^{un} = -\sum_{r=1}^{R} p_{u,i}^r \log p_{u,i}^r \tag{4}$$

We provisionally assume that the possible response for the item i to user u is the user $u's$ social friends' various ratings on the item i, denoted as r. For a specific item i, $p_{u,i}^r$ denotes the probability of each response r for user u, which is denoted as follow:

$$p_{u,i}^r = \frac{\exp(w_{u,i}^r)}{\sum_{r=1}^{R} \exp(w_{u,i}^r)} \tag{5}$$

where $w_{u,i}^r$ refers to user u' s weight of each possible response r for item i, formulated as follow:

$$w_{u,i}^r = \sum_{d=1}^{D}\left[e^{-\mu d} \times \frac{N_{v,i}^r}{N_{v,i}}\right], v \in U_u^d \qquad (6)$$

Considering that information can spread along social networks, we use D to represent the farthest social network layer, and d represents the current social network layer. U_u^d denotes user u's friends in layer d. $N_{v,i}$ and $N_{v,i}^r$ denote the number of user u's friends in layer d who give any response on item i, and the number of user u's friends in layer d who give response r on item i.

However, as D gets bigger, the number of user u's friend set U_u^d can grow exponentially and tends to be equal to U. To lower the time complexity, we restrict the max size of $U_{u,i}^d$ to κ where $\kappa \ll m$, and these users are the top κ similar users which can be computed by Pearson correlation coefficient (PCC) [30]. We also introduce an information spread coefficient μ. Term $e^{-\mu d}$ is used to capture the fact that the information spread is decaying with the social network layer d.

We use the uncertainty stimulus $si_{u,i}^{un}$ as the curiosity score. Moreover, we set a threshold γ to filter the incurious items and only select the items whose uncertainty stimulus is above the threshold into the curiosity feedback C_u. The users' curiosity feedback set as follows:

$$C_u = \{i \in (I - P_u)|si_{u,i}^{un} \geq \gamma\} \qquad (7)$$

4.3   Learning the CBPR

The purpose of Area Under Curve (AUC) [33] is to increase the gap between positive and negative samples. Since the AUC only contains two types of samples, which is inconsistent with the existing three types of samples in our mode. Therefore, this paper redefines the formula of AUC as follows, named AUC':

$$AUC' = \frac{\prod_{i \in P_u, j \in N_u} \sigma(x_{uij})}{|P_u||N_u|} + \frac{\prod_{k \in C_u, j \in N_u} \sigma(x_{ujk})}{|C_u||N_u|} \qquad (8)$$

where $\sigma(\cdot)$ is the sigmoid function. We take AUC' as our optimization objective and attempt to maximize it. Thus, for each user u, we have two preference orders as follows:

$$\prod_{(u,i),(u,j) \in (P_u \cup N_u)} P(x_{ui} \geq x_{uj})^{\delta(u,i,j)}[1 - P(x_{ui} \geq x_{uj})]^{1-\delta(u,i,j)} \qquad (9)$$

$$\prod_{(u,k),(u,j) \in (C_u \cup N_u)} P(x_{uk} \geq x_{uj})^{\epsilon(u,k,j)}[1 - P(x_{uk} \geq x_{uj})]^{1-\epsilon(u,k,j)} \qquad (10)$$

where $P(x_{ui} \geq x_{uj})$ denotes that the probability that user u likes item i is higher than the probability that user u likes item j. $P(x_{uk} \geq x_{uj})$ denotes that the probability that user u likes item k is higher than the probability that user u likes item j. $\delta(\cdot)$ and $\epsilon(\cdot)$ are indicator function:

$$\delta(u, i, j) = \begin{cases} 1 & if(u, i) \in P_u \ and \ (u, j) \in N_u \\ 0 & otherwise \end{cases}$$

$$\epsilon(u, k, j) = \begin{cases} 1 & if(u, k) \in C_u \ and \ (u, j) \in N_u \\ 0 & otherwise \end{cases}$$

For the convenience of calculation, we rewrite the formula above. The training process minimizes the objective function:

$$\mathcal{O} = \sum_u \left[\sum_{i \in P_u} \sum_{j \in N_u} -\ln \sigma(x_{uij}) + \sum_{k \in C_u} \sum_{j \in N_u} -\ln \sigma(x_{ukj}) \right] + \lambda_\Theta \|\Theta\|^2 \quad (11)$$

where $\sigma(x) = \frac{1}{1+e^{-x}}$, $x_{uij} = x_{ui} - x_{uj}$ and $x_{ukj} = x_{uk} - x_{uj}$. The preferences (x_{ui}, x_{uj}, x_{uk}) are modeled by matrix factorization. Parameter set $\Theta = \{W, V, b\}$ and $\lambda_\Theta \|\Theta\|^2$ is the L2-norm regulation.

4.4   Item Recommendation

We generate a candidate list of items for the target user u. It contains K items with the highest ranking score $\hat{x_{ul}}$. For any item $l \in I$, the rating of the user u for item l is $\hat{x_{ul}}$ which can be calculated as follow:

$$\hat{x_{ul}} = W_u V_l^T + b_l \quad (12)$$

where the vector W_u represents the user u, the vector V_l represents the item l, and b_l represents the bias of item l. $W \in \mathbb{R}^{m*f}$, $V \in \mathbb{R}^{n*f}$, $b \in \mathbb{R}^n$ and f is the latent factor numbers.

5   Experiments and Results Analysis

5.1   Experimental Settings

Datasets. Social datasets FilmTrust [8] and CiaoDVDs [7] are both used in our experiments, which are crawled from the FilmTrust website in 2011 and the Dvd Ciao website in 2013 respectively. As there are no model hyper-parameters, the validation set is not required. For both datasets, 80% of the data is randomly kept for training and the remainder for testing.

Baselines. Pop provides a list of non-personalized ranked items based on the frequency of the items that are selected by all the users. RankSGD [11] is a matrix decomposition method that utilizes an SGD optimizer to produce personalized ranked items. BPR [20] proposes a pair-wise assumption for item ranking. We apply a uniform sampling strategy for the item selection. SBPR [33] uses social networks to estimate item rankings. UC [26] combines the users' uncertainty stimulus with the matrix factorization [13] for item ranking. As far as we know, it is the only one method to calculate curiosity by quantifying the uncertainty stimulus. UCBPR is the first attempt to integrate curiosity into the BPR model framework, whose purpose is to test the feasibility of integrating curiosity into the Bayesian personalized ranking method. CBPR-1 and CBPR-2 are variants that correspond to Assumptions 1 and 2 respectively.

Parameter Settings. We use the LibRec library [6] to implement the Pop, RankSGD, BPR, SBPR, and UC methods. For each method, we use the parameter settings that achieve the best performance according to the public LibRec website[1]. We also fix the number of the latent factors $f = 10$ for all the matrix factorization methods as suggested in work [33].

Table 1. Comparison results on two datasets.

Model	FilmTrust			CiaoDVDs		
	P@10	R@10	F1@10	P@10	R@10	F1@10
Pop	0.1382	0.3245	0.1938	0.0105	0.0375	0.0164
RankSGD	0.1654	0.2706	0.2053	0.0070	0.0214	0.0105
BPR	0.3458	0.6066	0.4405	0.0120	0.0426	0.0188
SBPR	0.3362	0.5890	0.4281	0.0120	0.0402	0.0185
UC	0.3279	0.5512	0.4112	0.0117	0.0373	0.0178
UCBPR	0.3466	0.6098	0.4420	0.0144	0.0487	0.0222
CBPR-1	**0.3506**	**0.6248**	**0.4491**	**0.0152**	**0.0519**	**0.0235**
CBPR-2	0.3291	0.5895	0.4224	0.0133	0.0486	0.0209

---

[1] https://www.librec.net/.

5.2   Recommendation Performance

The comparison results are listed in Table 1. Comparing the experimental results between the UC and UCBPR, the UCBPR with user curiosity improves the F1 from 41.12% to 44.20% in the FilmTrust and from 1.78% to 2.22% in the CiaoDVD, which indicates the ranking list obtained by using the point-wise method combined with user curiosity is not as accurate as the ranking list obtained by using the pair-wise method combined with user curiosity, demonstrating that it is feasible to integrate the users' curiosity into the pair-wise model framework to improve recommendation performance.

Results show that CBPR-1 has significant improvement compared with other methods including the UC method on two datasets, especially in the CiaoDVDs. Particularly, the CBPR-1 outperforms the SBPR in most cases. One possible reason is that the SBPR only uses the friend relationship in social networks cannot fully use the potential information on social networks, which ignores that the users' curiosity on social networks can directly affect users' interest in items.

From the experiment results, we can find that the performance of the CBPR-1 is better than that of the CBPR-2, which indicates users will be more interested in the items from the curiosity feedback than the negative feedback. Therefore, Assumption 1 is better to capture users' behaviors. Furthermore, the results indicate that items from positive feedback and curiosity feedback can attract the users' interest which is consistent with the psychology of curiosity.

We also perform experiments on the CBPR and other baselines with different values of K from 5 to 30. The results are shown in Fig. 2. The proposed method achieves better performance across different values of K, which indicates the effectiveness of incorporating curiosity feedback in our model. Also, the gap between the CBPR and other baselines gets smaller as K increases. In the FilmTrust, a small K can better reflect the effect of the proposed model as the FilmTrust contains fewer items. Conversely, the CiaoDVDs has sufficient items, and the gap between the CBPR and other baselines becomes larger as K increases.

Table 2. CBPR-1 method results on the FilmTrust and the CiaoDVDs datasets with Precision, Recall, F1 metric by varying γ and D.

(a) P@5/@10

γ	P@5						P@10					
	FilmTrust			CiaoDVDs			FilmTrust			CiaoDVDs		
	$D=1$	$D=2$	$D=3$	$D=1$	$D=2$	$D=3$	$D=1$	$D=2$	$D=3$	$D=1$	$D=2$	$D=3$
0.0	0.4104	0.4120	0.4083	0.0157	**0.0173**	0.0147	0.3471	0.3474	0.3482	0.0142	0.0137	0.0127
0.4	0.4102	0.4128	0.4090	0.0166	0.0161	0.0167	0.3497	0.3494	0.3474	0.0138	0.0143	0.0140
0.8	0.4134	0.4138	0.4139	0.0157	0.0155	0.0145	0.3475	**0.3506**	0.3476	0.0141	0.0132	0.0130
1.2	0.4109	0.4141	0.4080	0.0169	0.0146	0.0153	0.3498	0.3474	0.3476	**0.0152**	0.0128	0.0126
1.6	0.4136	**0.4157**	0.4130	0.0156	0.0155	0.0144	0.3481	0.3486	0.3485	0.0148	0.0131	0.0118

(b) P@5/@10

γ	R@5						R@10					
	FilmTrust			CiaoDVDs			FilmTrust			CiaoDVDs		
	$D=1$	$D=2$	$D=3$	$D=1$	$D=2$	$D=3$	$D=1$	$D=2$	$D=3$	$D=1$	$D=2$	$D=3$
0.0	0.3878	0.3917	0.3852	0.0269	**0.0299**	0.0273	0.6165	0.6161	0.6197	0.0493	0.0490	0.0439
0.4	0.3913	0.3944	0.3876	0.0289	0.0290	0.0292	0.6240	0.6233	0.6164	0.0454	0.0495	0.0494
0.8	0.3952	0.3968	0.3938	0.0263	0.0261	0.0230	0.6208	**0.6248**	0.6172	0.0462	0.0432	0.0442
1.2	0.3864	0.3957	0.3863	0.0289	0.0239	0.0238	0.6217	0.6200	0.6175	**0.0519**	0.0418	0.0408
1.6	0.3935	**0.3977**	0.3948	0.0279	0.0258	0.0257	0.6211	0.6186	0.6195	0.0516	0.0448	0.0404

(c) R@5/@10

γ	F1@5						F1@10					
	FilmTrust			CiaoDVDs			FilmTrust			CiaoDVDs		
	$D=1$	$D=2$	$D=3$	$D=1$	$D=2$	$D=3$	$D=1$	$D=2$	$D=3$	$D=1$	$D=2$	$D=3$
0.0	0.3988	0.4016	0.3964	0.0199	**0.0219**	0.0191	0.4441	0.4443	0.4459	0.0221	0.0214	0.0197
0.4	0.4005	0.4034	0.3980	0.0211	0.0207	0.0212	0.4482	0.4478	0.4444	0.0212	0.0222	0.0218
0.8	0.4041	0.4051	0.4036	0.0197	0.0194	0.0178	0.4456	**0.4491**	0.4447	0.0216	0.0202	0.0201
1.2	0.3983	0.4047	0.3969	0.0213	0.0182	0.0186	0.4477	0.4453	0.4448	**0.0235**	0.0196	0.0193
1.6	0.4033	**0.4065**	0.4037	0.0200	0.0193	0.0184	0.4461	0.4459	0.4460	0.0230	0.0202	0.0182

5.3   Parameters Analysis

Table 2 shows the CBPR-1 results on the FilmTrust and the CiaoDVDs datasets by varying γ and D. Specifically, we vary D from 1 to 3 as more layers in the social networks are less effective. Given each fixed social network layer D and fixing the value of the information spread coefficient μ to be 1, we change the value of γ.

We respectively illustrate the results using P@5/10, R@5/10, and F1@5/10 that γ is tuned from 0.0 to 1.6 with step 0.4 on the FilmTrust and the CiaoD-VDs datasets (See results in Table 2). The results demonstrate that the optimal parameters settings on P@5/10, R@5/10, and F1@5/10 are $(D = 2, \gamma = 1.6)/(D = 2, \gamma = 0.8)$ for FilmTrust dataset, and $(D = 2, \gamma = 0.0)/(D = 1, \gamma = 1.2)$ for CiaoDVDs dataset. These results indicate that properly combining of social network layer depth D and properly using threshold parameters γ can improve item recommendation performance.

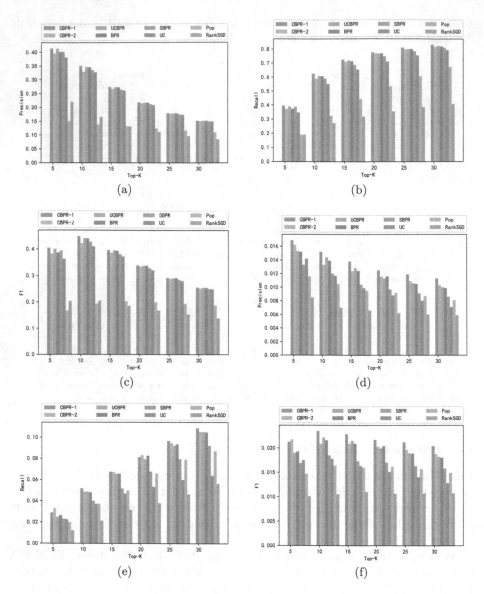

Fig. 2. Top-K Analysis. FilmTrust: (a) Precision, (b) Recall, (c) F1; CiaoDVDs: (d) Precision, (e) Recall, (f) F1.

6   Conclusion

The users' curiosity plays a crucial role in the users' selection. To integrate psychological theories on curiosity into the recommendation systems, we propose a Curiosity Enhanced Bayesian Personalized Ranking (CBPR) model. Particularly, we provide an approach to measure curiosity by modeling the uncertainty stimulus and incorporating it into the BPR model. It adopts two Bayesian pairwise preference rankings among items from the positive feedback, curiosity feedback, and negative feedback for users. The experimental results on two real-world social datasets show the effectiveness of our model.

Acknowledgement. This work was supported by the National Natural Science Foundation of China (62076100), Fundamental Research Funds for the Central Universities, SCUT (x2rjD2230080), the Science and Technology Planning Project of Guangdong Province (2020B0101100002), CAAI-Huawei MindSpore Open Fund, CCF-Zhipu AI Large Model Fund.

References

1. Berlyne, D.E.: Conflict, arousal, and curiosity (1960)
2. Butts, C.T.: Social network analysis: a methodological introduction. Asian J. Soc. Psychol. **11**, 13–41 (2007)
3. Cai, Y., Leung, H., Li, Q., Tang, J., Li, J.: TyCo: towards typicality-based collaborative filtering recommendation. In: ICTAI, pp. 97–104 (2010). IEEE Computer Society (2010)
4. Fan, W., Ma, Y., Yin, D., Wang, J., Tang, J., Li, Q.: Deep social collaborative filtering. In: RecSys, pp. 305–313 (2019). ACM (2019)
5. Gray, R.M.: Entropy and information theory (1990)
6. Guo, G., Zhang, J., Sun, Z., Yorke-Smith, N.: LibRec: a java library for recommender systems. In: UMAP 2015, CEUR Workshop Proceedings, vol. 1388. CEUR-WS.org (2015)
7. Guo, G., Zhang, J., Thalmann, D., Yorke-Smith, N.: ETAF: an extended trust antecedents framework for trust prediction. In: ASONAM 2014, pp. 540–547. IEEE Computer Society (2014)
8. Guo, G., Zhang, J., Yorke-Smith, N.: A novel Bayesian similarity measure for recommender systems. In: IJCAI 2013, pp. 2619–2625. IJCAI/AAAI (2013)
9. Hartung, F.M., Renner, B.: Social curiosity and gossip: related but different drives of social functioning. PLoS One **8**(7), e69996 (2013)
10. Huang, Z., Chen, H., Zeng, D.: Applying associative retrieval techniques to alleviate the sparsity problem in collaborative filtering. ACM Trans. Inf. Syst. **22**(1), 116–142 (2004)
11. Jahrer, M., Töscher, A.: Collaborative filtering ensemble for ranking. In: KDD Cup 2011 competition, 2011. JMLR Proceedings, vol. 18, pp. 153–167. JMLR.org (2012)
12. Jøsang, A.: A logic for uncertain probabilities. Int. J. Uncertainty Fuzziness Knowl.-Based Syst. **9**(03), 279–311 (2008)
13. Koren, Y., Bell, R.M., Volinsky, C.: Matrix factorization techniques for recommender systems. IEEE Comput. **42**(8), 30–37 (2009)
14. Lin, T., Gao, C., Li, Y.: Recommender systems with characterized social regularization. In: CIKM 2018, pp. 1767–1770. ACM (2018)
15. Lin, X., Zhang, M., Liu, Y., Ma, S.: A neural network model for social-aware recommendation. In: Sung, W.K., et al. (eds.) Information Retrieval Technology. AIRS 2017. LNCS, vol. 10648. Springer, Cham (2017). https://doi.org/10.1007/978-3-319-70145-5_10

16. Loewenstein, G.: The psychology of curiosity: a review and reinterpretation. Psychol. Bullet. **116**, 75–98 (1994)
17. Ma, H., Zhou, D., Liu, C., Lyu, M.R., King, I.: Recommender systems with social regularization. In: WSDM 2011, pp. 287–296. ACM (2011)
18. Rafailidis, D., Crestani, F.: Joint collaborative ranking with social relationships in top-n recommendation. In: CIKM 2016, pp. 1393–1402. ACM (2016)
19. Rafailidis, D., Crestani, F.: Recommendation with social relationships via deep learning. In: ICTIR 2017, pp. 151–158. ACM (2017)
20. Rendle, S., Freudenthaler, C., Gantner, Z., Schmidt-Thieme, L.: BPR: Bayesian personalized ranking from implicit feedback. In: UAI 2009, pp. 452–461. AUAI Press (2009)
21. Silvia, P.J.: Interest-the curious emotion. Curr. Dir. Psychol. Sci. **17**(1), 57–60 (2008)
22. Van, M., And, E., Schomaker, L.: Variants of the Borda count method for combining ranked classifier hypotheses. In: 7th International Workshop on Frontiers in Handwriting Recognition (2000)
23. Vega-Oliveros, D.A., Berton, L., Vazquez, F., Rodrigues, F.A.: The impact of social curiosity on information spreading on networks. In: ASONAM 2017, pp. 459–466 (2017)
24. Wang, X., Hoi, S.C.H., Liu, C., Ester, M.: Interactive social recommendation. In: CIKM 2017, pp. 357–366. ACM (2017)
25. Wu, L., Sun, P., Hong, R., Fu, Y., Wang, X., Wang, M.: SocialGCN: an efficient graph convolutional network based model for social recommendation. CoRR abs-1811-02815 (2018)
26. Wu, Q., Liu, S., Miao, C.: Modeling uncertainty driven curiosity for social recommendation. In: WI 2017, pp. 790–798. ACM (2017)
27. Wu, Q., Liu, S., Miao, C., Liu, Y., Leung, C.: A social curiosity inspired recommendation model to improve precision, coverage and diversity. In: WI 2016, pp. 240–247. IEEE Computer Society (2016)
28. Wu, Q., Miao, C.: Curiosity: from psychology to computation. ACM Comput. Surv. **46**(2), 1–26 (2013)
29. Wu, W., Yin, B.: Personalized recommendation algorithm based on consumer psychology of local group purchase e-commerce users. J. Intell. Fuzzy Syst. **37**(5), 5973–5981 (2019)
30. Xu, K., Mo, J., Cai, Y., Min, H.: Enhancing recommender systems with a stimulus-evoked curiosity mechanism. IEEE Transactions on Knowledge and Data Engineering, p. 1 (2019)
31. Xu, Y., Xu, K., Cai, Y., Min, H.: Leveraging distrust relations to improve Bayesian personalized ranking. Inf. **9**(8), 191 (2018)
32. Zhao, P., Lee, D.L.: How much novelty is relevant?: It depends on your curiosity. In: SIGIR 2016, pp. 315–324. ACM (2016)
33. Zhao, T., McAuley, J.J., King, I.: Leveraging social connections to improve personalized ranking for collaborative filtering. In: CIKM 2014, pp. 261–270. ACM (2014)

Modeling Online Adaptive Navigation in Virtual Environments Based on PID Control

Yuyang Wang[1]([envelope])[iD], Jean-Rémy Chardonnet[2][iD], and Frédéric Merienne[2][iD]

[1] The Hong Kong University of Science and Technology (Guangzhou), Guangzhou, China
yuyangwang@ust.hk

[2] Arts et Metiers Institute of Technology, LISPEN, HESAM Université, Paris, France
{jean-remy.chardonnet,frederic.merienne}@ensam.eu

Abstract. It is well known that locomotion-dominated navigation tasks may highly provoke cybersickness effects. Past research has proposed numerous approaches to tackle this issue based on offline considerations. In this work, a novel approach to mitigate cybersickness is presented based on online adaptive navigation. Considering the Proportional-Integral-Derivative (PID) control method, we proposed a mathematical model for online adaptive navigation parametrized with several parameters, taking as input the users' electro-dermal activity (EDA), an efficient indicator to measure the cybersickness level, and providing as output adapted navigation accelerations. Therefore, minimizing the cybersickness level is regarded as an argument optimization problem: find the PID model parameters which can reduce the severity of cybersickness. User studies were organized to collect non-adapted navigation accelerations and the corresponding EDA signals. A deep neural network was then formulated to learn the correlation between EDA and navigation accelerations. The hyperparameters of the network were obtained through the Optuna open-source framework. To validate the performance of the optimized online adaptive navigation developed through the PID control, we performed an analysis in a simulated user study based on the pretrained deep neural network. Results indicate a significant reduction of cybersickness in terms of EDA signal analysis and motion sickness dose value. This is a pioneering work which presented a systematic strategy for adaptive navigation settings from a theoretical point.

Keywords: Virtual reality · Cybersickness · Navigation · Neural computing

1 Introduction

Thanks to increasing computing power and the availability of many affordable head-mounted displays (HMDs) such as HTC Vive and Oculus Quest, the word "metaverse" has aroused profound discussion on the application of VR technologies among the public including mass media, industry, and academic community.

B. Luo et al. (Eds.): ICONIP 2023, CCIS 1964, pp. 325–346, 2024.
https://doi.org/10.1007/978-981-99-8141-0_25

Engineers are able to develop dozens of applications including training for medical operations, the organization of 3D virtual conferences, playing immersive games, visualizing 3D models, *etc.* [32]. When exposed to immersive environments, users can easily perform many tasks in a virtual world after wearing VR glasses, including walking/running, shooting, and fighting. The accomplishment of these tasks requires users to navigate through multiple virtual environments, which could be the most fundamental interaction process [42]. However, along with the navigation process, users usually experience cybersickness due to mismatched visual and vestibular information in the brain: users visually perceive objects moving while the body is still in position [36]. Therefore, this sensory conflict leads to sickness symptoms, such as headache, vomit, nausea, and sweating.

In this paper, we present a novel method to improve navigation experience in virtual environments. Our approach relies on online adaptation of navigation parameters based on pre-trained neural networks and laws from system control.

1.1   Cybersickness Evaluation

Due to the individual susceptibility to cybersickness, the severity of the symptoms experienced distributes differently among users [8]. Many subjective evaluation methods were proposed in past research, such as the well-known *Simulator Sickness Questionnaire* (SSQ) [24] that asks user to evaluate sixteen sickness symptoms after an immersive experience and classifies them into three categories: nausea, oculomotor and disorientation. However, researchers often stick to the overall cybersickness level. In this case, it might be meaningful to repeatedly ask participants a single question about their well-being, instead of asking about multiple symptoms. The *Misery Scale* (MISC) [6,7] and the Fast Motion Sickness Scale (FMS) [25] have been proposed accordingly for such convenience. However, subjective questionnaires are generally administered after users have performed an experiment; therefore they have to shift attention away from the experiment to personal feelings, leading to much disturbance on the resulting data [9].

Alternatively, biosignals are regarded as one of the most objective ways to represent individual differences (e.g., electrodermal activity (EDA), electroencephalography (EEG), heart rate variability (HRV), Eye tracking, *etc.*) [9]. Recent work has demonstrated success in assessing and predicting cybersickness by involving these signals, especially thanks to deep learning approaches [10,18, 37]. For example, a pioneering work presented the encoding of EEG signals to the cognitive representation relative to cybersickness, and by transferring it to VR video-based deep neural networks, the authors can predict cybersickness without any EEG signal [27]. This work subtly integrates individual EEG information into the visual information, making it possible to predict individually different cybersickness. Islam et al. [21,22] put forward a multimodal deep fusion neural network that can take stereoscopic videos, eye-tracking, and head-tracking data as inputs for predicting the severity of cybersickness. Particularly, when using a combination of eye-tracking and head-tracking data, the authors found that their network can predict cybersickness with an accuracy of 87.77%, which has out-performed state-of-the-art research. Their approach gains strong feasibility for being used in current consumer-level HMDs having already integrated eye and head tracking sensors. However, the adoption of measurement devices such as EEG and eye tracking may be hindered by their intrusiveness and inconvenience for collection and analysis. Typically, the collection of EEG signals requires the experimenter to put several electrodes on the user's head [31]. Despite EEG signals usually being noisy, such settings of real-time cybersickness evaluation may distract participants from the immersive experience. Recent progress in the development of wearable sensors provides insights for easy integration in immersive applications. One measurement that particularly attracted interest for several years in the VR community is the EDA (e.g., [38]). Indeed, the EDA signal can be easily recorded through a cheap electrical circuit, and is commonly regarded as a reflection of the sympathetic arousal [39]. In this work, we decided to opt accordingly for the EDA as a reliable cybersickness indicator.

EDA signals can be decomposed into two components: the skin conductance level (SCL) and the skin conductance response (SCR) [2]. SCL, associated with the tonic level of the EDA signal, changes slowly with a time scale of tens of seconds to minutes. Because of the differences in hydration, skin dryness, or autonomic regulation between respondents, SCL varies accordingly and can be significantly different among respondents. On the other hand, SCR, known as the phasic component of the EDA signal, rides on top of the tonic changes and demonstrates much faster variations. Alternations in SCR components of an EDA signal are observable as bursts or peaks in the signal. The phasic component is associated with specific emotionally arousing stimulus events (event-related SCRs, ER-SCRs). The rise of the phasic component can reach a peak within 1-5 s after the onset of stimuli [39].

The SCL at the forehead and finger area can demonstrate correlation with cybersickness occurrences but not during the recovery stage [16]: the SCL will increase after users are exposed to visual stimulation [30]. According to the spectrometer measurement of water vapor produced by sweating, an increased SCL is linked to an increased sweating. After the termination of the visual stimuli, the conductive path through the skin remains open despite reduced or absent sweat gland activity. Additionally, past research has found that the SCR presents strong correlation with both the onset of and the recovery from cybersickness [16]. SCRs collected from the forehead significantly indicate a sudden and sustained burst of activity preceding an increase in subjective cybersickness ratings. Though, SCRs gathered from the finger palmar site may not be related to the cybersickness, as the palm is less sensitive than the human forehead in both phasic and tonic levels [30]. Further past studies [15,17] confirmed that phasic changes of the skin conductance on the forehead can be used to measure the level of cybersickness. Last, it has been shown that the width of the SCR collected from the wristband could be an index of cybersickness [34], which further supports our approach to collect the EDA with a wristband sensor.

1.2   Adaptive Navigation

Adaptive navigation in virtual reality refers to the use of techniques and algorithms that adjust the users' virtual experience based on their behavior and preferences. It is a way to enhance the interaction experience by customizing the virtual environment to their needs and abilities. The severity of cybersickness symptoms increases with time, and after some time spent in the VR, the sickness severity either begins to stabilize or decrease; therefore, adaptive navigation in the VR environment appears to be meaningful and deserve investigating [11].

Many strategies have been proposed in past research to mitigate cybersickness by adapting navigation settings. For example, fuzzy logic has been used to integrate three user factors (gaming experience, ethnic origin, age) to derive an individual susceptibility index to cybersickness [45]. This work opens the possibility to adapt navigation settings based on individual characteristics. Fernandes and Feiner [12] explored the way to dynamically change the field of view (FOV) depending on the users' response to visually perceived motion in a virtual environment. As a result, users experience less cybersickness, without decreasing the sense of presence and minimizing the awareness to the intervention. However, Zielasko et al. failed to confirm the correlation between the reduced FOV and the severity of cybersickness, although a reduction of FOV allows the users to travel longer distances [47]. To reduce the risk of cybersickness, Argelaguet and Andujar [3] designed an automatic speed adaptation approach in which the navigation speed is computed on a predefined camera path using optical flow, image saliency and habituation measures. During navigation, users usually manipulate the speed based on the task and personal preferences, but they have to involuntarily adjust the speed frequently and unsmoothly, resulting in severe cybersickness. Hu et al. [19] carried out similar work for reducing cybersickness with perceptual camera control while maintaining original navigation designs. Considering the effect of speed on cybersickness, Wang et al. [44] proposed an online speed protector to minimize the total jerk of the speed profile considering both predetermined speed and acceleration constraints, leading users to report less severity of cybersickness. Additionally, Freitag et al. [13] developed an automatic speed adjustment method for travel in the virtual environment by measuring the informativeness of a viewpoint.

We believe that the evaluation and prediction of cybersickness is the first step, while the final objective is to reduce cybersickness through adaptive navigation. Therefore, through this work, we want to bridge the gap between evaluation and adaptation. A similar idea can be found in previous studies. Plouzeau et al. [38] created an innovative method to adapt the navigation acceleration in real time based on the EDA signal, resulting in a significant decrease of cybersickness levels among users while maintaining the same task performance. Similarly, Islam et al. [20] designed a closed-loop framework to detect the cybersickness severity and adapt the FOV during navigation. The framework can collect the user's physiological data (e.g., HR, BR, HRV, EDA) with which cybersickness can be predicted, and based on the sickness severity, the system can apply dynamic Gaussian blurring or FOV reduction to the VR viewer.

These studies usually assume that the sickness severity has a linear correlation with the adapted settings, and therefore the navigation settings are adapted with a linear proportional function which is not confirmed to the best of our knowledge. However, naturally, the perception of visual stimuli could be a nonlinear process, and presetting a linear adaptation strategy may prevent from finding the optimal adaptive settings. Accordingly, we propose to use a Proportional-Integral-Derivative (PID) control to adapt the navigation settings, without requiring any assumption beforehand. In this work, we propose to use physiological signals, and particularly the electrodermal activity (EDA) that can be obtained from a wristband sensor in real time (here an Empatica E4 wristband[1]), to adapt navigation in real time, and therefore mitigate cybersickness. By involving the EDA, we expect to further incorporate individual differences into a customized navigation experience. We firstly use the phasic component of the EDA signal as a measurement of the cybersickness level in real time. Secondly, we formulate a mathematical relation based on the PID model with several unknown parameters to adapt the navigation acceleration, taking the sickness severity as input. Optimization is performed to determine the optimal parameters to output a nonlinear adaptation strategy.

Past work proposed a similar idea as ours in which the navigation acceleration is adapted from the real-time evolution of EDA [38]:

$$a(t_i) = a(t_{i-1}) - 0.5 * \frac{dEDA(t_i)}{dt} \tag{1}$$

where $a(t_i)$ and $a(t_{i-1})$ are the accelerations at two successive frames, $EDA(t_i)$ is the magnitude of the EDA signal at time t_{i-1}. This formulation is designed for an acceleration-based control scheme in which the joystick can control the longitudinal or rotational acceleration directly until reaching a speed limit. However, the user might fail to control the acceleration when $EDA(t_i)$ keeps increasing or decreasing, which is the reason why this model is not sufficient to mitigate cybersickness efficiently. We therefore propose another model, based on the PID control scheme, adding a second term in Eq. 5 to stabilize the acceleration. Our contributions are summarized as follows:

- We develop a novel mathematical model for adapting the navigation acceleration based on the severity of cybersickness evaluated by the phasic component of the EDA signal.
- We propose and validate the use of deep neural networks (NN) in our studies. Deep NNs work as simulated users during the experiments, taking navigation as input and sickness as output. Such approach opens avenues to the development of intelligence in VR systems.

The paper is organized as follows. In Sect. 2, we will give a brief introduction to PID control and 1D convolutional neural networks, and we will present mathematically the proposed adaptive navigation model. In Sect. 3, we will demon-

---

[1] https://www.empatica.com/research/e4/.

strate a feasible approach to find the optimal parameters for the adaptive model. Further, we will discuss the results and the limitations of the work in Sect. 4, before concluding.

2   Adaptive Navigation Design

2.1   PID Controller

PID is the abbreviation of proportional-integral-derivative and is a control loop method with feedback widely used in industrial control systems and numerous applications needing constantly modulated control [41]. A PID controller continuously computes an error value $e(t)$ representing the difference between the expected setpoint and the measured process value, and applies a correction determined by proportional, integral and derivative terms (denoted P, I, and D respectively). The overall control function is given as,

$$u(t) = K_P * e(t) + K_I * \int_0^\tau e(t) + K_D * \frac{de(t)}{dt} \tag{2}$$

where, $u(t)$ is the output of the PID controller, K_p, K_i and K_d are non-negative coefficients for the proportional, integral and derivative terms respectively, τ is the time and t is the integration variable. This controller contains three terms with different control purposes:

- The proportional controller gives a feedback that is proportional to the error $e(t)$. If the error is large and positive, this term will also return a large and positive output, taking into account the coefficient K_p. However, it can not ensure that the system reaches the expected setpoint and maintains a steady-state error.
- The integral controller is involved to remedy the steady-state error. It integrates the historic cumulative value of the error until the error reduces to zero. However, the integral term decreases its output when a negative error appears, which will limit the response speed and influence the stability of the system.
- The derivative controller enables the system to estimate the future trend of the error based on its rate of change along time. It improves the stability of the system by compensating for a phase lag resulting from the integral term.

2.2   1D Convolutional Neural Network

There exist multiple models of neural networks. Among them, 1D convolutional neural networks (CNN) can achieve competitive performance compared to for example long short term memory (LSTM) on certain sequence-processing problems, usually at a significantly cheaper computational cost. Recently, 1D CNNs have been applied for audio generation and machine translation, obtaining great success [29]. In this work, 1D CNNs will be used to process sequential data.

2.3   Formulation of Adaptive Navigation

The objective of this section is to deduce the mathematical formulations for the adaptive navigation technique based on the PID control system.

Let $f(t)$ denote the phasic component of the EDA signal at time t, and the objective is to stabilize $f(t)$, given as,

$$E_f(t_i) = f(t_i) - f(t_{i-1}) \tag{3}$$

where $E_f(t_i)$ is the difference of the phasic component between the current time step t_i and the previous time step t_{i-1}. In idle state, $f(t_i)$ is expected to be 0 which means that there is no visual stimuli. Therefore, Eq. 3 can be simplified as,

$$E_f(t_i) = -f(t_{i-1}) \tag{4}$$

As the EDA signal is decomposed into the phasic (SCR) and the tonic (SCL) component, it implies that the variation of the EDA signal is associated with the phasic and the tonic component. Knowing that the tonic component usually varies slowly and the phasic component varies rapidly overlying the tonic component [5], in practice, we can use the phasic component to approximate the variation of the EDA signal. Mathematically, the variation of the EDA signal by time is noted as a temporal derivative, i.e., $f = \frac{dEDA}{dt}$. As depicted in Fig. 1, there exists a similar variation trend between the derivative of EDA and the phasic component of EDA.

As explained above, the SCR component is associated with arousing stimulus events. When a user is exposed to visual stimuli in immersive virtual environments, bursts or peaks appear. The severity of visual stimuli is associated with the salience of the burst or the signal's peak. In other words, in an idle situation in which the user does not receive any visual stimuli, there should not be any observable bursts or peaks. Furthermore, it means that we should design a navigation technique in which the visual stimuli should not arouse excessive physiological responses. Our goal is then to optimize navigation to stabilize the SCR component of the EDA signal.

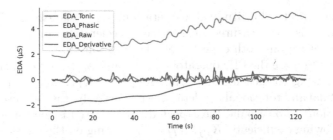

Fig. 1. Demonstration of one EDA signal including the phasic component, tonic component, and temporal derivative.

The navigation acceleration (both translational and rotational) **a** at time t_i can be parametrized by the following algebraic expression,

$$\mathbf{a}(t_i) = \mathbf{a}(t_{i-1}) + \psi_\mathbf{a}\mathbf{E_a}(t_i) + \text{diag}(\beta)\psi_f\mathbf{E_f}(t_i) \tag{5}$$

where

$$
\psi_\mathbf{a}(\cdot) = \begin{bmatrix} \psi_{a_l}(\cdot) & 0 \\ 0 & \psi_{a_r}(\cdot) \end{bmatrix}
$$
$$
= \begin{bmatrix} K_{Pl}(\cdot) + K_{Il}\int_0^\tau(\cdot) + K_{Dl}\frac{d(\cdot)}{dt} & 0 \\ 0 & K_{Pr}(\cdot) + K_{Ir}\int_0^\tau(\cdot) + K_{Dr}\frac{d(\cdot)}{dt} \end{bmatrix} \tag{6}
$$

$$
\psi_\mathbf{f}(\cdot) = \begin{bmatrix} \psi_f(\cdot) & 0 \\ 0 & \psi_f(\cdot) \end{bmatrix}
$$
$$
= \begin{bmatrix} K_{Pf}(\cdot) + K_{If}\int_0^\tau(\cdot) + K_{Df}\frac{d(\cdot)}{dt} & 0 \\ 0 & K_{Pf}(\cdot) + K_{If}\int_0^\tau(\cdot) + K_{Df}\frac{d(\cdot)}{dt} \end{bmatrix} \tag{7}
$$

$$
\mathbf{E_a}(t_i) = \begin{bmatrix} E_{a_l}(t_i) \\ E_{a_r}(t_i) \end{bmatrix} = \begin{bmatrix} a_{le}(t_i) - a_l(t_i) \\ a_{re}(t_i) - a_r(t_i) \end{bmatrix} \tag{8}
$$

$$
\mathbf{E_f}(t_i) = \begin{bmatrix} E_f(t_i) \\ E_f(t_i) \end{bmatrix} = \begin{bmatrix} -f(t_{i-1}) \\ -f(t_{i-1}) \end{bmatrix} \tag{9}
$$

$$
\text{diag}(\beta) = \begin{bmatrix} \beta_l & 0 \\ 0 & \beta_r \end{bmatrix} \tag{10}
$$

In Eq. 5, the second term $\psi_\mathbf{a}\mathbf{E_a}(t_i)$ ensures that the acceleration can vary around an expected value. Indeed, if the third term $\psi_f\mathbf{E_f}(t_i)$ keeps varying monotonously, the acceleration would also vary monotonously and reach an extremum. The third term implies then the adaptive quantity due to the visual stimulus or physiological response. $\text{diag}(\beta)$ represents diagonal coefficient matrices used to balance the importance between longitudinal and rotational motion in Eq. 5.

a_{le} is the expected longitudinal acceleration; a_{re} is the expected rotational acceleration; a_l is the measured longitudinal acceleration; a_r is the measured rotational acceleration. Both expected accelerations are 0.

$\psi_\mathbf{a}(\cdot)$ and $\psi_\mathbf{f}(\cdot)$ are the PID operators. $\psi_{a_l}(\cdot)$ and $\psi_{a_r}(\cdot)$ are elements of $\psi_\mathbf{a}(\cdot)$ with the following coefficients: K_{Pl}, K_{Il}, K_{Dl}, and K_{Pr}, K_{Ir}, K_{Dr}, acting on the longitudinal and rotational accelerations respectively; they are used to ensure the accelerations to be around the expected values. $\psi_f(\cdot)$ is the element of $\psi_\mathbf{f}(\cdot)$ with the following coefficients: K_{Pf}, K_{If}, K_{Df}, acting on the phasic component of the EDA signal. $\mathbf{E_a}(t_i)$ and $\mathbf{E_f}(t_i)$ are the errors between the expected steady states and the measured states at time step t_i. The errors will be substituted to the PID operators to compute the corrections.

To summarize, an adaptive navigation system can be described by Eq. 5 and the supplementary equations: Eq. 6, Eq. 7, Eq. 8, Eq. 9, and Eq. 10. The inputs of the system at time step t_i are the measured longitudinal and rotational accelerations, and the phasic component of EDA, and the outputs of the system are the corrected accelerations in both directions.

This parametrized model features the following eleven coefficients: K_{Pl}, K_{Il}, K_{Dl}, K_{Pr}, K_{Ir}, K_{Dr}, K_{Pf}, K_{If}, K_{Df}, β_l and β_r. As the objective of adaptation is to mitigate cybersickness, the nontrivial question is: can we find the optimal parameters to adapt the acceleration to reduce cybersickness?

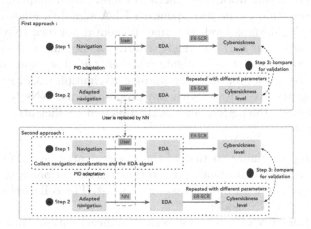

Fig. 2. Strategy to find the optimal parameters that can mitigate cybersickness in the adaptive model.

We can see two possibilities to answer this question, depicted in Fig. 2. The first approach is to assign different values to these coefficients and perform user studies to determine the best one that can mitigate cybersickness. As there are eleven coefficients and each of them is independent to the others, the combination of optimal coefficients reaches at least hundreds or thousands of groups. To prove that a specific group of coefficients can reduce cybersickness with a statistical significance, dozens of participants are required to evaluate the adaptation system, with much time needed to complete all the experiments. In general, three steps are required for this approach that we will call classical:

1. A group of users navigates through a virtual environment without adaptive navigation and evaluates the corresponding cybersickness level according to the EDA signal afterwards.
2. The same group of users navigates through the same virtual environment with the PID adapted navigation. In this step, we have to manually find appropriate values for the different parameters.
3. After evaluating the cybersickness level, a comparison between both approaches is performed to determine potential significantly reduced cybersickness levels.

The recent strong development of artificial intelligence in various domains has raised interest in developing intelligent systems that can predict phenomena at a price of less efforts for developers and end-users. Although considering human participants cannot be dismissed, the release of more and more efficient neural network (NN) algorithms represents a formidable opportunity to introduce them into VR applications and pave foundations for the development of more efficient and individualized VR. Particularly, as in the classical approach above, parameter tuning can be highly time-consuming and tiring for participants, using NNs to simulate users may be an interesting alternative to explore. In fact, in the issue considered in this work, the function of users is to map the motion profiles during navigation to the corresponding EDA signal through which we can compute the cybersickness level. Fortunately, neural network (NN) models can work for this purpose as they are widely regarded as nonlinear fitters. Therefore, in this paper, we propose to investigate whether, and if so, how, with NNs, we can determine the performance of the adapted navigation with less user studies; and here comes the second approach: we propose to use NNs to simplify user studies and evaluate the performance of the different coefficients. This approach includes three steps:

1. The first step is similar to that of the classical approach. We collect from past user studies longitudinal and rotational accelerations and EDA signals during navigation in an immersive environment. The collected data are then used to train the NN model.
2. With the trained NN model, any navigation acceleration can be ingested to output the corresponding EDA signal. Therefore, in this step we can replace users by the trained NN model, allowing not to conduct a vast number of user studies. The focus of the work becomes then to search for appropriate parameters that can mitigate cybersickness, which corresponds exactly to a parameter optimization problem. In mathematics, parametric optimization can be solved with different methods that have already been implemented in open-source softwares such as *Optuna* [1]. Once the objective of the optimization is determined, *Optuna* can find the optimal parameters of our model through a Bayesian optimization, especially sequential model-based optimization with Tree-Structured Parzen Estimator.
3. Similar to that in the classical approach, the third step investigates whether adapted navigation can reduce cybersickness by comparing the artificially generated ER-SCR feature of the EDA signal to that coming from non-adapted navigation.

Compared to the classical approach, the benefits of the second approach are: (1) replacing users by an NN, which alleviates the challenge of recruiting numerous participants; (2) transposing the adaptation problem to an optimization problem that can be solved with existing open-source software. In this case, the objective of the experiment is to collect enough data to train the NN model.

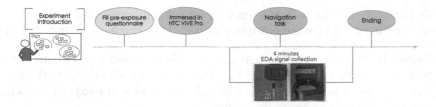

Fig. 3. Flowchart for the experiment including user navigation with HTC Vive Pro and data collection with Empatica E4 wristband.

Fig. 4. Virtual scenario in which the participants navigate along the highlighted path.

3 Data Collection and Parameters Computation

3.1 Data Acquisition

To avoid performing excessive user tests, we had to train an NN model that can map the navigation behavior (acceleration in this context) to the corresponding EDA signal. However, cutting down the number of user tests does not imply that we can completely get rid of them, we still need to collect enough data to train a high-quality NN model. Hence, we carried out a user experiment to collect the required data.

Participants. We invited 53 participants ($M_{age} = 26.3$, $SD_{age} = 3.3$, Females: 26) from the local city to participate in a navigation task in an immersive environment. To obtain much more samples, all participants were asked to participate three times on three different days, hence we collected 159 samples. They were rewarded with different gifts afterwards. Upon arrival, they were asked to fill one pre-exposure questionnaire to investigate on their health conditions and experience in playing games and using VR devices. From this questionnaire, no participants reported any health issues that would affect the experiment results. A consent form was signed by participants.

Task Design. The general experimental procedure is presented in Fig. 3. The whole experiment was carried out using an HTC Vive Pro head-mounted display.

1. Before the test, we gave the participants a brief introduction about how to control navigation with the HTC Vive Pro hand controllers. Due to the occurrence of cybersickness, they were allowed to terminate the experiment whenever they felt sick or severe discomfort.
2. The experimenter put an HTC Vive Pro on the participants' head and an Empatica E4 wristband on one participants' arm. The Empatica E4 can sample EDA at a frequency of 4 Hz and the EDA signal is sent during navigation to a processing computer through Bluetooth.
3. The participants were immersed in the virtual environment displayed in Fig. 4 and started to navigate following the trajectory highlighted in brown. The user could control the motion in different directions through the touchpad on the HTC Vive Pro hand controller. Together with the EDA signal, the longitudinal and rotational navigation accelerations were recorded synchronously.
4. The navigation task continued for four minutes, and the participants were removed from the head-mounted display at the end.

3.2 Model Architecture

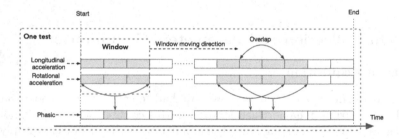

Fig. 5. Schematic representation of the data collected from one user session including the longitudinal and rotational accelerations, and the phasic component of EDA. Note that both accelerations were computed from the navigation speeds, and the phasic component of EDA was also preprocessed by the *Neurokit2* package.

During the experiment, we collected three signals with the same starting and ending times: the longitudinal acceleration, the rotational acceleration, and the EDA signal. As the phasic component is associated with arousing stimulus events, the NN model should link the navigating accelerations to the phasic component of EDA. Therefore, we extracted the phasic component from the original EDA thanks to *Neurokit2* [35], a Python toolbox for neurophysiological signal processing. Figure 5 represents a schematic representation of the data recorded from one participant session. Individual differences leading to different

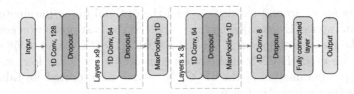

Fig. 6. Architecture of the deep neural network. The network is composed of fourteen 1D convolutional layers and one fully connected layer; the kernel size for the convolutional is 3.

magnitudes in the phasic response could make the model difficult to train; thus we normalized all the data to range between zero and one.

To allow the model to learn the local relationship between the acceleration and the phasic signal more easily, we used a moving window on the signal and reformulated the data structure. The current time clip of the phasic signal was associated with the accelerations from the previous time clip and the next time clip. For example, with a clip length of 1 s, to predict the phasic signal between 1 s and 2 s, we used the acceleration data from 0 s to 3 s; to predict the phasic signal between 2 s and 3 s, we used the acceleration data from 1 s to 4 s. The duration of the window clip was considered as a hyper-parameter for the NN model. In total, we collected 159 pairs of data[2] among which we used 119 pairs to train the model, and the rest of them (40) to test whether the model could reduce the level of cybersickness. With 119 pairs of data, we obtained 90530 clips as the training set, and 22633 as the testing set for the NN model.

Table 1. Settings of the hyper-parameters obtained from *Optuna*.

Phasic signal length	2.25 s
Dropout rate	0.00099
Learning rate	0.000288
# Convolutional layers	14
Epoch	1800
Batch size	256

The proposed model was implemented in the Tensorflow framework on an Nvidia GeForce RTX 2070 graphic card. Adam was chosen as the optimizer. Other hyper-parameters in the NN model (dropout rate, learning rate, the number of convolutional layers, the number of epochs, batch size) were optimally determined by *Optuna*. It took approximately three hours to train the model,

---

[2] One pair of data includes longitudinal and rotational accelerations and the corresponding EDA signal from one user session; one pair can be regarded as one data sample.

and *Optuna* spent around ten days to find the optimal hyper-parameters inside the searching space. Eventually, *Optuna* reported the best prediction accuracy in terms of the mean absolute error (MAE) is 0.015 (1D CNN) and 0.058 (LSTM). Therefore, we used 1D CNN in this work considering the lower loss error.

The hyper-parameters of the 1D CNN model are reported in Table 1. The model structure is given in Fig. 6.

3.3   Computing the Adaptive Coefficients

With the previously determined hyper-parameters, we obtained the best NN model that can map the accelerations to the phasic component of the EDA signal. At this stage, we can further find the optimal adaptive coefficients using the 40 pairs of data mentioned above. The idea is to employ *Optuna* again to use the number of ER-SCR as the optimization objective, and search for the best adaptive coefficients that can reduce the number of ER-SCR. For the non-adapted navigation, the phasic component of EDA is obtained from the data collected during the user study, while for the adapted navigation, the phasic component of EDA is predicted from the pre-trained NN model. The detailed steps of the process to find the optimal coefficients of the adapted navigation model (the eleven coefficients introduced in Sect. 2) are,

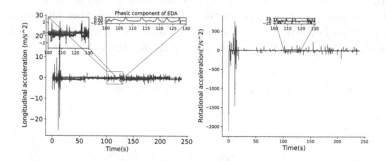

Fig. 7. Comparison between non-adapted (in red) and adapted (in blue) longitudinal (left) and rotational (right) accelerations. The adapted accelerations are determined by the phasic component of EDA (zoomed part in the left graph). (Color figure online)

1. With the 40 pairs of data, we can compute the number of ER-SCR for the non-adapted navigation of each sample, denoted by a vector $\mathbf{N}_{raw}$, used as a baseline to be compared with the adapted one.
2. *Optuna* will randomly choose values for the eleven unknown coefficients. With the model proposed in Eq. 5, we can compute a value for the adapted navigation accelerations based on the non-adapted one.

3. After obtaining the adapted accelerations, we can use the pre-trained NN model to compute the phasic component of EDA. As the pre-trained model only reads and returns the clipped data (as shown in Fig. 5), an additional step will reconstruct the phasic component obtained from multiple clipped data to a single sequence, and further compute the number of ER-SCR which is denoted by another vector $\mathbf{N}_{adapted}$.

4. We compute the difference $(\mathbf{N}_{raw} - \mathbf{N}_{adapted})$ to record the reduction of the number of ER-SCR, and the result is denoted by one vector $\mathbf{N}$ containing both positive and negative numbers. A positive number means that the person will experience reduced cybersickness, and a negative number denotes increased cybersickness. The goal during this optimization process is to have more positive numbers than negative ones, i.e., maximizing the percentage of positive numbers among the 40 samples. As the *Neurokit2* package provides diverse methods for computing the ER-SCR, we employed three methods from Kim [28], Gamboa [14], and *Neurokit2* [35] to compute the percentage of positive numbers. Next, another group of values will be randomly chosen for the coefficients and investigation will be performed to further increase the percentage of positive numbers, denoted by P_{pn}. The following formulation summarizes the optimization process operated by *Optuna* to find the optimal PID coefficients,

$$
\max_{\text{PID coefficients}} P_{pn} = \underbrace{\text{Percentage of positives}}_{\text{from Kim [28]}}
$$
$$
+ \underbrace{\text{Percentage of positives}}_{\text{from Gamboa [14]}} + \underbrace{\text{Percentage of positives}}_{\text{from \textit{Neurokit2} [35]}}
$$
(11)

5. Steps 2, 3 and 4 are repeated until the P_{pn} reaches a stable value. Once convergence is achieved, the computed coefficients are considered as the optimal ones, in our case, the ones reported in Table 2.

Table 2. Optimal coefficients of the adaptive model obtained from *Optuna*.

K_{Pl}	K_{Il}	K_{Dl}	K_{Pr}	K_{Ir}	K_{Dr}	K_{Pf}	K_{If}	K_{Df}	β_l	β_r
0.0113	0.0065	0.0137	0.0098	0.0012	0.0011	0.0730	0.2283	0.3724	0.0017	0.0012

3.4   Results

Based on the methodology presented above, a theoretical validation test was conducted in which the trained NN replaced users and adapted navigation was simulated numerically.

Figure 7 demonstrates the difference between non-adapted and adapted accelerations in the longitudinal and rotational directions for one test in the scenario presented in Fig. 4. During non-adapted navigation, the acceleration might reach extremely large magnitudes because the user moves rapidly. Adaptive navigation could compensate for such variation.

Table 3. The majority of samples (total is 40) report significantly decreased numbers of ER-SCR and $MSDV$ with our adaptive model.

Methods	Positive number	Percentage	χ^2	P-value	ϕ
Kim2004	36	90%	25.6	<.01	0.8
Gamboa2008	29	72.5%	8.1	<.01	0.45
Neurokit2	36	90%	19.6	<.01	0.7
$MSDV_l$	40	100%	40.0	<.01	1.0
$MSDV_r$	40	100%	40.0	<.01	1.0

To validate whether the adapted model with the coefficients computed in Table 2 can mitigate cybersickness, we compared the non-adapted navigation and the adapted one from a statistical viewpoint. We used the *Neurokit2* package to compute the number of ER-SCR in each navigation modality. The significance level was set to .05.

Results in Table 3 reveal that the adapted model can reduce the number of ER-SCR. According to the evaluation from the method of Kim [28] and *Neurokit2* [35], 90% of the total samples presented a significant reduced number of ER-SCR, $(\chi^2(1, N = 40) = 25.6, p < .01, \phi = .80)$ and $(\chi^2(1, N = 40) = 8.1, p < .01, \phi = .45)$ respectively, whereas only 72.5% from the method of Gamboa [14] provided significant results $(\chi^2(1, N = 40) = 19.6, p < .01, \phi = .70)$.

Additionally, to further validate the performance of the proposed adaptive navigation model, still using the samples above, we computed the motion sickness dose value $(MSDV)$ which is regarded as an objective cybersickness indicator [4, 26, 40]: a small $MSDV$ indicates less severity of cybersickness. The $MSDV$ can be computed with the following formula:

$$MSDV = \sqrt[n]{\int_0^T a^n(t)\mathrm{dt}} \tag{12}$$

where a is the navigation acceleration (m/s^2), T is the whole navigation time, and n equals to 2 here. The PID-adapted acceleration could significantly reduce the $MSDV$ along both longitudinal or rotational directions for all samples $(\chi^2(1, N = 40) = 40, p < .01, \phi = 1.0)$.

4   Discussion

Our model incorporates all accelerations (translational and rotational) and coefficients into an algebraic form. In Eq. 1, the coefficient 0.5 was derived empirically considering the physiological reaction time of the EDA [38], while in our model, all coefficients were optimized based on existing datasets constructed from past user studies.

Our adaptive navigation model was used to adapt the navigation acceleration, while it is worth noting that it could be used to adapt other navigation settings, e.g., field-of-view (FOV) vignetting [12] or geometry deformation [33]. To adapt different navigation parameters, we can replace the acceleration in Eq. 5 by the corresponding parameters and then follow the same procedure as shown in this study to find the best parameters. The objective of this work was not to compare the performance of different adaptive navigation settings for mitigating cybersickess, but to design a model of an adaptive strategy, and showcase it to one navigation parameter. For example, Islam et al. [20] developed a closed-loop framework to detect cybersickness and adapt the FOV accordingly; their work has gained advantage by using a deep LSTM neural network to predict real-time cybersickness with physiological data including heart rate, heart rate variability, EDA and breathing rate, while our work only used the phasic component of EDA to detect cybersickness. Therefore, we can legitimately wonder whether involving other physiological signals can improve the accuracy of adaptation. Even so, the main difference here is that after getting a feedback signal (i.e., the level of cybersickness), our model can make use of the proportional, integral and derivative components for adaptation, that can process problems with high nonlinearities.

We used a framework called *Optuna* twice in this work with different intentions. *Optuna* is an automatic hyperparameter optimization framework designed for optimizing an NN model. First, to avoid performing massive user studies, we needed an NN model to link navigation accelerations to the phasic component of EDA, and *Optuna* could help find the optimal parameters for the NN model. Although we came up with an NN model to reproduce human cognition, it could be a promising method in many user interaction design: since the NN model has been trained on the navigation data from real user experiments, our model can predict the user response (e.g., EDA) according to the visual stimuli, allowing a designer to improve the interaction interface based on the predicted response [46]. Second, after proposing the adaptive model, we had to determine the optimal parameters allowing to significantly mitigate the cybersickness level; therefore, we employed *Optuna* again for this parameter optimization problem.

Physiological measures have been praised as objective indicators for cybersickness and affective experience [23,43]. There was no need to employ subjective cybersickness measurements, such as SSQ, FMS, and MISC because the computational model evaluated cybersickness from an implicit point in real time instead of explicit subjective or verbal feedback which can only evaluate cybersickness for a certain duration. However, physiological signals have opened the possibility to continuously measure cybersickness, although the sensitivity of these signals to visual stimuli might limit the performance. For example, the duration of SCR windows usually varies from 1 to 5 s after the onset of stimulus, and there might be overlapping in two subsequent ER-SCRs if the recovery time is large than the inter-stimulus interval, leading to distortion of the ER-SCR [39].

Despite promising results achieved in simulation, more validation studies are required. The optimal parameters for adaptive navigation were found from a theoretical point, but we believe that an additional user study to compare the non-adapted and adapted navigation can further confirm the effectiveness of our approach. Put differently, the lack of user studies might weaken the reliability of the parameters computed. An experimental validation after the theoretical study not only can validate the performance of the adaptive parameters, but also can help define the search range in *Optuna* and strengthen the relationship between the model parameters, thus it can bring benefits to the tuning process. We intend to carry out such studies in future research, and at this stage we encourage readers to focus on the adaptive model apart from finding the optimal parameters. In addition, our model involved eleven parameters in order to adapt navigation. Future research can investigate deeply on the relationship between these parameters and simplify the model to have less parameters.

5   Conclusion

We proposed a pioneering mathematical model for adaptive navigation in virtual environments by integrating a PID controller, with a long-term vision that immersive experience should be individualized. The premise to run this model successfully requires the system to detect cybersickness accurately; otherwise the adaptive power from the PID controller is weakened. Many adaptive VR systems have been focusing on the detection and evaluation of cybersickness in immersive environments, while we paid more attention to utilizing the cybersickness to optimize the navigation settings backward. The work is a theoretical paper with a solid simulated validation based on the number of ER-SCR and $MSDV$. Our contribution is to lay down the foundations of intelligent VR, in which a VR system can act as an assistant to help users perform better, which justifies the need to build computational models of cybersickness involving the generation of artificial data through AI. The pandemics has further been a great facilitator to introduce such approach. Although we found optimal adaptive coefficients thanks to simulation in *Optuna*, we are planning to run more user studies to further improve its performance.

References

1. Akiba, T., Sano, S., Yanase, T., Ohta, T., Koyama, M.: Optuna: a next-generation hyperparameter optimization framework. In: Proceedings of the 25th ACM SIGKDD International Conference on Knowledge Discovery & Data Mining, pp. 2623–2631 (2019). https://doi.org/10.1145/3292500.3330701
2. Aqajari, S.A.H., Naeini, E.K., Mehrabadi, M.A., Labbaf, S., Rahmani, A.M., Dutt, N.: GSR analysis for stress: development and validation of an open source tool for noisy naturalistic GSR data (1) (2020). http://arxiv.org/abs/2005.01834
3. Argelaguet, F., Andujar, C.: Automatic speed graph generation for predefined camera paths. In: Lecture Notes in Computer Science, vol. 6133 LNCS, pp. 115–126 (2010). https://doi.org/10.1007/978-3-642-13544-6_11
4. Aykent, B., Merienne, F., Guillet, C., Paillot, D., Kemeny, A.: Motion sickness evaluation and comparison for a static driving simulator and a dynamic driving simulator. Proc. Institut. Mech. Eng., Part D: J. Automobile Eng. **228**(7), 818–829 (2014). https://doi.org/10.1177/0954407013516101
5. Benedek, M., Kaernbach, C.: A continuous measure of phasic electrodermal activity. J. Neurosci. Methods **190**(1), 80–91 (2010). https://doi.org/10.1016/j.jneumeth.2010.04.028
6. Bos, J.E., MacKinnon, S.N., Patterson, A.: Motion sickness symptoms in a ship motion simulator: effects of inside, outside, and no view. Aviat. Space Environ. Med. **76**(12), 1111–1118 (2005)
7. Bos, J.E., de Vries, S.C., van Emmerik, M.L., Groen, E.L.: The effect of internal and external fields of view on visually induced motion sickness. Appl. Ergon. **41**(4), 516–521 (2010). https://doi.org/10.1016/j.apergo.2009.11.007
8. Davis, S., Nesbitt, K., Nalivaiko, E.: A systematic review of cybersickness. In: Proceedings of the 2014 Conference on Interactive Entertainment - IE2014, pp. 1–9. ACM Press, New York, New York, USA (2014). https://doi.org/10.1145/2677758.2677780
9. Dennison, M.S., Wisti, A.Z., D'Zmura, M.: Use of physiological signals to predict cybersickness. Displays **44**, 42–52 (2016). https://doi.org/10.1016/j.displa.2016.07.002
10. Du, M., Cui, H., Wang, Y., Duh, H.: Learning from deep stereoscopic attention for simulator sickness prediction. IEEE Trans. Vis. Comput. Graph., 1–1 (2021). https://doi.org/10.1109/TVCG.2021.3115901
11. Dużmańska, N., Strojny, P., Strojny, A.: Can simulator sickness be avoided? a review on temporal aspects of simulator sickness. Front. Psychol. **9**, 2132 (2018). https://doi.org/10.3389/fpsyg.2018.02132
12. Fernandes, A.S., Feiner, S.K.: Combating VR sickness through subtle dynamic field-of-view modification. In: 2016 IEEE symposium on 3D user interfaces (3DUI), pp. 201–210. IEEE (2016). https://doi.org/10.1109/3DUI.2016.7460053
13. Freitag, S., Weyers, B., Kuhlen, T.W.: Automatic speed adjustment for travel through immersive virtual environments based on viewpoint quality. In: 2016 IEEE Symposium on 3D User Interfaces (3DUI), pp. 67–70. IEEE (2016). https://doi.org/10.1109/3DUI.2016.7460033
14. Gamboa, H.: Multi-modal behavioral biometrics based on HCI and electrophysiology. Ph.D. thesis, Universidade Técnica de Lisboa (2008)
15. Gavgani, A.M., Nesbitt, K.V., Blackmore, K.L., Nalivaiko, E.: Profiling subjective symptoms and autonomic changes associated with cybersickness. Auton. Neurosci. **203**, 41–50 (2017). https://doi.org/10.1016/j.autneu.2016.12.004

16. Golding, J.F.: Phasic skin conductance activity and motion sickness. Aviat. Space Environ. Med. **63**(3), 165–171 (1992)

17. Golding, J.F., Stott, J.R.: Comparison of the effects of a selective muscarinic receptor antagonist and hyoscine (scopolamine) on motion sickness, skin conductance and heart rate. Br. J. Clin. Pharmacol. **43**(6), 633–637 (1997). https://doi.org/10.1046/j.1365-2125.1997.00606.x

18. Hadadi, A., Guillet, C., Chardonnet, J.R., Langovoy, M., Wang, Y., Ovtcharova, J.: Prediction of cybersickness in virtual environments using topological data analysis and machine learning. Front. Virtual Reality 3 (2022). https://doi.org/10.3389/frvir.2022.973236

19. Hu, P., Sun, Q., Didyk, P., Wei, L.Y., Kaufman, A.E.: Reducing simulator sickness with perceptual camera control. ACM Trans. Graph. (TOG) **38**(6), 1–12 (2019). https://doi.org/10.1145/3355089.3356490

20. Islam, R., Ang, S., Quarles, J.: CyberSense: a closed-loop framework to detect cybersickness severity and adaptively apply reduction techniques. In: 2021 IEEE Conference on Virtual Reality and 3D User Interfaces Abstracts and Workshops (VRW), pp. 148–155. IEEE (2021). https://doi.org/10.1109/VRW52623.2021.00035

21. Islam, R., Desai, K., Quarles, J.: Cybersickness prediction from integrated HMD's sensors: a multimodal deep fusion approach using eye-tracking and head-tracking data. In: 2021 IEEE International Symposium on Mixed and Augmented Reality (ISMAR), pp. 31–40. IEEE, Bari, Italy (2021). https://doi.org/10.1109/ISMAR52148.2021.00017

22. Islam, R., et al.: Automatic detection and prediction of cybersickness severity using deep neural networks from user's physiological signals. In: 2020 IEEE International Symposium on Mixed and Augmented Reality (ISMAR), pp. 400–411. IEEE (2020). https://doi.org/10.1109/ISMAR50242.2020.00066

23. Kaneko, D., Stuldreher, I., Reuten, A.J., Toet, A., van Erp, J.B., Brouwer, A.M.: Comparing explicit and implicit measures for assessing cross-cultural food experience. Front. Neuroergonomics **2**, 5 (2021). https://doi.org/10.3389/fnrgo.2021.646280

24. Kennedy, R.S., Lane, N.E., Berbaum, K.S., Lilienthal, M.G.: Simulator sickness questionnaire: an enhanced method for quantifying simulator sickness. Int. J. Aviat. Psychol. **3**(3), 203–220 (1993)

25. Keshavarz, B., Hecht, H.: Validating an efficient method to quantify motion sickness. Hum. Fact. J. Hum. Fact. Ergon. Soc. **53**(4), 415–426 (2011). https://doi.org/10.1177/0018720811403736

26. Kilteni, K., Groten, R., Slater, M.: The sense of embodiment in virtual reality. Presence: Teleoperators Virtual Environ. **21**(4), 373–387 (2012). https://doi.org/10.1162/PRES_a_00124

27. Kim, J., Kim, W., Oh, H., Lee, S., Lee, S.: A deep cybersickness predictor based on brain signal analysis for virtual reality contents. In: 2019 IEEE/CVF International Conference on Computer Vision (ICCV), pp. 10579–10588. IEEE (2019). https://doi.org/10.1109/ICCV.2019.01068

28. Kim, K.H., Bang, S.W., Kim, S.R.: Emotion recognition system using short-term monitoring of physiological signals. Med. Biol. Eng. Compu. **42**(3), 419–427 (2004). https://doi.org/10.1007/BF02344719

29. Kiranyaz, S., Avci, O., Abdeljaber, O., Ince, T., Gabbouj, M., Inman, D.J.: 1D convolutional neural networks and applications: a survey. Mech. Syst. Signal Process. **151**, 107398 (2021). https://doi.org/10.1016/j.ymssp.2020.107398

30. Koohestani, A., et al.: A knowledge discovery in motion sickness: a comprehensive literature review. IEEE Access **7**, 85755–85770 (2019). https://doi.org/10.1109/ACCESS.2019.2922993

31. Krokos, E., Varshney, A.: Quantifying VR cybersickness using EEG. Virtual Reality **26**(1), 77–89 (2022). https://doi.org/10.1007/s10055-021-00517-2

32. Lee, L.H., et al.: All one needs to know about metaverse: a complete survey on technological singularity, virtual ecosystem, and research agenda (2021). arxiv:2110.05352

33. Lou, R., Chardonnet, J.R.: Reducing cybersickness by geometry deformation. In: 2019 IEEE Conference on Virtual Reality and 3D User Interfaces (VR), pp. 1058–1059. IEEE (2019). https://doi.org/10.1109/VR.2019.8798164

34. Magaki, T., Vallance, M.: Developing an accessible evaluation method of VR cybersickness. In: 2019 IEEE Conference on Virtual Reality and 3D User Interfaces (VR), pp. 1072–1073. IEEE (2019). https://doi.org/10.1109/VR.2019.8797748

35. Makowski, D., et al.: NeuroKit2: a Python toolbox for neurophysiological signal processing. Behav. Res. Methods **53**(4), 1689–1696 (2021). https://doi.org/10.3758/s13428-020-01516-y

36. Oman, C.M.: Motion sickness: a synthesis and evaluation of the sensory conflict theory. Can. J. Physiol. Pharmacol. **68**(2), 294–303 (1990). https://doi.org/10.1139/y90-044

37. Padmanaban, N., Ruban, T., Sitzmann, V., Norcia, A.M., Wetzstein, G.: Towards a machine-learning approach for sickness prediction in 360 stereoscopic videos. IEEE Trans. Vis. Comput. Graph. **24**(4), 1594–1603 (2018). https://doi.org/10.1109/TVCG.2018.2793560

38. Plouzeau, J., Chardonnet, J.R., Merienne, F.: Using cybersickness indicators to adapt navigation in virtual reality: a pre-study. In: 2018 IEEE conference on virtual reality and 3D user interfaces (VR), pp. 661–662. IEEE (2018). https://doi.org/10.1109/VR.2018.8446192

39. Sharma, V., Prakash, N.R., Kalra, P.: Audio-video emotional response mapping based upon electrodermal activity. Biomed. Signal Process. Control **47**, 324–333 (2019). https://doi.org/10.1016/j.bspc.2018.08.024

40. So, R.H.: The search for a cybersickness dose value. In: HCI (1), pp. 152–156 (1999)

41. Wang, L.: PID control system design and automatic tuning using MATLAB/Simulink. John Wiley & Sons (2020)

42. Wang, Y., Chardonnet, J.R., Merienne, F.: Design of a semiautomatic travel technique in VR environments. In: 2019 IEEE Conference on Virtual Reality and 3D User Interfaces (VR), pp. 1223–1224. IEEE, Osaka, Japan (2019). https://doi.org/10.1109/VR.2019.8798004

43. Wang, Y., Chardonnet, J.R., Merienne, F.: VR sickness prediction for navigation in immersive virtual environments using a deep long short term memory model. In: 2019 IEEE Conference on Virtual Reality and 3D User Interfaces (VR), pp. 1874–1881. IEEE, Osaka, Japan (2019). https://doi.org/10.1109/VR.2019.8798213

44. Wang, Y., Chardonnet, J.R., Merienne, F.: Development of a speed protector to optimize user experience in 3D virtual environments. International Journal of Human-Computer Studies 147, 102578 (dec 2021). https://doi.org/10.1016/j.ijhcs.2020.102578

45. Wang, Y., Chardonnet, J.R., Merienne, F., Ovtcharova, J.: Using fuzzy logic to involve individual differences for predicting cybersickness during VR navigation. In: 2021 IEEE Virtual Reality and 3D User Interfaces (VR), pp. 373–381. IEEE, Lisbon, Portugal (2021). https://doi.org/10.1109/VR50410.2021.00060

46. Yang, B., Wei, L., Pu, Z.: Measuring and improving user experience through artificial intelligence-aided design. Front. Psychol., 3000 (2020). https://doi.org/10.
3389/fpsyg.2020.595374
47. Zielasko, D., Meißner, A., Freitag, S., Weyers, B., Kuhlen, T.W.: Dynamic field
of view reduction related to subjective sickness measures in an HMD-based data
analysis task. In: Proceedings of IEEE VR Workshop on Everyday Virtual Reality
(2018)

Lip Reading Using Temporal Adaptive Module

Jian Huang[1] , Lianwei Teng[2] , Yewei Xiao[1(✉)] , Aosu Zhu[1],
and Xuanming Liu[1]

[1] Institute of Automation and Electronic Information, Xiangtan University,
Xiangtan, China
yeexiao2004@xtu.edu.cn
[2] College of Intelligent Science, National University of Defense Technology,
Changsha, China

Abstract. Lip reading is a fine-grained video understanding task that endeavors to recognize speech content by analyzing the movement of the speaker's mouth. In recent times, 3D-ResNet-18 has become the favored front-end network for most of the lip reading methods. However, a single 3D CNN layer within the 3D-ResNet-18-based front-end network might not have enough representation power to extract temporal features. To address this issue, we propose the incorporation of Temporal Adaptive Module (TAM) into the front-end network of lip reading methods. TAM is an uncomplicated temporal module that consists of two branches: a local branch that provides location-sensitive information, and a global branch that focuses on capturing long-term temporal dependencies. This combination of branches helps capture complex temporal structures and facilitates robust temporal modeling. Taking global and local relationships into consideration explicitly improves the feature representation. It can be easily used in classical building blocks of networks. We conducted ablation studies to determine the optimal TAM structure and compared our results with various related approaches on the LRW dataset. Our experimental outcomes prove the superiority of our approach.

Keywords: Lip Reading · TAM · Location Sensitive Information · Adaptive Temporal Aggregation

1 Introduction

Unlike speech recognition, lip reading is a task that predicts the speech content using only visual information, typically the speaker's lip movements. Lip reading is useful in the acoustic environments where audio signals are unreliable, and has wide-ranging applications in practice, for instance, aiding hearing-impaired people [1], analyzing silent movies [2], audio visual enhancement [3], face liveness detection [4], etc.

Deep learning has brought great progress in image recognition task, nonetheless, lip reading remains a challenging task, in part due to the high complexity of video data.

Current deep-learning based methods generally follow a two-step approach. The first stage is to use CNNs as the front-end network to extract the discriminative deep bottle features. Noda et al. [5] were the first group to use CNNs to lip reading works, they demonstrated that CNNs outperform traditional methods in extracting visual features. Whereas, 2D CNNs can only learn spatial features rather than learning temporal features when dealing with sequential inputs. To integrate temporal information into feature extraction, many researchers have introduced 3D CNNs to lip reading works to jointly learn spatial and temporal information. However, one drawback is that 3D CNNs require high computational cost. To solve this problem, Stafylakis et al. [6] proposed to use a combination of a shallow 3D convolutional layer and 2D ResNet-34 network as the front-end network to efficiently extract spatio-temporal information. They achieved the state-of-the-art performance on LRW with 83% accuracy. Due to its superior performance, the use of 3D-ResNet-18 (the combination of a spatio-temporal convolutional layer and 2D ResNst-18) has gradually become the mainstream front-end network.

The second is to use a sequence processing module as the back-end to capture temporal information. Current back-end networks for isolated words can be divided into two categories: Temporal Convolutional Networks (TCNs), Recurrent Neural Networks (RNNs). In fact, TCNs do have advantages over RNNs because they take less time to train and are more flexible in changing receptive field size. Martinez et al. [7] were the first to introduce Multi Scale TCN to lip reading task. In this TCN variant, each temporal convolution consists of several branches with different kernel size, and their outputs are simply concatenated. In this way, every convolution layers mix information on multiple temporal scales. Ma et al. [8] proposed a Densely Connected TCN to provide denser and more robust temporal features. They improved the recording performance of Martinez et al. on LRW from 85.3% to 88.4%.

Despite the recent progress, the mainstream lip reading methods suffer from the following shortcomings. For example, the front-end network based on 3D-ResNet-18 might lack sufficient representation power to extract the temporal features very well by using a single-layer 3D CNN. To address this issue, Xiao et al. [9] proposed a two-stream network, termed as DFTN. They introduced deformation flows that can capture the motion information of the faces. A 2D-ResNet-18 was used to extract features from the deformation flow, while a 3D-ResNet-18 was used for the raw videos. Extensive experiments demonstrated the effectiveness of their methods, but the computational cost was still huge. In addition, Hao et al. [10] inserted TSM inside residual branch of 2D-ResNet-18. Although the specific module can achieve the performance of 3D CNNs while maintaining the complexity of 2D CNNs. The channels in the features map are moved forward and backward along time dimensions, and the additional cost of data movement is non-negligible and will result in latency increased.

In this paper, we focus on capturing temporal information in a more flexible way, and introduce a novel lip reading model, termed as TAM-Net. Inspired by Liu et al. [11], we first insert Temporal Adaptive Module (TAM) into each residual branch of the 3D-Resnet-18 as the front-end. TAM is a lightweight temporal module consisting of two branches. The local branch focuses on short-term motion modeling by using temporal convolution to select location sensitive importance map, while the global branch uses fully connected layers to generate temporal adaptive aggregation kernels for capturing long-term temporal dependencies. Then we combine the local and global branch in a convolutional manner to adaptively aggregate temporal information.

The main contributions of this paper are summarized as follows. (1) In the task of lip reading, we propose a novel lip reading model for isolated words, termed as TAM-Net. We insert TAM into 3D-ResNet-18 network to enhance the temporal modeling capabilities. (2) We conduct ablation experiments on TAM-Net to find the optimal TAM structure. (3) We conduct comparative experiments with other related lip reading methods on LRW dataset, and our model achieves impressive performance.

2   The Proposed Work

In this section, we will first give the overall of TAM-Net, and then illustrate the design motivation of integrating TAM into the 3D-ResNet-18 network. Finally, we will describe the Temporal Adaptive Module in detail.

2.1   Lip Reading Models Based on TAM-Net Models

An overview of the pipeline is shown in Fig. 1. We choose the cropped mouth ROIs as the inputs, A 3D convolutional layer is first utilized to obtain the spatial-temporal features with shape of $C \times T \times H \times W$, where C is the channel numbers, T, H, W stand for the temporal spatial dimensions, respectively.

TAM can be easily integrated into the existing 2D CNNs to yield an efficient improvement with a low computational cost. Liu et al. [11] have demonstrated the superiority of TAM in the field of fined grained and motion-dominated actions. In this paper, we introduce TAM to the lip reading work, and we hope to enhance the temporal modeling capability by inserting TAM into 3D-ResNet-18. In the next, an average pooling operation is applied to squeeze the spatial knowledge and yield a 512-dimensional average-pooled vector for each frame. We choose PD-TCN as the back-end network, some detail will be discussed in the following subsections.

2.2   The Overview of TAM-Net

The structure of Temporal Adaptive Module can be seen in Fig. 2, formally, let $X \in R^{C \times T \times H \times W}$ denote the feature maps obtained by the 3D convolutional layers, where C represents the number of channels, and T, H, W stand for the temporal spatial dimensions, respectively. To ensure our model with only slightly increase in model complexity and computational burden, we propose to squeeze the global spatial information into 1D temporal signal by employing a global spatial pooling. Therefore, the aggregation spatial information $\widehat{X}_{c,t}$ can be generated by shrinking the spatial dimensions of the feature map:

$$\widehat{X}_{c,t} = \frac{1}{h \times w} \sum_{i,j}^{h \times w} X_{c,t,i,j} \tag{1}$$

Where $\widehat{X}_{c,t,i,j}$ is the i,j^{th} element of the activation tensor obtained from the tth frame, h and w are the height and width of the feature maps, respectively.

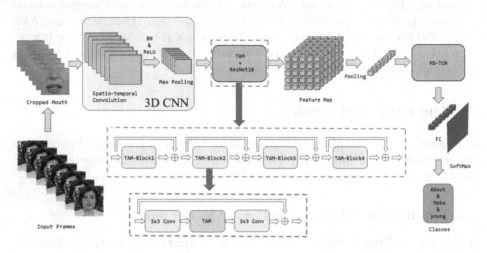

Fig. 1. The architecture of proposed method.

The proposed TAM is established based on this squeezed 1D temporal signal to keep the efficiency, and composed of two branches: a global branch and a local branch. Each of them focuses on a different aspect of temporal information. The local branch focuses on short-term motion modeling, while the global branch generates temporal adaptive aggregation kernels for capturing long-term temporal dependencies. We combine the local and global branch in a convolutional manner to adaptively aggregate temporal information, whereupon the TAM can be written as follows:

$$Y = G(X) \otimes (L(X) \odot X) \tag{2}$$

where $G(X)$ and $L(X)$ stand for the output of the global branch and local branch, $\otimes$ is convolution operation and $\odot$ denotes element-wise multiplication.

Local branch: We can find that the structure of local branch is similar to that of an SE block, applies global average pooling to the feature map to aggregate global information, and then uses an MLP with channel dimensionality reduction. It effectively enhances the expressive power of the learned features. Nevertheless, there are still some obvious differences between local branch and SENet. For instance, as shown in Fig. 2, the local branch is built by two temporal convolutional layers with ReLU non-linearity instead of fully connected layers as the basic layer across the temporal domain, which can produce location-sensitive importance maps for enhancing frame-wise features. The kernel size is set to 3 and the first Conv1D reduce the number of channels from C to C/β to control the model complexity. The other Conv1d increases the number of channels from C/β to C, and the following sigmoid activation generates the important weights W which are sensitive to temporal location. The final output of the local branch is formulated as follows:

$$L(X) = Re(W) \tag{3}$$

$$Z = L(X) \odot X = Re(W) \odot X \tag{4}$$

Where $Re(W)$ represents rescaling the weight $W \in R^{C \times T}$ to $W \in R^{C \times T \times H \times W}$, and denotes element-wise multiplication.

Global branch: Unlike the local branch, The global branch focuses on how to generate video adaptive kernels and aggregate temporal information in a convolutional manner. For efficiency, the global branch only models the temporal relations in a channel-wise manner and would not change the number of channels of the inputs. Just like the local branch, the squeezed feature map $\widehat{X}_c \in R^T$ is used as the input of the global branch without considering the spatial information. Two fully connected layers are stacked to generate adaptive kernels for aggregating temporal features from a global view, and the learned kernel is normalized with a Softmax function to yield a positive aggregation weight which can be written as follows:

$$\theta_c = G(X)_c = Softmax(F(W_2, \delta(F(W_1, \widehat{X}_c))) \tag{5}$$

Where $\theta_c \in R^k$ is the adaptive kernel for channel, K is the kernel size, F and δ stand for the FC layers and the ReLU function, respectively.

We can treat the global branch as an adaptive kernel generator, but it is still location invariant. The local branch provides the location sensitive information, and thereby solves this issue. The final procedures can be written as follows:

$$Y_{c,t,j,i} = G(X) \otimes Z = \theta \otimes Z = \sum_k \theta_{c,k} \cdot Z_{c,t+k,j,i} \tag{6}$$

Where $Y \in R^{C \times T \times H \times W}$ is the final output feature maps, $\theta = \{\theta_1, \theta_2, \ldots, \theta_c\}$ denote the adaptive kernels.

2.3 Partially Dense TCN

Densely connected network was designed to solve the vanishing-gradient problem by employing shallower layers in favor of benefiting feature propagation. In 2021, Ma et al. [8] have applied densely connected to TCNs for word-level lip reading. They found that DC-TCNs was able to cover the temporal scales in a denser fashion and thus more sensitive to words. Furthermore, to further improve the model's expression power, they explored two different DC-TCN

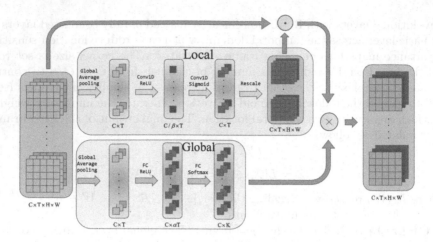

Fig. 2. Temporal Adaptive Module.

variants, and inserted Squeeze-and-Excitation (SE) Network [18] into both of them for additional channel-wise attention. The first variant is the fully dense TCN (FD-TCN) block which applies dense connections for all TC layers in a concatenated manner. The other DC-TCN variant is the partially dense (PD) block. In each PD block, the filters with identical dilation rates are employed in a multi-scale fashion.

In the work of [8], extensive experiments have demonstrated the superiority of PD-TCN, we thus choose PD-TCN as the back-end network with the following hyperparameters as the final PD-TCN model configuration: the kernel size K is set to $K = \{3, 5, 7\}$, the dilation rate D is set to $D = \{1, 2, 5\}$, the growth rate $C_0 = 128$ and the block number $B = 4$. Moreover, the reduction ratio in SE block is set to 16.

3   Experiments

3.1   Datasets and Implementation Details

Datasets: Our experiments are performed on one of the largest English word-level lip reading dataset, namely LRW [10]. This is a very challenging dataset, released in 2016 and still available to us today. LRW contains a total of 500 word-targets, each target has 800–1000 training samples, 50 testing samples, 50 validation samples, which was spoken by hundreds of different speakers. Furthermore, each sample consists of 29 frames, and the target word appears in the middle of the sample shot.

Implementation Details: In this experiment, we first use the mediapipe toolkit [13] to detect face and get the facial landmarks. The face is then aligned to a neural reference frame and the mouth ROIs with size of 96×96 is cropped. For efficiency, the cropped mouth ROIs are converted into grayscale. We train the word samples in the LRW dataset with a maximum sequence length of 29 frames to ensure that the word samples are placed in the center of the sequence. This setting can provide more contextual information to improve the training efficiency. The LRW dataset uses the target vocabulary "ACCESS" set by this method, and its position in the input sequence is shown in Fig. 3. The model is trained in an end-to-end fashion for 80 epochs with a batch of 32 on a single NVIDIA 3090 GPU with 24 GB memory. Our experiments are implemented with PyTorch, we use the Adam Optimizer, and the initial learning rate and the weight decay of LRW are set to 0.003 and 0.01, respectively. We decay the learning rate using a cosine scheduler without a warm-up phase. The calculation method of cosine learning rate is shown in formula (7), where t represents the number of training cycles, η is the initial learning rate, and η_t represents the cosine learning rate of t cycle. We use cross entropy loss as the loss function. During training, we use a random crop of 88×88 pixels and flip all the frames horizontally with a probability of 0.5 for data augmentation, and take a central crop for testing.

$$\eta_t = \frac{\eta}{2}(1 + cos(\frac{t}{4})) \tag{7}$$

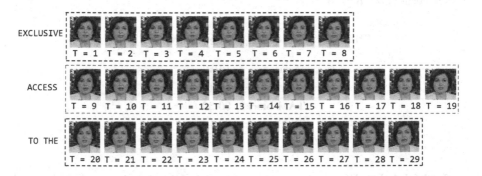

Fig. 3. The position in the input sequence of the target word "access".

Evaluation criteria: We take the recognition accuracy of overall word classes on all datasets as the evaluation criterion for this experiment.

3.2 Evaluation of TAM

In this section, we will perform ablation studies to find the optimal structure of TAM. We will first choose the best combination of parameters in TAM, then the most suitable adaptive kernel size and insertion position need to be discussed.

Parameter choices: We perform experiments with different combination of α and β to find out the optimal hyper-parameters in TAM. The results are shown in Fig. 4, we can get the conclusion that TAM Blocks with $\alpha = 2$, $\beta = 4$

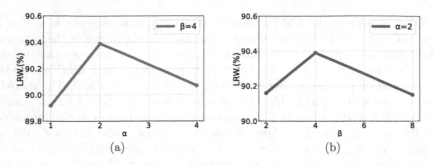

Fig. 4. Study on parameter choices of α and β.

achieve the highest performance on LRW, and this combination will be applied in the following experiments.

Adaptive kernel size: The adaptive kernel θ generated from the global branch with the global receptive field, thus could aggregate temporal information guided by the global context. K is the adaptive kernel size. When $K = 3$ and $K = 5$, the accuracy of our method is 90.16% and 90.39%, respectively. We can find that the larger K can bring better performance than the smaller one, thus we set kernel size $K = 5$.

Table 1. Study on the temporal receptive fields.

Position	LRW. (%)
TAM-block1	90.17
TAM-block2	**90.39**
TAM-block3	90.10

TAM in the different position: The TAM Module is inserted into each residual branch, and there are a total of three different positions. The TAM1 means that TAM is inserted before the first convolution of each residual block, the TAM2 is after the first convolution and the TAM3 is after the last convolution. The TAM2 have obvious improvements than other positions in Tab. 1, which will be the default position in subsequent experiments.

Compared to other temporal module: To verify the effectiveness of TAM, we make comparisons between TAM and TSM. TSM is another lightweight temporal module proposed by Lin et al. In the work of [8], Hao et al. introduce TSM to lip reading work and have achieved the performance of 3D CNNs while maintaining the complexity of 2D CNNs. For fair comparison, we choose the '3D-ResNet-18+PD-TCN' as the baseline model, we replace the back-end network in [8], as can be seen from Tab. 2, PD-TCN has brought a 3.46% improvement. In addition, both TAM and TSM can significantly improve the model performance, and TAM yields 0.7% higher accuracy than TSM.

Table 2. Study on the temporal receptive fields.

Methods	LRW. (%)
PBL [15]	82.80
GLMIM [19]	84.40
DFTN [9]	84.10
TSM [10]	86.23
PD-TCN [8]	88.36
WPCL [16]	88.30
MS-TCN+MSD [17]	87.90
TSM(PD-TCN)	89.69
PD-TCN (w/o pre-training) [14]	92.10
Ours (TAM-Net)	90.39

Comparison with the State of the Art: We report the performance of our model and other various relative approaches on the LRW dataset, and the results are shown in Tab. 2. We find that our method has achieved an accuracy of 90.39%, which is only 1.71% less than the Top-1 model. However, the method in [14] contains many other data augmentation methods including Time Mask, Mix Up, Word Boundary, etc. which lead to huge improvements, but consume a lot of computing resources.

4   Conclusion

In this paper, we propose a novel TAM-Net model that inserts TAM into 3D-ResNet-18 to enhance the temporal modeling capabilities. TAM is a lightweight model consisting of two branches, and each branch focuses on different temporal structures, which really contribute to more robust temporal modeling. This helps to capture rich discriminative feature representations at low computational cost. Due to its flexibility and lightweight design, TAM can be easily added to classical backbone networks seamlessly with negligible additional cost. We perform ablation studies to find the optimal structure of TAM, and conduct comparative experiments with the related methods on LRW dataset, and the experimental results demonstrate the superiority of TAM-Net.

References

1. Sun, K., Yu, C., Shi, W., Liu, L., Shi, Y.: Lip-interact: improving mobile device interaction with silent speech commands. In: Proceedings of the 31st Annual ACM Symposium on User Interface Software and Technology, pp. 581–593 (2018)
2. Jha, A., Namboodiri, V.P., Jawahar, C.V.: Word spotting in silent lip videos. In: 2018 IEEE Winter Conference on Applications of Computer Vision (WACV). IEEE (2018)

3. Afouras, T., et al.: Deep audio-visual speech recognition. IEEE Trans. Pattern Anal. Mach. Intell. (2018)
4. Rufai, S.Z., Selwal, A., Sharma, D.: On analysis of face liveness detection mechanisms via deep learning models. In: International Conference on Sustainable Computing and Data Communication Systems (ICSCDS), vol. 2022, pp. 59–64 (2022). https://doi.org/10.1109/ICSCDS53736.2022.9760922
5. Noda, K., Yamaguchi, Y., Nakadai, K., Okuno, H.G., Ogata, T.: Lipreading using convolutional neural network. In: Fifteenth Annual Conference of the International Speech Communication Association (2014)
6. Stafylakis, T., Tzimiropoulos, G.J.A.P.A.: Combining residual networks with LSTMs for lipreading (2017)
7. Martinez, B., Ma, P., Petridis, S., Pantic, M.: Lipreading using temporal convolutional networks. In: ICASSP 2020-2020 IEEE International Conference on Acoustics, Speech and Signal Processing (ICASSP), pp. 6319–6323. IEEE (2020)
8. Ma, P., Wang, Y., Shen, J., Petridis, S., Pantic, M.: Lip-reading with densely connected temporal convolutional networks. In: Proceedings of the IEEE/CVF Winter Conference on Applications of Computer Vision, pp. 2857–2866 (2021)
9. Xiao, J., Yang, S., Zhang, Y., Shan, S., Chen, X.: Deformation flow based two-stream network for lip reading. In: 2020 15th IEEE International Conference on Automatic Face and Gesture Recognition (FG 2020), pp. 364–370. IEEE (2020)
10. Hao, M., et al.: How to use time information effectively? Combining with time shift module for lipreading. In: ICASSP 2021-2021 IEEE International Conference on Acoustics, Speech and Signal Processing (ICASSP). IEEE (2021)
11. Liu, Z., et al.: TAM: temporal adaptive module for video recognition. In: Proceedings of the IEEE/CVF International Conference on Computer Vision (2021)
12. Chung, J.S., Zisserman, A.: Lip reading in the wild. In: Lai, S.-H., Lepetit, V., Nishino, K., Sato, Y. (eds.) ACCV 2016. LNCS, vol. 10112, pp. 87–103. Springer, Cham (2017). https://doi.org/10.1007/978-3-319-54184-6_6
13. Mediapipe. https://mediapipe.dev/
14. Ma, P., Wang, Y., Petridis, S., Shen, J., Pantic, M.: Training strategies for improved lip-reading. In: ICASSP 2022 - 2022 IEEE International Conference on Acoustics, Speech and Signal Processing (ICASSP), pp. 8472–8476 (2022). https://doi.org/10.1109/ICASSP43922.2022.9746706
15. Miao, Z., Liu, H., Yang, B.: Part-based lipreading for audio-visual speech recognition. In: 2020 IEEE International Conference on Systems, Man, and Cybernetics (SMC). IEEE (2020)
16. Tian, W., Zhang, H., Peng, C., Zhao, Z.-Q.: lipreading model based on whole-part collaborative learning. In: ICASSP 2022 - 2022 IEEE International Conference on Acoustics, Speech and Signal Processing (ICASSP), pp. 2425–2429 (2022). https://doi.org/10.1109/ICASSP43922.2022.9747052
17. Ma, P., Martinez, B., Petridis, S., Pantic, M.: Towards practical lipreading with distilled and efficient models. In: ICASSP 2021 - 2021 IEEE International Conference on Acoustics, Speech and Signal Processing (ICASSP), pp. 7608-7612 (2021). https://doi.org/10.1109/ICASSP39728.2021.9415063
18. Hu, J., Shen, L., Sun, G.: Squeeze-and-excitation networks. In: Proceedings of the IEEE Conference on Computer Vision and Pattern Recognition (2018)
19. Zhao, X., Yang, S., Shan, S., Chen, X.: Mutual information maximization for effective lip reading. In: 2020 15th IEEE International Conference on Automatic Face and Gesture Recognition (FG 2020), pp. 420-427 (2020). https://doi.org/10.1109/FG47880.2020.00133

AudioFormer: Channel Audio Encoder Based on Multi-granularity Features

Jialin Wang[ID], Yunfeng Xu[✉][ID], Borui Miao[ID], and Shaojie Zhao[ID]

Hebei University of Science and Technology, Shijiazhuang 050000, Hebei Province, China
hbkd_xyf@hebust.edu.cn

Abstract. To solve the problem of poor standardized feature extraction methods for speech emotion recognition tasks and insufficient depth representation capability for extracting acoustic samples, we first propose a Multi-granularity feature extraction method that takes into account the integrity of data features and overcomes the redundancy of existing feature extraction methods; secondly, we propose a Channel Audio Encoder Model that uses different Feature Encoders to extract High-order features. Experiments show that the proposed Multi-granularity feature-based Channel Audio Encoder achieves state-of-the-art performance in the IEMOCAP dataset. The method also experiments on a real-scene dataset to demonstrate its usability and provide a reference for aiding the diagnosis of mental illness.

Keywords: Speech Emotion Recognition · Channel Audio Encoder · Mental Illness

1 Introduction

Emotions play an important role in human communication [4]. With the continuous advancement of deep learning, Automatic Speech Recognition, Speech Emotion Recognition, and Multimodal Emotion Recognition have developed rapidly. Speech Emotion Recognition investigates to help machines understand human subjective emotions from audio information. In existing research methods, there are many data features, and for different data, the actual effect of data features is uneven. Moreover, acoustic feature extraction models generally rely on Convolutional Neural Networks, which use data feature embedding as the original input to reduce the temporal information relationship between data. Therefore, the in-depth extraction and exploitation of data features remain a challenging task.

In order to solve the above problems, many researchers have used deep learning methods to replace early emotion classification algorithms, and explored a series of End-to-End Speech Emotion Recognition technologies.

B. Luo et al. (Eds.): ICONIP 2023, CCIS 1964, pp. 357–373, 2024.
https://doi.org/10.1007/978-981-99-8141-0_27

Early speech emotion recognition tasks [25] used prosodic features as data features, and achieved certain results in emotion classification algorithms based on machine learning, such as Naive Bayesian algorithm and Support Vector Machine algorithm [15].

In recent years, with the continuous development of deep learning, the application of Deep Neural Network(DNN) emerges in an endless stream. Deep neural network is currently the basis of many artificial intelligence applications [8], and due to the breakthrough application of DNN in image recognition [6], the application volume of DNN has exploded. The outstanding performance of the Deep Neural Network stems from its ability to use statistical learning methods to extract deep features from the original sensory data [23]. The Deep Neural Network can be regarded as a neural network composed of multiple hidden layers. Combine shallow features to extract deep data features. Literature [18] proposed a feature extraction method combining Teager energy operator and Mel Frequency Cepstral Coefficient, which was applied to the Multilingual Speech Emotion Recognition system of English, German and Hindi. Literature [30] proposed a network that learns Multi-scale feature representations was used to fuse global features, but Convolutional Neural Networks(CNN) are unable to learn the contextual relationships of the data. Literature [22] applies Graph Convolutional Network (GCN) to the field of Speech Emotion Recognition, uses a window function to window the speech, and converts each windowed part into a graph node. The emergence and development of DNN provide technical support for the integration of different research fields. As demonstrated in the literature [1,13,16] attentional mechanisms can be used to good effect in computer vision, multimodal domains.

After that, the End-to-End speech emotion recognition technology improves the generalization ability of the model, enabling the model to output information directly from the original data. Current End-to-End speech recognition models include connectionist temporal classification (CTC) and RNN-Transducer. The purpose of the CTC technique is to map an input sequence of data samples to an output label sequence. RNN Transducer is widely welcomed by the industry due to its own streaming processing capabilities, even if it does not include the conditional independence assumption of CTC.

In response to the above challenges, we propose a Channel Audio Encoder Based on Multi-Granularity Features, the main purpose of which is how to better implement feature engineering and extract High-order features. The multi-granularity feature we propose refers to a variety of statistical features calculated based on the time-frequency features of acoustic data.

In summary, Our main contributions are as follows:

1. we proposed the conception of multi-granularity features and built an acoustic feature extraction method based on public datasets.
2. we proposed a Channel Audio Encoder Model to extract high-level features, solve the problem of model gradient disappearance, prevent model overfitting, and make the model converge faster.

3. We validated the validity of our proposed method on the IEMOCAP [2] dataset through comparative experiments and visualization of the results; we demonstrated the true usability of our proposed method on the Mental Illness Chinese(MIC) dataset, providing a theoretical basis for the auxiliary diagnosis of psychiatric disorders.

2   Related Work

Speech Emotion Recognition focuses more on the feature representation of emotional information in audio samples. Emotional features are divided into statistical domain, spectral domain and time domain features, these features are collectively referred to as low-level feature descriptors(LLDs) [12], many scholars will use LLDs features combined with high-level statistical functions features(HSFs) [12] when extracting acoustic features, this process can be seen as the first stage of feature extraction. Recurrent Neural Networks(RNN) [5] are widely used in the field of time series-based natural language processing, such as speech recognition, language modeling and machine translation. Variants of recurrent neural networks, such as long short-term memory networks and gated recurrent unit networks, are widely used in the field of Speech Emotion Recognition. This feature extraction process based on deep learning algorithms can be seen as the second stage of speech emotion recognition and is an emerging research area in machine learning that has received increasing attention in recent years [27]. The following will introduce from two aspects of Feature Engineering and High-order Feature Representation.

2.1   Feature Engineering

Features refer to information extracted from raw data that is helpful for predicting results. Feature engineering can be understood as a method of extracting data information. Feature engineering includes modules such as feature extraction, feature construction, and feature selection. Speech Emotion Recognition focuses more on the feature representation of emotional information in audio samples.

When building efficient classification models, audio feature-based models are widely used [28]. The HGFM model proposed by Xu [26] et al. combines spectral and time-domain features to extract three features of the audio data, which include the Zero Crossing Rate, the Mel Frequency Cepstrum Coefficients(MFCC), and the Constant-Q Transform. They use the openSMILE toolkit to extract the global acoustic features. Yoon [28] et al. extracted 12-dimensional MFCC features as well as 35-dimensional rhyme features, which contain F0 Frequency, Voicing Probabilities, and Loudness Contours. Sahu [20] et al. extracted five acoustic features, namely fundamental frequency features, harmonic features, speech energy features, root mean square energy features, and central moment features, and found through their study that the harmonic features directly related to the signal excitation contributed the most, the root mean square energy features contributed the same as the fundamental frequency features and the central moment features contributed the least. Schmid [21] et al. studied an audio embedding extractor that used short-time Fourier transform features with a window length of 25 ms and a frameshift of 10 ms as well as log-Melp features in its low-level features, and added fundamental frequency features by averaging the log-Melp over a certain period of time to finally obtain a 128-dimensional feature embedding and found the effectiveness of the fundamental frequency features. Mirsamadi [12] et al. applied different aggregation functions, such as mean, maximum, variance, and linear regression coefficients to LLDs features to generate HSFs features. These high-level statistical functions serve to roughly describe the temporal transformation and profile information of different LLDs features during speech under the assumption that emotional information depends on temporal changes and not on short-term LLDs statistical values. Tripathi [24] et al. proposed a method for speech emotion recognition based on speech features and speech transcription (text). The method shows that spectrograms and Mel frequency cepstral coefficients help to retain emotion-related features in speech.

This study found that LLDs features such as Mel Frequency Cepstral Coefficient, fundamental frequency, root mean square energy and speech energy can describe audio features more perfectly and effectively complete Speech Emotion Recognition tasks. At the same time, the application of HSFs features in LLDs features also provides a new idea for this paper.

2.2 High-Order Feature Representation

Classification features extracted by deep learning or machine learning algorithms can be called the High-order features. The deep learning models applied to time series data are mainly based on Recurrent Neural Network, Long Short-term Memory network(LSTM) and Gated Recurrent Unit(GRU). Lea [7] et al. proposed a Temporal Convolutional Network (TCN), which applies the idea of Recurrent Neural Networks to Convolutional Neural Networks to extract data features by spanning time steps. The main research work of most scholars is still centered on the Recurrent Neural Network and its variants. The Long Short-term Memory network is an improved Recurrent Neural Network, which can learn long-term dependent information and improve the gradient disappearance problem existing in recurrent neural networks. Its improvement is mainly reflected in two aspects, one is to introduce a new internal state for the transmission of historical information, and the other is to control the path of information transmission through the gating mechanism. The key of the long short-term memory network is the state of the cell unit, which is mainly controlled by the gating unit to selectively retain or discard information. The Gated Recurrent Unit network is based on the long short-term memory network, which has similar functions but optimizes the model structure. Zou [31] et al. used long short-term memory networks as encoders to receive speech spectral domain features in order to obtain information about specific features. Experiments on the IEMOCAP dataset show that their model achieves good performance under independent cross-validation.

Through the above research, we found that the Long-short-Term Memory network and the Gated Recurrent Unit as the feature extraction model have played an efficient role in the Speech Emotion Recognition task and achieved good performance. Based on the in-depth study of feature engineering and High-order features, we propose a Channel Audio Encoder based on Multi-granularity Features.

3 Methodology

This chapter introduces our proposed Channel Audio Encoder model based on Multi-granularity Features. The architecture of the model is shown in Fig. 1. The input of the model we propose is a speech emotion dataset and multi-granularity data features are obtained through various feature extraction methods, which are used as feature embedding and passed into feature encoders. The Channel Audio Encoder is composed of a bidirectional LSTM and a bidirectional GRU. Its main function to extract high-level features. Finally, the High-order features are classified by a Fully Connected Neural network(FCN). More details about the above architecture are presented in the following subsections.

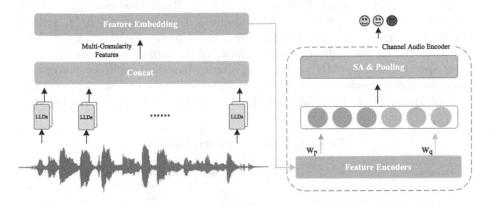

Fig. 1. Diagram of the overall architecture of AudioFormer. The green and purple circles in the right half of the diagram represent the output of different feature encoders. (Color figure online)

3.1 Problem Definition

Given a set of speech dataset $A = \{a_1, a_2, a_3, \cdots, a_n\}$, where n is the total number of samples in speech dataset A. For each speech sample a_n, we extract its acoustic features v_n generate a local feature embedding V_f. Define $E_f =$

$\{a_n, v_n, l_i\}$, $v_n \in V_f$, E_f denotes the n-dimensional feature embedding, l_i represents the a_n authentic sentiment label, where $L = \{l_1, l_2, l_3, l_4\}$, $l_i \in L$, the feature embedding matrix $V_f = \{v_1, v_2, v_3, \cdots, v_n\}$. By converting the real scenario problem into a mathematical problem and providing a theoretical basis for the solution, we realize that by inputting the dataset A, we generate the feature embedding V_f and finally get the prediction sample label l_i.

3.2   Multi-granularity Features Extraction

Acoustic features include frequency domain features, time domain features and statistical domain features. According to the different feature extraction processes, it can also be divided into features extracted from the original signal, features obtained by converting the original signal into frequencies and features obtained by changing the scale of quantized features. We utilize the LibROSA [11] toolkit to extract local features of acoustic samples from frequency, time domain and statistical perspectives.

Time-Frequency Features. We extracted a total of 11 features, including Mel Frequency Cepstrum Coefficient $f_{mfcc} \in \mathbb{R}^{1*a}$, Spectral Centroid $f_{cent} \in \mathbb{R}^{1*b}$, Spectral Flatness $f_{flat} \in \mathbb{R}^{1*c}$, Mel Spectrum $f_{mel} \in \mathbb{R}^{1*d}$, Chroma Feature $f_{chroma} \in \mathbb{R}^{1*e}$, Spectral Contrast $f_{cont} \in \mathbb{R}^{1*g}$, Zero Crossing Rate $f_{zcr} \in \mathbb{R}^{1*h}$, Pitch Feature $f_{pitch} \in \mathbb{R}^{1*k}$, Rmse Features $f_{rmse} \in \mathbb{R}^{1*o}$, Magnitude Feature $f_{mag} \in \mathbb{R}^{1*p}$ and Constant-Q Transform $f_{cqt} \in \mathbb{R}^{1*q}$, and the feature statistics are shown in Table 1. The Mel Frequency Cepstrum Coefficients take into account the auditory characteristics of the human ear by mapping the linear spectrum to the Mel nonlinear spectrum based on auditory perception, which is then converted to the cepstrum. That is, the spectrum is passed into a set of Mayer filters to obtain the Mayer spectrum. The Zero Crossing Rate

Table 1. Multi-granularity Features

Features	Feature representation	Dimensional representation	Dimensiona
Mel Frequency Cepstrum Coefficient	$mfcc$	a	50
Spectral Centroid	$cent$	b	1
Spectral Flatness	$flat$	c	1
Mel Spectrum	mel	d	128
Chroma Feature	$chroma$	e	12
Spectral Contrast	$cont$	g	7
Zero Crossing Rate	zcr	h	1
Pitch Feature	$pitch$	k	1025
Rmse Features	$rmse$	o	1
Magnitude Feature	mag	p	1025
Constant-Q Transform	cqt	q	12

is used to observe the number of changes in the sampling points of the speech signal and the changes in the time-domain waveform, which can reflect frequency information to a certain extent. The mel spectrum converts the frequency into a mel scale. The mel spectrum contains features in the time domain and frequency domain, and can perceive related frequency domain and amplitude information. The root mean square is also called an effective value in physics, and the root mean square energy expresses the comprehensive information of all sample points in a frame. The spectral centroid is a parameter describing the timbre properties of an acoustic signal, which explains the frequency distribution and energy distribution of the acoustic signal. A darker signal contains more low-frequency content and has a relatively low spectral centroid, whereas a bright signal has a relatively high spectral centroid.

Statistical Domain Characteristics. The characteristics of time series data based on the statistical domain include dozens of characteristics such as maximum value, minimum value, mean value, standard deviation, variance, skewness, and kurtosis. Machine learning models rely on the assumption of normality. Therefore, skewness is used to detect abnormal data to prevent it from affecting the predictive ability of machine learning models and kurtosis is used to detect outliers to determine whether it is similar to a normal distribution.

We calculate statistical domain features based on Spectral Flatness features, Zero Crossing Rate features, Rmse features, Pitch features, Magnitude features and Spectral Centroid features to obtain 15-dimensional statistical value features, which are comparable to 128-dimensional Mel Spectral features and 50-dimensional Mel Frequency Cepstrum Coefficient features, 12-dimensional Chroma features, 7-dimensional Spectral Contrast features and 12-dimensional Constant-Q Transform features together constitute 224-dimensional Multigranularity features.

3.3 Channel Audio Encoder Model Structure

The channel audio encoder consists of two feature encoders, which architecture diagram is shown in Fig. 2. The data features are passed through two feature encoders to obtain High-order Feature matrices w_p and w_q, and the High-order Feature matrices are concatenated to obtain the output of the Channel Audio Encoder CAE_{out}.

$$w_p = FE_p(V_f, F_w) \tag{1}$$

$$w_q = FE_q(V_f, F_w) \tag{2}$$

$$CAE_{out} = concat(w_p, w_q) \tag{3}$$

The feature representation is obtained through the fully connected layer of the ReLU activation function, and the Softmax activation function is used as a classifier to calculate the category probability.

$$f_{out} = ReLU(CAE_{out}) \tag{4}$$

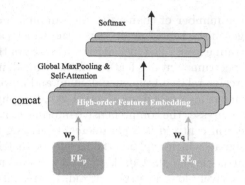

Fig. 2. Diagram of the overall architecture of Channel Audio Encoder. FE represent the Feature Encoder in the figure.

$$pred_i = \frac{\exp(f_{out})}{\sum_i \exp(f_{out})} \tag{5}$$

Among them, i belongs to the sample category, $pred_i$ is the predicted sample category probability. The model uses the cross-entropy loss function to optimize parameters, where l_i is the real label of the sample.

3.4   Feature Encoder

The feature encoder consists of a bidirectional LSTM network (or a bidirectional GRU network), a linear layer and a batch normalization layer, as shown in Fig. 3. The LSTM network solves the problem of gradient disappearance, and the GRU network has fewer parameters and reduces the risk of overfitting to a certain extent. LSTM and GRU can better solve the problems of gradient disappearance and historical information loss of RNN model, obtain the previous information and make the network converge faster, but the Unidirectional network lacks the perception of the following information. The introduction of a bidirectional network can better obtain context information and generate more comprehensive high-level features. For a Feature Encoder, its backbone network can be expressed as Eqs. 17 to 19.

$$\overrightarrow{h_t} = \overrightarrow{Bidirect}(E_f, \overrightarrow{h_{t-1}}) \tag{6}$$

$$\overleftarrow{h_t} = \overleftarrow{Bidirect}(E_f, \overleftarrow{h_{t-1}}) \tag{7}$$

$$h_t = [\overrightarrow{h_t}, \overleftarrow{h_t}] \tag{8}$$

Among them, E_f represents the n-dimensional feature embedding, $\overrightarrow{h_t}$ represents the forward hidden layer input of the bidirectional network at time t, $\overleftarrow{h_t}$ represents the reverse hidden layer input of the bidirectional network at time t, and h_t represents the hidden layer vector, including the context information of the bidirectional hidden layer. Use h_t as feature representation after normalization as input

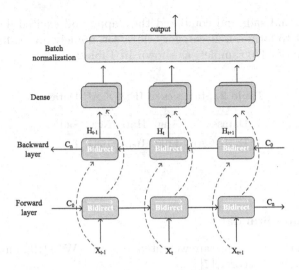

Fig. 3. Diagram of the overall architecture of Feature Encoder.

for downstream tasks. The Multi-granularity features we proposed can effectively represent sample information and fully describe sample features. Compared with the existing feature extraction network, the Channel Audio Encoder model we proposed can use the feature encoder to learn Multi-granularity features of samples, effectively combine context information and improve the classification accuracy of the model. The next chapter will verify the effect of the model in this paper through experiments.

4   Experiments and Dataset

We conducted comparative experiments on the IEMOCAP dataset and verified it in real scenarios on the Mental Illness Chinese dataset. For all experiments, we use an NVIDIA RTX 3090 GPU, the batch size is set to 512 to prevent the mode from overfitting, the learning rate is set to decay according to the number of epochs in the range [0.01,0.0003] and the weight decay range is set to [0.01, 0.0015]. The experiments compare some of the more advanced baseline codes. The following subsections introduce the dataset we used and some experimental results.

4.1   Dataset Introduction

IEMOCAP: The Interactive Emotional Dyadic Motion Capture (IEMOCAP) Database. The IEMOCAP dataset consists of 151 recorded conversation videos with 2 speakers per session, and a total of 302 videos across the entire dataset. The dataset is recorded in 5 sessions and 5 pairs of speakers. We selected the following five types of pure audio data for experiments, including angry, excited, happy, neutral and sad, and combined the happy and excited data sets as the happy data set to form a four-category data set, namely angry, happy, Sad and neutral, a total of 5470 samples, as shown in Table 2.

Table 2. Statistics of IEMOCAP Dataset

Dataset	Ang	Hap	Neu	Sad
IEMOCAP	1100	1614	1684	1072

4.2 Evaluation Standard

The evaluation metrics we use are weighted accuracy(WA) [19], unweighted accuracy(UA) [19] and F1 score [17].

$$WA = \sum_{j=1}^{C} p_j * a_j$$

$$p_j = \frac{N_j}{\sum_{j=1}^{C} N_j} \tag{9}$$

$$UA = \frac{1}{C} \sum_{j=1}^{C} a_j \tag{10}$$

$$F1 = \frac{2 * UA * R}{UA + R}$$

$$R = \frac{T_P}{T_P + F_N} * 100\% \tag{11}$$

Among them, C is the number of categories, p_j is the weight of category j, a_j is the accuracy rate of category j, N_j is the number of category j, R is the recall rate, T_P is the number of correct judgments as positive examples, F_N is the error judgment as the number of negative examples.

4.3 Compared Baselines

This section compares our method with some of the more advanced baselines.

GCN-Based SER [1]: The speech signal is modeled as a cyclic graph or a line graph, a graph convolutional network (GCN) based architecture is constructed, and the model is demonstrated to outperform standard GCN and other related deep graph architectures.

MWA-SER [14]: A new multi-window enhancement method is proposed to extract more audio features from speech signals by adopting multiple window sizes in the audio feature extraction process. Experiments show that its minimal extraction feature enhancement method combined with a deep learning model improves the performance of speech emotion recognition.

ECFW [10]: This study proposes an emotional-category-based feature weighting (ECFW) method, which aims to find the prominence of each feature under different emotions and apply this prominence as a priori knowledge. Furthermore, This study argues that different combinations of models and features result in large differences in the performance of SER. Features must be modeled with appropriate approaches to extract the most valuable information for emotional representation.

BLSTM-DSA [9]: Bidirectional Long Short-Term Memory with Directed Self-Attention (BLSTM-DSA) is proposed to mine the correlation of signals in audio, thereby increasing the diversity of information.

SpeechFormer [3]: It is used in cognitive speech signal processing, applying four stages to learn representations according to the natural structure of speech, and modeling the speech signal in conjunction with the statistical characteristics of speech.

4.4 Experimental Results

This section compares the experimental results of the above baseline models, as shown in Table 3. We compare five baseline models of IEMOCAP, as shown in Table 3, all of which are 4 classification models, consistent with our model. we achieved state-of-the-art results on the IEMOCAP dataset. On the IEMOCAP dataset, the unweighted accuracy and weighted accuracy have been significantly improved, achieving 5.32% and 4.74% improvement. Experiments demonstrate that our proposed multi-granularity feature extraction method and channel audio encoder architecture are effective. The confusion matrix and receiver operating curve are shown in Fig. 4 and Fig. 5. In the ROC curve, category 0 is the "angry" label, category 1 is the "happy" label, category 2 is the "neutral" label, and category 3 is the "sad" label. Figure 4 confusion matrix shows that the model proposed in this paper has good four-category performance. The ROC curve

Table 3. Overall performance of IEMOCAP dataset

Method	UA(%)	WA(%)
GCN-based SER(2021 ICASSP)	62.27	65.29
MWA-SER(2022)	65.0	–
ECFW(2021 INS)	–	60.80
BLSTM-DSA(2021 ESWA)	55.21	62.16
SpeechFormer-S(2022 INTERSPEECH)	64.5	62.9
AudioFormer(Our Method)	**69.74**	**70.61**

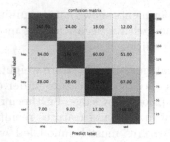
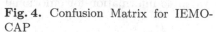

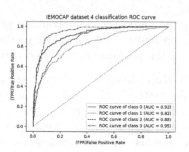

Fig. 4. Confusion Matrix for IEMO-CAP

Fig. 5. ROC Curve

in Fig. 5 shows that the classification performance of the "neutral" label is the best, and the area under the ROC curve reaches 0.95. Although the classification performance of the "happy" label is not as good as the other three categories, the AUC also reaches 0.82. AUC is an indicator used to measure the effect of the model. Usually, the value range of AUC is between 0.5 and 1. The larger the AUC, the better the classification effect of the model. The calculation method of AUC is as follows.

$$AUC = \frac{\sum_{i \in posCls} rank_i - \frac{M*(M+1)}{2}}{M*N} \tag{12}$$

Among them, $posCls$ is the positive case category,$rank_i$ is the sample sequence number of the rank position in ascending order according to the category probability, M is the number of positive samples, and N is the number of negative samples.

4.5 Real Scene Verification

The Speech Emotion Recognition task lacks practical applications in existing industrial scenarios. We delve into real scenarios and combine industry applications to apply our proposed model in the direction of psychologically assisted diagnosis. The Mental Illness Chinese dataset was obtained by collecting real patient voice samples from the industry, and the dataset was labeled by professional psychologists. Each patient reads the same article without significant affective bias and each sample is a piece of audio. The emotion category labels contain the following 3 categories: anxiety, depression and neutral. The statistical table of MIC dataset is shown in Table 4. The unweighted accuracy and F1 scores were 88.89% and 89.37, respectively, in the 3-classification experiments conducted on the Mental Illness Chinese dataset. the experimental results showed that our proposed model has a reasonable structure, sufficient theoretical basis and exists practical application value, which provides a reference basis for future related research.

Table 4. Statistics of MIC Chinese Dataset

Dataset	Anxiety	Depression	Neutral
MIC	2106	2106	2028

5 Ablation Experiments

Ablation experiments are commonly used methods in machine learning to help researchers verify the effects of the structure of each part of the overall network. In existing research, especially in complex Deep Neural Networks, ablation experiments are crucial to the study of deep learning. In this chapter, we analyze the effectiveness of our proposed Channel Audio Encoder based on Multi-granularity features through ablation experiments.

5.1 Channel Audio Encoder Ablation Experiment

The effectiveness of the Channel Audio Encoder is demonstrated by ablation experiments, as shown in Table 5. where L is a Unidirectional LSTM network, G is a Unidirectional GRU network, BL is a Bidirectional LSTM network, and BG is a Bidirectional GRU network. The ablation experiments demonstrate that our proposed Channel Audio Encoder achieves optimal performance in WA, UA and F1 scores, which indicates that the Channel Audio Encoder can closely link contextual information and learn global information more comprehensively for feature representation.

Table 5. Channel Audio Encoder Ablation Experiment

Method	WA(%)	UA(%)	F1-score
FE_L	66.73	67.41	66.81
FE_G	67.55	67.88	67.73
FE_{BL}	66.73	67.88	66.71
FE_{BG}	68.46	69.02	68.55
FE_{L+G}	68.01	69.09	68.02
FE_{BL+BG}	**69.74**	**70.61**	**69.91**

5.2 Multi-granularity Features Ablation Experiment

In this section, validity analysis of different dimensional features is performed in the same experimental setting. On the premise that the total number of samples remains unchanged, five multi-granularity feature modules are divided by the number of feature categories, as shown in Table 6, in which module A contains Zero Crossing Rate, Rmse Feature, mean MFCC, Spectral Flatness, Magnitude Feature, and Mel Spectrum features. module B adds Pitch Feature and Chroma Feature. module C adds Spectral Centroid and Spectral Contrast on the basis of module B. Module D is based on module C and adds the Constant-Q Transformation. Module E adds the Constant-Q Transform and the most valuable features of the Mel Frequency Cepstrum Coefficient on the basis of module D. Module A contains fewer data features and insufficient data feature representation, resulting in the worst experimental results. Module E has more data features, but the experimental results are poor due to feature redundancy. Module D, our proposed Multi-granularity feature extraction method, has the best experimental results.

Table 6. Multi-granularity Features Ablation Experiment

Module	Dimension	WA(%)	UA(%)	F1-score
A	192	66.08	67.0	66.25
B	208	67.36	68.12	67.47
C	219	68.0	68.83	68.14
D	224	**69.74**	**70.61**	**69.91**
E	286	67.64	68.24	67.73

6   Discussion

The traditional neural network can only pass the output of the neuron in the upper layer to the neuron in the next layer, so the traditional neural network cannot compute the time series data effectively, while the RNN is a recurrent neural network model in the form of chain connection. It can pass the output of the neuron to itself and at the same time output the hidden state to the next neuron in the current layer, which makes Recurrent Neural Networks have certain advantages in time series data processing. Unidirectional networks can only mine time series data information from front to back and retain past information, which may lead to a weaker learning ability of neural network models. Bidirectional networks consist of forward and backward networks, which can explore data information in both directions, better connect contextual information and improve the learning ability of neural network models. In this section, we analyze our proposed Channel Audio Encoder based on Multi-Granularity features based on ablation experiments.

Table 5 shows the experimental results of ablation for different methods. The experimental results of FE_{BL}, a feature encoder based on bidirectional LSTM, are slightly better than FE_L, a feature encoder based on unidirectional LSTM, in terms of unweighted accuracy, which indicates that although bidirectional LSTM increases the information perception of the inverse sequences, the model learning performance decreases due to the effect of the growth of the number of parameters brought by the inverse sequences, which is also corroborated by the literature [29]. The GRU network has the advantage of less number of parameters and faster model convergence and the feature encoder FE_{BG} based on bidirectional GRU is significantly better than the feature encoder FE_G based on GRU, which indicates that the bidirectional GRU network makes full use of the input of semantic information of the reverse sequence to increase the learning capability of the model while ensuring the number of parameters and the convergence performance of the model. The weighted and unweighted accuracies of FE_{L+G} using the LSTM and GRU based feature encoder obtained 1.28% and 1.68% improvement over FE_L as well as FE_G, which indicates that the method differs from the traditional sequential network structure by generating two different High-order feature matrices through the dual-channel feature encoder structure, which can compensate for the sequential network's data characterization deficiencies. The weighted and unweighted accuracies of the two-channel feature encoder FE_{BL+BG} based on bidirectional LSTM and bidirectional GRU are improved by 1.73% and 1.52% over FE_{L+G}, which indicates that the addition of inverse-order semantic information makes different models perceive the same inverse-order semantic information significantly differently, but this difference represents a positive difference, that is, different models perceive the feature embedding produces a different understanding and this understanding makes the individual sentiment classification more obvious, which leads to an improvement in model performance.

Table 6 shows the ablation experimental results of different feature modules, reflecting the differences in sensitivity between features and data. Module B adds Pitch Feature and Chroma Feature to improve the sensitivity of data features by adding features that describe the tonality of acoustic data. Module C has added Spectral Centroid and Spectral Contrast. By utilizing these two features, acoustic emotions are divided into high-frequency and low-frequency parts. The difference between high-frequency and low-frequency features is more pronounced, further enhancing the sensitivity of data features. Module D has added the Constant-Q Transform, which has the same frequency distribution as the acoustic data scale. It can directly obtain the amplitude value of the acoustic signal at the corresponding frequency, making it easier to distinguish similar notes. Module E has added the maximum value features of Constant-Q Transform and Mel Frequency Cepstrum Coefficient. Due to the inability of the maximum value features of both to represent global acoustic features, resulting in feature redundancy, the experimental effect is reduced.

7   Conclusion

In this paper, Multi-granularity feature and Channel Audio Encoder is set to solve some defects of Speech Emotion Recognition. Experiments show that our method achieves absolute improvements of 4.74% and 5.32% in WA and UA, respectively. Our approach can be used in various scenarios such as psychological diagnosis and treatment, intelligent customer service, and human-computer interaction. We plan to add micro-expression information based on the combination of speech and images to explore the impact of diseases on pronunciation and emotions.

Acknowledgements. This paper is founded by Supported projects of key R & D programs in Hebei Province(No. 21373802D) and Artificial Intelligence Collaborative Education Project of the Ministry of Education(201801003011).

References

1. Bahdanau, D., Cho, K., Bengio, Y.: Neural machine translation by jointly learning to align and translate. In: International Conference on Learning Representations (2014)
2. Busso, C., et al.: Iemocap: interactive emotional dyadic motion capture database. Lang. Resour. Eval. **42**(4), 335–359 (2008)
3. Chen, W., Xing, X., Xu, X., Pang, J., Du, L.: Speechformer: a hierarchical efficient framework incorporating the characteristics of speech (2022)
4. Cowie, R., Douglas-Cowie, E., Tsapatsoulis, N., Votsis, G., Taylor, J.: Emotion recognition in hci. Signal Process. Mag. IEEE (2001)
5. Elman, J.L.: Finding structure in time. Cogn. Sci. **14**(2), 179–211 (1990)
6. Krizhevsky, A., Sutskever, I., Hinton, G.: Imagenet classification with deep convolutional neural networks. Adv. Neural Inform. Process. Syst. **25**(2) (2012)
7. Lea, C., Flynn, M.D., Vidal, R., Reiter, A., Hager, G.D.: Temporal convolutional networks for action segmentation and detection. IEEE Computer Society (2016)
8. Lecun, Y., Bengio, Y., Hinton, G.: Deep learning. Nature **521**(7553), 436 (2015)
9. Li, D., Liu, J., Yang, Z., Sun, L., Wang, Z.: Speech emotion recognition using recurrent neural networks with directional self-attention. Expert Syst. Appl. **173**(3), 114683 (2021)
10. Li, D., Zhou, Y., Wang, Z., Gao, D.: Exploiting the potentialities of features for speech emotion recognition. Inf. Sci. **548**, 328–343 (2021)
11. Mcfee, B., Raffel, C., Liang, D., Ellis, D., Nieto, O.: librosa: audio and music signal analysis in python. In: Python in Science Conference (2015)
12. Mirsamadi, S., Barsoum, E., Zhang, C.: Automatic speech emotion recognition using recurrent neural networks with local attention. In: 2017 IEEE International Conference on Acoustics, Speech and Signal Processing (ICASSP) (2017)
13. Mnih, V., Heess, N., Graves, A., Kavukcuoglu, K.: Recurrent models of visual attention. Adv. Neural Inform. Process. Syst. **3** (2014)
14. Padi, S., Manocha, D., Sriram, R.D.: Multi-window data augmentation approach for speech emotion recognition (2020)
15. Pang, B.: Thumbs up? sentiment classification using machine learning techniques. In: Proceedings of EMNLP, Philadelphia. PA, USA, July 2002 (2002)
16. Peng, Z., Lu, Y., Pan, S., Liu, Y.: Efficient speech emotion recognition using multiscale cnn and attention (2021)
17. Powers, D.M.W.: Evaluation: from precision, recall and f-measure to roc, informedness, markedness and correlation (2020)
18. Qadri, S.A.A., Gunawan, T.S., Kartiwi, M., Mansor, H., Wani, T.M.: Speech emotion recognition using feature fusion of teo and mfcc on multilingual databases (2022)

19. Rozgi, V., Ananthakrishnan, S., Saleem, S., Kumar, R., Prasad, R.: Ensemble of svm trees for multimodal emotion recognition. In: Signal & Information Processing Association Summit & Conference (2012)
20. Sahu, G.: Multimodal speech emotion recognition and ambiguity resolution (2019)
21. Schmid, F., Koutini, K., Widmer, G.: Low-complexity audio embedding extractors. arXiv preprint arXiv:2303.01879 (2023)
22. Shirian, A., Guha, T.: Compact graph architecture for speech emotion recognition (2020)
23. Sze, V., Chen, Y.H., Yang, T.J., Emer, J.S.: Efficient processing of deep neural networks: a tutorial and survey. Proceedings of the IEEE **105**(12) (2017)
24. Tripathi, S., Kumar, A., Ramesh, A., Singh, C., Yenigalla, P.: Deep learning based emotion recognition system using speech features and transcriptions (2019)
25. Vinola, C., Vimaladevi, K.: A survey on human emotion recognition approaches, databases and applications. Elect. Lett. Comput. Vis. Image Anal. **2**(14), 24–44 (2015)
26. Xu, Y., Xu, H., Zou, J.: Hgfm : a hierarchical grained and feature model for acoustic emotion recognition. In: ICASSP 2020 - 2020 IEEE International Conference on Acoustics, Speech and Signal Processing (ICASSP) (2020)
27. Yazdani, A., Shekofteh, Y.: A persian asr-based ser: modification of sharif emotional speech database and investigation of persian text corpora. arXiv preprint arXiv:2211.09956 (2022)
28. Yoon, S., Byun, S., Jung, K.: Multimodal speech emotion recognition using audio and text. In: IEEE SLT 2018 (2018)
29. Yue Xibin, Hu Xiaolin, T.L.: The influence of the number of parameters in each layer of deep learning model on performance (in chinese). Comput. Sci. Appli. (2015)
30. Zhu, W., Li, X.: Speech emotion recognition with global-aware fusion on multi-scale feature representation (2022)
31. Zou, H., Si, Y., Chen, C., Rajan, D., Chng, E.S.: Speech emotion recognition with co-attention based multi-level acoustic information (2022)

A Context Aware Lung Cancer Survival Prediction Network by Using Whole Slide Images

Xinyu Liu, Yicheng Wang, and Ye Luo$^{(\boxtimes)}$

School of Software Engineering, Tongji University, Shanghai 201804, China
{1954090,2131490,yeluo}@tongji.edu.cn

Abstract. Lung cancer has caused enormous harm to human life and traditional whole slide image (WSI) based lung cancer survival prediction methods suffer from information loss and can not maintain the spatial context of the images, which may play the important roles into survival analysis. Meanwhile, the impact of the heterogeneity between the medical images and the natural images has been noticed for some pre-trained models on medical image representation learning. In this paper, we proposed a Context Aware Lung Cancer Survival Prediction Network (CA-SurvNet) by using the whole slide images, in which the survival prediction is decided by every patch of a WSI and its associated spatial context as well. Specifically, the representation of every WSI patch is first learned via a self-supervised learning based feature extractor, and then are sequentially concatenated followed by a channel-wisely dimensional reduction to preserve the significant information and maintain the spatial structure of the WSI simultaneously. Extensive experiments on two large benchmark datasets validate the superiority of the proposed method to its state-of-the-art competitors, and also its effectiveness of the WSI context preserving into the lung cancer survival prediction.

Keywords: Whole Slide Image · Lung Cancer Survival Prediction · Spatial Context · Contrastive Learning

1 Introduction

Lung cancer is a common malignant tumor and a leading cause of cancer-related deaths. Approximately 2.2 million people worldwide are diagnosed with lung cancer each year, resulting in 1.8 million deaths, accounting for 18% of all cancer-related deaths [1]. Survival prediction models are evaluation tools established using data analysis and machine learning technologies, which can assist doctors in providing individualized risk assessment and prognosis stratification for patients, and develop personalized diagnosis, treatment, and management plans [2]. Meanwhile, as the rapid development of computer technologies, the Whole Slide Pathological Images (WSI) [3] are able to be digitized and become one of the important ways for computer-aided disease diagnosis and survival prediction.

However, the characteristics of the pathological images including the large resolution and the heterogeneity to natural images are not well addressed and the performance of most WSI based survival prediction models are restricted. Traditional WSI based survival prediction analysis methods either manually labeled the region of interests (RoIs) or directly cropped the whole images into patches, and then performed the analysis on the selected regions or patches [4]. However, due to the extremely high resolution of WSI, manually labeling the RoIs is time-consuming and laborious [5]. Meanwhile, for the WSI patch based methods, some patch selection or filtering methods (e.g. random patch sampling [6], patch clustering [7], etc.) are first employed such that the number of WSI patches can be reduced into the reasonable level and important image patches can be selected to represent the entire WSI for survival prediction analysis [8]. However, approaches of this type suffer from information losses, as some patches containing critical survival prediction information may be overlooked and the spatial structure of the WSI is totally neglected, which could significantly impact the network's performance. Last but not least, WSIs significantly differ from natural images, and existing pre-trained feature extractors (usually pretrained on ImageNet dataset) are not appropriate to be directly used for WSI feature extraction.

In order to address the aforementioned issues, we propose a context aware lung cancer survival prediction network (CA-SurvNet) by using whole slide images. Specifically, we employ the self-supervised learning to train a specified WSI feature extractor such that the heterogeneity problem of the pre-trained feature extractor on the natural image to the WSIs is well addressed. Furthermore, instead of randomly selecting some important WSI patches or clustering the patches into sub-types, the proposed CA-SurvNet performed the survival analysis based on all the WSI patches and their spatial context. In other words, all the features of every WSI patch are first extracted by the pre-trained WSI_Encoder, and then concatenated horizontally or vertically. The following re-extraction module consists of multiple convolution and max-pooling layers and are used for further feature dimension reduction. By this way, the spatial context encoded by the order of patch arrangement can be maintained and the important information of each patch is kept as well. Later, the re-extracted feature is fed into the survival prediction module to make the final lung cancer survival prediction. At last, a multi-scaled CA-SurvNet (i.e. msCA-SurvNet) is explored to make the full use of the multi-scale WSIs on the survival prediction. Extensive experiments on two public datasets show the superiority of the proposed CA-SurvNet on the lung cancer survival prediction as well as the effectiveness of the spatial context maintained.

Our main contributions are as follows:

1. We propose a context aware lung cancer survival prediction network (CA-SurvNet) by using all the WSI patches and their inherited spatial context simultaneously.

2. A multi-scale version of the proposed CA-SurvNet is proposed, and better prediction result is obtained by fusing multi-scale WSI patches at feature level.

3. Extensive experiments on two public datasets show the superiority of our method to existing WSI based survival prediction methods, and also validate the effectiveness of the WSI context on lung cancer survival prediction.

2 Related Work

Due to limitations in computing capabilities, many histopathology studies have employed a two-stage patch-based workflow. This involves training a patch-level CNN using patches extracted from a WSI, followed by training a slide-level algorithm on the features extracted by the patch-level model to provide the final diagnosis. While these patch-based methods have been successful in identifying cancer [9], classifying cancer types [10], detecting cancer metastasis [11], and analyzing prognosis [12]. However, they often require significant annotation from experienced pathologists. Yao et al. [8]. divided each WSI into patches and treated the feature vectors of all image patches from the same patient as one bag, and used an attention mechanism to calculate the importance weight of each image patch for survival prediction. Li et al. [7] proposed a deep multiple-instance learning network with attention mechanism (AGMI-Net) for WSI analysis. The network consists of two main components: a feature extractor and an attention-guided multiple-instance aggregator. Their method is capable of effectively handling WSI with different sizes and resolutions. However, both of the aforementioned methods only selected a subset of patch clusters as inputs, and cannot obtain global feature information. Parker et al. [13] employed an unsupervised encoder to compress four types of data modalities, including clinical, genomic, and WSIs, into a patient feature vector. They then used a survival analysis network to predict the patient's survival time. However, this method incorporate data modalities beyond WSI. Chen et al. [14] proposed a multimodal co-attention transformer (MCAT) to predict survival outcomes from WSIs. Fan et al. [6] utilized a self-supervised learning (SSL) framework to predict cancer survival outcomes from WSIs. Zhu et al. [4] presented a whole slide tissue pathology image survival analysis framework (WSISA) for predicting survival outcomes from WSIs. However, due to the lack of guarantee for the orderedness of patch features, the aforementioned three methods may encounter challenges in utilizing spatial or contextual information in WSI, and may require further integration of other types of features or models to enhance their predictive accuracy and interpretability.

3 The Proposed CA-SurvNet

The overall structure of the proposed multi-scale CA-SurvNet (msCA-SurvNet) is illustrated in Fig. 1. For a given WSI, it is evenly cropped into patches at two magnification ratios, where P_1 and P_2 are the quantities of patches. For every

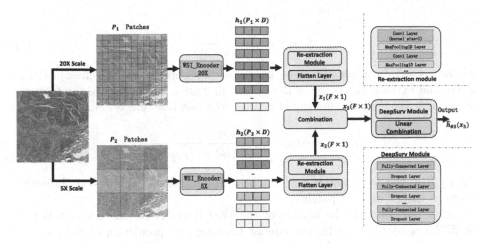

Fig. 1. Illustration the main idea of our proposed multi-scaled CA-SurvNet.

patch at a given magnification ratio, its representation is obtained by passing the WSI_Encoder module which is pre-trained via the self-supervised learning, where D is the dimensionality of the features obtained from a patch. Then, the patch representations (i.e. the feature vectors) are horizontally concatenated to be a feature matrix (i.e. H_1 and H_2) as the input of the Re-extraction Module followed by the Flatten Layer. The multi-scaled re-extracted features (i.e. x_1 and x_2) are then fused via a weighted linear combination to obtain x_3. The final prediction result y can be obtained by feeding the fused feature into the Survival Prediction Module. Detailed introductions of each module for the single scaled version of the CA-SurvNet is provided in the following Sections.

3.1 Pre-training the WSI_Encoder

In clinical practice, WSI often has high spatial resolution and the large difference between natural images, hence it is difficult but necessary to learn a WSI specified feature extractor. However, due to the high resolution of WSI, it is difficult to label it pixel-wisely hence directly learning the WSI feature extractor via the fully supervised learning is not easy either. Therefore, the self-supervised learning methods can be used for feature extraction training without the labeling issues. Inspired by this, we follow the framework of SimCLR [15] to pre-train a WSI feature extractor by using a large amount of unlabeled WSI patches. Specifically, given a WSI patch, the goal is to make similar images (i.e. the different versions of the same image patch via the data augmentation) closer in the feature space, while dissimilar images (two different image patches) are farther away. The enhanced images $W_{i,1}$ and $W_{i,2}$ will be encoded by the WSI_Encoder to obtain the image representations (i.e. $h_{i,1}$ and $h_{i,2}$). These representation vectors will then be further mapped to a latent space through a projection operation, resulting in new feature representations z_{i*2} and z_{i*2+1}. The cosine similarity [16] is used to define the similarity between image patchs as:

$$s_{i,j} = \frac{z_i^\top z_j}{\tau||z_i||||z_j||}, \tag{1}$$

where τ is an adjustable parameter known as the temperature, which adjusts the input and extends the range of the cosine similarity. z_i and z_j are the features of two images. For two augmented images, the loss function is defined as:

$$l(i,j) = -log\frac{exp(s_{i,j})}{\sum_0^{2P-1} 1[k! = i]exp(s_{i,k})}. \tag{2}$$

where P is the total number of patches. $s_{i,j}$ represents the cosine similarity between the ith and j^{th} patches.

After the training, the weights of the WSI_Encoder are saved and this pre-tr WSI_Encoder is used for subsequent training and prediction of the survival analysis models. It is worthy mentioning that, to distinguish the feature extractor pre-trained at different magnification ratio of WSIs, we use WSI_Encoder_5X, WSI_Encoder_20X to represent different feature extractors.

3.2 The Network Architecture of CA-SurvNet

Although the pre-trained WSI_Encoder can effectively extract the features from the WSI patches, there are still enormous patches remained (with 3000–10000 patches on average for each WSI at a 20× magnification ratio divided into patches), and it is still intractable for an end-to-end trained network to do the survival analysis. Moreover, previous methods by randomly sampling or clustering WSI patches to make the number of WSI patches tractable strictly restricted their performance due to some important WSI patches missing or the spatial structure of the WSI destroyed. In this paper, we propose a context aware lung cancer survival prediction network by using all patches of a whole slide images and maintaining the spatial structure of these patches simultaneously. The network architecture of the proposed CA-SurNet primarily consists of two main components, namely the Re-extraction Module and the Survival Prediction Module. The Re-extraction Module consists of multiple convolutional layers and max pooling layers. After the Re-extraction module, the Survival Prediction Module is employed to perform the final survival prediction task. The following content provides the detailed introductions to each module of the CA-SurvNet.

Re-extraction Module. The main aim of the Re-extraction Module is to further compress the features extracted by the WSI_Encoder. The re-extraction module consists of four pairs of the convolutional layer and a max pooling layer. Specifically, given the learned patch feature $h \in R^{D_0}$ (D_0 is the feature dimension.), for a patient, we pack the patches horizontally (from left to right) to be a feature matrix $H_0 \in R^{P_0 \times D_0}$, where P_0 is the number of WSI patches for a patient. When a patient has more than one WSI, these images are treated as distinct samples. When passing the first convolutional layer with the vertical depth of 1. That's $H_1 = Conv1D(H_0)$. The feature of each patch can be further compressed as $H_1 \in R^{P_0 \times D_1}$. A max pooling layer is followed after this convolutional layer. Thus, $H_1' = MaxPool(H_1)$ and $H_1' \in R^{P_0 \times D_1'}$. Here, D_1' is smaller than D_1 and D_1 is smaller than D_0. For the subsequent three convolutional layers, the depth is adjusted to be greater than 1. That's 3, 2, and 2, respectively. Thus, $H_2 = Conv2D_3(H_1')$ with depth of 3, $H_{i+1} = Conv2D_2(H_i')$, $i = 2, 3$ with depth of 2, and $H_i' = MaxPool(H_i)$, $i = 2, 3, 4$. The data flow of the Re-extraction Module is illustrated in Fig. 2.

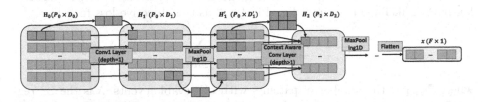

Fig. 2. Data flow of the proposed Re-extraction Module of our CA-SurvNet.

By setting the depth larger than 1, we can obtain a comprehensive feature by focusing on more than one patch feature (e.g. two patches as shown in Fig. 2.) and assign different weights to each patch feature. Moreover, the max pooling layers added between a pair of convolution layers can further compress the feature. Finally, a Flatten operation is performed on the last layer of the Re-extraction module. As each row of patch features is ordered, flattening into a one-dimensional feature representation preserves the spatial context of each image patch to some extent.

Survival Prediction Module. The second half of the CA-SurvNet is a feed-forward neural network used to predict the impact of patient covariates on hazard rates. The network architecture consists of four pairs of fully connected layers and dropout layers followed by a linear combination layer, and it is similar to DeepSurv [17]. The input of this module is the flattened feature output by the re-extraction module as x'. The network's output $\hat{h}_\theta(x')$ is a single node with linear activation used to estimate the logarithmic hazard function in the Cox proportional-hazards model [18].

The Cox Loss. To train the network, we adopt the traditional cox loss as the objective function. The Cox model estimates the logarithmic hazard function $h(x)$ using the linear function $\hat{h}_\theta(x) = \theta^T x$. To perform Cox regression, the weights θ must be adjusted to optimize the Cox partial likelihood. The partial likelihood is the product of probabilities for each event time t_i, which is the probability that individual i experiences the event given a set of individuals still at risk at time t_i. The Cox partial likelihood is parameterized by θ as:

$$L_c(\theta) = \prod_{i:e_i=1} \frac{exp(\hat{h}_\theta(x_i))}{\sum_{j \in R(t_i)} exp(\hat{h}_\theta(x_j))} \tag{3}$$

where t_i, e_i, and x_i are the values of the event time, event indicator, and baseline data, respectively, for the ith observation. This product is defined for the set of patients with observable events $e_i = 1$. The risk set $R(t_i)$ is the set of patients still at risk of dying at time t. Consequently, the regularization and average negative log partial likelihood are introduced in Eq. 3 as the objective loss function:

$$l(\theta) = -\frac{1}{N_{e=1}} \sum_{i:e_i=1} (\hat{h}_\theta(x_i) - log \sum_{j \in R(t_i)} e^{\hat{h}_\theta(x_j)}) + \lambda ||\theta||_2^2 \tag{4}$$

where $N_{e=1}$ is the number of patients with observable events, λ is the L2 regularization parameter, and gradient descent optimization is used to find the minimum network weights.

The Spatial Context Encoding. As mentioned before, all the WSI patches are horizontally packed together for the survival analysis. It is also reasonably to reorder the patches vertically. That's to pack the patches from top to bottom iteratively. It is worthy mentioning that the horizontal arrangement was considered the primary experimental setting, while the vertical arrangement was designed only for the ablation experiment purpose.

The Multi-scaled CA-SurvNet. In order to explore the effect of the multi-scaled WSI patches, we introduce the multi-scaled CA-SurvNet by the two shared-weights CA-SurvNets with dual-stream structure, named msCA-SurvNet. By passing two pre-trained WSI_Encoders, the patch features h_1 and h_2 with magnification ratios of $S_1 \times$ and $S_2 \times$ are obtained, respectively. Then h_1 and h_2 are feed into the Re-extraction modules to extract features x_1 and x_2 respectively. Since the number of nodes in the Flatten Layer is the same, the dimensions of x_1 and x_2 are the same. That makes the fusion of the multi-scaled WSI at feature level possible. Considering that, features magnified at large ratio may potentially retain more detailed information thus play more important roles, x_1 and x_2 can be fused via a weighted linear combination to be a multi-scale fused feature x_3. Thus, the weight can be set to be proportional to the area of the WSI, which is related to the square of the magnification ratio. The specific multi-scaled WSI feature fusion can be written as follows:

$$x_3 = \frac{S_1^2}{S_1^2 + S_2^2} x_1 + \frac{S_2^2}{S_1^2 + S_2^2} x_2 \tag{5}$$

where S_1 and S_2 represents the WSI magnification ratios corresponding to two different features respectively.

4 Experiment and Analysis

4.1 Dataset and Experiment Setup

Dataset. The proposed CA-SurvNet is validated on the TCGA-LUSC dataset [19] and NLST dataset [20]. The TCGA-LUSC dataset is a component of The Cancer Genome Atlas (TCGA) initiative, which aims to elucidate the genomic attributes and clinical information pertaining to lung squamous cell carcinoma (LUSC). The dataset comprises 1612 WSIs from 504 patients, 1569 samples of mRNA expression data specific to LUSC, and corresponding clinical information. The National Lung Screening Trial (NLST) is a lung cancer dataset collected by the National Cancer Institute's Division of Cancer Prevention (DCP) and Division of Cancer Treatment and Diagnosis (DCTD). It consists of 1104 WSIs from 404 patients. In this paper, only the WSI and corresponding survival clinical information is used. To ensure that different WSIs have the same number of patches, we resized the images and extracted 5000 patches for experimentation at the magnification ratio of 20x.

Experimental Setting. (1) To pre-train the WSI_Encoder, we employed resnet18 as the backbone. The training batch size was set to 4096 with a total of 100 epochs. The weight decay coefficient was set to 10e-6. The prediction head was used to output the feature dimensions of 256, while the encoder was set to output a feature dimension of 512. The size of the patch is fixed to 224×224. Magnification ratios of $5\times$ and $20\times$ were selected for the patch, respectively. (2) In the survival prediction experiment of CA-SurvNet, in order to optimize the model training, some common neural network training techniques are employed, including input standardization, the use of Scaled Exponential Linear Units (SELU) [21] as the activation function, the application of Adaptive Moment Estimation (Adam) [22] for gradient descent, the use of Nesterov momentum optimizer, and learning rate scheduling. A random hyperparameter optimization search was performed to adjust the hyperparameters of the network. The learning rate was set to 7.7e-4. The regularization parameter was set to 1.994. The parameter for dropout regularization was set to 0.147, and the learning rate decay was set to 6.494e-4.

Evaluation Metric. We utilized the consistency index C-index [23] as the evaluation metric for our model. The C-index, also known as the Concordance Index or Concordance Probability Estimate, is a common measure of the predictive ability of survival models. It is defined as the ratio of correctly predicted relative order of two events among all ordered pairs of data as:

$$C = \frac{\sum_{i<j}^{n}[t_i < t_j][y_i > y_j] + \frac{1}{2}\sum_{i<j}^{n}[t_i = t_j]}{\sum_{i<j}^{n}[t_i < t_j]} \tag{6}$$

where n represents the total number of patients. t_i and t_j represent the survival event times of the i-th and j-th samples, respectively, and y_i and y_j represent the corresponding survival function estimate of the model. The square brackets [·] denote the indicator function that takes a value of 1 or 0.

4.2 Comparison with State-of-the-Arts

To investigate the effectiveness of the msCA-SurvNet method proposed in this paper for survival prediction, We provide the five-fold cross-validation results to show the averaged results of the proposed multi-scale CA-SurvNet (msCA-SurvNet) in Table 1.

Table 1. The five-fold cross-validation results of the proposed msCA-SurvNet on the training and the testing datasets respectively. The bolded entries in the table represent the final obtained average C-index values.

Folds	C-index of Training Set		C-index of Testing Set	
	TCGA-LUSC	NLST	TCGA-LUSC	NLST
1	0.845	0.863	0.742	0.719
2	0.751	0.876	0.693	0.746
3	0.831	0.826	0.730	0.758
4	0.910	0.824	0.733	0.722
5	0.909	0.861	0.752	0.744
Avg±std	**0.849 ± 0.0621**	**0.854 ± 0.0221**	**0.729 ± 0.0249**	**0.738 ± 0.0149**

Moreover, the proposed msCA-SurvNet are also compared with four related methods, DL-MRPP [13], SSL [6], HANet [24], CapSurv [25],and WSISA [4]. These four methods also focus on survival prediction based on WSI images and have been tested on the TCGA-LUSC dataset in their respective papers. The results are presented in Table 2. The evaluation results indicate that, among all models, msCA-SurvNet exhibits superior performance on both TCGA-LUSC dataset and NLST dataset, with C-index metrics of 0.729 and 0.738, respectively. The main reasons may lie into the complete WSI patches used, the the spatial context information encoded, and the multi-scale WSIs contained. In addition, one branch of our CA-SurvNet model has approximately 70M parameters. For methods using msCA-SurvNet, the size of the model's parameter depends on the number of branches. Therefore, it is not a lightweight model and requires a server for reliable and accurate detection.

Table 2. Performance comparisons on TCGA-LUSC dataset and NLST dataset. Results from other methods are directly copied from the corresponding papers. The bolded entries in the table represent the best results.

Method	DL-MRPP [13]	SSL [6]	HANet [24]	CapSurv [25]	WSISA [4]	msCA-SurvNet
LUSC	0.670	0.679	0.668	0.673	0.638	**0.729**
NLST	0.723	0.711	0.734	0.730	0.703	**0.738**

4.3 Ablation Study

Effect of Spatial Context Encoding. In order to verify the importance of preserving the patch order or the spatial context in our method, an ablation experiment by randomly shuffling the patches was conducted to validate the effectiveness of the spatial context on the lung cancer survival prediction. Specifically, to ensure the comparability of the experiments, the same dataset and the same features were used. The main difference is that the patch features of each WSI in the counterpart experiments are packed with randomly shuffling or in a pre-defined order (e.g. horizontally, vertically, etc.). Here we use msCA-SurvNet for training in different conditions. Furthermore, to investigate the impact of both the horizontal and vertical arrangement methods proposed in this paper, we incorporated them into our experiments as well. The experimental results are shown in the first row of Table 3. Two methods for non-shuffling patch features achieve C-index scores of 0.729 and 0.726, respectively, surpassing the method utilizing shuffling patch features, which yields a score of 0.645. This indicates that preserving contour information also helps the performance of the model. The network can learn the associated information between adjacent patch blocks, which allows the WSI image to be processed as a complete entity rather than individual sampled image blocks. The results of this experiment demonstrate that the ordered patch encoding the spatial context did help to predict the survival rate. And the performance of the two ordered arrangement methods is relatively similar, indicating that both the horizontal and the vertical arrangements can effectively utilize context awareness to improve the model performance.

Effect of Different Multi-scale Fusion Strategy. To validate the effect of multi-scale WSI feature fusion strategy, we conduct the experiment by directly fusing the prediction results at different scales, and we name this fusion strategy as the decision-level-fusion. In other words, the predicted results of the CA-SurvNet model at 20× and 5× magnification ratios are first obtained and then directly fused. Specifically, the predicted results R_1 and R_2 are added together in a certain proportion to obtain R_3. The choice of the proportion is also related to the magnification ratio of the patch, and the specific formula for the combination is similar to Eq. 5 as: $R_3 = \frac{R_1 S_1^2}{S_1^2 + S_2^2} + \frac{R_2 S_2^2}{S_1^2 + S_2^2}$, where S_1 and S_2 represent the magnification ratios corresponding to two features respectively. We performed five-fold cross-validation to test this fusion strategy, and the results are presented in the second row of Table 3. From Table 3, we can see that feature fusion strategy performs better than fusion at decision, which may be due to the fact that feature fusion in the model can obtain image features of different scales more comprehensively, and the weight-sharing network can train the network completely. Thus, compared with simple fusion methods, it can more effectively utilize data features.

Table 3. Performance comparisons of different strategies used in our method. The bolded entries in the table represent the best results.

Spatial Context Encoding Strategies	Horizontally	Vertically	Randomly
	0.729 ± 0.0249	0.726 + 0.0301	0.645 ± 0.0475
Multi-scale Fusion Strategies	Feature-level-fusion	Decision-level-fusion	
	0.729 ± 0.0249	0.727 ± 0.0309	
Multi-sacle v.s. Single-scale	msCA-SurvNet	CA-SurvNet_20×	CA-SurvNet_5×
	0.729 ± 0.0249	0.717 ± 0.0319	0.684 ± 0.0193
w/ or w/o the Re-extraction Module	CA-SurvNet (w/)	DeepSurv (w/o)	
	0.717 ± 0.0319	0.624 ± 0.0398	

Multi-scale vs Single-scale. To show the effectiveness of our proposed mutliscaled WSI fusion approach, we conduct the experiments by training the proposed CA-SurvNet based on a single ratio of 20× and 5× WSI patches, respectively. Specifically, WSI_Encoder_20x and WSI_Encoder_5x were used to extract features at 20× and 5× magnification ratios, which were then input into our CA-SurvNet for training and prediction. The results of the experiments are presented in the third row of Table 3, which shows that the performance of the 20× magnification ratio is superior to that of the 5× one, with corresponding C-index metrics of 0.717 and 0.684, respectively. This may be due to the fact that the feature maps at higher magnification ratio have higher spatial resolution, allowing for the learning of more detailed features such as color and texture, while lower magnification ratio may result in some loss of information. Furthermore, the multi-scaled version (i.e. msCA-SurvNet) outperforms the single-scaled ones, with corresponding C-index metrics of 0.729 and 0.727, respectively. This indicates the multi-scale approach can learn more useful features for better results.

Effect of Re-extraction Module. In order to verify the importance of the Re-extraction module on the experimental results, an ablation experiment was conducted to explore whether the experimental results obtained with the presence of this module are more accurate and reliable. We compare our singlescaled CA-SurvNet with DeepSurv model, which has similar survival prediction pipeline as ours except for the feature ex-extraction part. Specifically, in order to ensure the comparability of the experiments, the same features extracted by WSI_Encoder is used as the input of the models. All WSI patches magnified at 20× ratio are used for comparisons. The experimental results are shown in the fourth row of Table 3. The performance of CA-SurvNet_20× is higher than that of the DeepSurv model, with C-index metrics of 0.717 and 0.624, respectively. This indicates that the Re-extraction module can extract features completely and improve the performance of the model to some extent.

5   Conclusion

In this paper, we propose a context aware lung cancer survival prediction network (CA-SurvNet) by using all the patches of a whole slide image and maintaining the spatial structure of these patches simultaneously. By fusing the multi-scaled WSIs into the proposed CA-SurvNet, more promising results are obtained onto the lung cancer survival prediction task. Additionally, we compare this method with other existing WSI survival prediction methods and design a series of ablation experiments to demonstrate the effectiveness of the model. Experimental results validate that our multi-scaled CA-SurvNet exhibits superior results by encoding the spatial context with all WSI patches on TCGA-LUSC dataset and NLST dataset.

Acknowledgements. This work was partially supported by the General Program of National Natural Science Foundation of China under Grant 62276189, and the Fundamental Research Funds for the Central Universities No. 22120220583.

References

1. Sung, H., Ferlay, J., Siegel, R.L., et al.: Global cancer statistics 2020: Globocan estimates of incidence and mortality worldwide for 36 cancers in 185 countries. CA: Cancer J. Clin. **71**(3), 209–249 (2021)
2. Pantanowitz, L., Valenstein, P.N., Evans, A.J., et al.: Clinical Statistics: Introducing Clinical Trials, Survival Analysis, and Longitudinal Data Analysis. CRC Press, 2nd ed. edn. (2018)
3. Pantanowitz, L., Valenstein, P.N., Evans, A.J., et al.: Review of the current state of whole slide imaging in pathology. J Pathol Inform **2**, 36 (2011)
4. Zhu, X., Yao, J., Zhu, F., et al.: WSISA: making survival prediction from whole slide histopathological images. In: Proceedings of the IEEE Conference on Computer Vision and Pattern Recognition, pp. 7234–7242 (2017)
5. Li, Y., Zhang, Y., Liang, X., et al.: Risk-aware survival time prediction from whole slide pathological images using deep learning. Sci. Rep. **12**(1), 1–13 (2022)
6. Fan, L., Sowmya, A., Meijering, E., et al.: Cancer survival prediction from whole slide images with self-supervised learning and slide consistency. IEEE Trans. Med. Imaging **42**(1), 1–14 (2023)
7. Li, Y., Yao, J., Xu, Z., et al.: Whole slide images based cancer survival prediction using attention guided deep multiple instance learning networks. In: Medical Image Computing and Computer Assisted Intervention, pp. 290–298 (2018)

8. Yao, J., Zhu, X., Huang, J.: Deep multi-instance learning for survival prediction from whole slide images. In: Medical Image Computing and Computer-Assisted Intervention, pp. 505–513 (2019)
9. Chuang, W.Y., Chang, S.H., Yu, W.H., et al.: Successful identification of nasopharyngeal carcinoma in nasopharyngeal biopsies using deep learning. Cancers 12(2), 507 (2020)
10. Coudray, N., Ocampo, P.S., Sakellaropoulos, T., et al.: Classification and mutation prediction from non-small cell lung cancer histopathology images using deep learning. Nat. Med. 24(10), 1559–1567 (2018)
11. Bejnordi, B.E., Veta, M., Van Diest, P.J., et al.: Diagnostic assessment of deep learning algorithms for detection of lymph node metastases in women with breast cancer. JAMA 318(22), 2199–2210 (2017)
12. Wang, S., Chen, A., Yang, L., et al.: Comprehensive analysis of lung cancer pathology images to discover tumor shape and boundary features that predict survival outcome. Sci. Rep. 8(1), 10393 (2018)
13. Parker, S.C.J., Khan, A., Talhouk, A., et al.: Deep learning with multimodal representation for pancancer prognosis prediction. Bioinformatics 35(14), i446–i454 (2019)
14. Chen, R.J., Lu, M.Y., Weng, W.H., et al.: Multimodal co-attention transformer for survival prediction in gigapixel whole slide images. In: Proceedings of the IEEE/CVF International Conference on Computer Vision, pp. 4015–4025 (2021)
15. Chen, T., Kornblith, S., Norouzi, M., et al.: A simple framework for contrastive learning of visual representations, pp. 1597–1607 (2020)
16. Salton, G., Wong, A.: Similarity measures. The SMART Retrieval Syst.: Exper. Autom. Document Process. 1, 145–159 (1970)
17. Katzman, J.L., Shaham, U., Cloninger, A., et al.: Deepsurv: personalized treatment recommender system using a cox proportional hazards deep neural network. BMC Med. Res. Methodol. 18(1), 24 (2018)
18. Cox, D.: Regression models and life-tables. J. Roy. Stat. Soc.: Ser. B (Methodol.) 34(2), 187–220 (1972)
19. Network, T.: Comprehensive molecular characterization of human colon and rectal cancer. Nature 487(7407), 330–337 (2012)
20. Team, N.L.S.T.R.: The national lung screening trial: overview and study design. Radiology 258(1), 243–253 (2011)
21. Klambauer, G., Unterthiner, T., Mayr, A., et al.: Self-normalizing neural networks. Adv. Neural. Inf. Process. Syst. 30, 971–980 (2017)
22. Kingma, D.P., Ba, J.: Adam: A method for stochastic optimization. ArXiv Preprint ArXiv:1412.6980, pp. 1–15 (2014)
23. Harrell, F.E., Jr., Lee, K.L., Mark, D.B.: Multivariable prognostic models: issues in developing models, evaluating assumptions and adequacy, and measuring and reducing errors. Stat. Med. 15(4), 361–387 (1996)
24. Chang, J.R., Lee, C.Y., Chen, C.C., Reischl, J., Qaiser, T., Yeh, C.Y.: Hybrid aggregation network for survival analysis from whole slide histopathological images. In: de Bruijne, M., et al. (eds.) Medical Image Computing and Computer Assisted Intervention – MICCAI 2021: 24th International Conference, Strasbourg, France, September 27 – October 1, 2021, Proceedings, Part V, pp. 731–740. Springer International Publishing, Cham (2021). https://doi.org/10.1007/978-3-030-87240-3_70
25. Tang, B., Li, A., Li, B., et al.: Capsurv: capsule network for survival analysis with whole slide pathological images. IEEE Access 7, 26022–26030 (2019)

A Novel Approach for Improved Pedestrian Walking Speed Prediction: Exploiting Proximity Correlation

Xiaohe Chen[1,2], Zhiyong Tao[1(✉)], Mei Wang[2], and Yuanzhen Zhou[1]

[1] Wuhan Research Institute of Posts and Telecommunications, Wuhan, China
taozhiyong@fhxy.net.cn
[2] National Engineering Research Center for Multimedia Software, School of Computer Science, Wuhan University, Wuhan, China

Abstract. Accurately predicting pedestrian speed is crucial for analyzing pedestrian behavior and optimizing intelligent transportation systems. This paper investigates the feasibility of modeling pedestrian walking speed as a time series. Building upon previous research highlighting the spatio-temporal nearest neighbor correlation in pedestrian walking speed, we propose a deep learning method that leverages this correlation. Experimental results demonstrate the superiority of our approach over traditional methods in accurately predicting pedestrian walking speed and capturing temporal characteristics and trends. The findings of this study have significant implications for enhancing pedestrian traffic flow management, improving the pedestrian travel experience, and enhancing overall traffic safety. Future research can focus on exploring advanced time series methods and deep learning models to further enhance the accuracy and practicality of pedestrian walking speed prediction.

Keywords: Pedestrian speed prediction · Time series modeling · Deep learning · Intelligent transportation systems

1 Introduction

Short-time pedestrian walking speed prediction predicts the future walking speed in a short period (e.g., seconds or minutes) by analyzing pedestrian trajectory data and environmental information. This technology is widely used in modern urban traffic management and intelligent transportation systems to optimize traffic flow, reduce congestion, and improve traffic efficiency and safety to provide convenient, efficient, and safe travel services [3]. Therefore, the research and development of short-time pedestrian walking speed prediction technology are significant for theory and practice.

However, pedestrian walking speed is affected by multiple factors [1], including the environment [1,3,7–10,14], road conditions [4–6,11,18,20], individual differences [6,17,21] and posture changes [12,16,27,28]. These factors contribute to the complexity of walking speed variations, making accurate prediction challenging. Additionally, there are significant inter-individual differences in walking speed, and the same pedestrian may exhibit varying speeds across different time periods and scenarios, further complicating the modeling process. Furthermore, pedestrian trajectory data often contain noise and outliers, which can disrupt prediction accuracy and complicate data processing.

The current research on short-time pedestrian walking speed still has significant limitations. Traditional research methods often acquire the physiological posture characteristics of pedestrians, such as gait, step frequency, and acceleration, through wearable devices or sensors and then model their posture and walking behavior patterns for speed prediction [13, 15, 19, 22, 23, 25, 26]. Although these prediction methods have been influential in short-time pedestrian walking prediction, they still need help with the problems of complex data collection, complex data processing, environmental effects, and poor model generalization ability. Based on these problems, we consider simplifying the data collection and feature extraction process, further analyzing pedestrian walking speed data features, and combining these features for prediction.

In our prior study, we observed that pedestrian walking speed exhibits a nearest-neighbor characteristic within contiguous time intervals, meaning that the current pedestrian walking speed is influenced by the speed observed in the immediately preceding time interval. This correlation can be effectively captured using an autoregressive model, wherein the pedestrian's walking speed at the current time step is modeled as a function of its previous time steps. Hence, this study investigates the viability of treating pedestrian walking speed as a time series and employs time series prediction methods to forecast pedestrian walking speed. We propose a walking speed prediction method incorporating the nearest neighbor principle to improve the deep learning prediction model accuracy by the weighted average of walking speed in the pedestrian nearest neighbor time interval. Compared to current methods, this research approach reduces dependence on factors such as pedestrian gait, thereby decreasing algorithmic complexity and cost. By incorporating the correlation and proximity patterns of pedestrian walking speed, we further enhance the accuracy and stability of the prediction model.

The main contributions of this study are as follows:

1) We propose and validate the feasibility and rationality of modeling pedestrian walking speed as a time series.
2) We identify and validate the correlation of pedestrian walking speed within adjacent time intervals and propose a prediction method based on proximity correlation.
3) We propose an innovative method for short-term pedestrian walking speed prediction that combines time series forecasting with the proximity of pedestrian walking speed. Compared to traditional baseline methods, this approach demonstrates significant performance improvement in pedestrian walking speed prediction.

2   Related Work

2.1   Pedestrian Walking Speed Analysis and Forecast

Pedestrian walking speed analysis plays a significant role in intelligent transportation systems, pedestrian behavior research, and health assessment. Previous studies have examined various factors influencing pedestrian walking speed, including weather and climate effects in severe cold regions [14], the relationship between gender and walking speed in different age groups [21], the impact of road conditions on pedestrian walking speed in winter [7], and the characteristics affecting walking speed and average walking speed [1]. These studies have generated valuable insights with practical applications in urban planning, traffic management, and pedestrian safety.

Early pedestrian speed prediction primarily relied on statistical methods. However, recent advancements in machine learning and deep learning techniques have enabled the utilization of these methods to analyze pedestrian walking data and predict walking speed patterns. [24] investigated the impact of sampling rate on pedestrian walking speed estimation using DualCNN-LSTM. [15] introduced a novel method for online pose classification and real-time walking speed estimation using handheld devices. [25] explored the influence of gait postures on walking speed estimation using smartphones and deep learning techniques. However, these methods often require strict data quality and feature selection criteria, making them more susceptible to data noise and outliers. Consequently, these limitations can lead to unpredictable prediction outcomes, posing challenges for accurate predictions in the presence of novel or unfamiliar pedestrian behavior patterns.

2.2   Time Series Forecasting

Various models have been employed for time series forecasting, including Autoregressive Models, Moving Average Models, Autoregressive Moving Average (ARIMA), Recurrent Neural Networks (RNN), Long Short-Term Memory (LSTM), and Transformer Models. Among these models, Recurrent Neural Networks and their variants, such as LSTM and GRU, are widely used in time series data forecasting [2]. RNNs and their variants are capable of processing sequential data and capturing temporal dependencies, making them suitable for time series prediction tasks. However, these models can be complex and require extensive training and tuning time. They may also face challenges such as gradient disappearance or explosion when dealing with long sequences. To optimize model performance and improve prediction accuracy, this paper introduces the concept of nearest neighbor correlation in pedestrian walking speed sequences.

3   Methodology

3.1   Problem Definition

In the study of pedestrian walking speed prediction, we model it using a time series model description. Assuming we have a series of discrete time steps indexed by t representing the time step. We use $v(t)$ to denote the observed pedestrian walking speed at step t. We predict the future sequence of pedestrian walking speeds $Y = \{v_t, v_{t+1}, \cdots, v_{t+h-1}\}$ using the past pedestrian walking speed observed sequence $X = \{v_{t-k}, v_{t-k+1}, \cdots, v_{t-1}\}$, where k represents the length of the past time steps, and h represents the length of the future time steps.

This paper aims to develop a time series model appropriate for predicting pedestrian walking speed from the current moment t to the next h time steps based on past observations of pedestrian walking speed. Therefore, we aim to establish a model f based on the available observation data: $Y = f(X)$, where X represents the series of past pedestrian walking speed observations, Y represents the series of future pedestrian walking speeds, and f represents the prediction model.

3.2   Framework Overview

The model framework of this paper is illustrated in Fig. 1. The model comprises three modules: the time series prediction module, the weighted average prediction module, and the fusion prediction module. The time series prediction module models the pedestrian walking speed sequence and learns its contextual information. The module considers the position of each data point in the sequence and its relationship with neighboring data points. It can effectively capture the time dependencies and patterns in the sequence.

However, relying solely on pedestrian walking speed as a one-dimensional feature may lead to information loss. Therefore, the weighted average prediction module assigns weights to each data point based on their temporal proximity, with weights increasing as the time proximity grows. This mitigates the potential information loss and enhances predictive accuracy.

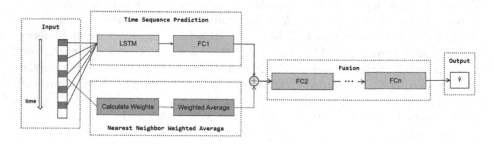

Fig. 1. Framework. The proposed model aims to enhance the accuracy and interpretability of pedestrian walking speed prediction by leveraging contextual information through the time series prediction module. Additionally, the model incorporates weighted average and fusion prediction modules to effectively capture relevant features and improve overall performance.

Finally, the results obtained from the time series prediction module and the weighted average prediction module are combined and analyzed in the fusion prediction module. This fusion process enables a comprehensive analysis of the evolution trend within the pedestrian walking speed sequences, leading to more accurate predictions and informed decisions.

3.3 Time Series Forecast Module

The functionality of the time series prediction module is to learn the contextual information of the pedestrian walking speed sequence. The input pedestrian walking speed sequence data undergoes preprocessing steps, including data cleaning and normalization, to ensure the accuracy and consistency of the data. Then, the time series data is fed into the time series prediction module for feature extraction and modeling.

In this study, the LSTM (Long Short-Term Memory) model is adopted as the underlying framework for the time series prediction module. The LSTM model is a variant of recurrent neural networks that captures the long-term dependency relationships in the pedestrian walking speed sequence. After processing by the base model, the obtained features are passed through a fully connected layer to generate the prediction results of the time series prediction module.

3.4 Weighted Average Forecast Module

Weighted average prediction module assigns weights to each data point in the sequence. Specifically, the nearest neighbor degree, represented by the time difference, is utilized to calculate the weight assigned to each data point for the target time point. The predicted value of pedestrian walking speed at the target time point can be obtained by taking a weighted average of the data points in the sequence.

In the calculation process, the nearest neighbor degree is calculated according to the following formula:

$$\Delta T = \{\Delta t_k\}_{k=1}^{K} \tag{1}$$

$$\Delta t_k = t - (t - k) \tag{2}$$

where K represents the length of the pedestrian walking speed sequence, and for each data point, Δt_k denotes the time difference between the target time point to be predicted and the time of the current data point. The corresponding weights based on the temporal proximity are calculated using the following formula:

$$W = \{w_k\}_{k=1}^{K} \tag{3}$$

$$w_k = \frac{\frac{1}{\Delta t_k}}{\sum_{k=1}^{K} \frac{1}{\Delta t_k}} \tag{4}$$

where Δt_k is the previously calculated temporal nearest neighbor degree, and the smaller the temporal nearest neighbor degree, the larger the weight, then the predicted value of walking speed corresponding to time t:

$$\widehat{v_t} = w_1 v_1 + w_2 v_2 + \cdots + w_K v_K \tag{5}$$

where v_i is the pedestrian walking speed at the i data point in the sequence, and w_i is the corresponding weight. By introducing the weighted average prediction module, this paper can consider the correlation between different time points in the series, making the pedestrian walking speed at the predicted time point more accurately reflect the evolution trend of the series.

3.5   Fusion Prediction Module

This paper employs a multilayer perceptron (MLP) as the fusion prediction module to integrate the outcomes of the time series prediction module and the weighted average prediction module. The output of the time series prediction module is $Y_s = \{v_{s_t}, v_{s_{t+1}}, \cdots, v_{s_{t+h-1}}\}$, and the output of the weighted average prediction module is $Y_k = \{v_{k_t}, v_{k_{t+1}}, \cdots, v_{k_{t+h-1}}\}$, where v_{s_t} and v_{k_t} denote the predicted pedestrian walking speed at time point t, respectively. Let W represents the weight parameter, and b denotes the bias parameter of the MLP. The predictions from both methods can be combined through the nonlinear transformation and weight learning of the MLP. More specifically, the computational process of the fusion prediction module can be formulated as follows:

$$\hat{Y} = \{\hat{v}_t, \hat{v}_{t+1}, \cdots, \hat{v}_{t+h-1}\} \tag{6}$$

$$\hat{v}_t = \sigma\left(W \bullet [v_{s_t}, v_{k_t}] + b\right) \tag{7}$$

where σ denotes the nonlinear activation function and $[v_{s_t}, v_{k_t}]$ denotes the stitching of the prediction results of the two methods into one vector. Through the calculation of the fusion prediction module, we can learn the weights and non-linear transformations of the fusion prediction module to obtain the prediction results of the final pedestrian walking speed sequence $\hat{Y} = \{\hat{v}_t, \hat{v}_{t+1}, \cdots, \hat{v}_{t+h-1}\}$.

4   Experiment

4.1   Dataset

We collected GPS data from 321 pedestrian walking trajectories, encompassing various individuals, different periods, and diverse environments. Each trajectory captures the location information of pedestrians over time, consisting of longitude, latitude, and corresponding timestamps. Considering the GPS positioning error, this study performs data pre-processing on the collected pedestrian GPS trajectory information. It uses the Kalman filtering method to eliminate the outliers in the trajectory, such as GPS drift points.

4.2   Correlation Verification

In order to validate the correlation between pedestrian walking speeds at adjacent time steps, we employed statistical techniques to assess the correlation coefficients across different time scales. This involved calculating the correlation coefficients between pedestrian walking speeds at adjacent data points within various time intervals. Hypothesis tests were then performed to determine the significance of any observed differences. Additionally, we utilized the autocorrelation function (ACF) to examine the autocorrelation of the pedestrian walking speed series. We further demonstrate the correlation between pedestrian walking speeds at adjacent time steps by comparing the autocorrelation at different time scales.

4.3 Hyperparameter Settings

We initially extract subsequences for each input trajectory data by employing a specific window length, thereby selecting a continuous segment from the trajectory data to serve as the observation sequence. The window length plays a crucial role in determining the length of these observation sequences. We employ three distinct window lengths to examine the impact of different time spans on the prediction task: 5, 10, and 15.

We set the initial learning rate to 0.001, conducted 60 training rounds, utilized a batch size of 32, and applied a dropout probability of 0.2. Moreover, we employ a learning rate decay strategy during the training process. Every five epochs, the learning rate decreases by 0.1 to facilitate convergence. Additionally, we employ the Dropout regularization method to accelerate convergence speed and mitigate overfitting risks.

5   Results

5.1   Correlation Verification

The results of the correlation validation for pedestrian walking speed within adjacent time steps are shown in Table 1. The correlation coefficients of pedestrian walking speed within adjacent time steps were calculated at different time steps, and the results using the t-test showed that the p-value was less than 0.05, implying a significant correlation of pedestrian walking speed within adjacent time steps.

Table 1. t-test for correlation coefficients of pedestrian walking speed within adjacent time steps at different time steps.

Time step	2 s	5 s	10 s	15 s	20 s	30 s
p-value	1.54×10^{-103}	3.54×10^{-59}	2.90×10^{-26}	9.02×10^{-14}	1.88×10^{-9}	4.14×10^{-7}

Figure 2 plots the autocorrelation function of the pedestrian walking speed sequence on a particular trajectory. The autocorrelation coefficients of the pedestrian walking speed series at different time steps are in the blue region, i.e., they have a significant linear relationship with previous values in the time series.

They can be used to predict future values. Therefore, the pedestrian walking speed series have significant autocorrelation and are suitable for modeling and prediction by time series analysis.

5.2   Model Performance

We use Mean Squared Error (MSE), Root Mean Squared Error (RMSE), Mean Absolute Error (MAE), and Coefficient of Determination (R^2) to measure the model performance. The smaller the MSE, RMSE, and MAE values, the smaller the difference between the model's prediction results and the actual values. The closer R^2 is to 1 means the model fits better, and the closer it is to 0, the worse.

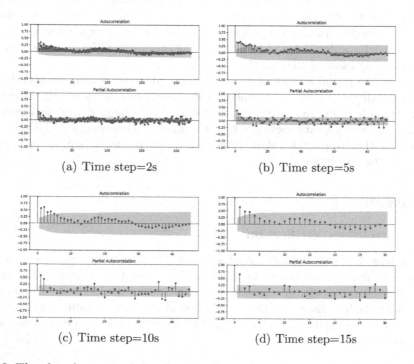

Fig. 2. The plot of autocorrelation function for a sequence of pedestrian walking speed on a specific trajectory at different time steps.

We choose LSTM as the base model and use it as the benchmark model for comparison with the method in this paper. We conducted experiments at different step times (2 s, 5 s, 10 s, 20 s) to evaluate the model's performance with a fixed sequence length of 10. The experimental results are shown in Table 2. The results show that the method in this paper significantly outperforms the LSTM benchmark model in all evaluation metrics. For example, the MSE of this method is 0.1687 at a step time of 5 s, while the MSE of the LSTM model is 0.1705, indicating that this method has a significant advantage in prediction accuracy. In addition, the results of RMSE, MAE, and R^2 metrics also support the superiority of this paper's method relative to other base models and LSTM models. We also compared the model inference times, and the inference times of this paper's model were significantly lower than those of the base model on the entire test data set, further demonstrating the superiority of this paper's method.

To further evaluate the method's effectiveness in this paper, we compared the ARIMA model, the four base models (Transformer, CNN, LSTM, GRU), and the method in this paper. We examined their performance metrics on different time scales. The experimental results are shown in Table 3. The results show that the method in this paper exhibits significant improvement relative to the base model at different time scales. In particular, the method based on the GRU base model significantly improves performance metrics, while the model inference time decreases significantly, and the prediction accuracy improves effectively.

To investigate the effect of the period on model performance, we conducted a series of experiments in which the sequence lengths were adjusted to 5, 10, and 15 with a fixed time step of 20 s. We evaluated the performance metrics of each model. The experimental results are shown in Table 4. The results show that the overall performance of the models improves as the sequence length increases. In particular, the performance metrics of each model reached the best level at the sequence length of 15, which indicates that longer sequence lengths help to provide more historical information, thus improving the prediction accuracy of the models. In addition, we observe that the method in this paper shows significant advantages over the base model for different sequence lengths. Both in MSE, RMSE, MAE, and R^2 metrics, the method in this paper achieves lower values, further demonstrating the better prediction performance of the method in this paper.

Table 2. Comparison of the performance of different models under different time step (sequence length of 10)

Time Step	Methods	Time	MSE	RMSE	MAE	R^2
2 s	LSTM	1.06	0.2212	0.4704	0.3396	0.2502
	LSTM+NWM(Ours)	1.2	0.2207	0.4698	0.339	0.2521
5 s	LSTM	0.19	0.1705	0.4129	0.2839	0.2752
	LSTM+NWM(Ours)	0.2	0.1687	0.4107	0.2836	0.2828
10 s	LSTM	0.15	0.1408	0.3752	0.2507	0.275
	LSTM+NWM(Ours)	0.16	0.14	0.3742	0.25	0.2789
15 s	LSTM	0.14	0.1177	0.3431	0.2281	0.3026
	LSTM+NWM(Ours)	0.14	0.1122	0.335	0.2252	0.3349
20 s	LSTM	0.13	0.1041	0.3227	0.2281	0.3026
	LSTM+NWM(Ours)	0.12	0.0966	0.3109	0.2252	0.3349

Table 3. Performance of different models on different time scales (series length k=10).

Time Step	Methods	Time	MSE	RMSE	MAE	R^2
2 s	ARIMA	273.81	0.2954	0.5435	0.3856	0.0010
	Transformer	1.29	0.2507	0.5007	0.3641	0.1505
	Transformer+NWM	1.34	0.2361	0.4859	0.3504	0.1997
	CNN	1.59	0.2244	0.4737	0.3434	0.2394
	CNN+NWM	1.60	0.22	0.469	0.3391	0.2546
	LSTM	1.06	0.2212	0.4704	0.339	0.2502
	LSTM+NWM	1.2	0.2207	0.4698	0.3396	0.2521
	GRU	0.81	0.2244	0.4737	0.3432	0.2394
	GRU+NWM	0.39	0.2243	0.4736	0.343	0.2399
5 s	ARIMA	108.98	0.2192	0.4682	0.3219	0.0653
	Transformer	0.83	0.2417	0.4917	0.346	0.0277
	Transformer+NWM	0.8	0.2212	0.4703	0.3492	0.0596
	CNN	0.1752	0.4186	0.2925	0.2549	0.1752
	CNN+NWM	0.174	0.4171	0.2889	0.2602	0.174
	LSTM	0.19	0.1705	0.4129	0.2839	0.2752
	LSTM+NWM	0.2	0.1687	0.4107	0.2836	0.2828
	GRU	0.19	0.1696	0.4119	0.2839	0.2788
	GRU+NWM	0.2	0.1686	0.4106	0.284	0.2833
10 s	ARIMA	49.75	0.1655	0.4068	0.275	0.1477
	Transformer	0.69	0.2473	0.4972	0.3851	0.019
	Transformer+NWM	0.71	0.1979	0.4448	0.3122	0.2733
	CNN	1.45	0.1752	0.4186	0.2925	0.2549
	CNN+NWM	1.45	0.174	0.4171	0.2889	0.2602
	LSTM	0.15	0.1408	0.3752	0.2507	0.275
	LSTM+NWM	0.16	0.14	0.3742	0.25	0.2789
	GRU	0.15	0.1377	0.3711	0.2477	0.2906
	GRU+NWM	0.16	0.1375	0.3708	0.2471	0.292
20 s	ARIMA	18.66	0.1124	0.3352	0.2273	0.3436
	Transformer	0.61	0.2157	0.4644	0.3709	0.0489
	Transformer+NWM	0.63	0.1412	0.3758	0.2591	0.2248
	CNN	1.37	0.1311	0.362	0.2432	0.2232
	CNN+NWM	1.4	0.1208	0.3475	0.2314	0.2843
	LSTM	0.13	0.1041	0.3227	0.2192	0.2811
	LSTM+NWM	0.12	0.0966	0.3109	0.213	0.3327
	GRU	0.13	0.0964	0.3105	0.2133	0.3343
	GRU+NWM	0.13	0.0947	0.3077	0.2114	0.3462

In summary, using a more complex model structure and specific algorithm optimization, the method in this paper can better capture the temporal information in the sequence data and provide more accurate prediction results. Therefore, the method in this paper has significant potential for practical applications and can provide a valuable reference for walking speed-related research and application areas.

Table 4. Performance of different models over different periods (time step = 20 s).

Seq length	Methods	Time	MSE	RMSE	MAE	R^2
5	ARIMA	434.82	0.3790	0.6156	0.4351	−0.2578
	Transformer	0.63	0.2372	0.487	0.3806	−0.3098
	Transformer+NWM	0.64	0.1741	0.4173	0.2982	0.0386
	CNN	1.5	0.1774	0.4212	0.3094	0.0206
	CNN+NWM	2.76	0.1254	0.3541	0.2317	0.3077
	LSTM	0.14	0.1338	0.3657	0.2374	0.2614
	LSTM+NWM	0.14	0.1223	0.3497	0.2281	0.3248
	GRU	0.14	0.1272	0.3567	0.2325	0.2976
	GRU+NWM	0.14	0.1228	0.3504	0.2278	0.3219
10	ARIMA	273.81	0.2954	0.5435	0.3856	−0.0010
	Transformer	1.29	0.2507	0.5007	0.3641	0.1505
	Transformer+NWM	1.34	0.2361	0.4859	0.3504	0.1997
	CNN	1.59	0.2244	0.4737	0.3434	0.2394
	CNN+NWM	1.60	0.22	0.469	0.3391	0.2546
	LSTM	1.06	0.2212	0.4704	0.339	0.2502
	LSTM+NWM	1.2	0.2207	0.4698	0.3396	0.2521
	GRU	0.81	0.2244	0.4737	0.3432	0.2394
	GRU+NWM	0.39	0.2243	0.4736	0.343	0.2399
15	ARIMA	240.68	0.2601	0.5100	0.3660	0.1146
	Transformer	0.61	0.2303	0.4799	0.4037	−1.2286
	Transformer+NWM	0.62	0.1054	0.3247	0.2358	0.0201
	CNN	1.4	0.152	0.3898	0.2575	0.161
	CNN+NWM	1.43	0.1247	0.3532	0.2309	0.3113
	LSTM	0.13	0.079	0.2811	0.2023	0.2355
	LSTM+NWM	0.13	0.0779	0.2791	0.2005	0.246
	GRU	0.13	0.0776	0.2786	0.2006	0.2487
	GRU+NWM	0.12	0.0762	0.2761	0.1985	0.2623

6   Conclusion

The experimental results show that the pedestrian walking speed of adjacent time steps is correlated, and it is reasonable to analyze walking speed as a time series. The time series prediction based on pedestrian trajectory data and the walking speed prediction model combined with the nearest neighbor weighted average method proposed in this study significantly outperformed the benchmark method and other comparison methods regarding prediction accuracy.

This paper demonstrates that pedestrian walking speed with adjacent time steps is correlated, providing more insight for further research. The prediction model proposed in this paper can be applied in pedestrian flow control and intelligent transportation in practical scenarios, which have practical application value. In addition, the method in this paper can provide reference and reference for time series-based data prediction in other fields, which has comprehensive promotion value.

References

1. Bohannon, R.W., Andrews, A.W.: Normal walking speed: a descriptive meta-analysis. Physiotherapy **97**(3), 182–189 (2011)
2. Cho, K., et al.: Learning phrase representations using RNN encoder-decoder for statistical machine translation. arXiv preprint arXiv:1406.1078 (2014)
3. El Hamdani, S., Benamar, N., Younis, M.: Pedestrian support in intelligent transportation systems: challenges, solutions and open issues. Transp. Res. Part C: Emerg. Technol. **121**, 102856 (2020)
4. Finnis, K.K., Walton, D.: Field observations to determine the influence of population size, location and individual factors on pedestrian walking speeds. Ergonomics **51**(6), 827–842 (2008)
5. Finnis, K., Walton, D.: Field observations of factors influencing walking speeds. In: 2nd International Conference on Sustainability Engineering and Science (2007)
6. Fitzpatrick, K., Brewer, M.A., Turner, S.: Another look at pedestrian walking speed. Transp. Res. Rec. **1982**(1), 21–29 (2006)
7. Fossum, M., Ryeng, E.O.: The walking speed of pedestrians on various pavement surface conditions during winter. Transp. Res. Part D: Transp. Environ. **97**, 102934 (2021)
8. Franěk, M.: Environmental factors influencing pedestrian walking speed. Percept. Mot. Skills **116**(3), 992–1019 (2013)
9. Franěk, M., Režný, L.: Environmental features influence walking speed: the effect of urban greenery. Land **10**(5), 459 (2021)
10. Franěk, M., Režný, L., Šefara, D., Cabal, J.: Effect of traffic noise and relaxations sounds on pedestrian walking speed. Int. J. Environ. Res. Public Health **15**(4), 752 (2018)
11. Fujiyama, T., Tyler, N.: Predicting the walking speed of pedestrians on stairs. Transp. Plan. Technol. **33**(2), 177–202 (2010)
12. Hausdorff, J.M.: Gait dynamics, fractals and falls: finding meaning in the stride-to-stride fluctuations of human walking. Hum. Mov. Sci. **26**(4), 555–589 (2007)
13. Li, Q., Young, M., Naing, V., Donelan, J.: Walking speed estimation using a shank-mounted inertial measurement unit. J. Biomech. **43**(8), 1640–1643 (2010)

14. Liang, S., Leng, H., Yuan, Q., Wang, B., Yuan, C.: How does weather and climate affect pedestrian walking speed during cool and cold seasons in severely cold areas? Build. Environ. **175**, 106811 (2020)
15. Park, J.G., Patel, A., Curtis, D., Teller, S., Ledlie, J.: Online pose classification and walking speed estimation using handheld devices. In: Proceedings of the 2012 ACM Conference on Ubiquitous Computing, pp. 113–122 (2012)
16. Perrin, O., Terrier, P., Ladetto, Q., Merminod, B., Schutz, Y.: Improvement of walking speed prediction by accelerometry and altimetry, validated by satellite positioning. Med. Biol. Eng. Compu. **38**, 164–168 (2000)
17. Pinna, F., Murrau, R.: Age factor and pedestrian speed on sidewalks. Sustainability **10**(11), 4084 (2018)
18. Silva, A.M.C.B., da Cunha, J.R.R., da Silva, J.P.C.: Estimation of pedestrian walking speeds on footways. In: Proceedings of the Institution of Civil Engineers-Municipal Engineer, vol. 167, pp. 32–43. Thomas Telford Ltd. (2014)
19. Soltani, A., et al.: Algorithms for walking speed estimation using a lower-back-worn inertial sensor: a cross-validation on speed ranges. IEEE Trans. Neural Syst. Rehabil. Eng. **29**, 1955–1964 (2021)
20. Tarawneh, M.S.: Evaluation of pedestrian speed in Jordan with investigation of some contributing factors. J. Safety Res. **32**(2), 229–236 (2001)
21. Tolea, M.I., et al.: Sex-specific correlates of walking speed in a wide age-ranged population. J. Gerontol. B Psychol. Sci. Soc. Sci. **65**(2), 174–184 (2010)
22. Wu, C.J., Kuo, C.H., Lin, Y.H., Liu, W.Y.: A feasible model training for LSTM-based dual foot-mounted pedestrian INS. IEEE Sens. J. **21**(12), 13616–13627 (2021)
23. Yang, S., Laudanski, A., Li, Q.: Inertial sensors in estimating walking speed and inclination: an evaluation of sensor error models. Med. Biolog. Eng. Comput. **50**, 383–393 (2012)
24. Yoshida, T., et al.: Sampling rate dependency in pedestrian walking speed estimation using DualCNN-LSTM. In: Adjunct Proceedings of the 2019 ACM International Joint Conference on Pervasive and Ubiquitous Computing and Proceedings of the 2019 ACM International Symposium on Wearable Computers, pp. 862–868 (2019)
25. Yoshida, T., Nozaki, J., Urano, K., Hiroi, K., Yonezawa, T., Kawaguchi, N.: Gait dependency of smartphone walking speed estimation using deep learning (poster). In: Proceedings of the 17th Annual International Conference on Mobile Systems, Applications, and Services, pp. 641–642 (2019)
26. Zihajehzadeh, S., Park, E.J.: Regression model-based walking speed estimation using wrist-worn inertial sensor. PLoS ONE **11**(10), e0165211 (2016)
27. Zijlstra, W.: Assessment of spatio-temporal parameters during unconstrained walking. Eur. J. Appl. Physiol. **92**, 39–44 (2004)
28. Zijlstra, W., Hof, A.L.: Assessment of spatio-temporal gait parameters from trunk accelerations during human walking. Gait Posture **18**(2), 1–10 (2003)

MView-DTI: A Multi-view Feature Fusion-Based Approach for Drug-Target Protein Interaction Prediction

Jiahui Wen, Haitao Gan$^{(\boxtimes)}$, Zhi Yang, Ming Shi, and Ji Wang

School of Computer Science, Hubei University of Technology, Wuhan 430068, China
htgan01@hbut.edu.cn

Abstract. Drug-Target protein Interaction (DTI) prediction is a crucial task in the field of drug discovery. Prediction methods based on deep learning have been demonstrated to significantly enhance the accuracy of DTI prediction. Existing approaches mainly extract features from drug molecular sequences and then utilize networks for learning and prediction. However, drug molecular images can clearly display features such as atoms, structures, and chemical bonds, which are difficult to capture in sequences. Therefore, this study introduces a deep learning approach based on multi-view feature fusion, leveraging Transformer to combine the graph structure and image features of drug molecules, thereby learning more comprehensive drug features. This enables the model to learn more intricate interaction features between amino acids and atoms during DTI simulation. The proposed model was evaluated on three benchmark datasets and demonstrated significant improvements over the latest baselines. Furthermore, to validate the efficacy of capturing drug image feature information, ablation experiments were conducted, indicating a notable enhancement in accuracy upon incorporating image data.

Keywords: Drug discovery · Drug-Target protein Interaction · Multi-view feature fusion · Transformer

1 Introduction

In recent years, DTI prediction has become a crucial research area in the field of bioinformatics. Its aim is to predict the likelihood of DTI and provide essential guidance for drug development. Predicting the effects of new drugs using known DTI datasets can save a significant amount of time and money in clinical experiments [1]. With the advancement of bioinformatics technology, the methods for predicting DTI have been evolving. Starting from virtual screening methods [2,3], to machine learning approaches based on chemogenomics, and more recently, to deep learning-based methods.

Inspired by deep learning, new methods for predicting DTI continue to emerge. Compared to traditional machine learning, these models generally do not require the definition and calculation of descriptors before modeling. Their

B. Luo et al. (Eds.): ICONIP 2023, CCIS 1964, pp. 400–411, 2024.
https://doi.org/10.1007/978-981-99-8141-0_30

network architecture mainly consists of drug feature extraction, protein feature extraction, and a classifier. The entire DTI prediction is considered a binary classification task, where features extracted from drug and protein data are used to infer the likelihood of their interaction through the classifier.

Lee et al. proposed a model named DeepConvDTI for predicting DTI, which uses convolutional layers to extract local residue features of the general protein class [4]. By detecting the fused convolutional results, the model has been shown to be able to identify protein binding sites. Öztürk H. et al. introduced a model named DeepDTA, which uses a convolutional layer to extract features and adopts a multi-task learning approach to simultaneously predict binding affinity and drug efficacy [5]. Wang et al. proposed a multiscale convolutional network, which uses different types of convolutional networks to extract local and global features of proteins as well as topological features of compounds [6].

Due to the inherent graph structure characteristics of drug molecules, the loss of spatial structural information may weaken the model's predictive ability. Tsubaki et al. first encoded the Simplified Molecular Input Line Entry System(SMILES) sequence of drug molecules into molecular graphs using the RDKit library, and then utilized both GNN and CNN to learn feature information from drug molecule sequences and protein sequences [7]. They used neural attention mechanisms to map regions with high weight values to known 3D protein-drug complex structures, which provided effective visualization for model analysis. Li et al. proposed a multi-objective neural network (MONN) that introduced a Gate Recurrent Unit (GRU) module to predict affinity and accurately determine the interactions and affinity between molecules [8].

The Transformer network is a deep learning model based on self-attention mechanisms [9], which has significant advantages in processing long sequences. Chen et al. proposed a more rigorous data set partitioning method, and mapped the weight of interaction features of drug molecules and proteins to molecular sequences using a Transformer decoder, and then predicted the molecular sequences with obtained attention weights [10]. Chen et al. proposed a Deep-Embedding network model that used a GNN with attention mechanisms and attention bidirectional long-term memory (BiLSTM) to perform DTI tasks [11]. Mutual-DTI [12] is a prediction model based on interaction features, which introduced an interaction feature extraction module to learn the complex internal interaction process between drugs and proteins during their mutual interactions.

The molecular image clearly displays information on the atoms, structure, and chemical bonds of the molecule. In comparison to SMILES sequences and molecular graphs, molecular images contain complex feature information. Qian et al. used images to represent drug molecules and utilized CNN convolutional aggregation operations to learn feature information from molecular images [13]. Qian et al. proposed a new image-based CAT-CPI model that uses Transformer to capture global features from the image [14].

Based on previous research, this paper proposes MView-DTI, a model that utilizes multi-view feature fusion to extract comprehensive drug features from the graph and image representations of drug molecules. In this study, we use

CNN to learn feature information from drug molecular images, and GNN to extract features from the embedded molecular graph. Through the use of self-attention mechanisms, we combine the two types of features to obtain more comprehensive drug molecule features. Additionally, we utilize a sliding window n-gram method to segment protein sequences, followed by using Conv1d and Gated Linear Unit (GLU) to extract protein features. Finally, the interaction feature extraction module is employed to learn from both drug molecule features and protein features. By comparing the performance of our model to the latest baseline on three datasets, we achieved significant improvements. Ablation experiments also demonstrated the effectiveness of fusing the features of drug molecule graphs and images.

2   Methodology

The model we propose is mainly composed of three parts: the drug molecule feature extraction module, the protein feature extraction module, and the interaction feature learning module. The overall framework of the model is shown in Fig. 1.

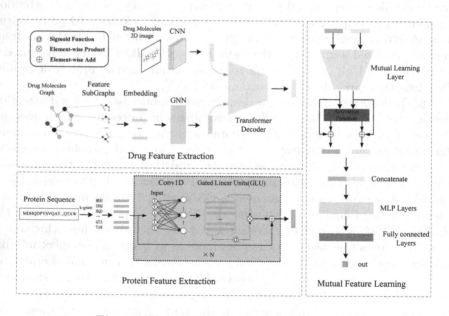

Fig. 1. MView-DTI Network Model Structure.

2.1   Drug Feature Extraction

The CNN network utilizes convolutional kernels to calculate the pixels in an image, which not only reduces the computational complexity but also effectively preserves the structural features of the image. The drug image provides a visualization of the molecular structure, with different structures highlighted in different colors. The drug image can provide clear atomic feature information. Using the RDKit library, we can obtain the 2D image corresponding to a drug molecule from its SMILES sequence $P \in R^{h \times w}$, where h and w represent the height and width of the image, respectively. Our CNN block consists of convolutional layers, batch normalization layers, activation layers, and pooling layers. We use the drug image as the input to the CNN convolutional layer, which is calculated as follows:

$$P_{out}^i = Conv\left(PW^i\right) + B^i \tag{1}$$

where i is network layer number, $Conv$ is the convolution operation, W and B are the network weights and biases respectively, and P_{out}^i is the output of the i-th layer. After passing through the CNN block, we obtain the feature information P_{drug} of the original image of the drug molecule.

A drug molecule consists of various types of atoms and chemical bonds, and its spatial structure is best represented by a graph structure. Thus, we can construct a drug molecule graph $G = \{N, E\}$, where N is the set of nodes representing atoms and E is the set of edges representing various chemical bonds. We consider an undirected graph G, where node $n_i \in N$ represents atom i of the drug molecule, and edge $e_{ij} \in E$ represents the chemical bond between atom i and atom j. Additionally, we adopt the r-radius subgraph method [15] to perform local search on the nodes in the graph, where the search radius r is determined by the hop count of a specific node. We define a subgraph $S_i^{(r)} = (N_i^{(r)}, E_i^{(r)})$ for node n_i within the radius r, as shown below:

$$N_i^{(r)} = \{n_j | j \in Sub(r, i)\} \tag{2}$$

$$E_i^{(r)} = \left\{ e_{ab}^{(r)} = \left\langle n_a^{(r)}, n_b^{(r)} \right\rangle | n_a^{(r)}, n_b^{(r)} \in Sub(r, i) \right\} \tag{3}$$

where $Sub(r, i)$ is the set of nodes connected to node n_i within a radius of r, including node n_i itself. $e_{ab}^{(r)}$ is the edge connecting nodes n_a and n_b. Using the defined subgraphs, we can obtain corresponding chemical features, such as atom type, atom degree, and aromaticity.

We initialize the encoded feature information by inputting it into the embedding layer, and then learn by inputting the obtained embedding vectors into the GNN layer. In the GNN, we represent the embedding of the k-th layer network of the i-th node v_i as $f_i^{(k)} \in R^d$, and update $f_i^{(n)}$ according to the following equation:

$$f_i^{(k)} = \sigma \left(f_i^{(k-1)} + \sum_{j \in Sub(r, i)} h_{ij}^{(k-1)} \right) \tag{4}$$

where σ is the sigmoid function defined as $\sigma(x) = 1/(1 + e^{-x})$, and $h_{ij}^{(k-1)}$ is the hidden vector between nodes v_j and v_i in the previous layer k − 1. The hidden vector can be computed using the following neural network:

$$h_{ij}^{(k)} = \varepsilon \left(\omega f_i^{(k)} + b \right) \tag{5}$$

where ε is a non-linear activation function $(ReLu : \varepsilon(x) = max(0, x))$, $\omega \in R^{d \times d}$ is a weight hyperparameter and $b \in R^d$ is a bias vector. After processing through the GNN layer, we obtain the feature vector G_{drug} for the drug molecule graph.

We flatten the obtained image features P_{drug} and together with the molecular graph feature vector G_{drug} as the input to the Transformer decoder layer, as shown in the Fig. 2. The output of the decoder layer is a sequence that includes both molecular image and graph feature information, with the same length as the molecular sequence.

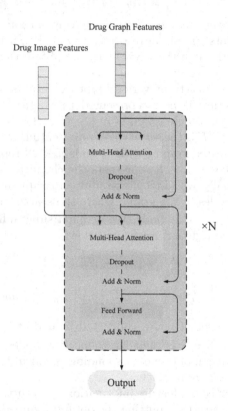

Fig. 2. Transformer Decoder used to fuse molecular graph features and image features. N indicates the number of layers of the module.

2.2   Protein Feature Extraction

We treat a protein sequence as a piece of text for learning, and for the "words" in the text, we define them as n-gram amino acids [16]. For a given amino acid sequence, we divide it into overlapping n-gram amino acids. There are 20 types of known amino acids, and the possible n-grams are 20^n. In this study, we set n to

3, which increases the diversity of features while avoiding the occurrence of low-frequency words in the learned representations. For example, a protein sequence $MVVMN \cdots QATP$ is split into $MVV, VVM, VMN, \cdots, QAT, ATP$. For a given protein sequence $S = a_1 a_2 a_3 \cdots a_L$, where L is the length of the protein sequence and a_i is the i-th amino acid, we split it into:

$$[a_1 a_2 a_3], [a_2 a_3 a_4], \cdots, [a_{L-2} a_{L-1} a_L]$$

We represent $[a_i a_{i+1} a_{i+2}]$ as a d-dimensional word embedding $a_{i:i+2} \in R^d$. We use the processed protein sequence obtained by the above method as the initialized d-dimensional embeddings, which are then inputted into a gated convolutional network with Conv1D and gated linear units [17]. We calculate the hidden layer as follows:

$$L_i(X) = (X \times \omega_1 + s) \otimes \sigma(X \times \omega_2 + t) \tag{6}$$

where L_i is the i-th layer of the gated convolutional network, $X \in R^{n \times d_1}$ is the input of the i-th layer, $\omega_1 \in R^{d_1 \times d_2}, s \in R^{d_2}, \omega_2 \in R^{d_1 \times d_2}, t \in R^{d_2}$ are learned parameters, n is the length of the sequence, d_1, d_2 are the dimensions of the input and hidden features, σ is the sigmoid function, and $\otimes$ is the matrix product. The output of the gated convolutional network is the final feature representation of the protein sequence.

2.3 Mutual Feature Learning

The interaction feature learning module consists of a mutual learning layer [12], a multi-layer perceptron layer, and a fully connected layer. We use the extracted drug feature vector and protein feature vector as input to the interaction feature learning module.

The mutual learning layer is a modified Transformer decoder that learns complex interactive feature information between internal amino acids and atoms during the drug molecule and protein interaction process. The key to capturing feature information is the Transformer's multi-head self-attention layer. The multi-head self-attention layer consists of several scaled dot-product attention layers, which are used to extract the interaction information between the encoder and decoder. The self-attention layer accepts three inputs: the key K, value V, and query Q, and calculates attention as follows:

$$attention(Q, K, V) = softmax\left(\frac{QK^T}{\sqrt{d_k}}\right) V \tag{7}$$

where d_k is a scaling factor that depends on the number of layers. This mechanism enables the decoder to dynamically focus on key parts of the encoder output, thereby directly capturing the interaction features of the given two sequences.

After the mutual learning layer, we obtain the interaction feature vectors for the drug and protein, represented as $D \in R^{b \times n_1 \times d}$ and $P \in R^{b \times n_2 \times d}$, respectively. Here, b is the batch size, n_1, n_2 are the number of words in the drug and protein sequences, respectively, and d is the feature dimension. The feature vectors are then normalized to obtain the weights of each word in the sequence:

$$W_D = F(D, 1) \tag{8}$$

$$W_P = F(P, 1) \tag{9}$$

where $F(input, dim)$ is a normalization function on the specified dimension, followed by an update to the potential feature vectors:

$$D_a = D \times 0.5 + D \odot W_D \tag{10}$$

$$P_a = P \times 0.5 + P \odot W_P \tag{11}$$

The feature vectors D_a and P_a undergo a maximum pooling operation and are then concatenated before being fed into the Multi-Layer Perceptron (MLP) layer for learning. The final structure is then fed back to the fully connected layer to obtain the predicted interaction result, denoted as $\hat{y}$. As a binary classification task, we train our model using binary cross-entropy loss:

$$Loss = -\frac{1}{N} \sum_{i=1}^{N} (y_i log \hat{y}_i + (1 - y_i) \log (1 - \hat{y}_i)) \tag{12}$$

where N is the total number of samples, $\hat{y}_i$ indicates the final prediction and y_i is the true label.

3   Experiment

3.1   Dataset

In the experiments, the learning rate is 0.0001 and batch size is 32. Each batch contains a set of drug molecules and the corresponding proteins. Our model was implemented using pytorch 1.10.0. All computations and training were done on GeForce RTX 3090.

We conducted DTI tasks on three datasets: Davis [18], DrugBank [19], and GPCR [10]. The Davis and DrugBank datasets were created by Zhao et al. [20] based on the original datasets. The Davis dataset includes 25772 samples, of which the total number of positive samples synthesized is 7320, which contains 68 unique drug with 379 unique proteins. The DrugBank dataset comprises 17511 positive interactions between 6655 unique drug and 4294 unique proteins. The GPCR dataset was partitioned according to a different criterion by Chen et al. [10]. The key to constructing GPCR dataset is that the drugs in its training set appear in only one class of samples (positive interaction or negative interaction DTI pairs), and in the test set appear in only the opposite class of samples. The specific details of data samples are shown in Table 1.

Table 1. Summary of the datasets.

Dataset	Drugs	Proteins	Interactions	Positive	Negative
Davis	68	379	25772	7320	18452
DrugBank	6655	4294	35002	17511	17511
GPCR	5359	356	15343	7989	7354

3.2 Results

We randomly divided the datasets into training, validation, and test sets in an 8:1:1 ratio. We repeated the experiments using three different random seeds, and for each experiment, we randomly partitioned the dataset using different random seeds. Each experiment was trained on the training set, and the network hyperparameters were fine-tuned using the validation set. Finally, the model's performance was evaluated on the test set. We employed the same preprocessing methods and compared our results with several state-of-the-art deep learning methods, which are listed below:

DeepEmbedding [11] used a bidirectional encoder representation from Transformer (BERT) pretraining method to extract substructure features from Transformer.

GNN-CPI [7] used molecular fingerprints and distance matrices as input features for compounds and learned to fuse them using a GNN network.

TransformerCPI [10] used molecular sequences and distance matrices as input features for drugs and learned the relationships between drug features and protein features using a Transformer encoder.

Mutual-DTI [12] used a mutual interaction feature extraction module to learn the complex interaction process between drugs and proteins.

The experimental results, as shown in Tables 2–4, demonstrate that the MView-DTI model outperformed other models in terms of AUC and Recall, regardless of whether the datasets were balanced (DrugBank), imbalanced (Davis), or strictly partitioned (GPCR). This shows that the MView-DTI model has good generalization properties. Meanwhile, the method (TransformerCPI, Mutual-DTI and MView-DTI) of considering internal features when interacting with each other performs better on the more strictly divided GPCR dataset. This suggests that it is necessary to re-learn the interaction features when predicting DTI (Tables 2–4).

Table 2. Comparison with other methods on Davis dataset

Methods	AUC	Precision	Recall
DeepEmbedding	0.869 ± 0.008	0.744 ± 0.025	0.607 ± 0.042
GNN-CPI	0.868 ± 0.004	0.748 ± 0.032	0.589 ± 0.038
TransformerCPI	0.889 ± 0.007	0.755 ± 0.013	0.610 ± 0.020
Mutual-DTI	0.895 ± 0.006	$\mathbf{0.773 \pm 0.016}$	0.645 ± 0.029
MView-DTI	$\mathbf{0.898 \pm 0.004}$	0.738 ± 0.018	$\mathbf{0.704 \pm 0.020}$

Table 3. Comparison with other methods on DrugBank dataset.

Methods	AUC	Precision	Recall
DeepEmbedding	0.809 ± 0.006	0.765 ± 0.016	0.681 ± 0.040
GNN-CPI	0.855 ± 0.006	0.755 ± 0.012	0.793 ± 0.029
TransformerCPI	0.873 ± 0.007	$\mathbf{0.782 \pm 0.017}$	0.806 ± 0.039
Mutual-DTI	0.854 ± 0.006	0.758 ± 0.032	0.791 ± 0.052
MView-DTI	$\mathbf{0.876 \pm 0.004}$	0.776 ± 0.018	$\mathbf{0.831 \pm 0.017}$

Table 4. Comparison with other methods on GPCR dataset.

Methods	AUC	Precision	Recall
DeepEmbedding	0.767 ± 0.022	0.675 ± 0.007	0.749 ± 0.027
GNN-CPI	0.771 ± 0.032	0.677 ± 0.021	0.750 ± 0.045
TransformerCPI	0.852 ± 0.004	0.770 ± 0.006	0.780 ± 0.023
Mutual-DTI	0.823 ± 0.007	$\mathbf{0.773 \pm 0.016}$	0.645 ± 0.029
MView-DTI	$\mathbf{0.862 \pm 0.007}$	0.772 ± 0.032	$\mathbf{0.781 \pm 0.058}$

3.3   Robustness Experiment

To ensure the reliability of MView-DTI in the presence of varying drug molecule images, we rotate the drug images at different angles prior to network learning. As show in Fig. 3. This is because the same chemical bond may be present at various locations in the images, which may interfere with the model's ability to learn features during image feature extraction. We conduct multiple experiments on the GPCR dataset using the same model parameters. Results are shown in Table 5. The results demonstrate that the model's performance fluctuates slightly after the rotations. In summary, MView-DTI exhibits robustness in processing geometrically rotated drug molecule images.

Table 5. Results of the geometric transformation on the GPCR dataset.

Methods	AUC	Precision	Recall
No Rotate	0.862 ± 0.007	0.772 ± 0.032	0.781 ± 0.058
Rotate 90°	0.856 ± 0.007	0.753 ± 0.006	0.806 ± 0.036
Rotate 180°	0.866 ± 0.003	0.764 ± 0.012	0.801 ± 0.020
Rotate 270°	0.862 ± 0.008	0.771 ± 0.010	0.781 ± 0.028

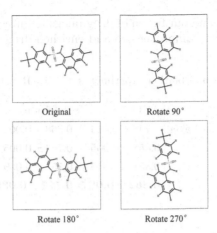

Original   Rotate 90°

Rotate 180°   Rotate 270°

Fig. 3. Different rotation angle of handling Drug images.

3.4 Ablation Study

We conducted ablation experiments on the GPCR dataset with the following settings:

No Mutual Learning Layer: We removed the mutual learning layer used in MView-DTI model to extract interaction features, and directly concatenated the extracted drug and protein features, which were then input to the subsequent learning module.

No MLP: We removed the MLP layer from the model and directly outputted the results through a fully connected layer.

No Image: We did not consider the image information of drug molecules and only used molecular graph features and protein features for mutual interaction learning.

For each setting, we used the same hyperparameters and conducted experiments with three different random seeds, taking the average value. As shown in Table 6, all three modules in our model contributed to its final performance. We observed a significant decrease in the model's prediction accuracy when the module for extracting interaction features was removed. We hypothesize that, after learning the features of drugs and proteins, the model only learns each sequence's respective feature information, while DTI is a process of mutual interaction. After introducing the Mutual learning module, the drug and protein features are treated as subjects, dynamically focusing on each other's key parts in the learning layer, directly capturing the interaction features of the given two sequences. By learning interaction features, the model can gain a deeper understanding of the DTI process, making it easier to capture the key parts that may be involved in the interaction when facing unknown drug and protein data, thus demonstrating higher performance in prediction results. Removing the drug molecule image information also led to a significant decrease in the model's performance. This suggests that drug molecule image information indeed contains unique features, and our fusion of drug image features and molecular graph features results in more comprehensive and enriched drug features.

Table 6. Ablation experiments on GPCR dataset.

Methods	AUC	Precision	Recall
No Mutual Learning Layer	0.748 ± 0.014	0.668 ± 0.004	0.693 ± 0.055
No MLP	0.858 ± 0.005	0.744 ± 0.005	**0.792 ± 0.018**
No Image	0.851 ± 0.015	0.754 ± 0.039	0.789 ± 0.057
MView-DTI	**0.862 ± 0.007**	**0.772 ± 0.032**	0.781 ± 0.058

4 Conclusions

This paper proposes a novel DTI prediction model based on multi-view feature fusion. In comparison to other methods, we take into account two perspectives of drugs: the image view and the molecular graph view. We utilize the CNN's ability to learn local features from images and the GNN's capability to effectively learn and represent complex topological structures of graph data to extract effective drug features from both perspectives. We then fuse the drug features from the two views and use the mutual learning module proposed in previous works to extract interaction features. Although MView-DTI has demonstrated promising performance improvements on various datasets, there is still room for enhancement. Three-dimensional protein data contains intricate binding site information, presenting a significant challenge in the development of network models tailored for proteins with three-dimensional structures.

Acknowledgements. This work is supported by the High-level Talents Fund of Hubei University of Technology under grant No. GCRC2020016, Open Research Fund Program of State Key Laboratory of Biocatalysis and Enzyme Engineering under grant No. SKLBEE2021020 and SKLBEE2020020.

References

1. Manoochehri, H.E., Nourani, M.: Drug-target interaction prediction using semi-bipartite graph model and deep learning. BMC Bioinformatics **21**, 1–16 (2020)
2. Maia, E.H.B., Assis, L.C., De Oliveira, T.A., Da Silva, A.M., Taranto, A.G.: Structure-based virtual screening: from classical to artificial intelligence. Front. Chem. **8**, 343 (2020)
3. Himmat, M., Salim, N., Al-Dabbagh, M.M., Saeed, F., Ahmed, A.: Adapting document similarity measures for ligand-based virtual screening. Molecules **21**(4), 476 (2016)

4. Lee, I., Keum, J., Nam, H.: DeepConv-DTI: prediction of drug-target interactions via deep learning with convolution on protein sequences. PLoS Comput. Biol. **15**(6), e1007129 (2019)
5. Öztürk, H., Özgür, A., Ozkirimli, E.: DeepDTA: deep drug-target binding affinity prediction. Bioinformatics **34**, I821–I829 (2018)
6. Wang, S., et al.: MCN-CPI: multiscale convolutional network for compound-protein interaction prediction. Biomolecules **11**(8), 1119 (2021)
7. Tsubaki, M., Tomii, K., Sese, J.: Compound-protein interaction prediction with end-to-end learning of neural networks for graphs and sequences. Bioinformatics **35**(2), 309–318 (2019)
8. Li, S., Wan, F., Shu, H., Jiang, T., Zhao, D., Zeng, J.: MONN: a multi-objective neural network for predicting compound-protein interactions and affinities. Cell Syst. **10**(4), 308–322 (2020)
9. Vaswani, A., et al.: Attention is all you need. In: Advances in Neural Information Processing Systems **30** (2017)
10. Chen, L., et al.: TransformerCPI: improving compound-protein interaction prediction by sequence-based deep learning with self-attention mechanism and label reversal experiments. Bioinformatics **36**(16), 4406–4414 (2020)
11. Chen, W., Chen, G., Zhao, L., Chen, C.Y.C.: Predicting drug-target interactions with deep-embedding learning of graphs and sequences. J. Phys. Chem. A **125**(25), 5633–5642 (2021)
12. Wen, J., Gan, H., Yang, Z., Zhou, R., Zhao, J., Ye, Z.: Mutual-DTI: a mutual interaction feature-based neural network for drug-target protein interaction prediction. Math. Biosci. Eng. **20**(6), 10610–10625 (2023)
13. Qian, Y., Li, X., Wu, J., Zhou, A., Xu, Z., Zhang, Q.: Picture word order compound protein interaction: predicting compound-protein interaction using structural images of compounds. J. Comput. Chem. **43**(4), 255–264 (2022)
14. Qian, Y., Wu, J., Zhang, Q.: CAT-CPI: combining CNN and transformer to learn compound image features for predicting compound-protein interactions. Front. Mol. Biosci. **9**, 963912 (2022)
15. Costa, F., De Grave, K.: Fast neighborhood subgraph pairwise distance kernel. In: Proceedings of the 26th International Conference on Machine Learning, pp. 255–262. Omnipress, Madison, WI, USA (2010)
16. Dong, Q.W., Wang, X., Lin, L.: Application of latent semantic analysis to protein remote homology detection. Bioinformatics **22**(3), 285–290 (2006)
17. Dauphin, Y.N., Fan, A., Auli, M., Grangier, D.: Language modeling with gated convolutional networks. In: Proceedings of the 34th International Conference on Machine Learning, pp. 933–941. PMLR, Sydney, NSW, Australia (2017)
18. Davis, M.I., et al.: Comprehensive analysis of kinase inhibitor selectivity. Nat. Biotechnol. **29**(11), 1046–1051 (2011)
19. Wishart, D., et al.: DrugBank: a comprehensive resource for in silico drug discovery and exploration. Nucleic Acids Res. **34**, D668–D672 (2006)
20. Zhao, Q., Zhao, H., Zheng, K., Wang, J.: HyperAttentionDTI: improving drug-protein interaction prediction by sequence-based deep learning with attention mechanism. Bioinformatics **38**(3), 655–662 (2022)

User Multi-preferences Fusion
for Conversational Recommender Systems

Yi Zhang[1], Dongming Zhao[2], Bo Wang[3(✉)], Kun Huang[2], Ruifang He[3],
and Yuexian Hou[3]

[1] Tianjin International Engineering Institute, Tianjin University, Tianjin, China
zhangyi_0125@tju.edu.cn
[2] AI Lab, China Mobile Communication Group Tianjin Co., Ltd., Tianjin, China
huangkun1@tj.chinamobile.com
[3] College of Intelligence and Computing, Tianjin University, Tianjin, China
{bo_wang,rfhe,yxhuo}@tju.edu.cn

Abstract. Conversational recommender systems (CRS) aim to provide recommendations by inferring user preferences during conversations. Many current CRS models utilize third-party information, such as reviews, to supplement the extraction of user preferences. Consequently, users develop preferences for third-party information and their own preferences extracted from original dialog data. However, the prevailing approach of combining these preferences as a unified whole for self-attention at the element level compromises their independence. In real-life decision-making, we refer to third-party information and it is important to distinguish whether the reference is from a third party or from the original dialog data. This paper emphasizes the independence of users' own preferences and third-party information. To effectively integrate multiple user preferences, we propose an **A**ttentive **W**ide&**D**eep Conversational **R**ecommender (**AWDCore**). Specifically, we design an attentive wide linear module and an attentive deep neural network to capture the low-order linear and high-order nonlinear relationships between the user's own preference and third-party information, respectively. To highlight the significance of the user's current preference, we incorporate attention mechanisms and a SENet layer in the wide module and deep neural networks, respectively. The learned user preferences are then employed for recommendation and dialogue generation. Extensive experiments have demonstrated the effectiveness of our approach in both recommendation and conversation tasks.

Keywords: Conversational Recommender System · Multi-preference fusion · Wide&Deep

1 Introduction

In conversational recommender systems (CRS), the first party represents the system, and the second party represents the current user. Third-party information refers to external information that the user considers when making decisions, extending beyond the current dialog context. To address the limited information within the dialog context, CRS methods incorporate third-party information. This includes introducing external

B. Luo et al. (Eds.): ICONIP 2023, CCIS 1964, pp. 412–425, 2024.
https://doi.org/10.1007/978-981-99-8141-0_31

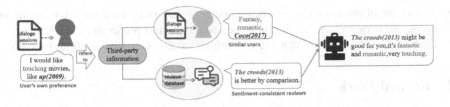

Fig. 1. An example of a user referring to third-party information in conversation recommendation. The fused user multi-preferences are in red, and Items are in italics. (Color figure online)

reviews to enrich item information and generate informative responses, as well as referencing similar users for user preference learning.

However, current approaches in conversational recommender systems (CRS) combine third-party information and the user's own preference extracted from the dialogue context into a unified representation. Self-attention is then applied at the element level to capture user preference. This approach compromises the independence of third-party information and the user's own preference. In real-life decision-making, individuals consciously consider third-party information. For instance, as illustrated in Fig. 1, when users seek *touching* movies similar to *"Up"*, they refer to similar users and sentiment-consistent reviews, selectively incorporating this third-party information. By capturing the fused user preference, the system provides convincing recommendations. Therefore, it is crucial to differentiate between third-party information and original dialogue data used for user reference. Preserving the independence of the user's own preference and third-party information is of utmost importance. While UCCR [14] maintains this independence by employing a linear combination of the two, it falls short in modeling the complex relationship between them. Thus, a nonlinear approach is also necessary.

In this work, we emphasize the independence of the user's own preference and third-party information and propose an **A**ttentive **W**ide&**D**eep **C**onversational **Re**commender (**AWDCore**) model to fuse user's multi-preferences. Specifically, we regard similar users and sentiment-consistent reviews as third-party information. AWDCore contains three parts: (1) We first design a user's own preference learner to encode the user's current preference and historical preference from the current session and historical sessions, respectively. (2) For third-party information learner, inspired by user-based and item-based collaborative filtering [23], we further subdivide similar users into historical similar users and current similar users and encode them based on the learned user's own preference. Following RevCore [16], we encode sentiment-aware retrieved reviews through Transformer. (3) The attentive wide linear module and attentive deep neural networks are used to learn the linear and nonlinear relationship between the user's own preference and third-party information respectively. To sum up, our contributions are as follows:

1. In the user preference learning of CRS, we are the first to emphasize the independence of the user's own preference and third-party information.
2. We apply the Wide&Deep idea [3] to CRS. Considering that user's current preference is the main basis for the user's preference learning, both the wide and deep modules contain attentive components to highlight the user's current preference.

3. To the best of our knowledge, it is the first time that users' own preferences, similar users, and reviews have been jointly used to model user preferences in CRS.
4. Experiments show that AWDCore outperforms state-of-the-art baselines on CRS.

2   Related Work

2.1   Conversational Recommender Systems

Recent CRS works can be divided into two categories: attribute-based approaches and chit-chat-based approaches. The attribute-based approaches obtain user preferences by asking questions about item attributes. Many works adopt multi-armed bandit [4] and reinforcement learning [12] to achieve accurate recommendations in fewer conversation rounds. They typically have a simplified dialogue module to fill recommendation results into systematic response text [12, 19, 27].

In contrast to attribute-based approaches, chit-chat-based approaches focus on creating a more human-like dialogue experience by seamlessly integrating the recommendation task within the dialogue module [2, 13, 14, 16, 29, 30]. These approaches typically employ sequence-to-sequence models for response generation. To compensate for the limited contextual information in conversations, existing works incorporate third-party information. RevCore [16] and C^2-CRS [30] leverage external reviews, while UCCR [14] utilizes similar users for learning user preferences.

2.2   Wide&Deep

As a powerful structure in the field of recommendation systems, Wide&Deep [3] is widely used in various recommendation scenarios. Its core idea is to combine Wide and Deep components to balance the model's memorization and generalization. To improve these two components, there are a series of extended works on Wide&Deep.

On the one hand, considering that the Wide component relies on artificially designed feature crosses, Guo et al. [6] propose DeepFM to replace the Wide component with Factorization Machine (FM). Since the FM structure is only suitable for explicit second-order feature crosses, Wang et al. [25] design DCN with a Cross Network to achieve explicit high-order feature crosses while accelerating calculation. xDeepFM [15] is then proposed with a CIN structure to further improve the Cross Network, it performs feature crosses vector-wise to enhance the concept of domain.

On the other hand, since the Deep component of Wide&Deep is simply stacking features and completely depends on the subsequent DNN to learn feature crosses, He et al. [7] proposes an NFM model with the Bi-Interaction to capture second-order feature crosses before hidden layers, making it easier for subsequent hidden layers to learn useful higher-order feature crosses. AFM [26] and AutoInt [21] introduce the Attention mechanism to explicitly learn the differences between different feature crosses, enhancing model interpretability.

This paper extends the Wide&Deep idea and the second category of CRS works by introducing multiple third-party information and fuse user multi-preferences.

3   Preliminary

CRS contains three objects, namely the user, entities mentioned in conversation, and words mentioned in conversation. Considering that a user may have multiple sessions, we reorganize the sessions in chronological order. Each user has two dialogue attributes: the current session and the historical sessions. Then, we have the following definitions:

User's Own Preference. The user's own preference has two parts: current preference and historical preference, both of which are learned from the dialogue data. *Users' current preference*: For a user u from the user set $\mathcal{U}$, the T-th session is the current dialogue session. When the current dialogue session reaches t-th turn, all entities mentioned before this turn $\mathcal{C}_e = \{e_1^T, ..., e_t^T\}$ are regarded as the current entities, the current words $\mathcal{C}_w$ is defined in the same way as $\mathcal{C}_e$, User's current preference $\mathcal{P}^c$ is learned from $\mathcal{C}_e$ and $\mathcal{C}_w$. *User's historical preference*: historical dialogue sessions are all sessions before the current dialogue session. Historical entities $\mathcal{H}_e = \{H_e^1, ..., H_e^{T-1}\}$ and historical words $\mathcal{H}_w$ are extracted from $T-1$ sessions, we can capture user's historical preference P^h based on $\mathcal{H}_e$ and $\mathcal{H}_w$.

The Third-Party Information. We divide third-party information into two categories: Similar users and Sentiment-consistent reviews, as they are commonly referred to by users in decision-making [16]. Inspired by user-based and item-based collaborative filtering [23], we further classify Similar users into two subcategories: *Historical similar users* and *Current similar users*. In addition, *Sentiment-consistent reviews* refer to item-related reviews that maintain sentimental consistency with the user's current dialogue.

Task Definition. Our task is: Taking User's current preference P^c, User's historical preference P^h, Historical similar users P^{lh}, Current similar users P^{lc} and Sentiment-consistent reviews P^r as input, we learn the linear and nonlinear relationships among them and obtain final user preference $P(u)$, to accurately recommend Items $\mathcal{I}_{t+1}$ and generate appropriate response $\mathcal{S}_{t+1}$ to the user u at the t-th turn of a conversation.

4   Method

In this section, we present the **A**ttentive **W**ide&**D**eep **Co**nversational **Re**commender (**AWDCore**) to the CRS. We first introduce how to learn the representations of users' own preferences and third-party information. Then we propose the Attentive Wide&Deep model to learn the low-order linear and high-order nonlinear relationships between the user's own preference and third-party information via feature crosses. Based on the learned user preference, we finally describe our solutions for both recommendation and conversation. The overview of AWDCore is shown in Fig. 2.

4.1   User's Own Preference Learner

Given the dialogue sessions, we encode the user's own preference in a separate entity and word semantic spaces. Next, we present the encoding modules for the user's current preference and the user's historical preference respectively.

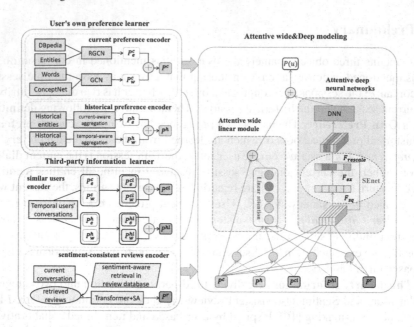

Fig. 2. The overview of our model AWDCore, where "SA" denotes self-attention. First, the representations of the user's own preference and third-party information are encoded by corresponding learners. Then, these multi-preferences are fused by the Attentive Wide&Deep modeling module.

Encoding User's Current Preference. In Entity semantic space, we use DBPedial [1] to enrich the representations of entities. To capture the semantic relations between entity nodes, we use RGCN [20] to encode DBPedial. We obtain the node representations matrix N_e of entities in the current conversation $C_e = \{e_1^T, ..., e_t^T\}$ on the top RGCN layer. As each entity is of different importance for user u, the self-attention mechanism is adopted to generate the user's current entity representation P_e^c. The processing approach of word space is consistent with the entity space, except that before self-attention we use ConceptNet [22] and GCN [11] to enhance the word's representation.

As there is a natural semantic gap between the two semantic spaces, we adopt contrastive learning [5] to bridge the semantic gap. Specifically, given a batch of sample B, we suppose that two representations h_1 and h_2 of a user will be more relevant than others. Therefore, the objective for aligning the semantic spaces is as follows:

$$\mathcal{L}_C(h_1, h_2) = \log \frac{\sum_{u \in B} e^{\mathrm{sim}(h_1^u, h_2^u)/\tau}}{\sum_{u,u' \in B} e^{\mathrm{sim}(h_1^u, h_2^{u'})/\tau}} \tag{1}$$

where $\mathrm{sim}(\cdot)$ is the cosine similarity and u' stands for other users in B except for u. Finally, we perform a gate mechanism to combine the words and entities representations to derive the user's current preference P^c.

Encoding User's Historical Preference. We encode the user's historical preference in a separate semantic space. In the entity space, considering historical sessions may confuse the current recommendation, following [14], for all historical entities $\mathcal{H}_e = \{H_e^1, ..., H_e^{T-1}\}$, we first use RGCN and self-attention to learn i-th session's entity representation h_e^i, and then aggregate all $T-1$ sessions according to their similarity with the current dialogue session. The final historical entity representation P_e^h is built as:

$$P_e^h = \sum_{i=1}^{T-1} sim(h_e^i, P_e^c)\, h_e^i,$$
$$sim(h_e^i, P_e^c) = \text{Softmax}(h_e^i W_s P_e^c\, /\, \lambda_e) \tag{2}$$

In the word space, we identify that the closer the session to the current one, the more effective it will be. And in a session, the utterance with the latter turn will be more important. Thus, for all historical words $\mathcal{H}_w = \{H_w^1, ..., H_w^{T-1}\}$, where $\mathcal{H}_w^i = \{w_1^i, ...w_t^i\}$ we first use the GCN and self-attention to obtain i-th session's word representation $\mathcal{V}_w^i = \{v_{w_1^i}, ..., v_{w_t^i}\}$. Then we further assign temporal weights to the representations at the word level and session level respectively, which is defined as follows:

$$P_w^h = \sum_{i=1}^{T-1} \Big(SM(T-1, i) \sum_{n=1}^{t} SM(t, n)\, v_{w_n^i} \Big) \tag{3}$$

where $SM(a, b) = \text{SoftMax}(1, ..., a)[b]$. After obtaining the historical representations of entity and word, similar to the user's current preference, we use contrastive learning as Eq. 1 to bridge the semantic gap, and then the gate mechanism is applied to combine the two representations to get the user's historical preference representation P^h.

4.2 Third-Party Information Learner

Encoding Similar Users. *For historical similar users*, Considering that users' preferences dynamically change over time, we compute the cosine similarity between users in units of historical recommendation turn. Concretely, for target user u, assuming that user u' has N historical recommendation turns, the definition of historical similar users of u in entity space is as follows:

$$P_e^{hl}(u, u') = \sum_{n=1}^{N} \phi(\text{sim}(P_e^h(u), P_e^h(u_n'))) \, P_e^c(u_n') \tag{4}$$

where $\phi(x) = \max(0, x - \phi_e)$, we avoid introducing too much noise via this truncation. $P_e^{hl}(u, u')$ is the similarity between u and u', and all these similarities are aggregated to obtain historically similar users of user u in entity space $P_e^{hl}(u)$. The processing of word space is the same as an entity. Finally, the gate mechanism is applied to combine two spatial representations to obtain historical similar users representation P^{lh}. *For current similar users*, entities or words mentioned in the current conversation are considered as the comparison object, we apply the same method as Eq. 4 to calculate the similarity between u' and target user u. Then, the two spatial representations are combined via a gate to obtain current similar users representation P^{lc}.

Encoding Sentiment-Consistent Reviews. Given the dialogue context C, considering that the reviews with sentiment consistency can be more valuable for reference [16], we firstly analyze the sentiment of dialogue context through analysis tool (e.g. NLTK), the score interval is set to $[0, 4]$. Then for the items mentioned in the context, we retrieve reviews from the reviews database with keeping their sentiment matching the context (e.g., they should be both positive or negative). Note that we use the same tool to conduct sentimental analysis for context and reviews. Since Transformer [18,24] has a good performance in NLP tasks, here we adopt Transformer to encode the retrieved reviews, so as to obtain the sentiment-consistent reviews representation P^r.

4.3   Attentive Wide&Deep Modeling

We design an Attentive Wide&Deep module to encode the low-order linear and high-order nonlinear relationships between the user's own preference and third-party information, highlighting the prior feature of the user's current preference. In this section, we will introduce the Attentive wide linear module and Attentive Deep neural networks. Note that the Attentive Wide Linear Module cannot be implicitly replaced by the Attentive Deep Neural Networks, as they learn the linear and nonlinear relationships at different level (vector and bit level respectively). By performing calculations at the vector level, the feature fields can be emphasized.

Attentive Wide Linear Module. Each representation is regarded as a feature of user preference, we define the second-order feature crosses as the operational combination of two representations. In this module, we attempt to learn the low-order (below the third-order) linear relationship between the user's own preference and third-party information via low-order feature crosses on the vector level.

Firstly, we perform a second-order feature cross between the user's current preference P^c and any other representations through the Hadamard product. We believe that the linear relationship between two features can be captured by this vector-level operation. Considering the different importance of feature crosses in different combinations, after the second-order feature crosses we use a linear attention mechanism to learn the weight values of different feature crosses, which is defined as:

$$\hat{y} = h^T \sum P^c \circ P' + b \tag{5}$$

where $P' \in \{P^h, P^{lh}, P^{lc}, P^r\}$, and $h \in \mathbb{R}^k, b \in \mathbb{R}$ denote the weights and bias respectively. We suppose the original feature closer to the final output will be more prominent since it can effectively avoid the information loss caused by the bottom layers' transmission. Based on the above consideration, we add the user's current preference representation to the output of Eq. 5 to highlight its prior premise. Accordingly, the final representation of the Attentive wide linear module is learned as:

$$\hat{y}_w = W_0 + W_1 P^c + h^T \sum P^c \circ P'$$ (6)

Via Eq. 6 the low-order linear relationship between the user's own preference and third-party information can be obtained, and the user's current preference can be highlighted at the same time.

Attentive Deep Neural Networks. Here, we study how to capture the high-order nonlinear relationship between the user's own preference and third-party information. It is inefficient to directly cross high-order features as the low-order since there are more than 2^n high-order feature combinations for n features. For efficiency, we first use a SENet [8] layer on the vector level to emphasize informative features for user preference learning, and then the DNN is conducted to perform high-order feature crosses on the bit-level.

Firstly, the SENet is divided into two steps: Squeeze and Excitation. The Squeeze operation is designed to compress each feature to extract a more refined feature descriptor Z, which is achieved by using a global average pooling. Formally, for the feature map $P_s \in \mathbb{R}^{(N \times bs \times embed)}$ formed by stacking N features with embedding dim (bs, C), global average pooling is adopted as follows, where $Z \in \mathbb{R}^{(1 \times 1 \times C)}$:

$$Z = F_{sq}(P_s) = \frac{1}{N \times bs} \sum_{i=1}^{N} \sum_{j=1}^{bs} P_s(i, j)$$ (7)

According to the feature descriptor Z, then we use the Excitation operation to obtain the weight of each feature and excite the corresponding feature:

$$S = F_{ex}(Z) = \sigma(W_2 \delta(W_1 Z))$$ (8)

where δ is the ReLU function, σ is the sigmoid function, and $W_1 \in \mathbb{R}^{\frac{c}{r} \times C}$ is the learnable weight matrix of the first Full Connection Layer (FC), which is used to compress the feature of the C-dim and fully capture feature-wise dependencies, $W_2 \in \mathbb{R}^{C \times \frac{c}{r}}$ is another weight matrices of the second FC to restore dimension. The function of these two FCs is to fuse the information of each feature, to learn the importance of each feature. The output of the SENet is obtained by rescaling P_s with the activations S:

$$\hat{y}_{se} = F_{scale}(P_s, S) = P_s S$$ (9)

Via Eq. 7, 8, 9 we believe that the correlation among features can be explicitly modeled, and the prior feature of the user's current preference can be automatically highlighted at the same time. Considering that DNN can calculate the nonlinear relationship among representations on the bit-level, and it has the advantages of fast speed with simple structure, we adopt the DNN to autonomously perform high-order feature crosses. The $l-$ hidden layer of DNN is defined as:

$$y^{l+1} = ReLU(BN(W^l y^l + b))$$ (10)

where W^l and b are two learnable parameters, BN is Batch Normalization. We set up 4 such hidden layers to get the final output of this module $\hat{y}_d$. Finally, we obtain the learned user's preference representation as:

$$P(u) = \hat{y}_w + \lambda\hat{y}_d \tag{11}$$

4.4 Optimization

Given the final user preference $P(u)$, We learn how to leverage it to both recommendation and response generation tasks. Firstly, **Objective for Recommendation**. The probability that recommends an item i to user u is calculated as: $p_{rec}(i, u) = $ Softmax$(P(u)^\top r_i)$, where r_i is the representation of item i. Then we apply a cross-entropy loss as the optimization objective for recommendation:

$$\mathcal{L}_{rec} = -\sum_{u \in \mathcal{U}} \sum_{i=1}^{M} \log p_{rec}(i, u) + \lambda_{cl} \sum_{j \in \{c,h\}} \mathcal{L}_{\mathcal{C}}(P_w^j, P_e^j) \tag{12}$$

where M is the number of items, $\mathcal{L}_{\mathcal{C}}(P_w^j, P_e^j)$ denotes the aligning semantic spaces loss.

Objective for Response Generation. Here for the response generation task, we use Transformer [24] as the encoder-decoder framework following [14,29]. The input utterances are first encoded to capture the global semantic feature via multi-head attention and fully connected feed-forward layers, then the decoder outputs a predicted representation R_d for the next token generation. In order to improve users' satisfaction with the recommendations and generate more informative responses, here we introduce the user preference into token generation as a bias:

$$p_{res}(y_t|y_1, ..., y_{t-1}) = \text{Softmax}(W^r R_d + \mathcal{F}(P(u))[y_t] \tag{13}$$

Where $\mathcal{F}$ is a linear function used to align the dimension of users and vocabulary. Then we use the cross-entropy loss to train the response generation:

$$\mathcal{L}_{res} = -\sum_{u \in \mathcal{U}} \sum_{t=2}^{N_t} \log(p_{res}(y_t|y_1, ..., y_{t-1})) \tag{14}$$

5 Experiment

5.1 Experimental Setting

Datasets. We evaluate our model on TG-Redial [28] and Redial [13] datasets, as both datasets associate every dialogue session with a unique user identity, allowing us to obtain historical sessions about each user. In order to obtain the user's historical dialogue information, the two datasets are arranged in chronological order. We randomly select several users' two consecutive rounds (for TG-Redial) or three consecutive rounds (For Redial) of dialogue sessions as validation and test sets, and the remaining is train sets. The training, validation, and test sets are split with a ratio of 8:1:1. Since the above datasets do not contain the review data, the reviews for movies in TG-ReDial and ReDial are retrieved from douban[1] and IMDB[2] respectively.

Baselines. We consider two major subtasks for evaluation in CRS, namely recommendation and conversation. Therefore, We not only compare our model with CRS models but also select several classic recommendation or conversation models as baselines: (1) *TextCNN* [10]utilizes a CNN-based model to encode current conversational context to learn user preferences. (2) *SASRec* [9] exploits the user's historical information for recommendation. (3) *Transformer* [24] utilizes a encoder-decoder structure to generate conversational responses. (4) *KBRD* [2] adopts an external KG to strengthen the understanding of user preferences. (5) *KGSF* [29] utilizes both entity-oriented and word-oriented KGs to enhance the representation of conversation data. (6) *RevCore* [16] incorporates reviews for the first time to enhance the understanding of user preferences. (7) *UCCR* [14] makes a linear combination of the user's current preferences, historical preferences, and similar user preferences to capture user preferences.

Evaluation Metrics. For recommendation tasks, we want to know whether AWDCore can accurately model user preference to enable the system to provide accurate recommendations, thus we adopt NDCGG@k and MRR@k ($k = 10, 50$). For the conversation task, we use Distinct n-gram ($n = 2, 3, 4$) to measure the diversity. BLEU-2,3 [17] is adopted for testing the generations' accuracy.

Implementation Details. The embedding dimensions for recommendation and conversation are 128 and 300 respectively. The layer numbers of both RGCN and GCN are 1. We use Adam optimize with a learning rate of 0.001, where the batch size is 128. In experiments, we consider the semantic fusion (Eq. (1)) as a pre-training task, whose epochs are 3, and the epochs for recommendation and dialogue tasks are 16 and 20 respectively. The λ_e in historical entities encoder (Eq. (2)) equals 0.1, the λ in user final preference representation (Eq. (11)) is 0.35.

5.2 Evaluation on Recommendation Task

In this part, we conduct a series of experiments to verify the effectiveness of our proposed method on the recommendation task, the results are shown in Table 1. In general, CRS models perform better than non-CRS models (e.g., SASRec and TextCNN). Since these CRS models emphasize the natural integration of recommendation results into the conversation, which mutually improves both each other. In CRS models, firstly, KGSF achieves better performance than KBRD. It aligns the word and entity semantic spaces through mutual information maximization to enhance data representation. Secondly, Revcore performs better than KGSF. As it introduces review information to help better

---

[1] https://movie.douban.com/.
[2] https://www.imdb.com/.

Table 1. The recommendation results. The marker * indicates that the improvement is statistically significant compared with the best baseline.

Dataset	TG-ReDial				ReDial			
Models	NDCG@10	NDCG@50	MRR@10	MRR@50	NDCG@10	NDCG@50	MRR@10	MRR@50
SASRec	0.0017	0.0042	0.0012	0.0013	0.0324	0.0617	0.0320	0.0401
TextCNN	0.0025	0.0055	0.0021	0.0027	0.0496	0.0788	0.0431	0.0487
KBRD	0.0075	0.0145	0.0055	0.0063	0.1010	0.1328	0.0794	0.0858
KGSF	0.0096	0.0175	0.0073	0.0088	0.1110	0.1556	0.0837	0.0932
Revcore	0.0110	0.0183	0.0080	0.0092	0.1145	0.1611	0.0861	0.0956
UCCR	0.0122	0.0214	0.0088	0.0107	0.1182	0.1642	0.0883	0.0981
AWDCore	0.0205*	0.0341*	0.0120*	0.0174*	0.1210*	0.1723*	0.0903*	0.1054*

model user preference. Finally, UCCR achieves better performance than other baselines. It supplements the current session with historical sessions and look-alike users, modeling these user diverse preferences in a linear manner.

Our model AWDCore outperforms all the baselines since AWDCore utilizes both user's own preferences and third-party information (similar users, sentiment-consistent reviews) to help understand user preference. To fuse these user multi-preferences, we maintain the independence of the user's own preferences and third-party information and model the linear and nonlinear relationships between the two. Such a way can be beneficial to the recommendation module by fully fusing these user multi-preferences.

5.3   Evaluation on Conversation Task

Table 2. Results of conversation. The marker * indicates that the improvement is statistically significant compared with the best baseline.

Dataset	TG-ReDial					ReDial				
Models	Bleu-2	Bleu-3	Dist-2	Dist-3	Dist-4	Bleu-2	Bleu-3	Dist-2	Dist-3	Dist-4
Transformer	0.0274	0.0089	0.2638	0.5167	0.7936	0.0215	0.0061	0.0548	0.2018	0.4325
KBRD	0.0423	0.0119	0.3482	0.6911	0.9972	0.0238	0.0088	0.0712	0.2883	0.4893
KGSF	0.0461	0.0135	0.4447	1.0450	1.5792	0.0249	0.0091	0.0756	0.3024	0.5177
Revcore	0.0467	0.0136	0.4513	1.0932	1.6631	0.0252	0.0098	0.0769	0.3065	0.5283
UCCR	0.0494	0.0145	0.5365	1.2783	1.9376	0.0257	0.0106	0.0818	0.3289	0.5635
AWDCore	0.0679*	0.0261*	0.6323*	1.5540*	2.2072*	0.0312*	0.0146*	0.1061*	0.4064*	0.6238*

The conversation results are shown in Table 2, we can note that KBRD performs better than the basic Transformer model. Since it adopts KG-based representations of entities to generate responses. KGSF outperforms KBRD as it performs KG-based Transformer attention layers to interact with the contextual information with the generated response. Besides, Revcore achieves better performance than KGSF on Dist metrics, It introduces reviews as a richer external source, bringing more diversity. Finally, among all the baselines, UCCR achieves the best performance. It enhances user representation with user historical preference and look-alike users and adds the enhanced representation bias into the generated responses.

Compared with these baselines, Our AWDCore performs consistently better in all evaluation metrics. The main reason is that AWDCore integrates user self-preferences, similar users, and reviews to better model user preference, and serves this preference representation as a vocabulary bias to generate proper tokens. In addition, the two external knowledge of reviews and KG can also assist in generating informative responses.

5.4 Ablation Study

To understand the effectiveness of each component, we study the performances of three variants of AWDCore: (1) AWDCore w/o Wide removes the Attentive wide linear module; (2) AWDCore w/o Deep removes the Attentive deep neural networks; (3) AWD-Core w/o Attentive removes the attentive components in both wide and deep modules.

Table 3. Results of ablation and variation study on the recommendation task.

Dataset	TG-ReDial				ReDial			
Models	NDCG@10	NDCG@50	MRR@10	MRR@50	NDCG@10	NDCG@50	MRR@10	MRR@50
AWDCore	**0.0205**	**0.0341**	**0.0120**	**0.0174**	**0.1210**	**0.1723**	**0.0903**	**0.1054**
w/o Wide	0.0198	0.0321	0.0113	0.0162	0.0928	0.1427	0.0707	0.0892
w/o Deep	0.0126	0.0205	0.0091	0.0106	0.0812	0.1313	0.0675	0.0741
w/o Attentive	0.0173	0.0296	0.0098	0.0158	0.0854	0.1201	0.0621	0.0695

As shown in Table 3, Firstly, we can see that removing any component would result in a decrease in model performance, which demonstrates the effectiveness of all components in our method. In particular, removing attentive deep neural networks leads to the largest performance decrease, since attentive deep neural networks are used to model the complex nonlinear relationships among user multi-preferences, it is the key to fully fuse user multi-preferences. Besides, the variants that remove attentive components also lead to a sharp performance decrease, which verifies that user current preference is the main basis for user preference learning.

6 Conclusion and Future Work

In this paper, we proposed a novel method AWDCore for user preference modeling in CRS. By maintaining the independence of the user's own preference and third-party information, we model the linear and nonlinear relationships between the two, such that user multi-preferences can be fully fused. Extensive experiments verify the effectiveness of our approach on CRS both recommendation and conversation tasks. In future work, We will investigate how to better use user preference representation for dialogue generation to make the response more persuasive and explainable for the recommendation results.

Acknowledgements. This work was supported by the National Natural Science Foundation of China (62376188, 62272340, 61876129, 62276187, 61976154).

References

1. Auer, S., Bizer, C., Kobilarov, G., Lehmann, J., Cyganiak, R., Ives, Z.: DBpedia: a nucleus for a web of open data. In: Aberer, K., et al. (eds.) ASWC/ISWC-2007. LNCS, vol. 4825, pp. 722–735. Springer, Heidelberg (2007). https://doi.org/10.1007/978-3-540-76298-0_52
2. Chen, Q., et al.: Towards knowledge-based recommender dialog system. In: EMNLP, pp. 1803–1813 (2019)
3. Cheng, H.T., et al.: Wide & deep learning for recommender systems. In: DLRS, pp. 7–10 (2016)
4. Christakopoulou, K., Radlinski, F., Hofmann, K.: Towards conversational recommender systems. In: SIGKDD (2016)
5. Gao, T., Yao, X., Chen, D.: SimCSE: simple contrastive learning of sentence embeddings. arXiv arXiv:2104.08821 (2021)
6. Guo, H., Tang, R., Ye, Y., Li, Z., He, X.: DeepFM: a factorization-machine based neural network for CTR prediction. In: IJCAI, pp. 1725–1731 (2017)
7. He, X., Chua, T.S.: Neural factorization machines for sparse predictive analytics. In: SIGIR, pp. 355–364 (2017)
8. Hu, J., Shen, L., Sun, G., Albanie, S.: Squeeze-and-excitation networks. In: CVPR, pp. 7132–7141 (2018)
9. Kang, W.C., Mcauley, J.: Self-attentive sequential recommendation. In: ICDM, pp. 197–206 (2018)
10. Kim, Y.: Convolutional neural networks for sentence classification. In: EMNLP, pp. 1746–1751 (2014)
11. Kipf, T.N., Welling, M.: Semi-supervised classification with graph convolutional networks. arXiv arXiv:1609.02907 (2016)
12. Lei, W., et al.: Estimation-action-reflection: towards deep interaction between conversational and recommender systems. In: WSDM, pp. 304–312 (2020)
13. Li, R., Kahou, S., Schulz, H., Michalski, V., Charlin, L., Pal, C.: Towards deep conversational recommendations. In: NeurIPS, pp. 9748–9758 (2018)
14. Li, S., Xie, R., Zhu, Y., Ao, X., Zhuang, F., He, Q.: User-centric conversational recommendation with multi-aspect user modeling. In: SIGIR, pp. 223–233 (2022)
15. Lian, J., Zhou, X., Zhang, F., Chen, Z., Xie, X., Sun, G.: xDeepFM: combining explicit and implicit feature interactions for recommender systems. In: SIGKDD, pp. 1754–1763 (2018)
16. Lu, Y., et al.: RevCore: review-augmented conversational recommendation. In: ACL, pp. 1161–1173 (2021)
17. Papineni, K., Roukos, S., Ward, T., Zhu, W.J.: BLEU: a method for automatic evaluation of machine translation. In: ACL, pp. 311–318 (2002)
18. Rahali, A., Akhloufi, M.A.: End-to-end transformer-based models in textual-based NLP. AI **4**, 54–110 (2023)
19. Ren, X., Yin, H., Chen, T., Wang, H., Zheng, K.: Learning to ask appropriate questions in conversational recommendation. In: SIGIR, pp. 808–817 (2021)
20. Schlichtkrull, M., Kipf, T.N., Bloem, P., Van Den Berg, R., Titov, I., Welling, M.: Modeling relational data with graph convolutional networks. In: 15th International Conference on the Semantic Web, ESWC, pp. 593–607 (2018)
21. Song, W., Shi, C., Xiao, Z., Duan, Z., Tang, J.: AutoInt: automatic feature interaction learning via self-attentive neural networks. In: ACM (2019)

22. Speer, R., Chin, J., Havasi, C.: ConceptNet 5.5: an open multilingual graph of general knowledge. In: AAAI, pp. 4444–4451 (2017)
23. Thakkar, P., Varma, K., Ukani, V., Mankad, S., Tanwar, S.: Combining user-based and item-based collaborative filtering using machine learning. In: Satapathy, S.C., Joshi, A. (eds.) Information and Communication Technology for Intelligent Systems. SIST, vol. 107, pp. 173–180. Springer, Singapore (2019). https://doi.org/10.1007/978-981-13-1747-7_17
24. Vaswani, A., et al.: Attention is all you need. In: NIPS, pp. 6000–6010 (2017)
25. Wang, R., Fu, B., Fu, G., Wang, M.: Deep & cross network for ad click predictions. In: ADKDD (2017)
26. Xiao, J., Ye, H., He, X., Zhang, H., Wu, F., Chua, T.S.: Attentional factorization machines: Learning the weight of feature interactions via attention networks. In: IJCAI, pp. 3119–3125 (2017)
27. Xie, Z., Yu, T., Zhao, C., Li, S.: Comparison-based conversational recommender system with relative bandit feedback. In: SIGIR (2021)
28. Zhou, K., Zhou, Y., Zhao, W.X., Wang, X., Wen, J.R.: Towards topic-guided conversational recommender system. In: COLING, pp. 4128–4139 (2020)
29. Zhou, K., Zhao, W.X., Bian, S., Zhou, Y., Wen, J.R., Yu, J.: Improving conversational recommender systems via knowledge graph based semantic fusion. In: SIGKDD, pp. 1006–1014 (2020)
30. Zhou, Y., Zhou, K., Zhao, W.X., Wang, C., Jiang, P., Hu, H.: C^2-crs: Coarse-to-fine contrastive learning for conversational recommender system. In: WSDM (2022)

Debiasing Medication Recommendation with Counterfactual Analysis

Pei Tang, Chunping Ouyang$^{(\boxtimes)}$, and Yongbin Liu

University of South China, Hengyang, China
ieptsky@gmail.com, ouyangcp@gmail.com, yongbinliu@usc.edu.cn

Abstract. The AI-driven medication recommendation has emerged as a crucial undertaking in the field of healthcare research. Recent literature has focused on leveraging patients' diagnoses, procedures, and historical visit information for medication recommendation. However, this approach can lead to recommendation biases due to spurious correlations among the historical visit information. Previous studies have either failed to address this bias issue or attempted to mitigate recommendation biases through dataset manipulation, albeit at the expense of increased computational costs. In this study, we propose CAMeR (Counterfactual Analysis based Medication Recommendation), which is a novel debiasing model based on counterfactual analysis. The model preserves medications information while emphasizing the primary influence of diagnoses and procedures. Unlike traditional factual reasoning approaches that address biases before or during training, counterfactual reasoning mitigates the impact of post-training spurious correlations. Additionally, we incorporate contrastive loss computation in the embedding module of our model to calibrate the feature construction for patients with multiple visit information. We validate the CAMeR on widely adopted datasets, MIMIC-III and MIMIC-IV, and experimental results unequivocally demonstrate its superiority over state-of-the-art methods.

Keywords: Counterfactual Analysis · Medication Recommendation · Contrastive Learning · Causal Inference

1 Introduction

With the emergence of a vast amount of medical data, particularly electronic health records (EHRs), leveraging deep learning techniques and EHRs for medication recommendation has become a current research focus. The existing studies mainly revolve around recommending drug combinations based on patients' clinical information [1]. Taking the patient depicted in Fig. 1 as an example, during each visit, the patient is diagnosed with a series of diseases and undergoes a range of medical procedures. Diagnoses and procedures, among others, are considered clinical information by physicians and are used to prescribe medications. The combination of medications and clinical information constitutes the information

© The Author(s), under exclusive license to Springer Nature Singapore Pte Ltd. 2024
B. Luo et al. (Eds.): ICONIP 2023, CCIS 1964, pp. 426–438, 2024.
https://doi.org/10.1007/978-981-99-8141-0_32

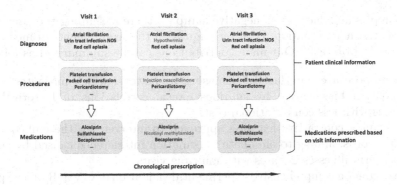

Fig. 1. An example of electronic health records.

for each visit. The physician recommends a series of medications for the patient based on information from previous visits and current clinical information, a process known as medication recommendation. To ensure comprehensive medication recommendations, physicians need to consider various clinical information and assess potential drug-drug interactions (DDIs), making the recommendation process complex and time-consuming. Hence, simulating the medication recommendation process using deep learning techniques to provide improved recommendations for physicians has become a crucial task.

Existing deep learning medication recommendation methods [2,9,10,16–18] have made significant progress, but there are still two unavoidable influences. First, the current work would introduce historical visit information to assist with recommendations in addition to using clinical information such as current diagnoses and procedures. However, due to spurious correlation between historical visit information and current symptoms, there is a recommendation bias in using visit information containing historical information for medication recommendation. Second, in cases where a patient has multiple visits, one of the visits (as shown in Fig. 1, Visit 2) may contain visit information that does not match the overall characteristics of the patient due to an transient illness symptoms. These visit information can disrupt the medication recommendation process and further misleading the recommendation outcomes.

We propose a Counterfactual Analysis based Medication Recommendation (CAMeR) to address the aforementioned issues. Firstly, we construct a causal graph for factual reasoning. By considering the prediction process from a counterfactual perspective, we assume unobserved clinical information at the current time to maximize the model's tendency for erroneous predictions. We then eliminate this predictive bias using causal intervention, truncating the backdoor paths between input data and the model, thereby alleviating bias. Secondly, to address the insufficient information encoding between multiple visits, we employ contrastive learning loss to constrain the relationships among diagnoses, procedures, and medications across multiple visits. Specifically, we treat diagnoses, procedures, and medications that appear multiple times in visit records as positive samples and the rest as negative samples, thereby shaping the embedding process of visit information to better align with the requirements of medication recommendation tasks. Our main contributions can be summarized as follows:

- We propose a counterfactual debiasing framework, which mitigates recommendation bias caused by spurious causal correlations of historical visit records with less computational cost.
- We employ contrastive learning approach to intervene in the embedding of visit records, aiming to eliminate information interference caused by sudden short-term illnesses in cases with multiple visits.
- We conduct a comprehensive experimental validation of CAMeR on the public datasets MIMIC-III and MIMIC-IV, and the experimental results show that its performance exceeds that of the current optimal methods.

2   Problem Formulation

In EHRs, each patient can be represented as a sequence of medical visit information $R(n) = [V_1^n, V_2^n, \ldots, V_n^n]$, where $n \in \{1, 2, \ldots, N\}$ denotes the patient number, $t \in \{1, 2, \ldots, T\}$ denotes the visit number. N represents the total number of patients, and T represents the maximum number of visits for the corresponding patient R_n. To simplify the description, we will omit the index n, and subsequent work will use the longitudinal visit information R for a single patient, given by $R = [V_1, V_2, \ldots, V_t]$, where each visit's information is represented as $V(t) = [M_t, D_t, P_t]$. For the t-th visit of a single patient, $M_t = [m_1^t, m_2^t, \ldots, m_i^t]$ represents the set of medications, $D_t = [d_1^t, d_2^t, \ldots, d_j^t]$ represents the set of diagnoses, and $P_t = [p_1^t, p_2^t, \ldots, p_k^t]$ represents the set of procedures. Here, $i \in \{1, 2, \ldots, I\}$, $j \in \{1, 2, \ldots, J\}$, and $k \in \{1, 2, \ldots, K\}$, where I, J, and K indicate the maximum number of disease diagnoses and procedures for the t-th visit record. In addition, prior to considering the recommendation of drug combinations, we need to take into account the EHRs graph and DDIs graph. We represent the EHRs graph and DDIs graph using binary adjacency matrices A_e and A_d respectively, $A_e, A_d \in A^{|M| \times |M|}$. where M represents the collection of medications, $A_e[i, j] = 1$ indicates that the i-th drug and the j-th drug form a drug combination, and $A_d[i, j] = 1$ indicates the presence of adverse effects between the i-th and j-th drugs.

The objective of medication recommendation is to generate the current medication M_t based on the patient's current diagnosis D_t, procedure P_t, historical medical records $[V_1, V_2, \ldots, V_{t-1}]$, the EHRs graph A_e, and the DDIs graph A_d.

3   The CAMeR

Figure 2 illustrates the architecture of CAMeR. The model consists of three components: a historical scoring network with contrastive loss embedding, a Transformer-based generation network, and a counterfactual debiasing framework based on causal reasoning.

3.1   Historical Score Computation

The module determines the likelihood of retrieving drugs from the prescription history based on the similarity between current and historical information.

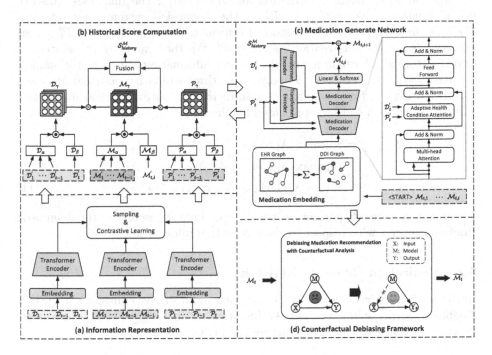

Fig. 2. An overview of our proposed CAMeR: (a) Information Representation; (b) Historical Score Computation; (c) Medication Generate Network; and (d) Counterfactual Debiasing Framework.

Information Representation. According to Sect. 2, we understand that the patient denote as $R = [[M_1, \ldots, M_{t-1}], [D_1, \ldots, D_t], [P_1, \ldots, P_t]]$. For these data, we apply the same treatment, utilize embedding layers and Transformer encoding modules for embedding. Then, the impact of unexpected conditions in the visit data on the construction of patient features with multiple visits is considered. For example, in Fig. 1, the red font represents the emergent conditions in the visit information, which, if incorporated into the construction of patient features, will affect the recommendation of the later visit as a disturbing factor. To solve this problem, we introduce contrast learning in the process of patient information embedding, and obtain the contrast loss $\mathcal{L}_{\text{contrastive}}$ through the cosine similarity between features. Contrastive learning can constrain the embedding vectors [3, 12, 19], bringing positively paired vectors closer and negatively paired vectors apart. Ultimately, we obtain the embedded vectors corresponding to our visit data, denoted as $H_r = [[M_1', \ldots, M_{t-1}'], [D_1', \ldots, D_t'], [P_1', \ldots, P_t']]$. It is worth noting that we do not obtain $M_{t,i}'$ from the input.

Compute Historical Score. The historical scoring network takes the embedding vector and the intermediate output $M'_{t,i}$ of the generation network as inputs. Here, we define the medication's historical information as $\mathcal{M}_\alpha = [M'_1, \ldots, M'_{t-1}]$, the medication's current information as $\mathcal{M}_\beta = M'_{t,i}$; the diagnosis's historical information as $\mathcal{D}_\alpha = [D'_1, \ldots, D'_{t-1}]$, the diagnosis's current information as $\mathcal{D}_\beta = D'_t$; and the procedure's historical information as $\mathcal{P}_\alpha = [P'_1, \ldots, P'_{t-1}]$, the procedure's current information as $\mathcal{P}_\beta = P'_t$. We then multiply the aforementioned historical information with the current information to obtain the similarity matrix $\mathcal{M}_\gamma$, $\mathcal{D}_\gamma$, and $\mathcal{M}_\gamma$. Finally, these three similarity matrices are fused together using the fusion module to obtain the ultimate historical medication score $\mathcal{S}^{\mathcal{M}}_{\text{history}}$, which can be represented by the following formula:

$$
\begin{aligned}
\mathcal{S}^{\mathcal{M}}_{\text{history}} &= \text{Softmax}\left(\lambda_1 \mathcal{M}_\gamma \times \mathcal{D}_\gamma + (1 - \lambda_1)\,\mathcal{M}_\gamma \times \mathcal{P}_\gamma\right) \\
&= \text{Softmax}\left(\lambda_1 \left(\mathcal{M}_\alpha \odot \mathcal{M}_\beta\right) \times \left(\mathcal{D}_\alpha \odot \mathcal{D}_\beta\right) \right. \\
&\quad \left. + (1 - \lambda_1)\left(\mathcal{M}_\alpha \odot \mathcal{M}_\beta\right) \times \left(\mathcal{P}_\alpha \odot \mathcal{P}_\beta\right)\right)
\end{aligned}
\tag{1}
$$

where λ_1 is a learnable parameter, which represents the weight of the diagnostic similarity matrix with respect to the medication information.

3.2 Medication Generate Network

In the medication generate network, for each iteration, we consider the current diagnosis D_t, procedure P_t, and the historical rating $\mathcal{S}^{\mathcal{M}}_{\text{history}}$ of the previously generated medication $M_{t,i}$, to generate the next medication $M_{t,i+1}$.

Medication Graph Embedding. We construct a medication embedding layer based on the EHRs graph A_e and DDIs graph A_d. We utilize GCN (Graph Convolutional Network) to learn features. The GCN process is as follows:

$$
\text{GCN}(\boldsymbol{A}, \boldsymbol{W}) = \text{ReLU}\left(\tilde{\boldsymbol{D}}^{-\frac{1}{2}}\left(\boldsymbol{A} + \boldsymbol{I}\right)\tilde{\boldsymbol{D}}^{-\frac{1}{2}}\boldsymbol{W}\right)
\tag{2}
$$

where $\tilde{\boldsymbol{D}}$ is a diagonal matrix such that $\tilde{\boldsymbol{D}}_{ii} = \sum_j \boldsymbol{A}_{ij}$ and $\boldsymbol{I}$ are identity matrices. We employed a two-layer GCN to learn features on both graphs, with specific processing steps for A_e and A_d as follows:

$$
\boldsymbol{B}_e = \text{GCN}(\boldsymbol{A}_e, \text{GCN}(\boldsymbol{A}_e, \boldsymbol{W}_e)\boldsymbol{W}_1)
\tag{3}
$$

$$
\boldsymbol{B}_d = \text{GCN}(\boldsymbol{A}_d, \text{GCN}(\boldsymbol{A}_d, \boldsymbol{W}_d)\boldsymbol{W}_2)
\tag{4}
$$

$$
\boldsymbol{E}_m = \boldsymbol{B}_e + \lambda_2 \boldsymbol{B}_d
\tag{5}
$$

where $\boldsymbol{W}_1$, $\boldsymbol{W}_2$ and λ_2 are learnable parameters, while $\boldsymbol{W}_e$ and $\boldsymbol{W}_d$ represent embeddings derived from the EHRs graph and the DDIs graph, respectively. The parameter λ_2 is utilized to allocate weights for graph fusion, ultimately resulting in $\boldsymbol{E}_m$, which serves as one of the inputs for the decoder.

Generate Network. We have incorporated a two-layer medication decoder based on Transformer decoder layers. Specifically, we have named them decoder1 and decoder2. It is noteworthy how we integrate the diagnosis D'_t and procedure P'_t, as illustrated in Fig. 2(c). We treat them separately as inputs to a multi-head attention layer and another Transformer encoder. The simplified process of decoder1 is as follows:

$$E_1 = E_m + \text{MultiHeadAttention}(E_m, E_m, E_m) \tag{6}$$
$$E_2 = E_1 + \text{MultiHeadAttention}(E_1, D'_t, P'_t) \tag{7}$$
$$E_3 = E_2 + \text{FNN}(E_2) \tag{8}$$

The output E_3 here and D''_t, P''_t, which has undergone another encode, are then used as input to decode2.

$$E_4 = \text{MedDecoder2}(E_3, D''_t, P''_t)$$
$$= \text{MedDecoder2}(\text{MedDecoder1}(E_m, D'_t, P'_t), D''_t, P''_t) \tag{9}$$
$$M'_{t,i} = \text{Softmax}(\text{Linear}(E_4)) \tag{10}$$

The symbol $M'_{t,i}$ represents the generation probability of each medication in the list, and ultimately, this result is fused with the historical score $\mathcal{S}^M_{\text{history}}$.

$$M_{t,i+1} = \lambda_3 \mathcal{S}^M_{\text{history}} + (1 - \lambda_3) M'_{t,i} \tag{11}$$

The parameter λ_3 is trainable, and $M_{t,i+1}$ represents the selection probabilities of all medications in the medication table, After multiple generations, the final recommended set of medications M_t is obtained.

3.3   Counterfactual Debiasing Framework

Figure 3(a) shows the factual inference causal diagram for the medication recommendation task, where X includes the D_t, P_t, and historical visit information. We note that the focus of the medication recommendation task is on D_t, P_t itself, although historical medical history contains more potential information,

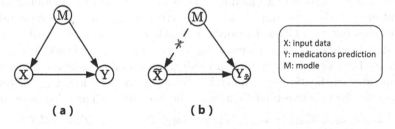

(a)   **(b)**

Fig. 3. The original causal graph (a) and its counterfactual alternatives to the recommendation bias (b). $\tilde{X}$ represents the removal of D_t and P_t from input data, and $Y_{\tilde{x}}$ is the corresponding output.

but also introduces data information bias due to its spurious correlations. Using Bayesian inference rules [13], we can view the inference process as follows:

$$Y_x = P(Y \mid X) = \sum_c P(Y \mid X, c) P(c \mid X) \tag{12}$$

The variable c represents any confounding factor captured by the model trained on a biased training dataset. In the presence of these confounding factors, the model incorporates spurious correlations between data during the inference process. Consequently, when the dataset undergoes changes, the Recommended Performance often suffer a decline due to the influence of these spurious correlations. This situation affected by spurious causality is known as the backdoor path [6] in causal reasoning: $M \rightarrow X$.

Causal inference encourages us to directly manipulate the nodes in the causal graph and observe the outputs, a behavior known as causal intervention [6], denoted as $do(\cdot)$. We employ the concept of counterfactual thinking to imbue the model with imagination, enabling the analysis and intervention of recommendation bias. Specifically, we assume that the model cannot observe D_t and P_t in the input x, thereby maximizing the model's recommendation bias. In Fig. 3(b), we depict the predicted outcomes after intervention as follows:

$$Y_{\tilde{x}} = Y(do(X = \tilde{x})) \tag{13}$$

where $\tilde{x}$ represents the result of counterfactual processing applied to the input data X. Our ultimate objective is to strengthen the direct influence of D_t and P_t on the predicted outcomes, thereby mitigating data bias: $Y_x \backslash Y_{\tilde{x}}$. To accomplish this goal, we employ a simple yet empirically effective subtraction operation:

$$Y_{\text{last}} = Y_x - \delta Y_{\tilde{x}} = Y_x - \delta Y(do(X = \tilde{x})) \tag{14}$$

where δ is a hyperparameter to balance the relationship between bias distillation and result recommendation. Our proposed CAMeR replaces the conventional one-time prediction with Y_{last}, generating debiased medication recommendation. This essentially entails a "second round" of prediction: one targeting the original observation Y_x, and another targeting the counterfactual assumption $Y_{\tilde{x}}$.

3.4   Training Objective

We utilize the Negative Log-Likelihood Loss $\mathcal{L}_{\text{nll}}$ to train the model. Similarly, as mentioned earlier, we employ a contrastive loss $\mathcal{L}_{\text{contrastive}}$ to constrain the relationships between medical records. Following certain rules, we obtain three contrastive losses $\mathcal{L}_{\text{m2p}}$, $\mathcal{L}_{\text{m2d}}$, and $\mathcal{L}_{\text{d2p}}$ for pairwise comparisons. Additionally, to enhance the safety of the model and reduce potential drug interactions among recommended medication combinations, we apply the DDI loss $\mathcal{L}_{\text{ddi}}$ to regulate the adverse effects between medications. The specific configuration is as follows:

$$\mathcal{L} = \begin{cases} \mathcal{L}_{\text{nll}} + \omega_{\text{m2p}} \mathcal{L}_{\text{m2p}} + \omega_{\text{m2d}} \mathcal{L}_{\text{m2d}} + +\omega_{\text{d2p}} \mathcal{L}_{\text{d2p}} + \omega_{\text{ddi}} \mathcal{L}_{\text{ddi}}, DDI > \psi \\ \mathcal{L}_{\text{nll}} + \omega_{\text{m2p}} \mathcal{L}_{\text{m2p}} + \omega_{\text{m2d}} \mathcal{L}_{\text{m2d}} + +\omega_{\text{d2p}} \mathcal{L}_{\text{d2p}}, \qquad\qquad DDI \leq \psi \end{cases} \tag{15}$$

where ψ represents the threshold for DDI, allowing us to confine the DDI rate within the desired threshold.

4   Experiments

4.1   Dataset

We utilized the MIMIC-III [5] and MIMIC-IV [4] datasets, which are publicly available critical care medical information datasets released on PhysioNet. Table 1 summarizes the statistical information of the datasets. Furthermore, we incorporated DDIs knowledge from the TWOSIDES dataset [11].

Table 1. Statistics of the data.

Items	MIMIC-III	MIMIC-IV
# of patients/# of visits	6350/15032	6136/17813
med./prod./diag. space size	131/1430/1958	121/4001/1851
avg. # of visits	2.367	2.90
avg. # of med./prod./diag. per visit	11.44/3.84/10.51	6.68/2.18/11.78
total # of DDI pairs	448	337

4.2   Baselines and Metrics

Under the same data processing conditions [18], we compared our method with mainstream baselines.

- **LR** is the standard Logistic Regression.
- **ECC** [8] employing multiple SVM classifiers for medication recommendation.
- **LEAP** [20] is a sequential generation system that utilizes LSTM to construct models for medication recommendation.
- **GameNet** [9] integrates drug-drug interaction knowledge as a memory module implemented through graph convolutional networks.
- **MICRON** [17] is a recurrent residual learning model proposed in the paper for drug change prediction problems.
- **SafeDrug** [18] Equipped with a Message Passing Neural Network (MPNN) module and a local binary learning module.
- **COGNet** [16] works within an encoder-decoder framework and introduces a copying or predicting mechanism for medication recommendation.
- **DrugRec** [10] designs a causal graph model that identifies and eliminates recommendation bias through front-door adjustment in scenarios with multiple visits.

To evaluate prediction accuracy, we employ the number of drugs, DDI Rate (DDI), Jaccard similarity score (Jaccard), average F1 (F1), and Precision Recall AUC (PRAUC) as evaluation metrics.

4.3 Results

Table 2 presents the experimental results of various methods on the MIMIC-III and MIMIC-IV datasets. The experimental findings demonstrate that models considering historical patient records, such as GAMENet, COGNet, and DrugRec, outperform instance-based methods like LR, ECC, and LEAP in overall performance. Particularly, specifically designed approaches like DrugRec, MICRON, and SafeDrug exhibit lower DDI rates due to their controlled DDI loss. DrugRec achieves a fine balance between DDI control and model performance through a dedicated DDI control module, although its results fall slightly short of the baseline model's optimal performance. In contrast, our CAMeR model surpasses other methods in terms of performance, and even in scenarios where DDI is given special consideration, the performance of the $CAMeR_{ddi}$ model, while experiencing a slight decrease, still outperforms all baseline models.

Table 2. Experimental results on MIMIC-III / MIMIC-IV.

Model	Jaccard	F1	PRAUC	DDI	Avg. # of Drugs
LR	0.4877/0.4429	0.6439/0.6121	0.7512/0.6452	0.0820/0.8430	16.1846/11.1565
ECC	0.4901/0.4625	0.6521/0.6345	0.6849/0.6668	0.0834/0.0815	18.0455/18.1526
LEAP	0.4513/0.4326	0.6124/0.6061	0.6566/0.6456	0.0742/0.0689	18.7348/13.2568
GAMENet	0.5089/0.4899	0.6634/0.6537	0.7629/0.7494	0.0864/0.0928	27.5146/20.6529
MICRON	0.5159/0.4937	0.6712/0.6571	0.7701/0.7554	0.0650/0.0938	18.4944/16.9344
SafeDrug	0.5212/0.5013	0.6731/0.6610	0.7650/0.7511	0.0603/0.0864	20.9894/19.5764
COGNet	0.5289/0.5081	0.6836/0.6637	0.7713/0.7545	0.0850/0.0802	26.9528/24.0359
DrugRec	0.5274/0.5052	0.6801/0.6593	0.7710/0.7506	0.0605/0.0618	21.8962/15.2592
$CAMeR_{ddi}$	0.5331/0.5100	0.6861/0.6646	0.7718/0.0.7579	**0.0599/0.0597**	25.1315/23.4006
CAMeR	**0.5392/0.5188**	**0.6906/0.6720**	**0.7739/0.7624**	0.0828/0.0799	25.4923/22.0691

4.4 Ablation Study

Following that, we conducted ablation studies to gain a better understanding of how each component in the CAMeR model influences its performance. We explored the following variants:

- CAMeR *w/o causal*: This model excludes the counterfactual analysis framework, thereby retaining recommendation biases.
- CAMeR *w/o m2p*: This model disregards the contrastive loss between medications and procedures.
- CAMeR *w/o m2d*: This model disregards the contrastive loss between medications and diagnoses.
- CAMeR *w/o d2p*: This model disregards the contrastive loss between diagnoses and procedures.
- CAMeR *w/o contrastive*: This model does not incorporate any contrastive learning loss, including $\mathcal{L}_{m2p}$, $\mathcal{L}_{m2d}$ and $\mathcal{L}_{d2p}$.

Table 3 presents the results of our ablation experiments, and as anticipated, our contrastive learning embedding module and counterfactual debiasing framework demonstrate their effectiveness in improving performance. Notably, the counterfactual debiasing framework has the most significant impact on the model, highlighting its substantial contributions. Overall, the complete CAMeR model outperforms all other ablation models in terms of performance, further affirming the effectiveness of our work.

Table 3. Ablation study for CAMeR on MIMIC-III dataset.

Model	Jaccard	F1	PRAUC	DDI	Avg. # of Drugs
- w/o causal	0.5340 ± 0.0012	0.6864 ± 0.0008	0.7713 ± 0.0018	0.0856 ± 0.0009	25.9671 ± 0.0586
- w/o m2p	0.5368 ± 0.0015	0.6888 ± 0.0017	0.7726 ± 0.0014	0.0862 ± 0.0007	25.8743 ± 0.0924
- w/o m2d	0.5384 ± 0.0010	0.6893 ± 0.0019	0.7732 ± 0.0011	0.0852 ± 0.0007	26.4273 ± 0.0672
- w/o d2p	0.5381 ± 0.0018	0.6891 ± 0.0010	0.7733 ± 0.0013	0.0861 ± 0.0009	26.0658 ± 0.0951
- w/o contrastive	0.5359 ± 0.0017	0.6879 ± 0.0014	0.7722 ± 0.0020	0.0832 ± 0.0010	25.5874 ± 0.0731
Ours	**0.5392 ± 0.0012**	**0.6906 ± 0.0011**	**0.7739 ± 0.0017**	**0.0828 ± 0.0011**	25.4923 ± 0.0672

4.5 Case Study

To assess the impact of spurious correlations between models and data on medication recommendation, Table 4 presents an example of medication recommendation using a patient's visit information from the MIMIC-III dataset. We selected COGNet and the ablation model CAMeR *w/o causal*, both with spurious correlations, as baseline models. In the table, we provide a detailed overview of the patient's clinical information for each visit, including ICD codes for diagnoses and procedures, as well as the actual prescribed drugs and drugs recommended by different methods. It can be observed that the baseline models, COGNet and CAMeR *w/o causal*, which are affected by causal spurious correlations, consistently predict the frequently occurring medication "J01D" across the dataset, even though it is unnecessary in this particular instance. However, our CAMeR model captures and mitigates the spurious correlations between historical visit information, reducing recommendation bias. Consequently, it avoids inertia in recommending incorrect drugs and improves the accuracy of recommendations.

Table 4. Example recommended medication for a patient with three visits. Here "IoU" refers to the intersection over union ratio between the recommended medication group and the actual medication group, where a higher ratio indicates better recommendation performance. Additionally, the critical medications that contribute to prediction bias are highlighted in black font with an underline.

Patient Clinical Information	Method	Recommended Medications (ATC3)	IoU
1st Visit Diag: 2724, 4280, 00845, 4019, 42822, 2851, 72992, V1251, 28981, 99812, 5571, E8844, V1301, 28860, 78559, 3481, 44489, V1588, 87343 Prod: 2751, 5732, 4513, 4523, 9904	Ground Truth	N02B, A02B, A06A, A12C, C07A, N02A, B01A, C01B, N05B, D06A	-
	COGNet	N02B, A01A, A02B, A06A, B05C, A12A, A12C, C01C, A07A, C07A, A12B, N02A, B01A, C10A, C01B, N05C, <u>**J01D**</u>, B02B, N05A, A04A, A11G	0.3478
	CAMeR w/o causal	N02B, A01A, A02B, A06A, B05C, A12C, C01C, A07A, C07A, C03C, A12B, N02A, B01A, <u>**J01D**</u>, B02B, N05A, A04A	0.3500
	CAMeR	N02B, A01A, A02B, A06A, B05C, A12A, A12C, C01C, A07A, C07A, A12B, N02A, B01A, N05A, A04A, J01C	0.3684
2st Visit Diag: 5119, 78551, 42731, 2720, 78552, 99592, 4019, 42789, 0389, E8798, 78959, 5990, E9289, 51881, 2866, 2800, 99674, 42732, 5921, 8670, 28984, 41519, 44489, 28959, 5750, 42989, 59970 Prod: 9671, 9604, 3891	Ground Truth	N02B, A01A, A02B, A06A, B05C, A12A, A12C, C01C, A07A, C07A, A12B, N02A, J01M, B01A, C01B, N05C, N05B, C01E, C01A, C05A	-
	COGNet	N02B, A01A, A02B, A06A, B05C, A12A, A12C, C01C, C07A, A12B, J01M, B01A, C10A, C01B, N05C, <u>**J01D**</u>, A04A, R03A, D06A, R01A	0.5365
	CAMeR w/o causal	N02B, A01A, A02B, A06A, B05C, A12A, A12C, C01C, A07A, C07A, C03C, A12B, B01A, C01B, N05C, <u>**J01D**</u>, R03A, D06A, R01A	0.5600
	CAMeR	N02B, A01A, A02B, A06A, B05C, A12A, A12C, C01C, A07A, C07A, C03C, A12B, J01M, B01A, C01B, N05C, R03A, D06A, R01A	0.625
3st Visit Diag: 2724, 4271, 00845, 4019, 42789, V5861, 2761, V1251, E9426, 4275, E9289, 51881, 70703, 430, 34982, 8670, 3481, 41519, 5920, 4441, 70722, 44489, 4440, E9420 Prod: 3893, 966, 9672, 3734	Ground Truth	A01A, A02B, A06A, B05C, A12C, C01C, N01A, C07A, A12B, B01A, C01B, N05C, B02B, A04A, A07D	-
	COGNet	N02B, A01A, A02B, A06A, B05C, A12A, A12C, C01C, A07A, N01A, C07A, C03C, A12B, N02A, B01A, C10A, C01B, N05C, C09A, <u>**J01D**</u>, N03A, N05A, R03A, N05B, A03F, R01A, C01A	0.4000
	CAMeR w/o causal	N02B, A01A, A02B, A06A, B05C, A12A, A12C, C01C, A07A, N01A, C07A, C03C, A12B, B01A, C10A, C01B, N05C, <u>**J01D**</u>, N05A, R03A, D07A, R01A, C05A	0.4615
	CAMeR	N02B, A01A, A02B, A06A, B05C, A12A, A12C, C01C, A07A, N01A, C07A, C03C, A12B, B01A, C10A, C01B, N05C, N05A, R03A, D07A, R01A, C05A	0.4800

5   Related Work

Medication Recommendation. Deep learning techniques hold immense potential and advantages in the field of medication recommendation. For instance, Zhang et al. [20] encode patient diagnoses and procedures, formulate medication recommendation as a multi-label classification problem, and employ a recurrent decoder for medication advice. Shang et al. [9] further introduce GameNet, which combines RNN and memory networks, enhancing the performance of the medication recommendation model. However, these works do not consider DDIs. Yang et al. [17,18] consider the medication side effects in DDIs, and successively proposed SafeDrug and MICRON models to reduce the risk of medication administration. Wu et al. [16] combine the transformer framework and introduce a copying or predicting mechanism to generate sets of drugs. Nevertheless, these methods overlook the issue of recommendation bias.

Causal Inference. Causal Inference plays a significant role in the field of medication recommendation. By employing causal interventions, it becomes possible to eliminate confounding biases caused by spurious correlations or data imbalances. Many existing approaches [6,7,14,15] construct causal graphs and utilize causal reasoning techniques to alleviate confounding bias [7,13][21–23, 29]. Sun et al. [10] were the first to introduce causal reasoning techniques in the field of medication recommendation. They modeled multiple visit scenarios and mitigated the impact of latent recommendation biases through front-door adjustment. Counterfactual analysis represents a novel method within causal reasoning, capable of distinguishing between spurious and genuine correlations. For the first time in the context of medication recommendation, we have employed counterfactual reasoning to address recommendation biases.

6   Conclusion

In this paper, we propose the CAMeR model, which effectively mitigates the recommendation bias due to spurious correlations between historical visit information after training by performing counterfactual analysis and interventions on the constructed causal graphs. In addition, we apply contrast learning in the embedding module to restrict the feature construction for patients with multiple visits, eliminating the interference from sudden short-term illnesses on the recommendation task. Finally, we conduct comprehensive experiments on the MIMIC-III and MIMIC-IV datasets. The experimental results demonstrate the outstanding performance of our proposed CAMeR model in the task of medication recommendation, outperforming other baseline models. Additionally, through ablation study and case study, we further validate the effectiveness of the counterfactual analysis module and contrastive learning loss.

References

1. Ali, Z., et al.: Deep learning for medication recommendation: a systematic survey. Data Intell. **5**(2), 303–354 (2023)
2. Bhoi, S., Lee, M.L., Hsu, W., Fang, H.S.A., Tan, N.C.: Personalizing medication recommendation with a graph-based approach. ACM Trans. Inf. Syst. (TOIS) **40**(3), 1–23 (2021)
3. He, K., Fan, H., Wu, Y., Xie, S., Girshick, R.: Momentum contrast for unsupervised visual representation learning. In: Proceedings of the IEEE/CVF Conference on Computer Vision and Pattern Recognition, pp. 9729–9738 (2020)
4. Johnson, A., Bulgarelli, L., Pollard, T., Celi, L.A., Mark, R., Horng IV, S.: MIMIC-IV-ED. PhysioNet (2021)
5. Johnson, A.E., et al.: MIMIC-III, a freely accessible critical care database. Sci. Data **3**(1), 1–9 (2016)
6. Pearl, J.: Causal inference in statistics: an overview. Stat. Surv. **3**, 96–146 (2009)
7. Qian, C., Feng, F., Wen, L., Ma, C., Xie, P.: Counterfactual inference for text classification debiasing. In: Proceedings of the 59th Annual Meeting of the Association for Computational Linguistics and the 11th International Joint Conference on Natural Language Processing (Volume 1: Long Papers), pp. 5434–5445 (2021)
8. Read, J., Pfahringer, B., Holmes, G., Frank, E.: Classifier chains for multi-label classification. Mach. Learn. **85**, 333–359 (2011)
9. Shang, J., Xiao, C., Ma, T., Li, H., Sun, J.: GAMENet: graph augmented memory networks for recommending medication combination. In: Proceedings of the AAAI Conference on Artificial Intelligence, vol. 33, pp. 1126–1133 (2019)
10. Sun, H., Xie, S., Li, S., Chen, Y., Wen, J.R., Yan, R.: Debiased, longitudinal and coordinated drug recommendation through multi-visit clinic records. In: Advances in Neural Information Processing Systems, vol. 35, pp. 27837–27849 (2022)

11. Tatonetti, N.P., Ye, P.P., Daneshjou, R., Altman, R.B.: Data-driven prediction of drug effects and interactions. Sci. Transl. Med. **4**(125), 125ra31 (2012)
12. Velickovic, P., Fedus, W., Hamilton, W.L., Liò, P., Bengio, Y., Hjelm, R.D.: Deep graph infomax. In: ICLR (Poster), vol. 2, no. 3, p. 4 (2019)
13. Wang, T., Huang, J., Zhang, H., Sun, Q.: Visual commonsense R-CNN. In: Proceedings of the IEEE/CVF Conference on Computer Vision and Pattern Recognition, pp. 10760–10770 (2020)
14. Wang, Y., et al.: Should we rely on entity mentions for relation extraction? Debiasing relation extraction with counterfactual analysis. arXiv preprint arXiv:2205.03784 (2022)
15. Wang, Z., Culotta, A.: Identifying spurious correlations for robust text classification. arXiv preprint arXiv:2010.02458 (2020)
16. Wu, R., Qiu, Z., Jiang, J., Qi, G., Wu, X.: Conditional generation net for medication recommendation. In: Proceedings of the ACM Web Conference 2022, pp. 935–945 (2022)
17. Yang, C., Xiao, C., Glass, L., Sun, J.: Change matters: medication change prediction with recurrent residual networks. arXiv preprint arXiv:2105.01876 (2021)
18. Yang, C., Xiao, C., Ma, F., Glass, L., Sun, J.: SafeDrug: dual molecular graph encoders for recommending effective and safe drug combinations. arXiv preprint arXiv:2105.02711 (2021)
19. Zbontar, J., Jing, L., Misra, I., LeCun, Y., Deny, S.: Barlow twins: self-supervised learning via redundancy reduction. In: International Conference on Machine Learning, pp. 12310–12320. PMLR (2021)
20. Zhang, Y., Chen, R., Tang, J., Stewart, W.F., Sun, J.: LEAP: learning to prescribe effective and safe treatment combinations for multimorbidity. In: Proceedings of the 23rd ACM SIGKDD International Conference on Knowledge Discovery and Data Mining, pp. 1315–1324 (2017)

Early Detection of Depression and Alcoholism Disorders by EEG Signal

Hesam Akbari and Wael Korani(✉) ⓘ

Department of Information Science, University of North Texas, Texas, USA
Wael.Korani@unt.edu

Abstract. The World Health Organization reported that more than 264 and 80 million patients worldwide suffer from depression and alcoholism, respectively. Depression and alcoholism might cause severe negative repercussions on a patient's life and relationships, such as self-harm and suicide. A person can lead a normal life after these brain disorders are timely and accurately diagnosed and cured. In order to recognize the brain's activity and identify different mental disorders, Electroencephalography (EEG) is often employed. The EEG signals in our study are separated into rhythms in the empirical wavelet transform domain, and then linear and nonlinear features are extracted. Significant features are selected by a feature selection method, and the output of the feature selection method is fed into a classifier. In this paper, a fast and effective diagnostic tool is proposed to detect and recognize depression and alcoholism disorders. The proposed diagnostic tool is built on the Salp Swarm Algorithm and the Tree Growth Algorithm as feature selection methods and Cascade Forward Neural Network and Feed-forward Neural Network classifiers. The diagnostic tool is evaluated on two datasets for depression and alcoholism, and the results show that the classification accuracies are 100% and 99.58% for depression and alcoholism, using 10-fold cross-validation strategy, respectively. The proposed diagnostic tool can be used in hospitals and clinics for fast and accurate detection of depression and alcoholism. In addition, we introduce a novel depression diagnostic index and alcoholism diagnostic index, which can be used as biomarkers for healthcare provider to diagnose depression and alcoholism without using machine learning approaches.

Keywords: Feature selection · EEG · depression · alcoholism

1 Introduction

After the industrialization of societies and the increase in urbanization, an increase in the rate of psychological stress of urban life in the cities of the world has been witnessed, which causes to prevalence of depression and alcoholism disorders in most of the countries [6, 28].

Both depression and alcoholism disorders can affect the daily routine of patients, which leads the patient to lack control of emotions and failure in

© The Author(s), under exclusive license to Springer Nature Singapore Pte Ltd. 2024
B. Luo et al. (Eds.): ICONIP 2023, CCIS 1964, pp. 439–452, 2024.
https://doi.org/10.1007/978-981-99-8141-0_33

education or loss of jobs; these are the short-term effects of depression and alcoholism disorders on patients' lives and the long-term effects can be even committed dangerous behaviors such as self-harm or suicide [6,7]. According to the World Health Organization (WHO), more than 264 and 80 million people around the world are living with depression and alcoholism, respectively [28]; on the other hand, about 700,000 people die by suicide every year, and this number is increasing every year (https://www.who.int/news-room/fact-sheets/detail/suicide). The problem is that although the medications can treat depression and alcoholism disorders, the diagnosis takes time and is prone to human error [22].

Electroencephalography (EEG) signals are recorded to show brain activity, but the EEG signals are nonlinear and non-stationary, making it difficult for medical professionals to decode the information visually, even though they can be useful instruments for illuminating cerebral activity [24]. Therefore, it is desirable to provide a computer-based method for the accurate detection of depressed and alcoholic patients based on EEG signals.

The previous works that have been conducted for depression and alcohol EEG signals detection can be categorized based on extracted features from EEG signals into four main groups, including time (T) domain, frequency (F) domain, time-frequency (TF) domain methods, and deep learning (DL).

In T-domain methods, features were extracted directly from EEG signals, and in F-domain methods, features were extracted from the frequency spectrum of EEG signals. In TF methods, the EEG signal were separated into its sub-bands, and then features have been extracted. In DL methods, the EEG signals were fed into a deep neural network to decide which class the EEG signal belongs to. The main problem of T domain and F domain methods is that the extracted features cannot explain the complexity and nonlinearity of EEG signals, which is why TF methods have been developed to analyze the frequency components of EEG signals. In fact, in TF methods, the EEG signals are separated into several sub-bands with specific frequency bandwidths, but the number of sub-bands can be changed by changing the frequency sampling, which means TF-based methods are non-adaptive to frequency sampling, which is the main weakness of TF methods. Additionally, although the DL approaches such as convolutional neural networks (CNNs) and Long short-term memory (LSTM) have demonstrated outstanding performance in biomedical processing applications, the outcomes for the identification of alcoholic and depressed EEG signals were not very promising [1,15].

These reasons motivate us to decompose the EEG signals into five brain rhythms instead of using T-domain, F-domain, TF-domain, or DL methods. The medical team describes the behavior of EEG signals based on five main rhythms, including delta, theta, alpha, beta, and gamma. The bandwidths of these five rhythms are fixed, which means any method based on these rhythms is not sensitive to frequency sampling.

The best features are chosen before being implemented in a machine learning algorithm since doing so helps to increase classification accuracy (ACC) while also minimizing the method's complexity and the associated computational cost. In the majority of earlier research, best features were selected using the conventional feature selection (FS) methods, which are based on statistical p-values. In p-value-based algorithms, features having a p-value less than 0.05 are called statistical significant and can be utilized for classification. However, the fundamental drawback of conventional FS approaches is that they are unable to choose the best features when all features have a p-value lower than 0.05.

To find the optimum answer for engineering problems, metaheuristic algorithms were developed based on a group behavior of living organisms; more recently, metaheuristic algorithms have been utilized as FS approaches to choose the best features. The effectiveness of 12 different FS approaches is evaluated in the current work to select the best FS method for detecting alcoholism and depression instead of using traditional p-value based methods.

Another drawback of earlier research is that the majority of them classified the features using straightforward classification methods like support vector machines (SVM), least-squares SVM (LS-SVM), and K-nearest neighbor (KNN), probabilistic neural network (PNN) and linear regression (LR). Further study is needed to determine how well different classifiers work for detecting alcoholic and depressive EEG signals application. Due to this, the performance of seven different classifiers is examined in the current work to select the most effective classifier to detect alcoholism and depression using EEG signals.

In the proposed method, the five brain rhythms are separated from the EEG signal using the Empirical Wavelet Transform (EWT) filter bank, then 17 linear and nonlinear features are extracted from each rhythm. After that, the best features are chosen using proper FS methods, and these selected features are fed

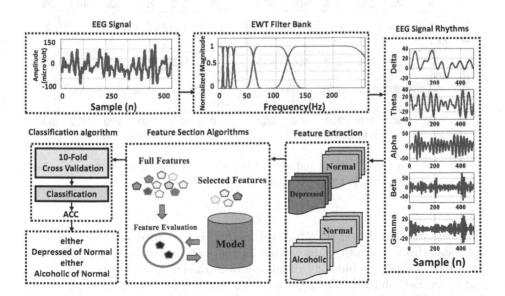

Fig. 1. Block diagram of the proposed method

into a classifier to classify the EEG signals. In order to assess the effectiveness of the proposed solution, two distinct classification tasks: depression vs. normal and alcoholism vs. normal are defined in the current study. Figure 1 shows the block diagram of the proposed method.

2   Materials and Method

2.1   Used Databases

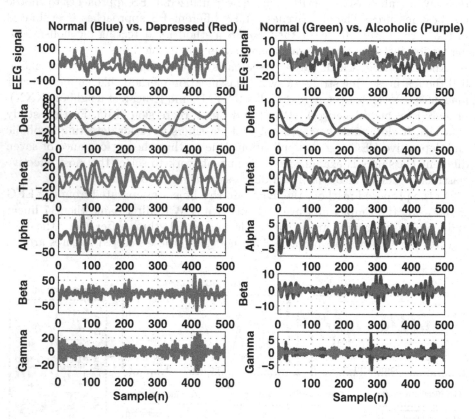

Fig. 2. A sample of EEG signal and its rhythms for normal and depressed groups (Left) and normal and alcoholic groups (right).

Depression Database: the proposed method is evaluated using self-recorded EEG data from 22 healthy subjects and 22 depressive patients [6]. The EEG data were recorded using a bipolar montage on both the left and right hemispheres of the brain. Each individual's EEG data were recorded for ten minutes. The power line incursion and muscular artifacts were removed by visually analyzing the signal and using a notch filter with a 0.5 to 150 HZ bandpass, respectively [7]. The sampling rate was set at 256 Hz. In the current study, the EEG data are divided into segments of 500 samples. The data recording experiments were approved by the AJA University of Medical Sciences Research Ethics Committee in Iran with authorization ID: IR.AJAUMS.REC.1399.049. In the current study, 1000 EEG data for the depressed and 1000 EEG signals for the normal groups are used to assess the performance of the proposed framework in the classification task between depressed and normal groups. Figure 2-left shows a sample of normal and depressed EEG signals.

Alcoholism Database: The University of California (UCI) has a public database that has EEG signals from both alcoholic and healthy patients available online at UCI website. 122 EEG recordings of participants who were either alcoholic or normal were obtained from the 10/20 worldwide montage [8]. EEG signals were recorded using the 90 photos of variously chosen things, and those signals are discussed in detail in [15]. 64 electrodes on the human scalp were used for collecting the EEG data. Each individual accomplished 120 trials for a variety of stimuli. There are two kinds of EEG signals in the dataset: alcoholic and normal. The EEG signals were recorded at a frequency rate of 256 Hz for a length of 32 s, with a resolution of 12 bits. After removing trails that contain undesired eye and body movements, the EEG signals of each class were recovered. The resulting EEG recording is then divided into four segments, each of which lasts for eight seconds. There are in total 2048 samples in each section. In the current study, 120 EEG signals from each normal and alcoholic group are employed to evaluate the performance of the proposed method in the alcoholic vs. normal classification task. Figure 2-right shows the normal and alcoholic EEG signals.

2.2 Empirical Wavelet Transform

The Empirical wavelet transform (EWT) is an adaptive time-frequency transformation that is effective for analyzing nonlinear and non-stationary signals like EEG signals. In EWT, the signal is decomposed into its intrinsic mode functions by considering the frequency components. In fact, with proper segmentation of the frequency spectrum, the frequency boundaries for the filter bank are defined, and then the signal is filtered.

In the current study, the frequency boundaries of the EWT filter bank are set to [0,4], [4,8], [8,13], [13,30], [30,60] to extract delta, theta, alpha, beta, and gamma rhythms; any frequency component higher than 60 Hz is considered a noise. More details about EWT and how to set up the parameters to separate the EEG signal rhythms can be found in [6]. Figure 3 shows the designed filter bank by EWT to separate the rhythms and Fig. 2 shows the separated rhythms for EEG signals of depression and alcoholism database.

2.3 Feature Extraction

The ability of machine learning to accurately classify different classes is improved by the extraction of relevant features. To decode the complicated functioning of

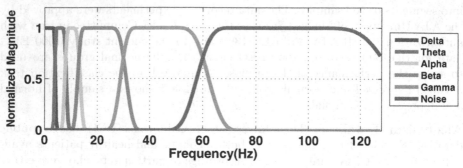

Fig. 3. Shows the designed filter band by EWT and separated EEG rhythms.

the brain, previously, several features have been defined for EEG signals. In the current work, 17 linear and nonlinear features including Mean (F1), Standard Deviation (F2), Variance (F3), Skewness (F4), Kurtosis (F5), Maximum (F6), Minimum (F7), Range (F8), Median (F9), Interquartile Range (F10), Correlation Coefficient (F11), Sum of Power Spectral Density (F12), Shannon Entropy (F13), Log Energy Entropy (F14), Threshold Entropy (F15), Sure Entropy (F16), Norm Entropy (F17) are extracted to distinguish between the classes of EEG signals.

2.4 Feature Selection

A FS method generally seeks to choose a set of extremely unique characteristics from the original dataset [14]. The overlay techniques are the most used FS algorithms, primarily due to their great performance in resolving this challenging issue [4]. A vector of size S, where S is the total number of original features, is used in this approach to represent features [20]. The learning algorithm is used as an index to gauge the effectiveness of the solution during the evaluation of the subset of features [4]. The best solution (the optimum feature subset) is ultimately achieved.

In this work, we evaluate the performance of 12 FS algorithms for depression and alcoholism detection applications using metaheuristic algorithm as a FS approach [4,20]. These FS techniques are known by the acronyms ACS (Ant Colony System), ASO (Atom Search Optimization), DE (Differential Evolution), GA (Genetic Algorithm), GWO (Grey Wolf Optimization), HGSO (Henry Gas Solubility Optimization), HHO (Harris Hawk Optimization), PSO (Particle Swarm Optimization), SCA (Sine Cosine Algorithm), SSA (Salp Swarm Algorithm), TGA (Tree Growth Algorithm) and WOA (Whale Optimization Algorithm). In this study, each approach uses a population size = 10 and maximum number of iterations = 100. The classification error in the 10-fold cross-validation (CV) strategy is set as the fitness function in all FS methods. The parameter settings for the employed techniques are shown in Table 1.

2.5   Classifiers

The selected features are utilized as the input to the classification procedures. To quantify the error rate during validation, a classifier using the 10-fold CV strategy is used. In this study, seven different classification methods are implemented: the generalized regression neural network (GRNN), the cascade forward neural network (CFNN), KNN, the feed-forward neural network (FFNN), the neural network with multiple layers (MNN), the neural network (NN) and SVM [28]. Table 2 lists the specific classifier parameter settings.

Table 1. Parameter settings of used FS methods

Method	Parameter	Value
All	Population size	10
	Maximum iterations	100
	K-fold CV	10
ASO	Depth weight(α)	50
	Multiplier weight(β)	0.2
DE	CR	0.9
HHO	Number of hawks	10
TGA	θ	0.8
	λ	0.5
	Size of the first group	3
	Size of the second group	5
	Size of the fourth group	3
GA	CR	0.8
	MR	0.01
GWO	Number of wolves	10
PSO	c1 and c2	2
	w	[0.9, 0.4]
HGSO	Number of gas types	3
SCA	α	2
SSA	Number of salps	10
WOA	Number of whales	10
ACS	Number of ants	10
	$\alpha, \beta, \tau, \eta$	1
	ρ	0.2
	ϕ	0.5

Table 2. Parameter settings of classifiers

Method	Parameter, value
All	CV, 10-fold
NN, MNN, FFNN, CFNN	Hidden layer size, 10
	Maximum epochs allowed, 50
GRNN	The spread of radial basis functions, 1
SVM	Kernel function, Radial basis function
	Sigma, 1
KNN	distance metric, Euclidean
	k-value, 5

3   Results and Discussion

At the first step of the proposed framework, the EEG signals are separated into five brain rhythms: delta, theta, alpha, beta, and gamma in the EWT domain. Figure 2 shows the designed filter bank by EWT to separate rhythms, and Fig. 3 shows the separated rhythms in the EWT domain. In the current study, any frequencies higher than gamma frequencies are considered a noise and disregarded.

In the second step, 17 features are extracted from each rhythm to distinguish between the classes; in other words, a total of 85 features (5 rhythms × 17 features) are extracted in the feature extraction part. In the third step, the best features for each classification task are selected. Finally, the selected features are fed into classifiers.

In order to evaluate the performance of the proposed framework, the classification ACC is computed, which shows the ability of the classifier to correctly classify EEG signals. The p-values for all extracted features are lower than 0.05, which indicates that in terms of statistical science, all of them can be used in classifiers as discriminant features for designed classification tasks.

In order to select the best classifiers, the performance of seven different classifiers: GRNN, CFNN, KNN, FFNN, MNN, NN, and SVM, is evaluated for two defined classification tasks. Table 3 shows the classification accuracies before selecting the best group of features.

Table 3. The classification ACC for seven different classifiers in Depression vs. normal and Alcoholic vs. normal classification tasks

classification task	Classification ACC (%)						
	GRNN	CFNN	KNN	FFNN	MNN	NN	SVM
Depression vs. normal	96.35	**99.85**	97.25	99.75	99.45	96.35	84.10
Alcoholic vs. normal	91.25	95.83	90.42	**98.75**	95.83	95.42	76.35

Table 4. The resulting classification ACC and the number of selected features for each FS method in Depression vs. normal and alcoholic vs. normal classification tasks.

	Depression vs. Normal		Alcoholism vs. Normal	
	ACC of CFNN (%)	Number of selected features	ACC of FFNN (%)	Number of selected features
ASO	99.95	42	98.75	41
DE	99.95	73	99.17	51
HHO	99.95	50	99.17	59
TGA	**100**	**40**	99.58	45
GA	99.95	37	98.75	47
GWO	99.95	48	99.58	51
PSO	99.95	47	99.58	46
HGSO	99.95	38	99.17	53
SCA	99.95	48	98.75	48
SSA	99.9	52	**99.58**	**44**
WOA	99.9	40	98.75	44
ACS	96.7	15	87.08	15

The results show that the classification ACC of 99.85% by CFNN and 98.75% by FFNN classifiers are higher than other classification methods in depression and alcoholism classification tasks, respectively. On the other hand, the classification ACCs for the SVM classifier in the depression and alcoholism tasks are 84.10% and 76.35%, respectively, which are lower than the ACC for the CFNN and FFNN classifiers. These results show the importance of choosing the significant classifier in each particular classification task.

The selection of significant features improves the performance of the classifier and decreases the computational cost. As a result, the proposed model becomes faster and more accurate after selecting the best group of features. In the current study, the performance of 12 different FS algorithms: ACS, ASO, DE, GA, GWO, HGSO, HHO, PSO, SCA, SSA, TGA, and WOA, is evaluated to choose the best FS method in two defined classification tasks. The fitness function of the FS methods for depression and alcoholism classification tasks are selected based on the performance of CFNN and FFNN classifiers, respectively.

Table 4 shows the classification ACC and the number of selected features for each FS method and for each classification task, respectively. In case of the depression, TGA achieves the perfect classification ACC of 100%. Although the improvement of classification ACC is only 0.15%, the number of selected features is 40, which means that the TGA drops 45 features out of the 85 features. The number of features is decreased by 52.94% using TGA in the depression task.

Similarly, in the Alcoholic vs. normal classification task, the SSA utilizes 44 features, which decreases the number of features by 48.23%. In addition, the classification ACC improves from 98.75 up to 99.58, which means 0.83% difference after selecting the best features.

The selected features by TGA and SSA for depressed vs. normal and alcoholic vs. normal classification tasks are recorded in Table 5, respectively. Table 5 shows that the number of features selected from beta rhythm is more than other brain rhythms in both classification tasks, which indicates the importance of beta rhythm in depression and alcoholic detection. In previous research, the importance of beta rhythm in diagnosis of depression [13] and alcoholism [26] was emphasized, because the beta rhythms are the states in the brain that occur during normal consciousness.

Table 5 also shows that F15 (threshold entropy) and F16 (sure entropy) for depression and F3 (variance) for alcoholism detection are extracted from all rhythms, which shows the importance of these features. Table 6 shows a comparison between the proposed framework and previous research. In depressed vs. normal classification tasks, the reported classification ACC by the proposed framework is better than others.

Although the proposed framework's ACC of 99.58% for the alcoholic vs. normal classification task is lower than the reported ACC of 99.98% [23], our results are reported using a 10-fold CV strategy, whereas theirs results were reported using a leave-one-out CV technique (LOO), which shows that our classification ACC was reported under more stringent conditions than the other study.

The development of a depression diagnostic index (DDI) and an alcoholism diagnostic index (ADI) using features derived from the EWT domain is one of the important contributions in our work. These diagnostic indexes are utilized as biomarkers by medical providers to identify disorders without using AI tool. The proposed indexes facilitate for medical providers to make a final decision. These indexes shows a clear distinction between the two classes as shown in Fig. 4. The two indexes are proposed after many tries as follows:

$$DDI = AF_{16}^{\gamma} + BF_{10}^{\gamma} + CF_{14}^{\gamma} + DF_{15}^{\gamma} \tag{1}$$

$$ADI = LF_{2}^{\delta} + MF_{3}^{\delta} + NF_{12}^{\delta} \tag{2}$$

where A, B, C, D, L, M, and N is -0.12, 0.45, 0.23, 0.09, 0.95, -0.06 and 0.02, respectively. Figure 4 shows the variation of our proposed indexes DDI and ADI, and they can separate the groups perfectly. These indexes will help neurologists to diagnose depression and alcoholism disorders.

The contribution of the current study can be listed below:

– We introduce a novel index for depression and alcoholism called ADI and DDI, respectively which can assist medical providers in making fast and accurate diagnoses.
– The proposed framework provides a better classification ACC compared with previous works in depression and alcoholism detection.
– The majority of previous research works were designed to identify one disorder, but we develop a model for two brain disorders.
– the proposed framework is based on the EEG signal rhythms, which means it is not sensitive to sampling frequency. Thus, the frequency bandwidth of rhythms are constant and the designed filter bank in EWT domain can separate rhythms in one step process without considering the sampling frequency.

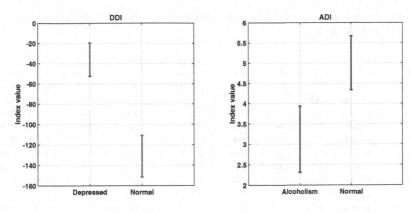

(a) Caption for the first figure   (b) Caption for the second figure

Fig. 4. Ranges of DDI and ADI for classes.

- The high-frequency noise is canceled during the processing stage and by discarding the frequencies greater than 60 Hz as noise, unlike to the majority of previous works that used pre-processing approaches to do so.
- The suggested framework is fast and useful for real-time applications that requires less than 0.06 s for rhythm separations and about 1 s for feature extraction.
- The suggested framework is entirely automatic, and it does not dependent on a psychiatrist's experience or psychiatric counseling. Doctors and nurses can utilize it easily. Although doctors make the final diagnosis, this system may help them make an accurate diagnosis of depression.

Table 5. Feature number of the selected features by FS methods for application of depression and alcoholic.

	Depression vs. Normal	Alcoholic vs. Normal
Delta	1, 4, 9, 12, 14, **15, 16**	2, **3**, 4, 6, 12, 13, 14, 17
Theta	3, 6, 10, 14, **15, 16**	1, **3**, 4, 5, 6, 9, 11, 13
Alpha	1, 3, 7, 10, 12, **15, 16**	1, **3**, 5, 7, 9, 11, 13, 15, 16
Beta	3, 4, 5, 6, 8, 9, 10, 13, 14, **15, 16**	1, 2, **3**, 4, 5, 8, 10, 12, 13, 14, 15, 16
Gamma	5, 7, 8, 9, 10, 11, 12, **15, 16**	2, **3**, 6, 11, 12, 14, 16

Table 6. Feature number of the selected features by FS methods for application of depression and alcoholic.

Depression Vs. Normal					Alcoholic Vs. Normal				
Method	FS	Classifier	CV	ACC (%)	Method	FS	Classifier	CV	ACC (%)
[27], TF	p-value	NN	No	98.11	[34], TF	Not used	NN	10-fold	82.98
[5], TF	Not used	PNN	No	91.30	[2], T	p-value	SVM	3-fold	91.70
[18], TF	p-value	LR	LOO	90.05	[17], TF	p-value	KNN	10-fold	95.80
[16], TF	p-value	PNN	No	99.50	[33], TF	Not used	SVM	10-fold	94.29
[3], T	p-value	SVM	No	98.00	[25], TF	p-value	LS-SVM	10-fold	97.02
[24], T	Not used	SVM	10-fold	98.00	[12], T	Not used	SVM	5-fold	90.00
[21], T	Not used	SVM	LOO	81.23	[31], TF	p-value	SVM	10-fold	97.91
[19], F	p-value	No	No	91.30	[30], TF	p-value	LS-SVM	10-fold	97.08
[10], T	p-value	SVM	No	94.00	[32], TF	Not used	KNN	10-fold	98.91
[11], T	p-value	LR	LOO	92.00	[9], F	Not used	LS-SVM	LOO	98.80
[2], DL	Not used	CNN	10-fold	95.96	[8], TF	p-value	LS-SVM	LOO	98.75
[29], TF	p-value	SVM	10-fold	99.54	[15], DL	Not used	LSTM	10-fold	93.00
[6], TF	p-value	SVM	10-fold	98.76	[28], T	HGSO	FFNN	10-fold	99.16
[7], T	p-value	SVM	10-fold	99.30	[23], TF	p-value	SVM	LOO	99.98
Present Work	TGA	CFNN	10-fold	100.00	Present Work	SSA	FFNN	10-fold	99.58

4   Conclusion

The two brain disorders that have the biggest prevalence worldwide are depression and alcoholism. In the present research, we introduce a diagnostic tool to identify both of these disorders based on EEG signals. EEG signals were divided into five rhythms in the EWT domain, and then significant features were extracted and fed to the classifier in a 10-fold CV strategy.

This method detected depression and alcoholism with 100% and 99.58% classification ACC, respectively. The results show that in the identification of depression and alcoholism, respectively, TGA and SSA perform better than other FS approaches. In addition, the CFNN and the FFNN classifiers perform superior to others in detecting alcoholism and depression, respectively.

Because the beta rhythm represents the brain states that take place throughout typical awareness, the results show that depression and alcoholism had a greater impact on it than other rhythms. On the other hand, the variance of rhythms is a good feature in alcoholism detection. In addition, the threshold and Sure entropies are important features that can be derived from EEG signal rhythms in the identification of depression.

In future research, the framework will be used to detect different mental diseases such as Parkinson's, Alzheimer's, and stress.

References

1. Acharya, U.R., Oh, S.L., Hagiwara, Y., Tan, J.H., Adeli, H., Subha, D.P.: Automated EEG-based screening of depression using deep convolutional neural network. Comput. Methods Programs Biomed. **161**, 103–113 (2018)
2. Acharya, U.R., Sree, S.V., Chattopadhyay, S., Suri, J.S.: Automated diagnosis of normal and alcoholic EEG signals. Int. J. Neural Syst. **22**(03), 1250011 (2012)
3. Acharya, U.R., et al.: A novel depression diagnosis index using nonlinear features in EEG signals. Eur. Neurol. **74**(1–2), 79–83 (2015)
4. Agrawal, P., Abutarboush, H.F., Ganesh, T., Mohamed, A.W.: Metaheuristic algorithms on feature selection: a survey of one decade of research (2009–2019). IEEE Access **9**, 26766–26791 (2021)
5. Ahmadlou, M., Adeli, H., Adeli, A.: Fractality analysis of frontal brain in major depressive disorder. Int. J. Psychophysiol. **85**(2), 206–211 (2012)
6. Akbari, H., Sadiq, M.T., Rehman, A.U.: Classification of normal and depressed EEG signals based on centered correntropy of rhythms in empirical wavelet transform domain. Health Inf. Sci. Syst. **9**, 1–15 (2021)
7. Akbari, H., et al.: Depression recognition based on the reconstruction of phase space of EEG signals and geometrical features. Appl. Acoust. **179**, 108078 (2021)
8. Anuragi, A., Sisodia, D.S.: Empirical wavelet transform based automated alcoholism detecting using EEG signal features. Biomed. Signal Process. Control **57**, 101777 (2020)
9. Anuragi, A., Sisodia, D.S., Pachori, R.B.: Automated alcoholism detection using Fourier-Bessel series expansion based empirical wavelet transform. IEEE Sens. J. **20**(9), 4914–4924 (2020)
10. Bachmann, M., Lass, J., Suhhova, A., Hinrikus, H.: Spectral asymmetry and Higuchi's fractal dimension measures of depression electroencephalogram. Comput. Math. Methods Med. **2013**, 251638 (2013)
11. Bachmann, M., et al.: Methods for classifying depression in single channel EEG using linear and nonlinear signal analysis. Comput. Methods Programs Biomed. **155**, 11–17 (2018)
12. Bae, Y., Yoo, B.W., Lee, J.C., Kim, H.C.: Automated network analysis to measure brain effective connectivity estimated from EEG data of patients with alcoholism. Physiol. Meas. **38**(5), 759 (2017)
13. Cai, H., Sha, X., Han, X., Wei, S., Hu, B.: Pervasive EEG diagnosis of depression using deep belief network with three-electrodes EEG collector. In: 2016 IEEE International Conference on Bioinformatics and Biomedicine (BIBM), pp. 1239–1246. IEEE (2016)
14. Dokeroglu, T., Deniz, A., Kiziloz, H.E.: A comprehensive survey on recent metaheuristics for feature selection. Neurocomputing. **494**, 269–296 (2022)
15. Farsi, L., Siuly, S., Kabir, E., Wang, H.: Classification of alcoholic EEG signals using a deep learning method. IEEE Sens. J. **21**(3), 3552–3560 (2020)
16. Faust, O., Ang, P.C.A., Puthankattil, S.D., Joseph, P.K.: Depression diagnosis support system based on EEG signal entropies. J. Mech. Med. Biol. **14**(03), 1450035 (2014)
17. Faust, O., Yu, W., Kadri, N.A.: Computer-based identification of normal and alcoholic EEG signals using wavelet packets and energy measures. J. Mech. Med. Biol. **13**(03), 1350033 (2013)
18. Hosseinifard, B., Moradi, M.H., Rostami, R.: Classifying depression patients and normal subjects using machine learning techniques and nonlinear features from EEG signal. Comput. Methods Programs Biomed. **109**(3), 339–345 (2013)

19. Knott, V., Mahoney, C., Kennedy, S., Evans, K.: EEG power, frequency, asymmetry and coherence in male depression. Psych. Res. Neuroimaging 106(2), 123–140 (2001)
20. Korani, W., Mouhoub, M.: Review on nature-inspired algorithms. Oper. Res. Forum. 2, 1–26 (2021). https://doi.org/10.1007/s43069-021-00068-x
21. Liao, S.C., Wu, C.T., Huang, H.C., Cheng, W.T., Liu, Y.H.: Major depression detection from EEG signals using kernel eigen-filter-bank common spatial patterns. Sensors 17(6), 1385 (2017)
22. McHugh, R.K., Weiss, R.D.: Alcohol use disorder and depressive disorders. Alcohol Res. Curr. Rev. 40(1), 1–8 (2019)
23. Mehla, V.K., Singhal, A., Singh, P.: A novel approach for automated alcoholism detection using Fourier decomposition method. J. Neurosci. Methods 346, 108945 (2020)
24. Mumtaz, W., Xia, L., Ali, S.S.A., Yasin, M.A.M., Hussain, M., Malik, A.S.: Electroencephalogram (EEG)-based computer-aided technique to diagnose major depressive disorder (MDD). Biomed. Signal Process. Control 31, 108–115 (2017)
25. Patidar, S., Pachori, R.B., Upadhyay, A., Acharya, U.R.: An integrated alcoholic index using tunable-q wavelet transform based features extracted from EEG signals for diagnosis of alcoholism. Appl. Soft Comput. 50, 71–78 (2017)
26. Propping, P., Krüger, J., Mark, N.: Genetic disposition to alcoholism. An EEG study in alcoholics and their relatives. Human Genet. 59, 51–59 (1981)
27. Puthankattil, S.D., Joseph, P.K.: Classification of EEG signals in normal and depression conditions by ANN using RWE and signal entropy. J. Mech. Med. Biol. 12(04), 1240019 (2012)
28. Sadiq, M.T., Akbari, H., Siuly, S., Li, Y., Wen, P.: Alcoholic EEG signals recognition based on phase space dynamic and geometrical features. Chaos Solitons Fractals 158, 112036 (2022)
29. Sharma, M., Achuth, P., Deb, D., Puthankattil, S.D., Acharya, U.R.: An automated diagnosis of depression using three-channel bandwidth-duration localized wavelet filter bank with EEG signals. Cogn. Syst. Res. 52, 508–520 (2018)
30. Sharma, M., Deb, D., Acharya, U.R.: A novel three-band orthogonal wavelet filter bank method for an automated identification of alcoholic EEG signals. Appl. Intell. 48, 1368–1378 (2018)
31. Sharma, M., Sharma, P., Pachori, R.B., Acharya, U.R.: Dual-tree complex wavelet transform-based features for automated alcoholism identification. Int. J. Fuzzy Syst. 20, 1297–1308 (2018)
32. Thilagaraj, M., Rajasekaran, M.P.: An empirical mode decomposition (EMD)-based scheme for alcoholism identification. Pattern Recogn. Lett. 125, 133–139 (2019)
33. Upadhyay, R., Padhy, P., Kankar, P.: Alcoholism diagnosis from EEG signals using continuous wavelet transform. In: 2014 Annual IEEE India Conference (INDICON), pp. 1–5. IEEE (2014)
34. Zhong, S., Ghosh, J.: HMMs and coupled HMMs for multi-channel EEG classification. In: Proceedings of the 2002 International Joint Conference on Neural Networks. IJCNN 2002 (Cat. No. 02CH37290), vol. 2, pp. 1154–1159. IEEE (2002)

Unleash the Capabilities of the Vision-Language Pre-training Model in Gaze Object Prediction

Dazhi Chen and Gang Gou(✉)

State Key Laboratory of Public Big Data, College of Computer Science and
Technology, Guizhou University, Guiyang 550025, Guizhou, China
gs.dzchen21@gzu.edu.cn, gougang_v_h@126.com

Abstract. In a retail environment, it is valuable to evaluate the products of interest to perform accurate recommendations. However, the existing method (gaze following) only predicts the gaze area, and the prediction problem of gaze objects has not been fully explored. To this end, this paper proposes a new visual language model based on pre-trained large language models for the gaze object prediction framework, named EdgeCLIP. Primarily, we employ a set of adaptable and instructive cues to judiciously infuse instructional cues into the extensive language model, while proficiently retaining its pre-training knowledge. Secondly, we introduce a multi-head pooled attention block, MPATB, to achieve semantic enhancement and extract the joint representation of multimodal components, thereby mitigating the discrepancy in fixation points and subsequently reducing inaccurate predictions of gaze objects. Furthermore, we introduce a regulatory loss function that effectively governs the gaze heatmap within the stared box. A large number of experiments have proved that our model outperforms previous models. The code will be available in: https://github.com/fadaishaitaiyang/EdgeCLIP.

Keywords: Deep learning approaches · Human-centered computing · Applied computing

1 Introduction

In recent times, the swift advancement of computer vision and machine learning has brought about substantial upheaval within the realm of retail. Nevertheless, the predicament of prognosticating gaze objects remains inadequately resolved. For instance, within a retail setting, the salesperson often finds themselves unable to ascertain whether the customer's gaze fixates upon the product they are presenting, thus placing the salesperson in a quandary as to whether they should approach and engage in salesmanship. At present, the task of gaze object prediction can offer a solution to this predicament. Unlike gaze estimation, the gaze

B. Luo et al. (Eds.): ICONIP 2023, CCIS 1964, pp. 453–466, 2024.
https://doi.org/10.1007/978-981-99-8141-0_34

object prediction task allows intelligent systems to learn to classify and predict the boundaries of objects that people are looking at.

In the present era, however, the paucity of datasets utilized for the prognostication of gaze objects renders research arduous, owing to the costly endeavor of data collection and annotation. Recently, expansive visual language models such as CLIP [12] have flourished across the expanse of computer vision, owing to their formidable semantic information. This resounding triumph serves as an impetus for us to delve into its applicability within the realm of gaze object prediction. The application of CLIP models to gaze object prediction presents a challenge, given that CLIP models undergo training via image-level contrast learning. This mode of learning signifies a deficiency in the pixel-level recognition capabilities requisite for gaze object prediction. One solution to bridge the chasm in representation granularity involves fine-tuning the model on the dataset specific to gaze object prediction. However, the scale of the gaze object prediction dataset pales in comparison to that of the visual language pre-training dataset, thus frequently impinging upon the feasibility of fine-tuning the model for gaze object prediction.

To surmount this formidable challenge, adapters emerge as the sole panacea. Adapters, when employed in conjunction with LLMs (Large Language Models), represent neural modules that assimilate into LLMs, housing a modest assemblage of supplementary trainable parameters. This ingenious integration facilitates the streamlined fine-tuning of specific tasks while leaving the pre-training parameters of LLMs unscathed. Currently, the prevailing adapter configurations (as depicted in Fig. 1) encompass the Series Adapter [6], the Parallel Adapter [5], and the illustrious LoRA (Low-Rank Adaptation) [7]. LoRA, renowned for its parameter-efficient fine-tuning (PEFT) [4], embeds this approach into Alpaca, enabling models to achieve performance levels akin to full fine-tuning with a reduced number of trainable parameters. The triumph and promise showcased by Alpaca and Alpaca-LoRA have spawned a multitude of adaptations and applications, thereby prompting us to ponder: Can Adapters facilitate the utilization of visual language pre-training models within the domain of gaze object prediction tasks?

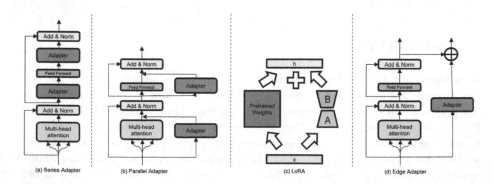

Fig. 1. A detailed illustration of the model architectures of four different adapters.

In order to address the aforementioned quandaries, we present an adapter tailored specifically for gaze object prediction. This adapter ingeniously infuses instructional cues into the LLM (Large Language Model), deftly preserving its pre-training knowledge. Unlike its predecessors, our adapter eschews integration within a clip block and instead adopts a decoupled design, enabling the effective utilization of the language model's formidable semantic information. Simultaneously, drawing inspiration from the [Side Adapter], we embrace a single-forward design that minimizes the computational overhead of CLIP. To harness the head location information more effectively, we introduce a multi-head pooled attention block, known as MPATB, which not only augments semantic comprehension but also unearths the collaborative representation of multimodal elements. This concerted effort serves to alleviate fixation point discrepancies and mitigate erroneous gaze-object predictions. Furthermore, we introduce a regulatory loss function that effectively governs the gaze heatmap within the stared box.

This work aims to fully unleash the capabilities of visual language pre-trained models in gaze object prediction. To achieve this, we propose a new framework, EdgeCLIP, to estimate gaze heatmap, detect retail objects, as shown in Fig. 2. Scene and Face images are first jointly processed by the improved clip feature extractor. Then, the object detection head discovers bounding boxes, and the gaze prediction head predicts gaze heatmap, so we can jointly consider gaze prediction and object detection results to carry out our contributions are summarized below:

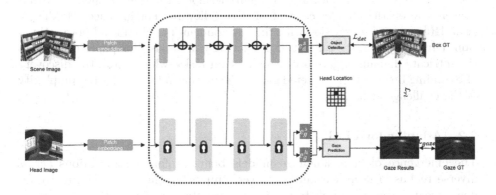

Fig. 2. Overview of the proposed method

1: We first discuss the current mainstream adapter method and provide a new decoupled adapter for the gaze object prediction task. The adapter can adaptively inject instructional cues into LLM while effectively retaining its pre-training knowledge.

2: We propose a multi-headed pooled attention block MPATB to achieve semantic enhancement and mine the joint representation of multimodal elements, thereby reducing the deviation of fixation points, and thus reducing incorrect gaze-object prediction.
3: We also propose a regulatory loss, which can effectively regulate the stared heatmap in the stared box.
4: A large number of experiments have proved that our method is superior to the previous model.

2   Related Work

2.1   Gaze Following

Gaze following, as a practical technique, the gaze following task, proposed by Recasens et al., is a well-studied branch of gaze estimation. Existing gaze estimation work can be divided into three categories according to different scenes, namely fixation point estimation, gaze following, and 3D gaze estimation. This article is related to gaze following tasks. In the initial stages, Recasens et al. [13] prognosticated the gaze area by extracting head posture and gaze direction through depth models. Parks et al. [11], on the other hand, amalgamated saliency maps with human head posture and gaze direction, effectively predicting the observer's gaze area. Tonini et al. [16] uses additional depth maps to effectively solve gaze following tasks. Furthermore, there exist endeavors that concentrate on the broader concept of gaze following. For instance, Mukherjee et al. [10] resumed interaction with the environment based on head pose estimation. Diverging from the pursuit of predicting gaze areas, this paper delves into the intricate domain of gaze object prediction tasks, which entail the discovery of bounding boxes for the targeted gaze objects, thus intensifying the complexity of the challenge at hand.

2.2   Adapter for LLM

The emergence of large language models bears profound implications across diverse realms of deep learning. However, employing these models directly for downstream tasks often fails to yield optimal outcomes. Currently, two prevailing solutions have garnered attention: fine-tuning and adapters. Yet, as language models continue to grow in size and potency, fine-tuning proves to be an increasingly inefficient approach, potentially compromising the inherent capabilities of the pre-training process. Hence, researchers have embarked upon the quest for a novel model tuning methodology, with initial forays observed in the realm of NLP [6]. As large-scale visual language models have emerged, the exploration of computer vision has undergone a surge of intensity. Zhou et al. [18] fine-tunes the CLIP model for the image classification task by training only the input prompts of the text encoder of the CLIP. Zhang et al. [17], on the other hand, embed trainable adapter modules into fixed CLIP models and fine-tune these adapters with limited supervision. However, the majority of these endeavors concentrate on image-level tasks or visual language tasks, lacking direct applicability to the Gaze Object Prediction task at hand.

3 Method

Given the scene image I_s and head position mask H, the head image I_h is usually generated by cropping the scene image I_s. The goal of the Gaze Object Prediction task is to predict the bounding box of a human-initiated object.

3.1 EdgeCLIP

To capture the holistic essence of the scene and the intricate details of the head, conventional gaze following tasks often rely on two distinct networks to process the scene image, denoted as I_s, and the head image, denoted as I_h, separately. However, if we were to apply this approach to address the gaze object prediction task, an additional branch dedicated to object detection would need to be incorporated, resulting in escalated computational demands imposed on the model. Thus, we opt for a shared backbone strategy, leveraging head-specific features that effectively cater to the distinct requirements of gaze prediction and object detection tasks.

As shown in Fig. 2, we first use the patch embedding layer to extract the input-specific features of the scene image and head image before sharing the backbone:

$$e_s^{S\star} = \psi^s\left(I_s\right), e_s^{h\star} = \psi^h\left(I_h\right) \tag{1}$$

where $\psi^s\left(\cdot\right)$ and $\psi^h\left(\cdot\right)$ represent the convolutional layers of the scene picture and the head image, respectively. $e_s^{S\star}$ and $e_s^{h\star}$ represent extracted features. We then feed $e_s^{S\star}$ and $e_s^{h\star}$ into the shared backbone network Edge CLIP and produce the features in a general way:

$$e_g^s, e_d^s = \psi^c\left(e_s^{S\star}\right), e_g^h = \psi^c\left(e_s^{h\star}\right) \tag{2}$$

where e_g^s and e_g^h represent general features from scene images and head images, e_d^s represents the features generated by the first three blocks for simple feature pyramids, and ψ^c represents a shared backbone network. By sharing the backbone network, this can effectively reduce the computing cost compared to the previous paradigm. After that, we feed into a simple feature pyramid using the general scene e_g^s as input to generate specific object features for detecting locations:

$$e_{det} = \phi^{det}\left(e_g^s\right) \tag{3}$$

where ϕ^{det} refers to a simple pyramid of features, and e_{det} represents features used for object detection. Then, the general scene feature e_g^s, the general head feature e_g^h, and the head position H are sent into the multi-head pooled attention layer to generate gaze-specific features that predict gaze outcomes:

$$e_{gaze} = \phi^{pool}\left(e_g^s, e_g^h, H\right) \tag{4}$$

where e_{gaze} represents the features used for gaze prediction networks, and ϕ^{pool} refers to the multi-headed pooled attention layer.

Through the above design, the computational cost can be effectively reduced and satisfactory performance can be maintained, forming an effective and efficient gaze target detection framework.

3.2   Gaze Prediction

As shown in Fig. 3, the gaze prediction network predicts gaze outcomes using two gaze-specific features (i.e., e_g^s and ee_g^h) and a head position map (i.e., H) as inputs.

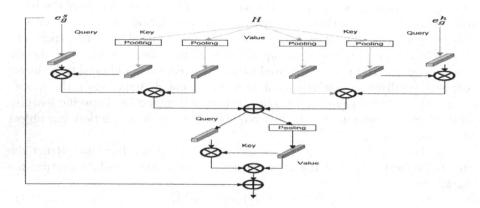

Fig. 3. Detailed network architecture of our multi-headed pooled attention block

The conventional gaze following task conventionally intertwines the head position with the scene image, thereby extracting comprehensive scene features. In an endeavor to augment the semantic richness of the overall scene feature, [multimodel] incorporates depth images into the aforementioned approach. Nonetheless, employing this technique for the gaze object prediction task may inadvertently misdirect the object detection process. Hence, we employ the 'head-delay' methodology, offering head position cues to the gaze prediction network, thereby circumventing such potential pitfalls.

Drawing upon the aforementioned enhancements, we adhere to the approach proposed by [2] to undertake gaze object prediction. In this endeavor, we simultaneously introduce two distinct gaze-specific features and head position maps into the Multihead Pool Attention Module without undergoing any further processing. Our multi-head pooled attention module comprises two primary components. In the initial component, we subject the general scene feature e_g^s and the general head feature e_g^h to convolutional filtering, subsequently employing the resulting feature as the Query. Concurrently, we leverage the pooling treatment head position H and employ the obtained features as the Key and Value, the pooling referencing the SPPF module of yolov5. Subsequently, self-attention is performed on the features acquired from the two branches. Finally, the features from both branches are fused by means of addition. The precise formula can be represented as follows:

$$e_q^s = Conv\left(e_g^s\right), e_q^h = Conv\left(e_g^h\right)$$

$$e^s = \left(e_q^s \top \bigotimes pool\left(H\right)\right) \bigotimes pool\left(H\right)$$

$$e^h = \left(e_q^h \top \bigotimes pool\left(H\right)\right) \bigotimes pool\left(H\right) \tag{5}$$

$$e = e^s + e^h$$

The resulting features e are fed into the following part to enhance the semantic information, which can be written as:

$$e' = pool\left(e\right)$$

$$e_{out} = \left(e^\top \bigotimes e'\right) \bigotimes e' \tag{6}$$

Finally, the output heatmap e_{out} is fed into the gaze prediction module to obtain the final heatmap M. Specifically, the gaze prediction module reference [2] is an encoder-decoder structure that proves its effectiveness in multiple gaze followings.

3.3 Object Detection

When presented with an image, the backbone network proves adept at comprehending the visual composition, thereby yielding a generalized feature e_g^s. Conventionally, this feature is processed through a feature pyramid to facilitate fusion, subsequently being fed into the detection head. However, the Vision Transformer [3] produces feature maps of equal size, rendering this approach unsatisfactory, as shown in Fig. 4. Recently, [8] surmounted these challenges with minimal modifications, constructing a concise feature pyramid solely from the final block of the trunk. This method efficiently extracts high-resolution features. Nonetheless, within a retail setting, numerous items may occupy only a few pixels, rendering object detection accuracy compromised by adopting the

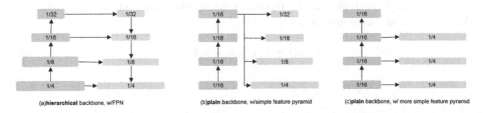

Fig. 4. A typical hierarchical-backbone detector (left), plain-backbone detector (middle) and our more simple feature pyramid.

aforementioned technique. Hence, we have made slight adjustments. Instead of employing the last block for upsampling, we exclusively utilize the initial three blocks to generate a potent object detection feature e_{det}. This approach circumvents the need for feature pyramid structure fusion, significantly reducing the computational burden of the model.

3.4 Regulatory Loss

Gaze object prediction research endeavors demand superior gaze heatmaps to attain precise outcomes. For instance, in the case of the Lian model, it achieved an impressive AUC of 84%; however, the Avg.Dist measure stood at 0.321. While this error magnitude may appear minute, such a discrepancy can result in substantial deviations from the actual gaze point, ultimately leading to erroneous gaze object prediction. A superior-quality heatmap has the capacity to precisely pinpoint the gaze point. Hence, we put forth a novel regulatory LOSS to address the aforementioned challenges.

Precisely, we stochastically generate a matrix G, mirroring the dimensions of the predicted heat map, with all elements set to 0. Simultaneously, employing the actual box $b = (x_1, y_1, x_2, y_2)$ as a mask, we overlay it onto the corresponding positions of the predicted heat map, where (x_1, y_1) and (x_2, y_2) denote the coordinates of the upper left and lower right corners. Then, we calculate the pixel-wise disparity between the matrix and the predicted heat map. Thus, our loss function can be expressed as follows:

$$\mathcal{L}_{rl} = 1 - \left(\frac{2 \sum_i^N p_i g_i + 1}{\sum_i^N (p_i^2 + g_i^2) + 1} \right) \tag{7}$$

where N represents the number of elements in box b, the predicted volume $p_{i \subseteq P}$ and the ground volume $g_{i \subseteq G}$.

Given that our detection branch adopts a YOLOv4-based detection head, we employ a methodology akin to YOLOv4 to compute the detection loss, denoted as mathcalLdet. This loss encompasses the amalgamation of detection confidence score, category classification score, and bounding box regression. In the gaze prediction branch, we commence by generating a ground truth heatmap derived from the actual gaze points. Subsequently, we evaluate the heat map loss, mathcalLgaze, by quantifying the mean squared error between the predicted heat map and the ground truth heat map. To ensure comprehensive training, the total loss comprises these three components:

$$\mathcal{L}_{total} = \mathcal{L}_{rl} + \mathcal{L}_{gaze} + \mathcal{L}_{det} \tag{8}$$

4 Experiments

4.1 Setups

Datasets. In this paper, we employ the GOO dataset [15] to assess the efficacy of our proposed approach. The GOO dataset comprises comprehensive annotations encompassing foveated points, gaze objects, and bounding boxes for a diverse range of 24 categories. Notably, GOO stands as the pioneering dataset in the domain of gaze target detection that amalgamates both real and synthetic data. The GOO-Synth subset showcases a collection of 192,000 composite images, while the GOO-Real subset comprises 9,552 real-world images. The GOO dataset inherently presents numerous challenges inherent to gaze object prediction tasks, including the presence of diminutive objects and the existence of multiple objects within each image.

Implementation Details. Our proposed methodology incorporates the ViT-B/16CLIP model as the shared backbone network. For the object detection branch, we employ Non-Maximum Suppression (NMS) with a threshold of 0.3 to eliminate redundant bounding boxes while retaining the first 100 boxes per image. In the gaze estimation branch, we apply a Gaussian blur with a kernel size of 3 to seamlessly fuse it with the actual box's gaze point. To optimize the network, we employ the Adam optimization algorithm, conducting a total of 100 epochs. The batch size is set to 64, and the initial learning rate is set to 10^{-4}.

Evaluation Metrics. For object detection, we use Average Prediction (AP) as our metric following the previous method. For the target forecast, we used two metrics as per previous work, Heatmap Area Under Curve (AUC %), and Average distance (Avg.Dist.). AUC is the confidence level of the heat map to evaluate the predicted heat map versus the real map. Avg.Dist predicts the Euclidean distance between gaze positioning and the true gaze point.

4.2 Comparison with State-of-the-Arts

We compare our method with the current method, as shown in Table 1, Table 2, where Table 1 uses the GOO-Synth subset and Table 2 uses the GOO-Real subset. These comparisons also included the standard gaze analysis baseline, or random. Random represents generating a heat map of each pixel by sampling values from a Gaussian distribution.

Our method obtains better results than all corresponding methods and is the most advanced method for all datasets in terms of AUC. In particular, its relative performance improvement in the GOO-Synth and GOO-Real datasets is abrupt (0.3–2.6% and 0.68–12.88% AUC, respectively). In the case of Avg.Dist., our approach lagged behind Tonini et al. [16] while outperforming others. It's worth noting that Tonini et al. [16] is more complex than our model and may not work in a retail environment by using additional depth maps. On the other hand We can say that the use of depth maps by Tonini et al. [16] helped improve Avg.Dist, and this method performed worse than us in terms of AUC.

Table 1. Gaze estimation performance on the GOO-Synth

Methods	GOO-Synth	
	AUC↑	Avg.Dist↓
Random	49.7	0.454
Recasens [13]	92.9	0.162
Lian [9]	95.4	0.107
Chong [2]	95.2	0.075
Ours	**95.5**	**0.071**

Moreover, one can observe that following the work in the traditional gaze [7] [9], two separate networks are used to process scene images and head images. Under this paradigm, Lian et al. [9] and Chong et al. [2] achieved promising performance, as shown in Table 1. However, we hired a shared backbone to perform gaze object detection, and the gaze object detection performance confirmed that it was indeed superior to the above paradigm.

At the same time, Table 2 considers the different settings for object detection performance. First, initializing the original YOLOv4 [2] of CSPDarknet53 yielded 43.69% APs. For example, the Yolov4 object detector can only handle this single task, and cannot solve the gaze estimation task. However, our approach integrates gaze estimation and object detection into a unified framework. Since two tasks exist at the same time, we can only keep data augmentation methods that are suitable for both tasks, such as random cropping and color transformation. Eventually, our model reached the 48.008%AP.

Table 2. Gaze estimation performance and object detection on the GOO-Real

Method	gaze estimation		object detection		
	AUC	Avg.Dist	AP	AP50	AP75
Recasens et al. [13]	85	0.22	-	-	-
Lian et al. [9]	84	0.321	-	-	-
Chong et al. [2]	79.6	0.252	-	-	-
Tonini et al. [16]	91.8	**0.164**	-	-	-
Faster-RCNN [14]	-	-	-	25.47	-
YOLOv4 [1]	-	-	43.69	84.02	43.59
ours	**92.48**	0.1989	**48.008**	**86.15**	**48.17**

4.3   Ablation Studies

In EdgeCLIP, we introduce the concept of EdgeAdapter, an intricate multi-head pooled attention block (MPATB), and a meticulously crafted regulatory loss mechanism, all of which synergistically contribute to the pursuit of precise gaze object prediction. The comprehensive results of our ablation experiments are presented in Table 3. Primarily, we observe a notable decline in performance when the EdgeAdapter is substituted, thereby substantiating its efficacy in harnessing the potential of expansive language models for eye evaluation tasks. Moreover, the significant deterioration in performance when the traditional splicing operation replaces the multi-head pooled attention module serves as compelling evidence attesting to the efficacy of the latter. Notably, upon removal of the regulatory loss, a discernible increase in Avg.Dist is observed, emphasizing the pivotal role played by our proposed loss in accurately homing in on the intended gaze point.

Table 3. Ablation studies on GOO-Real, we report the performance of gaze estimation.

Setups	gaze estimation	
	AUC	Avg.Dist
#a w/o edgeadapter	88.28	0.2026
#b w/o MPATB	70.72	0.418
#c w/o loss	91.95	23.08
EdgeCLIP	92.48	0.1989

Table 4 elucidates the performance analysis of diverse adapters depicted in Fig. 1, specifically concerning the GOO-Real dataset. Evidently, both the Series Adapter and the Parallel Adapter exhibit a significant decline in performance, signifying the inadequacy of previous adapters in seamlessly integrating the semantic nuances of extensive language models into the gaze evaluation task. While LoRA demonstrates commendable performance in terms of AUC, it falls noticeably short when compared to our EdgeAdapter in the aspect of Avg.Dist. In conclusion, the resounding success of our EdgeAdapter is unequivocally established.

Table 4. Ablation experiments with different adapters on the Goo-Real dataset.

Setups	Gaze Estimation	
	AUC↑	Avg.Dist↓
#a w/o Series Adapter	60.14	0.7453
#b w/o Parallel Adapter	59	0.2992
#c w/o LoRA	89.6	0.2607
EdgeCLIP	**92.48**	**0.1989**

4.4 Qualitative Analysis

Figure 5 presents the exquisite visual outcomes achieved by the innovative Edge-CLIP approach. Consequently, this culminates in an accurate forecast of the gaze objects. The top panel of the figure showcases the qualitative outcome of object detection using our method, while the bottom panel illustrates a comparative analysis between our approach and the Chong's method.

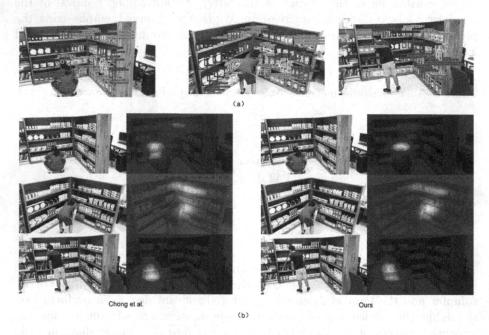

Fig. 5. (a) the qualitative outcome of object detection. (b) Sample predicted points and heatmaps between our method and the Chong's method. Green line represents the ground truth gaze vector and gaze object bounding box, while the red line is the model prediction. (Color figure online)

5 Conclusion

This article explores how to construct a unified framework for the Gaze Object Prediction Task. In order to perform the two tasks of object detection and gaze evaluation at the same time, we first propose an edgeadapter decoupling integrated into the CLIP model and use a backbone network that adaptively injects instructional cues into LLM, while effectively retaining its pre-training knowledge. Then, a multi-headed pooled attention block MPATB is proposed to achieve semantic enhancement and mine the joint representation of multimodal elements, thereby reducing the deviation of fixation points, and thus reducing incorrect gaze-object prediction. It also proposes a regulatory loss that can effectively regulate the gaze heatmap in the real box. T The remarkable efficacy of EdgeCLIP, as demonstrated by its impressive performance on the GOO dataset, unequivocally affirms the efficacy of our approach. It is poised to invigorate related endeavors in the realm of multimodal learning and small object detection.

Acknowledgements. This paper was supported by National Natural Science Foundation of China (NSFC) (62162010); Guizhou Provincial Science and Technology Support Program Project, Qianke He Support [2022] General 267.

References

1. Bochkovskiy, A., Wang, C.Y., Liao, H.Y.M.: Yolov4: optimal speed and accuracy of object detection. arXiv preprint arXiv:2004.10934 (2020)
2. Chong, E., Wang, Y., Ruiz, N., Rehg, J.M.: Detecting attended visual targets in video. In: Proceedings of the IEEE/CVF Conference on Computer Vision and Pattern Recognition, pp. 5396–5406 (2020)
3. Dosovitskiy, A., et al.: An image is worth 16 × 16 words: transformers for image recognition at scale. arXiv preprint arXiv:2010.11929 (2020)
4. Fu, Z., Yang, H., So, A.M.C., Lam, W., Bing, L., Collier, N.: On the effectiveness of parameter-efficient fine-tuning. arXiv preprint arXiv:2211.15583 (2022)
5. He, J., Zhou, C., Ma, X., Berg-Kirkpatrick, T., Neubig, G.: Towards a unified view of parameter-efficient transfer learning. arXiv preprint arXiv:2110.04366 (2021)
6. Houlsby, N., et al.: Parameter-efficient transfer learning for NLP. In: International Conference on Machine Learning, pp. 2790–2799. PMLR (2019)
7. Hu, E.J., et al.: Lora: low-rank adaptation of large language models. arXiv preprint arXiv:2106.09685 (2021)
8. Li, Y., Mao, H., Girshick, R., He, K.: Exploring plain vision transformer backbones for object detection. In: Avidan, S., Brostow, G., Cissé, M., Farinella, G.M., Hassner, T. (eds.) Computer Vision. ECCV 2022. LNCS, vol. 13669, pp. 280–296. Springer, Cham (2022). https://doi.org/10.1007/978-3-031-20077-9_17
9. Lian, D., Yu, Z., Gao, S.: Believe it or not, we know what you are looking at! In: Jawahar, C.V., Li, H., Mori, G., Schindler, K. (eds.) ACCV 2018. LNCS, vol. 11363, pp. 35–50. Springer, Cham (2019). https://doi.org/10.1007/978-3-030-20893-6_3
10. Mukherjee, S.S., Robertson, N.M.: Deep head pose: gaze-direction estimation in multimodal video. IEEE Trans. Multimed. **17**(11), 2094–2107 (2015)
11. Parks, D., Borji, A., Itti, L.: Augmented saliency model using automatic 3d head pose detection and learned gaze following in natural scenes. Vision. Res. **116**, 113–126 (2015)
12. Radford, A., et al.: Learning transferable visual models from natural language supervision. In: International Conference on Machine Learning, pp. 8748–8763. PMLR (2021)
13. Recasens, A., Khosla, A., Vondrick, C., Torralba, A.: Where are they looking? In: Advances in Neural Information Processing Systems, vol. 28 (2015)
14. Ren, S., He, K., Girshick, R., Sun, J.: Faster R-CNN: towards real-time object detection with region proposal networks. In: Advances in Neural Information Processing Systems, vol. 28 (2015)

15. Tomas, H., et al.: GOO: a dataset for gaze object prediction in retail environments. In: Proceedings of the IEEE/CVF Conference on Computer Vision and Pattern Recognition, pp. 3125–3133 (2021)
16. Tonini, F., Beyan, C., Ricci, E.: Multimodal across domains gaze target detection. In: Proceedings of the 2022 International Conference on Multimodal Interaction, pp. 420–431 (2022)
17. Zhang, R., et al.: Tip-adapter: training-free clip-adapter for better vision-language modeling. arXiv preprint arXiv:2111.03930 (2021)
18. Zhou, K., Yang, J., Loy, C.C., Liu, Z.: Learning to prompt for vision-language models. Int. J. Comput. Vision **130**(9), 2337–2348 (2022)

A Two-Stage Network for Segmentation of Vertebrae and Intervertebral Discs: Integration of Efficient Local-Global Fusion Using 3D Transformer and 2D CNN

Zhiqiang Li[1,2], Xiaogen Zhou[1,2], and Tong Tong[1,2,3](✉)

[1] College of Physics and Information Engineering, Fuzhou University, Fuzhou, China
ttraveltong@gmail.com
[2] Fujian Key Lab of Medical Instrumentation and Pharmaceutical Technology, Fuzhou University, Fuzhou, China
[3] Imperial Vision Technology, Fujian, China

Abstract. In the field of computer-aided diagnosis (CAD) for spinal diseases, the fundamental task of multi-label segmentation for vertebrae and intervertebral discs (IVDs) assumes a significant role. However, the distinctive characteristics inherent to the spinal structure pose considerable challenges to the segmentation process, impeding its practical applicability in clinical settings. Convolutional neural networks have been widely used in this task; however, their limited receptive field restricts their capacity to capture extended-range spatial correlations. Consequently, the model's ability to accurately delineate vertebral boundaries is compromised, leading to a notable deterioration in the quality of segmentation outputs. To address this limitation, we propose a novel two-stage convolutional neural network (CNN) framework that incorporates both 3D Transformers and 2D CNNs. By synergistically leveraging the advantages of Transformers in facilitating the integration of long-range dependencies and the ability of CNNs to learn global and local features, our proposed approach exhibits promising potential in enhancing the segmentation performance for vertebrae and intervertebral discs. Moreover, we introduce a graph convolution module into our network architecture to exploit the inherent spatial dependencies present in MRI scans of spinal structures, thereby extracting semantic feature representations and further augmenting the efficacy of segmentation. The evaluation of our proposed method is conducted on the MRSpineSeg Challenge dataset, encompassing T2-weighted MR images.

Keywords: Two-stage · Combined with 3D Transformers and 2D CNN · multi-label · Graph Convolution module · Segmentation of Vertebrae and Intervertebral Discs · deep learning

---

Supported by National Natural Science Foundation of China under Grant 62171133.

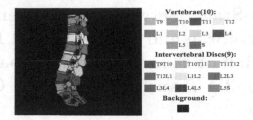

Fig. 1. The task involves the multi-label segmentation of volumetric MR images depicting the vertebrae and intervertebral discs, encompassing 10 distinct labels for vertebrae and 9 for intervertebral discs. It is worth noting that the labels correspond to vertebrae located in the thoracic (T), sacral (S), and lumbar (L) regions.

1   Introduction

The spinal serving as the central axis of the skeletal structure, assumes a vital role in protecting essential organs, blood vessels, and nerves [1]. As the population ages, the incidence of spinal disorders has witnessed a significant increase. In the domain of computer-aided diagnosis and treatment of spine-related diseases, the multi-label segmentation of volumetric magnetic resonance (MR) images pertaining to vertebral bones and intervertebral discs assumes a critical significance. Accurate segmentation of the spinal region, as depicted in Fig. 1, empowers medical practitioners to assess the structural characteristics and overall health of vertebrae and intervertebral discs, thereby facilitating early detection, diagnosis, and surgical planning for various spinal conditions, including deformities, traumas, tumors, and fractures.

Currently, with the progress of artificial intelligence, contemporary medical image spinal segmentation techniques are predominantly built upon two predominant strategies:1) Traditional machine learning based methods. Bao et al. [2] employed a linear iterative clustering algorithm to acquire superpixel MRI images of the spine, enabling the subsequent segmentation of the spinal region. Viji et al. [3] applied a probabilistic boosting tree (PBT) approach in conjunction with fuzzy support vector machine segmentation to achieve automated detection of the spinal canal.2) Deep learning-based methods. In contrast to conventional methodologies, deep learning techniques have demonstrated remarkable efficacy in the domain of spinal segmentation.

Particularly, convolutional neural networks (CNNs) [4–8] have been widely adopted, yielding significant advancements in spinal MR image segmentation. Noteworthy models such as the fully convolutional neural network (FCNNs) [4,5] and U-Net [6,7] have played a prominent role in these advancements. However, the effectiveness of FCNNs is limited by the restricted spatial range of the convolutional layers, impeding the model's ability to capture long-range spatial correlations. Despite the increasing diversity of models employed in spinal segmentation, they often overlook the distinctive chain structure of the spine and neglect the structural interdependencies among neighboring vertebrae and

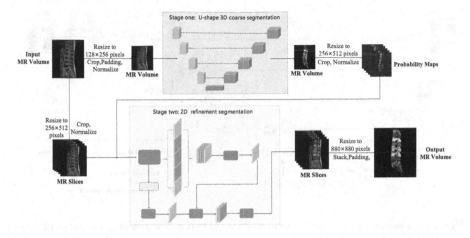

Fig. 2. Our proposed segmentation network consists of two stages, namely 3D coarse segmentation and 2D refinement segmentation.

lumbar discs. These approaches overlook the holistic architecture of the spine, the persistent long-range dependencies between vertebrae, and the inherent relationships among them. Furthermore, the significant computational and memory requirements associated with these methods impose limitations on their adaptability in diverse spinal segmentation scenarios.

Our work presents the following main contributions:

1. We propose a novel two-stage network architecture designed specifically for the segmentation of biomedical 3D MR images. Our approach involves the integration of a coarsely segmented 3D Transformer to capture long-distance dependencies, along with a finely segmented 2D CNN to capture local high-level features effectively.
2. The incorporation of both 3D and 2D networks enables our model to assimilate a broader range of feature information from images with varying dimensions, thus enhancing its ability to learn diverse representations.
3. To further augment the segmentation performance of our proposed two-stage network, we introduce graph convolution modules within both the 3D and 2D networks. This integration harnesses the power of graph convolution to exploit spatial relationships, leading to improved segmentation outcomes.

2   Methods

2.1   Overall Architecture Design

We presents an innovative methodology for multi-class segmentation, employing a two-stage approach. In particular, we introduce a U-shaped 3D coarse segmentation network, leveraging Transformers as the foundation for the initial segmentation stage, followed by a refinement segmentation network based

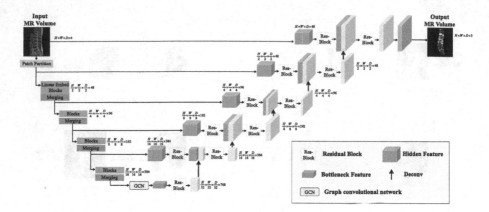

Fig. 3. An outline of the architecture of the 3D coarse segmentation network is presented. The input to the initial segmentation stage consists of 3D multi-modal MRI images with 4 channels. The encoded feature representations in the Swin transformer are transmitted to a CNN-decoder via skip connections at multiple resolutions. The final segmentation output comprises 3 output channels.

on DeepLabv3+ in the subsequent stage. The 3D coarse segmentation network utilizes Swin Transformers as the encoder, which is connected to FCNN-based decoders via skip connections. The decoder generates probability maps for the coarse segmentation task. Subsequently, during the refinement segmentation stage, the volumetric MR image and the probability map derived from the 3D coarse segmentation network serve as inputs for the 2D refinement segmentation network, aiming to achieve more precise and intricate segmentation results. Our proposed two-stage network is specifically tailored for multi-category segmentation of vertebrae and intervertebral discs in volumetric MR images. Figure 2 provides a visual depiction of the network architecture, offering an overview of its structural components.

2.2   3D Coarse Segmentation Stage

Inspired by the effectiveness of the "U-shaped" network architecture, we present a U-shaped 3D coarse segmentation network built upon the Swin Transformer. This network is designed for application during the coarse segmentation stage. The structural configuration of the coarse segmentation network is illustrated in Fig. 3 of this study.

Our coarse segmentation network follows a contracting-expanding pattern, incorporating a stack of transformers as the encoder and establishing connections with the decoder through skip connections. The input token $X \epsilon R^{H \times W \times D \times S}$ to the coarse segmentation network exhibits a patch resolution of $(\hat{H}, \hat{W}, \hat{D})$ and a dimension of $\hat{H} \times \hat{W} \times \hat{D} \times S$. To facilitate the projection of a 3D token sequence with a dimensional parameter $[\frac{H}{\hat{H}}] \times [\frac{W}{\hat{W}}] \times [\frac{D}{\hat{D}}]$ onto an embedding space of dimensional parameter C, we employ a patch partition layer. This layer enables the transformation of the input token sequence into an embedded representation.

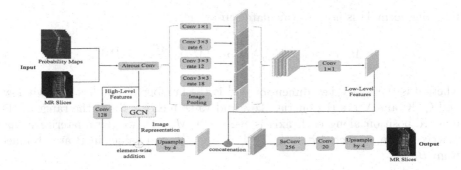

Fig. 4. The architecture of the 2D refinement segmentation network is delineated. The inputs of this network include both the 2D MR sagittal slice and its corresponding coarse probability map, which is produced by the 3D coarse segmentation network.

In order to capture token interactions effectively, we incorporate a self-attention mechanism that operates across non-overlapping windows generated during the partitioning phase. Within the transformer encoder architecture, at a specific layer denoted as l, we employ windows of size M×M×M to evenly divide a 3D token sequence into $[\frac{\hat{H}}{M}] \times [\frac{\hat{D}}{M}] \times [\frac{\hat{W}}{M}]$ regions. These partitioned window segments are subsequently shifted by $([\frac{M}{2}], [\frac{M}{2}], [\frac{M}{2}])$ voxels in layer l+1. Instead of the conventional multi-head self-attention (MSA) module, the Swin Transformer utilizes a shifted windows module, which constrains self-attention calculations to non-overlapping local windows using the shifted windows strategy. This approach not only facilitates efficient computation but also enables the modeling of token dependencies across the entire sequence.

The Swin Transformer module consists of a multi-head self-attention (MSA) module with a shifted window and a two-layer MLP, embedded between Gaussian Error Linear Units (GELU) nonlinearities. Prior to each MSA and MLP module, a LayerNorm (LN) layer is applied. Moreover, residual connections are established between two Swin Transformer modules, enhancing information flow within the network. The introduction of the shifted window division method in the Swin Transformer module optimizes its computational efficiency. The calculation process of the Swin Transformer module, employing this method, can be outlined as follows:

$$\hat{Z}^l = W - MSA(LN(Z^{l-1})) + Z^{l-1} \tag{1}$$

$$Z^l = MLP(LN(\hat{Z}^l)) + \hat{Z}^l \tag{2}$$

$$\hat{Z}^{l+1} = SW - MSA(LN(\hat{Z}^l)) + Z^l \tag{3}$$

$$Z^{l+1} = MLP(LN(\hat{Z}^{l+1})) + \hat{Z}^{l+1} \tag{4}$$

where $\hat{Z}^l$ and Z^l stand for the (S) W-MSA modules and the MLP module's respective block l output characteristics. Similar to other studies [9,10], the following formula is used to calculate self-attention:

$$Attention(Q, K, V) = SoftMax(\frac{QK^T}{\sqrt{d}} + B)V \qquad (5)$$

where d is the query/key dimension, M^2 is the number of patches in a window, and Q, K, and V are the queries, key, and value metrics. Since the range of the relative position along each axis is $[-M + 1, M - 1]$, we parameterize a bias matrix with a smaller size, $B \in R^{(2M-1) \times (2M-1)}$, and values in B are obtained from $\hat{B}$.

2.3   2D Refinement Segmentation Stage

During the 2D segmentation stage, our methodology is primarily guided by the design principles of DeepLabv3+ [?]. In the encoder phase, we employ parallel atrous convolution at multiple rates, commonly referred to as Atrous Spatial Pyramid Pooling (ASPP) [11], to effectively encode multi-scale context information. For the segmentation task, we adopt the Xception architecture and incorporate depthwise separable convolution to enhance both the efficiency and precision of network training. Furthermore, to refine the segmentation outcomes, we introduce a straightforward yet highly efficacious decoder module, which builds upon the aforementioned foundation. The architectural details of the refinement segmentation network are presented in Fig. 4.

The 2D refinement segmentation network takes as input the 2D MR sagittal slice and the coarse probability map corresponding to that slice, which is generated by the 3D coarse segmentation network. Incorporating the coarse probability map enables the 2D refinement segmentation network to leverage the implicit 3D semantic information of the image. By effectively integrating the semantic features of the spinal structure with detailed information, the network achieves accurate segmentation. The high-resolution MR slices contain detailed information, and the 2D refinement segmentation network combines this information with the 3D semantic information to produce fine segmentation.

2.4   Graph Convolution Module

The graph convolution module consists of three consecutive stages: Pooling, Graph Convolutional Network (GCN), and Unpooling. In the Pooling stage, the input image representation is transformed into a graph-based representation to facilitate subsequent processing by the GCN stage. The GCN stage aims to generate graph representations enriched with semantic information through the application of graph convolution operations. In the final Unpooling stage, the obtained semantic graph representation is mapped back to the semantic image representation and passed to the convolution layer for further processing.

Table 1. The mean DSC (%) for the proposed method and other methods on the MRSpineSeg Challenge dataset.

Methods	T9	T10	T11	T12	L1	L2	L3	L4	L5	S
nnUNet	0.00±0.00	3.07±15.02	70.71±28.28	79.86±18.10	80.55±16.81	79.17±18.17	80.59±14.55	84.44±8.43	85.58±8.83	86.71±2.07
VNet	0.00±0.00	28.96±29.87	72.45±27.34	80.64±20.83	81.98±19.47	84.33±13.57	86.63±4.70	86.69±3.31	86.16±4.03	85.54±2.49
UNETR	0.00±0.00	26.17±35.85	66.99±26.52	75.66±18.59	79.23±16.82	79.23±15.32	79.21±15.25	80.80±13.49	82.80±7.99	83.43±4.59
3D Graphonomy	20.78±30.97	44.32±38.50	75.67±23.83	82.14±14.62	83.56±14.73	82.22±13.82	82.65±12.51	82.80±12.88	84.48±10.82	82.52±4.28
3D Deeplabv3+2D ResUNet	24.18±25.39	49.91±37.25	78.86±23.43	86.59±13.35	88.20±9.62	87.67±7.88	87.27±6.78	86.76±7.04	86.93±6.11	87.58±3.45
3D Graphonomy+2D Deeplabv3	23.59±23.12	44.77±35.39	77.09±23.52	84.78±13.90	86.27±13.09	86.07±12.69	86.35±11.52	85.92±11.57	85.87±9.96	85.82±3.33
Ours	31.12±21.99	56.90±34.25	80.75±20.34	87.34±9.74	88.19±9.76	87.68±9.64	88.54±3.83	88.31±3.15	87.83±3.08	87.53±2.54

Table 2. The mean DSC (%) for the proposed method and other methods on the MRSpineSeg Challenge dataset.

Methods	T9T10	T10T11	T11T12	T12L1	L1L2	L2L3	L3L4	L4L5	L5S
nnUNet	0.00±0.00	0.00±0.00	74.78±26.38	81.19±19.41	80.44±19.75	81.02±19.13	85.42±13.41	85.37±9.59	85.07±10.09
VNet	0.00±0.00	44.97±33.05	78.68±24.01	83.21±21.37	86.17±14.78	87.17±13.47	89.11±4.05	86.52±7.10	84.83±8.12
UNETR	0.00±0.00	42.97±38.16	73.31±28.29	76.48±21.08	78.53±22.19	80.74±18.55	81.39±16.30	80.69±15.22	82.44±7.76
3D Graphonomy	22.55±39.85	61.39±30.74	80.01±20.09	83.07±15.86	83.84±17.79	83.54±15.16	83.77±15.31	82.42±12.32	82.33±12.38
3D Deeplabv3 + 2D ResUNet	27.15±36.07	74.05±28.64	84.22±20.74	87.78±13.74	89.09±10.75	88.07±12.83	88.34±8.22	85.86±7.39	85.65±11.95
3D Graphonomy + 2D Deeplabv3	26.42±35.10	73.52±25.93	84.33±18.65	87.07±14.31	87.30±15.28	87.11±14.12	87.37±13.96	85.81±11.43	85.72±10.57
Ours	28.35±34.11	76.92±22.48	86.03±16.54	88.91±10.06	89.37±9.94	88.83±10.12	89.99±4.02	87.47±5.45	86.73±7.62

3 Experiments

3.1 Dataset

Our proposed method was evaluated on the MRSpineSeg Challenge dataset, which comprises a total of 215 T2-weighted MR volumetric images. During the experiment, 172 images were utilized, and they were partitioned into training, validation, and testing sets in a ratio of 7:2:1. The volumetric images encompassed 10 vertebrae, 9 intervertebral discs (IVDs), and backgrounds, resulting in a total of 20 distinct categories. The original images exhibited varying dimensions, with widths and heights ranging from 512 to 1024, while the number of slices along the coronal axis ranged from 12 to 20.

3.2 Implementation Details

For 3D networks, the pre-processing stage comprises a series of steps aimed at preparing the input data. These steps include cropping, resizing, padding, and normalization. To begin with, the cropping step involves center-cropping the images along the depth direction to eliminate the non-spine portion, as half of the image does not contain spinal information. Subsequently, the cropped image is resized to a dimension of $18 \times 256 \times 128$ pixels, with zero filling applied in the depth direction to ensure uniformity. Lastly, the normalization process involves computing the mean and variance values across all the images. These values are then utilized to subtract the mean from each pixel and divide by their standard deviation, resulting in a normalized representation of the data.

Our methodology was implemented using the Python programming language based on the PyTorch deep learning framework. The model was trained on an Nvidia RTX 3090 GPU with 24 GB of RAM. During the 3D segmentation stage, a preliminary probability map of dimensions $20 \times 18 \times 256 \times 128$ was generated,

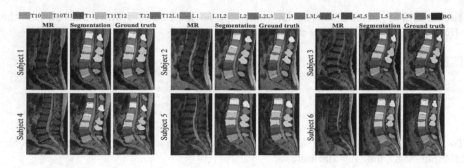

Fig. 5. Using our segmentation method, sagittal slices depicting vertebrae and intervertebral discs from six subjects were acquired. The label BG means the background in these slices.

with an MR volume of size $16 \times 256 \times 128$ serving as input. The Adam [13] optimizer was employed for optimization, with a weight decay of 0.0001. We initiated the learning rate at 0.001, and reduced it by a factor of 5 every 33 epochs. The batch size was set to 2, which was limited by the available GPU memory.

3.3 Evaluation Metrics

To assess the segmentation performance, several metrics were employed in our experiment, including the Dice similarity coefficient, precision(DSC), and recall. These metrics are computed as follows:

$$Dice = \frac{2TP}{FP + 2TP + FN}. \tag{6}$$

Table 3. Ablation experiments were conducted on the MRSpineSeg Challenge dataset to assess the effectiveness of each component in segmentation of the ten classes of vertebrae T9-S. The mean DSC (%) was used to validate the components.

| Coarse Segmentation | | Refinement Segmentation | | T9 | T10 | T11 | T12 | L1 | L2 | L3 | L4 | L5 | S |
3D Swin Transform	3D GCM	2D Deeplabv3+	2D GCM										
✓				21.70±29.97	47.68±34.94	77.39±23.17	84.89±16.34	85.04±15.56	85.48±13.24	86.22±4.78	86.12±3.40	85.53±4.16	85.27±2.82
✓	✓			25.41±24.71	50.19±36.35	78.87±23.12	85.11±16.35	86.14±15.57	86.53±13.26	87.28±4.77	87.15±3.41	86.62±4.13	86.36±2.82
✓		✓		27.11±21.65	53.99±34.63	79.25±22.11	86.49±11.21	87.64±10.37	87.23±10.17	87.81±6.91	87.55±7.26	87.34±4.53	87.27±2.75
✓		✓	✓	28.23±21.47	54.78±33.63	80.37±22.14	86.23±11.20	87.73±10.4	87.29±10.16	87.86±6.90	87.59±7.25	87.41±4.50	87.33±2.72
✓	✓	✓		28.59±21.39	54.39±33.25	80.28±20.44	87.12±9.76	88.10±9.77	87.63±9.61	88.49±3.85	88.28±3.15	87.74±3.13	87.44±2.54
✓	✓	✓	✓	31.12±21.99	56.90±34.25	80.75±20.34	87.34±9.74	88.19±9.76	87.68±9.64	88.54±3.83	88.31±3.15	87.83±3.08	87.53±2.54

Table 4. Ablation experiments were conducted on the MRSpineSeg Challenge dataset to assess the effectiveness of each component in segmentation of the nine classes of IVDs T9T10-L5S. The mean DSC (%) was used to validate the components.

| Coarse Segmentation | | Refinement Segmentation | | T9T10 | T10T11 | T11T12 | T12L1 | L1L2 | L2L3 | L3L4 | L4L5 | L5S |
Coarse Segmentation	3D GCM	2D Deeplabv3+	2D GCM									
✓				20.58±25.94	67.33±29.91	81.44±21.70	85.70±16.57	86.96±14.31	86.71±13.70	87.86±4.11	85.37±5.50	84.29±8.15
✓	✓			24.81±35.02	69.92±29.24	82.88±21.65	86.68±16.56	87.92±14.31	87.78±13.73	88.88±4.13	86.37±5.49	85.36±8.11
✓		✓		25.18±26.91	71.08±30.00	83.73±24.41	87.59±10.52	88.23±10.07	88.03±11.61	89.57±6.35	86.88±8.78	86.57±7.66
✓		✓	✓	26.35±26.91	72.92±27.34	84.27±21.00	88.60±10.51	89.26±10.07	88.21±11.61	89.26±10.07	86.95±8.78	86.62±7.66
✓	✓	✓		27.11±34.66	74.33±24.10	85.59±16.69	88.94±10.07	89.41±9.92	88.75±10.09	89.97±4.00	87.44±5.47	86.65±7.67
✓	✓	✓	✓	28.35±34.11	76.92±22.48	86.03±16.54	88.91±10.06	89.37±9.94	88.83±10.12	89.99±4.02	87.47±5.45	86.73±7.62

$$Pre = \frac{TP}{TP + FP}. \tag{7}$$

$$Recall = \frac{TP}{TP + FN}. \tag{8}$$

where TP, FP, FN, and TN denote the number of true positives, false positives, false negatives, and true negatives, respectively.

3.4 Experiment Results

The Table 5 displays the precise values of Mean Recall, Mean Precision, and Dice Similarity Coefficient (DSC) achieved by the two-stage segmentation network for vertebrae, intervertebral discs (IVD), and all 19 spinal structures. We have presented some exemplary images with well-performing segmentation results in Fig. 5.

We conducted a comparative analysis of our proposed spinal segmentation method with several other methods, including nnUNet [14], VNet [15], UNETR [16], 3D Graphonomy [12], 3D Deeplabv3 [11] + 2D ResidualUNet [17], and 3D Graphonomy [12] + 2D Deeplabv3 [11]. The evaluation of the segmentation performance across these methods was based on three crucial metrics: the Dice similarity coefficient (DSC), Precision, and Recall. Tables 1 and 2 present the DSC evaluation indexes specifically for the segmentation of each vertebra and intervertebral disc (IVD). Our proposed segmentation network demonstrated superior performance compared to the other methods, achieving excellent segmentation results for the seven categories of vertebrae T12-S (DSC > 87.34%) and the seven categories of IVDs T11-S (DSC > 86.03%). These quantitative comparison results highlight the notable superiority of our proposed methodology. Furthermore, Fig. 6 showcases specific segmentation results obtained by applying different algorithms to the aforementioned dataset, providing visual evidence of the superior segmentation outcomes achieved by our proposed method.

Table 5. The average values of Recall, Precision, and Recall were computed for the segmentation of vertebrae, intervertebral discs (IVDs), and all 19 spinal structures using our proposed two-stage segmentation network.

	Mean Recall	Mean Precision	Mean Dice
Background	98.96 ± 0.30	98.99 ± 0.50	98.97 ± 0.21
Vertebrae	85.87 ± 8.46	86.16 ± 6.54	85.61 ± 6.68
IVDs	88.76 ± 8.31	85.95 ± 6.71	86.96 ± 7.40
Overall	87.19 ± 7.84	86.04 ± 5.85	86.21 ± 6.57

The performance of our network on the segmentation of T9-T11 vertebrae (DSC $\leq 80.75\%$) and T9-T12 IVDs (DSC $\leq 76.92\%$) is unsatisfactory due to several factors. Firstly, the dataset contains very limited samples of T9-T11 vertebrae and T9-T12 IVDs, with most of them being incompletely shaped. Secondly, the top of the image contains three types of vertebrae (T9-T11) and two types of IVDs (T9-T12), making segmentation difficult due to the limited receptive field at the top. These factors contribute to the suboptimal segmentation results of our network in these regions.

3.5   Ablation Study

To assess the efficacy of each constituent element within our network architecture, a series of ablation experiments were conducted, yielding results that have been presented in Tables 3 and 4. The evaluation process encompassed six distinct configurations involving the integration of the 3D Swin Transform and 3D GCM during the 3D Coarse Segmentation stage, as well as the utilization of the 2D Deeplabv3+ and 2D GCM during the 2D Refinement Segmentation stage. These meticulous experiments effectively demonstrated the augmentation of segmentation performance for both Vertebrae and IVDs through the inclusion of the graph convolutional module and the employment of a dual network strategy.

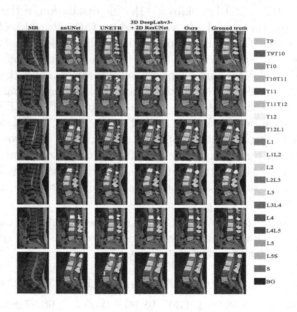

Fig. 6. Visualized comparison of results using different segmentation networks.

3.6   Effect of the Two-Stage Framework

The incorporation of 2D refinement stages into 3D segmentation tasks has demonstrated considerable effectiveness in enhancing the performance of segmentation algorithms. This enhancement is substantiated by the findings presented in Tables 1 and 2, which elucidate the improvements attained through the integration of 2D refinement in the 3D Graphonomy and 3D Graphonomy+2D Deeplabv3 frameworks, respectively. In comparison to the sole utilization of 3D Graphonomy, the inclusion of a 2D refinement stage within the 3D Graphonomy+2D Deeplabv3 approach yielded a notable increase in the average Dice similarity coefficient (DSC) across the eight classes of vertebrae T11-S and T10T11-L5S, as well as the eight classes of intervertebral discs (IVDs). Similarly encouraging results were obtained from the ablation experiments conducted, as evidenced by the outcomes presented in Tables 3 and 4. The incorporation of high-resolution images within the 2D networks contributed to a more comprehensive representation of detailed information, enabling the model to acquire a richer understanding of the underlying features. Consequently, the synergistic combination of 3D and 2D networks facilitated the assimilation of more contextual information from images with varying dimensions, culminating in a discernible enhancement in segmentation performance.

3.7   Effect of the GCM

The quantitative findings elucidated in Tables 3 and 4 provide compelling evidence that the incorporation of the graph convolution module into either the 3D or 2D network yields notable advantages in enhancing the segmentation performance of the model. Notably, it should be acknowledged that during the training phase of both the 3D and 2D networks, the segmentation results may not strictly adhere to the spatial order of the spinal structure. Consequently, the inclusion of the graph convolution module in both the 3D and 2D networks emerges as a more favorable approach for boundary position segmentation.

4   Conclusion

This paper presents a novel two-stage framework designed for achieving precise multi-label segmentation of vertebrae and intervertebral discs. The proposed framework integrates 3D transformers and 2D convolutional neural networks (CNN) to attain accurate and reliable segmentation outcomes. In the initial stage of the framework, 3D transformers are employed to generate preliminary probability graphs, thereby establishing a foundation for subsequent processing. Subsequently, in the second stage, the 2D MR sagittal slice and the corresponding rough probability graph derived from the 3D rough segmentation network are jointly inputted into the 2D network to achieve refined segmentation results with heightened precision. Notably, the integration of graph convolution modules within both the 3D and 2D networks plays a crucial role in addressing pertinent challenges associated with pixel labeling isolation, as well as rectifying errors pertaining to shape and positional segmentation outcomes. These modules contribute to the enhancement of segmentation accuracy by effectively resolving issues related to isolation and correction within the segmentation process. Through comprehensive comparisons with state-of-the-art spinal segmentation methodologies utilizing publicly available datasets, the proposed framework has exhibited superior performance, underscoring its efficacy and potential for advancing the field of spinal segmentation.

References

1. Lopez, I.B., Benzakour, A., Mavrogenis, A., Benzakour, T., Ahmad, A., Lemee, J.-M.: Robotics in spine surgery: systematic review of literature. Int. Orthop. **47**(2), 447–456 (2023)
2. Bao, X.-X., et al.: Recognition of necrotic regions in MRI images of chronic spinal cord injury based on Superpixel. Comput. Methods Programs Biomed. **228**, 107252 (2023)
3. Viji, C., Rajkumar, N., Suganthi, S., Venkatachalam, K., Kumar, T.R., Pandiyan, S.: An improved approach for automatic spine canal segmentation using probabilistic boosting tree (PBT) with fuzzy support vector machine. J. Ambient. Intell. Humaniz. Comput. **12**, 6527–6536 (2021)
4. Pang, S., et al.: SpineParseNet: spine parsing for volumetric MR image by a two-stage segmentation framework with semantic image representation. IEEE Trans. Med. Imaging **40**(1), 262–273 (2020)
5. Pang, S., et al.: DGMSNet: spine segmentation for MR image by a detection-guided mixed-supervised segmentation network. Med. Image Anal. **75**, 102261 (2022)
6. Yang, Z., Wang, Q., Zeng, J., Qin, P., Chai, R., Sun, D.: RAU-Net: u-net network based on residual multi-scale fusion and attention skip layer for overall spine segmentation. Mach. Vis. Appl. **34**(1), 10 (2023)
7. Wang, B., Qin, J., Lv, L., Cheng, M., Li, L., Xia, D., Wang, S.: MLKCA-Unet: multiscale large-kernel convolution and attention in UNet for spine MRI segmentation. Optik **272**, 170277 (2023)
8. Tao, R., Zheng, G.: Spine-transformers: vertebra detection and localization in arbitrary field-of-view spine CT with transformers. In: de Bruijne, M., et al. (eds.) MICCAI 2021. LNCS, vol. 12903, pp. 93–103. Springer, Cham (2021). https://doi.org/10.1007/978-3-030-87199-4_9
9. Hu, H., Gu, J., Zhang, Z., Dai, J., Wei, Y.: Relation networks for object detection. In: Proceedings of the IEEE Conference on Computer Vision and Pattern Recognition, pp. 3588–3597 (2018)
10. Hu, H., Zhang, Z., Xie, Z., Lin, S.: Local relation networks for image recognition. In: Proceedings of the IEEE/CVF International Conference on Computer Vision, pp. 3464–3473 (2019)
11. Chen, L.-C., Papandreou, G., Kokkinos, I., Murphy, K., Yuille, A.L.: DeepLab: Semantic image segmentation with deep convolutional nets, Atrous convolution, and fully connected CRFs. IEEE Trans. Pattern Anal. Mach. Intell. **40**(4), 834–848 (2017)
12. Gong, K., Gao, Y., Liang, X., Shen, X., Wang, M., Lin, L.: Graphonomy: universal human parsing via graph transfer learning. In: Proceedings of the IEEE/CVF Conference on Computer Vision and Pattern Recognition, pp. 7450–7459 (2019)
13. Kingma, D.P., Ba, J.: Adam: a method for stochastic optimization. arXiv preprint arXiv:1412.6980 (2014)
14. Isensee, F., Jaeger, P.F., Kohl, S.A., Petersen, J., Maier-Hein, K.H.: nnU-net: a self-configuring method for deep learning-based biomedical image segmentation. Nat. Methods **18**(2), 203–211 (2021)

15. Milletari, F., Navab, N., Ahmadi, S.-A.: V-net: Fully convolutional neural networks for volumetric medical image segmentation. In: 2016 Fourth International Conference on 3D Vision (3DV), pp. 565–571. IEEE (2016)
16. Hatamizadeh, A., et al.: UNETR: transformers for 3d medical image segmentation. In: Proceedings of the IEEE/CVF Winter Conference on Applications of Computer Vision, pp. 574–584 (2022)
17. He, K., Zhang, X., Ren, S., Sun, J.: Deep residual learning for image recognition. In: Proceedings of the IEEE Conference on Computer Vision and Pattern Recognition, pp. 770–778 (2016)

Integrating Multi-view Feature Extraction and Fuzzy Rank-Based Ensemble for Accurate HIV-1 Protease Cleavage Site Prediction

Susmita Palmal[✉], Sriparna Saha, and Somanath Tripathy

Indian Institute of Technology Patna, Bihar, India
{susmita_2121cs34,sriparna,som}@iitp.ac.in

Abstract. Acquired immunodeficiency syndrome (AIDS) continues to be a significant cause of mortality, disability, and economic repercussions, especially in underdeveloped countries. Extensive research has been conducted to develop effective therapies for human immunodeficiency virus (HIV) infection, including the prediction of HIV-1 protease cleavage sites. Accurate prediction of these sites can expedite the discovery of new HIV-1 protease inhibitors. Motivated by this, we propose a novel approach for HIV-1 protease cleavage site prediction using numerical descriptors based on octapeptide sequences. Our method incorporates multi-view feature extraction, combining sequence order effects of amino acids with physicochemical features. To capture important information, we utilize a convolutional neural network for feature extraction. For the classification task, we employ a fuzzy rank-based ensemble method, utilizing Random Forest, Logistic Regression, and Support Vector Machine as base classifiers. The ensemble combines their predictions to make the final prediction. Experimental evaluation on benchmark datasets demonstrates the effectiveness of our approach, achieving average Accuracy, AUC, Precision, Recall, and F-measure of 0.93, 0.95, 0.85, 0.76, and 0.80, respectively. Comparisons with existing studies confirm the potential of our proposed technique. The source code can be downloaded from Github: https://github.com/SusmitaPalmal/RC_CNN_HIV.

Keywords: Protease cleavage site · Multi-view features · Fuzzy rank-based ensemble · Latent feature extraction

1 Introduction

Human Immunodeficiency Virus (HIV) is a kind of lentivirus that infects and attacks the immune system, leading to Acquired Immunodeficiency Syndrome (AIDS) [10]. It stays hidden for a long time after entering the body, making it a great challenge for the treatment. The virus has two different types: HIV-1 and HIV-2. HIV-1 is the more life-threatening and common of these two. Therefore, several studies have focused on this particular strain of HIV. One of

B. Luo et al. (Eds.): ICONIP 2023, CCIS 1964, pp. 480–492, 2024.
https://doi.org/10.1007/978-981-99-8141-0_36

the approaches is the utilization of HIV-1 protease inhibitors. HIV-1 protease (PR) is a necessary enzyme that accomplishes a crucial role in HIV replication by cleaving or dismantling the polyproteins. Therefore, inhibiting the activity of HIV-1 PR is an effective way to treat AIDS. Consequently, understanding the HIV protease-cleavable peptide sites can help scientists develop more potent HIV protease inhibitors. To this end, several inhibitors have been invented to prevent the cleaving process of peptide substrates. However, the effectiveness of these inhibitors depends on their ability to bind tightly to the PR. To design reliable and effective PR inhibitors, it is necessary to understand the substrate specificity of HIV-1 PR [2]. Moreover, identifying HIV protease cleavage sites can be challenging due to the time and effort required for laboratory-based investigations.

In recent years, machine learning-based computational methods have been pitched to address the limitations of laboratory-based experiments in detecting HIV-1 protease cleavage sites. The prediction of cleavage sites is handled by these computational methods as a standard prediction problem. These techniques combine many classifiers with diverse properties of protein or substrate sequences. Naani [11] suggested that combining classifiers trained in different feature spaces can significantly reduce errors in HIV-1 PR cleavage site prediction. Furthermore, HIVcleave [19] utilized a discriminant function method and a sequence-coupling model to make accurate predictions. The encoding technique of protein sequence also plays an important role in HIV-1 Protease cleavage site prediction. Gok and Ozcerit [3] have established this by combining Taylor's Venn diagram with orthonormal encoding. On the other hand, DeepCleave [7] employed one-hot encoding for input protein sequence feature extraction. Rognvaldsson et al. [18] proposed another effective approach by using Orthonrmal Encoding for feature extraction from protein sequences in combination with an SVM classifier. On the other hand, Li et al. [8] treated unknown octapeptides as unlabeled sets for the prediction task and performed the classification using a biased SVM. Furthermore, Hu et al. [4] proposed the EvoCleave method by integrating the co-evolutionary pattern of substrate sequences. They used SVM classifier for the final classification. Onah et al. [13] adopted a hybrid approach that combines multiple input variables. Moreover, Hu et al. [5] introduced EM-HIV, a biased SVM classifier ensemble learning algorithm. Recent studies have also focused on deep learning-based approaches to achieve improved performance in various research objectives [14,15]. Based on that, in another recent study [16], a deep learning-based stacked auto-encoder is applied for latent feature extraction followed by classification using a majority voting-based classifier ensemble. Furthermore, S. Palmal et al. [17] propose a multi-objective optimization with a classifier ensemble to improve accuracy and AUC in HIV-1 protease cleavage site prediction.

We can see in the above studies that previous researchers have employed various feature extraction techniques and applied different machine-learning approaches or deep learning-based approaches or their ensembles to predict cleavage sites. However, there exists a lot of scope for improving the perfor-

482   S. Palmal et al.

mance of this study. Inspired by this, we apply Convolutional Neural Network
(CNN) [1] based feature extraction from the multi-view features of the octapeptide sequence. It works by applying convolutional filters to input multi-view
data, allowing it to learn and extract important features automatically which
was unexplored in the previous studies. Here, we have used physicochemical and
sequential features of the octapeptide sequence as multi-view features. Instead
of directly using multi-view features, we calculated the Quasi Residue Couple
[12] based on the combined representation of these two types of features. After
extracting important features, we applied a fuzzy rank-based ensemble [9] of
classifiers for the final classification. This ensemble technique incorporates the
confidence levels of the base classifiers, which was never explored for the HIV-1
protease cleavage site prediction task.

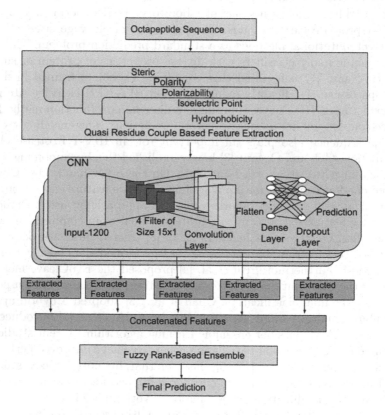

Fig. 1. Architecture for the fuzzy rank-based ensemble model

The key contributions of the suggested method are enumerated below:

1. The suggested study makes use of multi-view features, a combined representation of physicochemical and indirect sequence order features for the prediction
of HIV-1 protease cleavage site. This is extracted from octapeptide sequences

using the Quasi Residue Couple (RC) [12] method rather than just one kind of feature collection.

2. Here, five important physicochemical features, namely Hydrophobicity, Iso-electric Point, Polarizability, Steric Property, and Polarity, have been used with the indirect sequence order features. These multi-view features are capable of grasping various data characteristics. We will call this feature set as RCPhySeq feature set onwards.

3. The proposed study is the first to utilize Convolutional Neural Network [1] to extract informative latent feature vectors from multi-view features. The main idea behind using a CNN is to learn complex and abstract features by applying convolutional filters across the input data.

4. The final classification is applied using a Fuzzy rank-based ensemble [9] of classifiers where Random forest, Logistic regression, and SVM(Linear) are utilized as base classifiers. Unlike existing fusion methods found in research, this ensemble technique incorporates the confidence levels of the base classifiers when making predictions on test samples.

5. The proposed approach attains state-of-the-art performance for the majority of the cases across four benchmark data sets. A detailed description of the data is given in Sect. 2.1.

2   Proposed Work

In the proposed work (Fig. 1), we have created a Convolutional Neural Network (CNN) [1] based feature extraction from multi-view features. We further have applied a fuzzy rank-based ensemble of classifiers [9] for the prediction of HIV-1 protease cleavage site using four benchmark datasets.

We have used multiple views of feature sets, which consist of physicochemical features with indirect sequence order of octapeptide. This feature encoding is done by applying the Quasi Residue Couple [12] method. Additionally, the utilization of a CNN to discover latent feature representations, which offers more informative insights than the uncompressed feature set, is a novel exploration in this research. Finally, the fuzzy rank-based ensemble [9] of classifiers for the final classification incorporates the confidence levels of the base classifiers which was never explored for the HIV-1 protease cleavage site prediction task. We followed a ten-fold cross-validation method for individual datasets.

2.1   Data Set

Our proposed method utilizes four benchmark datasets obtained from the UCI machine learning repository (https://archive.ics.uci.edu/ml/datasets/HIV-1+protease+cleavage). These datasets consist of both cleaved and uncleaved octamers, with each octamer comprising a combination of eight amino acids. The descriptions of different benchmark data sets are given in Table 1.

2.2   Feature Extraction

Initially, we have taken octapeptide sequences from each dataset. There are twenty unique amino acids. Each of the octapeptides' eight places can be filled by any of these twenty different amino acids. However, only octapeptide sequences are insufficiently informative to build a prediction model. Thus, multi-view feature extraction using Quasi Residue Couple (RC) has been investigated here. After that, using the multi-view features, we applied CNN to learn complex and abstract features by using convolutional filters across the input data.

Table 1. Details of four benchmark data sets

	Data746	Data1625	Data Schilling	Data Impens
#Octamers	746	1625	3272	947
#Cleavage sites	401	374	434	149
#Non-Cleavage site	345	1251	2838	798

Quasi-Residue Couple (RC): This technique presents an effective method for feature encoding of octapeptide sequences by incorporating multiple views (physicochemical and indirect sequence order) to obtain an informative feature set. The physicochemical characteristics, which encompass molecular properties of the twenty amino acids, play a crucial role in determining their intrinsic reactivity. Our study focuses on the five important physicochemical properties which are retrieved from AAindex [6] database.

A small description of these properties is given below.

– Hydrophobicity: Refers to the propensity of water to divert non-polar molecules.
– Isoelectric Point: Influences cleavage specificity based on charge distribution.
– Polarizability: Affects stabilizing contacts with the protease active site.
– Steric Property: Determines molecular arrangement and interactions within the substrate.
– Polarity: Impacts solubility, water contact, and protein interactions.

Further, the RC model is constructed considering the indirect sequence order effect of peptide residues while taking into account physicochemical features. In Algorithm 1, the entire operation is described. Here, the order is represented by the mathematical constant n, where n is given the values $1, 2$, and 3 to take into account the indirect sequence order influence of the current amino acid on the first, second, and third next position of it in the octapeptide sequence. For each amino acid couples up to order $n \leq 3$, P outputs are concatenated into a $1,200$-dimensional vector.

Algorithm 1: RC-based Feature Extraction
Input:

- *AminoAcid[1 : 20]*: Array of amino acids
- *Physico_ Che[1 : 20]*: Array of physicochemical properties of amino acids
- *n*: Order
- *residue[1 : 8]*: Residue of octapeptide
- *itr1, itr2*: Iterations

Procedure:

1. **For** $n \leq 3$ **do**
2.   **For** $itr1 \leq 20$ **do**
3.   **For** $itr2 \leq 20$ **do**
4. $h1, h2 \leftarrow 0$
5.   **For** $k \leq 8 - n$ **do**
6.   **If** *residue[k] = AminoAcid[itr1]* $\wedge$ *residue[k+n] = AminoAcid[itr2]* **then**
7. $h1 \leftarrow h1 + Physico\_ Che[itr1]$
8. $h2 \leftarrow h2 + Physico\_ Che[itr2]$
9.   **End If**
10.   **End For**
11. $p1 \leftarrow \frac{1}{8-n} \times h1$
12. $p2 \leftarrow \frac{1}{8-n} \times h2$
13. $P \leftarrow p1 + p2$
14.   **End For**
15.   **End For**
16. **End For**

Note: In this algorithm, the variables $h1$ and $h2$ are used to calculate the sum of the physicochemical properties of amino acids, and $p1$ and $p2$ are used to calculate the respective probabilities. The variable P represents the sum of probabilities.

A Convolutional Neural Network Based Feature Extraction: The proposed model uses a convolutional neural network (CNN) for the HIV-1 cleavage site prediction and extracts the features from the multiview features (RCPhySeq Features) for the next stage of the model. The architecture of CNN is shown in Fig. 1. A CNN takes inputs from the individual feature set (RCPhySeq Features) of data, and each input data is processed through a set number of filters or kernels in convolution layers. The feature map is generated through a convolution process by combining the input and filter matrix values. This combination is achieved by performing element-wise multiplication and then adding the corresponding elements. To initialize the filter matrix, the Glorot normal initializer is utilized. This initializer assigns suitable initial values to the filter matrix to promote effective learning in the convolutional neural network. The parameter configurations of CNN are as follows. We used a single convolution layer with four filters of size 15. The convolution layer has a stride of 2 and the 'same' padding. Each layer's activation function is 'TANH'. It has two hidden layers, each holding 700 and 200 units. The 'batch size' that we utilized was 8, and the training epoch was 25. As a loss function (see Eq. 1), the binary cross entropy with L2 regularizer was employed.

$$L(y, \hat{y}) = -\frac{1}{N} \sum_{i=0}^{N} [y_i \log(\hat{y}_i) - (1 - y_i) \log(1 - \hat{y}_i)] + \frac{1}{2} \lambda \sum_{k=1}^{K} \sum_{j=1}^{n_k} \sum_{i=1}^{m_k} w_{ij}^{k^2} \tag{1}$$

In this equation, y_i represents the class label, and $\hat{y}_i$ represents the prediction value. N denotes the batch size. The kth weight matrix is denoted by $W^k = (w_{ij}^k)_{m_k \times n_k}$ with dimensions $m_k \times n_k$, where k ranges from 1 to K, representing the total number of weight matrices. The regularization parameter is denoted by λ.

2.3   Fuzzy Rank-Based Ensemble

Let us consider, for the base classifier c, the confidence scores for M number of classes are denoted as $(P_1^c, P_2^c, P_3^c, \ldots, P_M^c)$, where $c = 1, 2, 3$. We start by accumulating the confidence scores of base classifiers. As $(P_1^c, P_2^c, P_3^c, \ldots, P_M^c)$ represents probabilities, the following equation must be true:

$$\sum_{k=1}^{M} P_k^c = 1, \quad \forall c \in \{1, 2, 3\} \tag{2}$$

Now, we denote the fuzzy ranks generated using two non-linear functions as $(R_1^{c_1}, R_2^{c_1}, R_3^{c_1}, \ldots, R_M^{c_1})$ and $(R_1^{c_2}, R_2^{c_2}, R_3^{c_2}, \ldots, R_M^{c_2})$. The calculation of fuzzy ranks is achieved through the following equations:

$$R_k^{c_1} = 1 - \tanh\left(\frac{(P_k^c - 1)^2}{2}\right) \tag{3}$$

$$R_k^{c_2} = 1 - \exp\left(-\frac{(P_k^c - 1)^2}{2}\right) \tag{4}$$

In Eq. 3, a classification is rewarded. The reward amount grows as the value of Eq. 3 increases and gets closer to 1. On the other hand, Eq. 4 calculates the deviation from 1, where smaller values indicate a greater deviation. Let $(RS_1^c, RS_2^c, RS_3^c, \ldots, RS_M^c)$ be the fused rank scores, where RS_k^c is given by the equation:

$$RS_k^c = R_k^{c_1} \times R_k^{c_2} \tag{5}$$

The rank score (RS_k^c) is the product of the reward and deviation for a given base learner's confidence score. This fused rank score, which combines the fuzzy ranks produced by the two different types of functions, indicates the degree of confidence in a certain class. Now, the fused score tuple is $(FS_1, FS_2, FS_3, \ldots, FS_M)$, where FS_k is given by the equation:

$$FS_k = \sum_{c=1}^{L} RS_k^c, \quad \forall k = 1, 2, \ldots, M \tag{6}$$

$$\text{class}(I) = \min_k FS_k \tag{7}$$

This combined score can be used to determine the end result for each class. The winning class is determined by finding the class with the lowest fused score given in Eq. 7. Here, class(I) represents the winning class, which is determined by finding the class with the minimum fused score.

2.4 Steps of Proposed Technique

The step-by-step description of the proposed method is enumerated below. We applied ten-fold cross-validation for the proposed study.

1. We used four Benchmark datasets containing the Octapeptide sequences. A detailed description is available in Sect. 2.1.
2. We extracted multi-view features, a combined representation of physicochemical and indirect sequence order features, using the Quasi Residue Couple(RC) [12] method from each dataset.
3. We considered five types of physicochemical features, namely, Hydrophobicity, Isoelectric Point, Polarizability, Stearic property, and Polarity.
4. We constructed a total of five combined representations of physicochemical and indirect sequence order features for individual datasets using RC. [12]. Let us consider it as a RCPhySeq feature set having 1200 features in it. Finally, the total number of features that we are considering is 60,000 ($1200 \times 5 = 60,000$).
5. We further applied the CNN model on each type of extracted feature set (RCPhySeq) with respect to individual datasets. Then, the informative latent feature was collected from the last layer of CNN having dimension 200.
6. We concatenated all the five different output features ($dimension : 200 \times 5 = 1000$) extracted using CNN from the RCPhySeq feature set.
7. Further, we performed the final classification using a fuzzy rank-based ensemble [9] of classifiers where Random forest (RF), Logistic regression (LR), and SVM(Linear) are utilized as base classifiers.

3 Experimental Results

The proposed method is implemented using Python (version 3.9). In this section, we present the experimental results obtained. The evaluation metrics considered include Accuracy (Acc), Area Under the Curve (AUC), Precision (Pre), Recall, F-Measure, and PR_AUC (area under the precision-recall curve). Accuracy is calculated as the ratio of correctly classified samples (true positives and true negatives) to the total number of data points. The AUC represents the overall discriminative ability of the model in distinguishing between positive and negative samples. The F-Measure, also known as the harmonic mean of recall and precision, provides a balanced measure of the model's performance. A higher PR_AUC or area under the precision-recall curve indicates improved classifier performance. It is useful to address unbalanced classification issues. All these metrics are frequently employed for performance evaluation in several bioinformatics applications [4].

Comparison with Base Classifiers: Ten-fold cross-validation result for the prediction of HIV-1 Protease cleavage site using the proposed model is depicted in Table 2. We have compared our model with respect to different baseline classification approaches. After executing all the steps of feature extraction techniques, we have applied the proposed Fuzzy rank-based ensemble classifiers and all the base classifiers (RF, LR, SVM). Further, we have reported the results in terms of five metrics, namely: ACC, AUC, Precision, Recall and F-measure. To this end, 20 results have been reported using four benchmark datasets. Among these, in 10 cases, our proposed model with a fuzzy rank-based ensemble has achieved better performance. Whereas the RF, LR, and SVM models have yielded better results in 3 case, 6 cases, and 3 cases, respectively. So, it is visible that the pro-

Table 2. Result with 10 Fold Cross validation

Model	ACC	AUC	F-measure		
			Pre	Recall	F-mea
Data746					
Proposed Model with Fuzzy Rank-based Ensemble	**0.934**	**0.981**	**0.941**	0.938	**0.939**
Proposed Model with RF	0.930	0.979	0.937	0.935	0.936
Proposed Model with LR	0.920	**0.981**	0.938	**0.943**	**0.939**
Proposed Model with SVM	0.921	0.978	0.932	0.923	0.926
Data1625					
Proposed Model with Fuzzy Rank-based Ensemble	0.954	0.976	0.929	0.867	0.896
Proposed Model with RF	**0.956**	0.975	0.926	**0.883**	**0.902**
Proposed Model with LR	0.954	0.973	**0.934**	0.864	0.896
Proposed Model with SVM	0.951	**0.978**	0.922	0.861	0.890
Data Schilling					
Proposed Model with Fuzzy Rank-based Ensemble	**0.921**	**0.919**	**0.738**	**0.638**	**0.682**
Proposed Model with RF	0.917	0.895	0.721	0.617	0.663
Proposed Model with LR	0.913	0.887	0.695	0.618	0.653
Proposed Model with SVM	0.917	0.896	0.713	0.629	0.667
Data Impens					
Proposed Model with Fuzzy Rank-based Ensemble	0.915	0.913	**0.810**	0.592	0.676
Proposed Model with RF	0.911	0.924	0.775	0.617	0.684
Proposed Model with LR	0.914	**0.926**	0.790	**0.631**	0.696
Proposed Model with SVM	**0.920**	0.904	0.804	0.601	**0.702**

Table 3. Comparison with state-of-the-art techniques

Dataset	Metrices	Proposed Method	Palmal et al. [17]	EM_HIV [5]	Evo Cleave [4]	Rdgnvalds-son et al. [18]	HIV Cleave [19]	Deep Cleave [7]
Data746	AUC	**0.98**	0.96	–	0.93	0.92	0.74	0.44
	PR_AUC	**0.98**	0.86	–	0.92	0.91	0.81	0.49
	Precision	**0.94**	0.89	–	0.90	0.85	0.92	0.41
	Recall	**0.94**	0.92	–	0.80	0.90	0.70	0.14
	F-Measure	**0.94**	0.90	–	0.85	0.87	0.80	0.21
Data1625	AUC	**0.98**	**0.98**	**0.98**	0.93	0.97	0.73	0.46
	PR_AUC	**0.95**	0.81	0.94	0.84	0.90	0.61	0.21
	Precision	**0.93**	0.92	0.82	0.85	0.85	0.69	0.13
	Recall	0.87	0.84	**0.91**	0.74	0.80	0.67	0.14
	F-Measure	**0.90**	0.87	0.86	0.80	0.83	0.68	0.13
Data Schilling	AUC	0.92	**0.96**	**0.96**	0.78	0.93	0.59	0.52
	PR_AUC	0.75	0.51	**0.80**	0.36	0.68	0.34	0.13
	Precision	0.74	**0.82**	0.54	0.50	0.66	0.31	0.13
	Recall	0.64	0.56	**0.91**	0.20	0.66	0.41	0.43
	F-Measure	**0.68**	0.65	**0.68**	0.28	0.66	0.35	0.20
Data Impens	AUC	0.91	**0.94**	0.92	0.88	0.90	0.56	0.45
	PR_AUC	**0.78**	0.40	0.73	0.64	0.70	0.29	0.14
	Precision	**0.81**	0.78	0.51	0.77	0.69	0.29	0.14
	Recall	0.59	0.39	**0.81**	0.42	0.62	0.45	0.34
	F-Measure	**0.68**	0.48	0.62	0.54	0.65	0.35	0.20

posed method has achieved better performance in most cases. Here, the obtained average ACC, AUC, Precision, Recall, and F-measure values are 0.93, 0.95, 0.85, 0.76, and 0.80, respectively.

Comparison with State-of-Art Models: We have compared our model performance (see Table 3) with six different state-of-art models, namely, Palmal et al. [17], EM_HIV [5], EvoCleave [4], Rdgnvaldsson et al. [18], HIVCleave [19], and DeepCleave [7] with respect to AUC, PR_AUC, Precision, Recall and F-Measure values respectively. Here also, 20 different results have been reported for each of the models. Among them, the proposed method has attained the best results in 13 cases, whereas Palmal et al. [17] and EM_HIV [5] work have attained the best results in 4 and 7 cases, respectively. The rest of the reported models have underperformed. However, EM_HIV [5] have not reported any result using Data746 in their work. Thus, we have not reported the comparison of this model with our work in terms of Data746. Here, the results of the proposed model have been rounded off to two decimal point values, aligning with the convention followed by the majority of state-of-the-art models.

4 Discussion

Our proposed model outperforms both base classifiers and state-of-the-art models in most scenarios, showcasing its superior performance. This improvement is primarily attributed to incorporate multi-view features encompassing physico-chemical and indirect sequence order information. This has been performed using the Quasi-Residue Couple method, which is capable of giving high-dimensional feature representation from the octapeptide sequence for each sample. Further, by leveraging a CNN for feature extraction, our model can learn intricate and abstract features through the application of convolutional filters. To further enhance classification accuracy, we opted for an ensemble approach rather than relying solely on single classifiers. Through the implementation of a fuzzy rank-based ensemble, we compared the performance of our proposed model with that of individual classifiers, as shown in Table 2. Consistently, our ensemble approach demonstrated superior performance in the majority of cases. The utilization of multi-view features and the integration of a CNN contribute to the improved performance of our model. Combining the predictions of multiple classifiers through the ensemble approach enhances accuracy and robustness in classification tasks. These results validate the our proposed model's effectiveness and ability to effectively handle unbalanced classification issues.

5 Conclusion

In this proposed study, we employed a fuzzy rank-based ensemble of classifiers with multi-view features to predict HIV-1 protease cleavage sites. Our model achieved impressive average scores for ACC (0.93), AUC (0.95), Precision (0.85), Recall (0.76), and F-measure (0.80). Notably, our proposed model not only outperformed baseline models but also surpassed other state-of-the-art models across multiple benchmark datasets. This demonstrates the effectiveness of using Quasi Residue Couple (QR)-based feature encoding to generate multi-view features. Additionally, our CNN-based feature extraction approach yielded superior performance compared to alternative feature extraction methods employed in other state-of-the-art models. Furthermore, the fuzzy rank-based ensemble proved to be an effective approach as it incorporated confidence levels from the base classifiers when making predictions on test samples. Future research can explore alternative feature encoding techniques to further enhance the classification performance of our proposed model.

Acknowledgment. Dr. Sriparna Saha gratefully acknowledges the Young Faculty Research Fellowship (YFRF) Award, supported by Visvesvaraya Ph.D. Scheme for Electronics and IT, Ministry of Electronics and Information Technology (sMeitY), Government of India, being implemented by Digital India Corporation (formerly Media Lab Asia) for carrying out this research.

References

1. Arya, N., Saha, S.: Multi-modal classification for human breast cancer prognosis prediction: proposal of deep-learning based stacked ensemble model. IEEE/ACM Trans. Comput. Biol. Bioinf. **19**(2), 1032–1041 (2020)
2. Brik, A., Wong, C.H.: HIV-1 protease: mechanism and drug discovery. Organ. Biomolecul. Chem. **1**(1), 5–14 (2003)
3. Gök, M., Özcerit, A.T.: A new feature encoding scheme for HIV-1 protease cleavage site prediction. Neural Comput. Appl. **22**, 1757–1761 (2013)
4. Hu, L., Hu, P., Luo, X., Yuan, X., You, Z.H.: Incorporating the coevolving information of substrates in predicting HIV-1 protease cleavage sites. IEEE/ACM Trans. Comput. Biol. Bioinf. **17**(6), 2017–2028 (2019)
5. Hu, L., Li, Z., Tang, Z., Zhao, C., Zhou, X., Hu, P.: Effectively predicting HIV-1 protease cleavage sites by using an ensemble learning approach. BMC Bioinformatics **23**(1), 447 (2022)
6. Kawashima, S., Ogata, H., Kanehisa, M.: Aaindex: amino acid index database. Nucleic Acids Res. **27**(1), 368–369 (1999)
7. Li, F., et al.: Deepcleave: a deep learning predictor for caspase and matrix metalloprotease substrates and cleavage sites. Bioinformatics **36**(4), 1057–1065 (2020)
8. Li, Z., Hu, L., Tang, Z., Zhao, C.: Predicting hiv-1 protease cleavage sites with positive-unlabeled learning. Front. Genet. **12**, 658078 (2021)
9. Manna, A., Kundu, R., Kaplun, D., Sinitca, A., Sarkar, R.: A fuzzy rank-based ensemble of CNN models for classification of cervical cytology. Sci. Rep. **11**(1), 14538 (2021)
10. Miller, R.J., Cairns, J.S., Bridges, S., Sarver, N.: Human immunodeficiency virus and aids: insights from animal lentiviruses. J. Virol. **74**(16), 7187–7195 (2000)
11. Nanni, L.: Comparison among feature extraction methods for HIV-1 protease cleavage site prediction. Pattern Recogn. **39**(4), 711–713 (2006)
12. Nanni, L., Lumini, A., Gupta, D., Garg, A.: Identifying bacterial virulent proteins by fusing a set of classifiers based on variants of chou's pseudo amino acid composition and on evolutionary information. IEEE/ACM Trans. Comput. Biol. Bioinf. **9**(2), 467–475 (2011)
13. Onah, E., et al.: Prediction of HIV-1 protease cleavage site from octapeptide sequence information using selected classifiers and hybrid descriptors. BMC Bioinformatics **23**(1), 1–20 (2022)
14. Palmal, S., Arya, N., Saha, S., Tripathy, S.: A Multi-modal graph convolutional network for predicting human breast cancer prognosis. In: Tanveer, M., Agarwal, S., Ozawa, S., Ekbal, A., Jatowt, A. (eds.) Neural Information Processing: 29th International Conference, ICONIP 2022, Virtual Event, 22–26 November 2022, Proceedings, Part VII, pp. 187–198. Springer, Singapore (2023). https://doi.org/10.1007/978-981-99-1648-1_16
15. Palmal, S., Arya, N., Saha, S., Tripathy, S.: Breast cancer survival prognosis using the graph convolutional network with choquet fuzzy integral. Sci. Rep. **13**(1), 14757 (2023)
16. Palmal, S., Saha, S., Tripathy, S.: HIV-1 protease cleavage site prediction using stacked autoencoder with ensemble of classifiers. In: 2022 International Joint Conference on Neural Networks (IJCNN), pp. 1–8. IEEE (2022)
17. Palmal, S., Saha, S., Tripathy, S.: Multi-objective optimization with majority voting ensemble of classifiers for prediction of HIV-1 protease cleavage site. Soft Comput. 1–11 (2023)

18. Rögnvaldsson, T., You, L., Garwicz, D.: State of the art prediction of HIV-1 protease cleavage sites. Bioinformatics **31**(8), 1204–1210 (2015)
19. Shen, H.B., Chou, K.C.: HIVcleave: a web-server for predicting human immunodeficiency virus protease cleavage sites in proteins. Anal. Biochem. **375**(2), 388–390 (2008)

KSHFS: Research on Drug-Drug Interaction Prediction Based on Knowledge Subgraph and High-Order Feature-Aware Structure

Nana Wang[1,2,3] (ID), Qian Gao[1,2,3](✉) (ID), and Jun Fan[4]

[1] Key Laboratory of Computing Power Network and Information Security, Ministry of Education, Shandong Computer Science Center, Qilu University of Technology (Shandong Academy of Sciences), Jinan, China
gq@qlu.edu.cn

[2] Shandong Engineering Research Center of Big Data Applied Technology, Faculty of Computer Science and Technology, Qilu University of Technology (Shandong Academy of Sciences), Jinan, China

[3] Shandong Provincial Key Laboratory of Computer Networks, Shandong Fundamental Research Center for Computer Science, Jinan, China
10431210633@stu.qlu.edu.cn

[4] China Telecom Digital Intelligence Technology Co., Ltd., No. 1999, Shunhua Road, Jinan 250101, Shandong, China
fanjun.sd@chinatelecom.cn

Abstract. Effective drug-drug interaction (DDI) prediction can prevent adverse reactions and side effects caused by taking multiple drugs at the same time. However, most methods that obtain drug information through large-scale biomedical knowledge graphs (KGs), ignore the problem of high noise and complexity, and have certain limitations in obtaining rich neighborhood information for each entity in the KG. Therefore, this paper proposes an end-to-end method called Knowledge Subgraph and High-order Feature-aware Structure (KSHFS) to address DDI prediction. In KSHFS, this paper first designs a subgraph extraction module to reduce the noise caused by the KG, remove irrelevant information, and effectively utilize the entity information in external knowledge graphs to assist DDI prediction. Then, a high-order feature-aware module is designed to aggregate entity information propagated from high-order neighbors, learn high-order structural embeddings for each entity, and effectively capture potential semantic neighborhood features of drug pairs. Finally, in binary DDI prediction, a self-attention mechanism is used for feature fusion to predict drug interaction events. The experimental results demonstrate that the KSHFS model outperforms the baseline models in binary and multi-relation DDI prediction based on various evaluation metrics, including AUC, AUPR, and F1.

Keywords: Drug-drug interaction · Knowledge graph · Graph neural networks · Prediction

Supported by the Natural Science Foundation of Shandong Province (ZR2022MF333).

B. Luo et al. (Eds.): ICONIP 2023, CCIS 1964, pp. 493–506, 2024.
https://doi.org/10.1007/978-981-99-8141-0_37

1   Introduction

A study shows that 67% of elderly Americans have taken five or more medications in the past decade, which could increase the risk of adverse drug reactions and even death due to potential harmful interactions between drugs [1]. Therefore, establishing a computer model for simulating drug-drug interactions (DDIs) is an urgent issue to be addressed. The rapid development of current technology offers the possibility of computational drug discovery and drug safety research. The introduction of many computational prediction methods can help provide patients with safer and more effective prescriptions and reduce the time and cost of clinical trials.

Indeed, a number of artificial intelligence-based models for DDI event prediction have been proposed, including Perozzi et al. [2] using random wandering to learn drug embeddings in networks, Ryu et al. [3] designing a deep neural network (DNN) framework for multi-relational DDI prediction, and on the other hand, due to the prevalence of knowledge graphs (KGs), it has widely led to relational inference and recommendation influx of research, especially the recent study by Yu et al. [4] using KG for DDI prediction via multi-relational embedding models and neighbourhood relationship propagation mechanisms. They both applied KG to machine learning models to extract drug features using various embedding methods. However, these methods learn direct node potential embeddings but are limited in obtaining rich neighbourhood information for each entity in KG and ignore the noise introduced by large biological knowledge graphs.

In order to address the noise problem and the limitation of neighbourhood information in KG, the design goal of this paper is to reduce the noise in KG and capture the higher-order structure and semantic relationships in KG. This study proposes a drug interaction prediction model based on knowledge subgraphs and high-order feature-aware structure (KSHFS). The design goal of this paper is to reduce the noise in the KG and to capture the higher-order structure and semantic relationships in the KG. Briefly, given a multi-relationship network of DDIs, this paper first uses knowledge graph extraction subgraphs to anchor the relevant subgraphs in the KG, captures multi-relationship information between drugs through the embedding model RotatE [5], and learns the potential embeddings of each entity. Then, this paper proposes a feature-aware module which firstly extracts the first-order relationship-aware network structure information of entities by propagating the neighbourhood information of entities under different relationships through the knowledge graph attention network KGAT [6]. In addition, this paper proposes a high-order feature-aware structure and innovatively applies the graph convolution network NGCF in recommender systems to drug interaction prediction to extract the drug pair higher-order structural information and semantic relationships to learn drug representations. Finally, this paper combines the acquired first-order neighbourhood information and higher-order feature information embedding to obtain the final entity representation that can be used for binary and multi-relational DDI prediction. The contributions of this paper are summarised as follows:

- This paper proposes a new high-order relation-aware network, which comprises a subgraph extraction module, a feature information perception module, and a DDI prediction module as an end-to-end framework. It can be used to explore the topological structure of drugs in a knowledge graph and predict potential drug-drug interactions. The KSHFS framework has the following advantages:

  (a) KSHFS utilizes drug knowledge graph (KG) subgraph anchoring to relevant subgraphs in KG to reduce noise and then uses the topological relationships of each entity in KG to obtain rich neighbourhood information, capturing multi-relation information between drugs, which is beneficial for DDI prediction.

  (b) KSHFS designs higher-order feature-aware structure and innovatively applies NGCF to drug interaction prediction by learning embedding representations in graph structures to recursively propagate embeddings along higher-order connectivity, aggregating all topological neighbourhood information from them and extracting higher-order structures and semantic relationships to learn drug representations.

  (c) KSHFS utilizes self-attention networks for feature fusion, which can reduce dependence on external information and better capture internal correlations of data or features. Additionally, self-attention mechanisms can focus more on important features.

- This study conducted comparative experiments, ablation experiments, and hyperparameter setting experiments on the DrugBank dataset and the DeepDDI dataset. The experimental results demonstrate that the proposed drug-drug interaction (DDI) prediction model is more effective than more advanced neural network-based methods, with an improvement of approximately 1% in ACC, AUC, AUPR, and F1.

2 Method

This section will provide a detailed description of the framework and implementation details of the proposed KSHFS model. Specifically, given a multi-relational DDI network $G = \{(\mathbf{d}_i, r, \mathbf{d}_j) | \mathbf{d}_i, \mathbf{d}_j \in D, r \in R\}$, where D consists of drugs and R consists of relations, each tuple $(\mathbf{d}_i, r, \mathbf{d}_j)$ describes the biomedical relation r between entities $\mathbf{d}_i$ and $\mathbf{d}_j$. The primary objective of KSHFS is to forecast whether there exists an interaction or a particular kind of interaction between pairs of drugs $(\mathbf{d}_i, \mathbf{d}_j)$.

The structure of KSHFS is shown in Fig. 1, which consists of three modules. Specifically, in the Knowledge Graph-based Subgraph Extraction Module (a), a knowledge graph is used to extract subgraphs, which can reduce interference from many unnecessary entities and relations and lower noise. Then, the RotatE embedding model is used to capture multi-relational information among drugs. In the Feature Information Perception Module (b), a relation-aware structure (right) and a high-order feature-aware Structure (left, red box) are designed separately. In the relation-aware structure, the KGAT algorithm can adaptively adjust the weights of entities and relations between drug pairs to capture the feature information between drug pairs better. In the high-order feature-aware structure, graph structure data is utilized to better capture the high-order structure and semantic relations between drug pairs and mine their potential relationships. In the Prediction Module (c), binary and multi-relational DDI predictions are performed. In particular, in the binary DDI prediction (red box), a self-attention mechanism is designed for feature fusion to learn entity representations of drug pairs better.

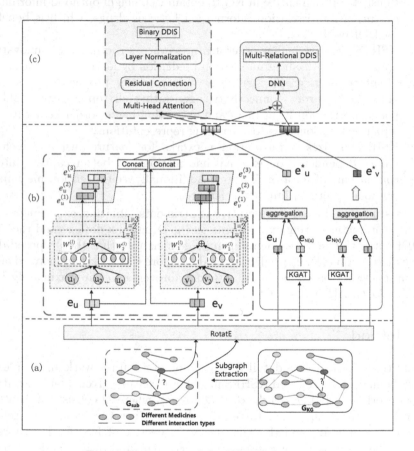

Fig. 1. Illustration of the KSHFS framework.

2.1   Extracting Knowledge Graph Subgraph

To obtain the complex relationships between drugs and learn the potential features of each entity through the multi-relational network of DDI, this paper initializes all entity embeddings using the KG embedding model RotatE.

The external biomedical knowledge graph G_{KG} [4] is defined as $G_{KG} = \{(u, r, v) \mid u, v \in \varepsilon, r \in R\}$, where each triple (u, r, v) describes the biomedical relationship r between entities u and v. G_{KG} is calculated using the negative distance score function [7], where a higher score implies a higher possibility of having an edge with relation r between u and v in the complex embedding space. The negative distance score function is shown in Formula 1:

$$\text{score}(u, r, v) = -\|u \circ r - v\|_2^2 \tag{1}$$

where $\|x\|_2$ is the l_2 parametrization of x and $\circ$ denotes the Hadamard (elementwise) product.

The biomedical information knowledge graph (KG) is complex and large, and many unimportant edge information may introduce noise to the prediction of drug interactions. Focusing on local subgraphs can extract rich information while reducing noise and irrelevant information. Zhang [8] emphasized the importance of nodes in the subgraph relative to the central nodes u and v.

Specifically, for entities u and v, this paper first extracts the k-hop neighbor nodes of entities u and v, as shown in Formula 2:

$$N_k(u) = \{s \mid d(s, u) \leq k\}, N_k(v) = \{s \mid d(s, v) \leq k\} \tag{2}$$

where $d(,)$ represents the distance between two nodes in the G_{KG} network, and according to the experimental results analysis in Sect. 3.4.1, the value of k in this paper is 2. Then, based on the intersection of these nodes, a closed subgraph $G_{\text{sub}} = \{(u, r, v) \mid u, v \in N_k(u) \cap N_k(v), r \in R\}$ is obtained.

This paper uses the BPR loss [7] to optimize nested loops:

$$\mathcal{L}_{KG} = - \sum_{(u,r,v)\in\mathcal{G},(u',r',v')\notin G} \ln \sigma \left(\text{score}(u, r, v) - \text{score}(u', r', v') - \gamma_1 \right) \tag{3}$$

2.2 Feature Information Perception Module

Relation-Aware Structure. To obtain the network structure embedding of entities, inspired by the literature [4,6], we designed a relation-aware structure to learn drug representations and constructed a feature information-aware module for knowledge graph attention networks. The main idea of this module is to assign different weights to neighboring nodes in the subgraph, considering that different edge relationships propagate different information, thus avoiding high-cost matrix operations, reducing time complexity, and obtaining more information from the relation-aware structure in cases of sparse data.

For an entity u, $N(u)$ denotes its neighbors. We extract information from $N(u)$ [4] and define it as:

$$e_{N_r(u)} = \sum_{(u,r,N_{(u)})\in N_r(u)} \left(\alpha_{N_{(u)}} N_{(u),r} + \mathcal{F}\left(N_{(u)}\right) \right) \tag{4}$$

Using the basis decomposition method [4] to solve the problem of parameter overfitting, according to Eq. 4, use r and the original information to propagate drug $N(u)$ information, accumulate the information propagated by u under each relationship to acquire the first-order neighborhood embedding of u: $e_{N(u)} = \frac{1}{|N(u)|} \sum_{r \in R} e_{N_r(u)}$. Using the GraphSage aggregator [9], perform a non-linear transformation one$_u$ and $e_{N(u)}$ to obtain the final embedding: $e_u^* = \text{ReLU}\left(S * \mathcal{F}\left(e_u \| e_{N(u)}\right)\right)$.

Higher-Order Feature-Aware Structures. Although the relationship-aware structure constructed above has achieved good results, it still has certain limitations due to its propagation mechanism that only considers obtaining the first-order neighbourhood features of each node, resulting in incomplete embedding information. Therefore, this paper proposes a high-order feature-aware structure that further updates the high-order connectivity between drugs by adding high-order neighbourhood information of drugs, refines embedding propagation through high-order connectivity relationships, measures the cooperative similarity between drugs, and obtains hidden information between drug pairs to predict DDI events accurately.

In addition to the nearest neighbours, KSHFS is extended to two layers of embedding propagation to extract high-order structures and semantic relationships. For drug u, in the l-th layer, its recursive representation is [10]:

$$e_u^{(l)} = \text{LeakyReLU}\left(m_{u \leftarrow u}^{(l)} + \sum_{i \in N_u} m_{u \leftarrow i}^{(l)}\right) \tag{5}$$

$$m_{u \leftarrow i}^{(l)} = p_{ui}\left(\mathbf{W}_1^{(l)} e_i^{(l-1)} + \mathbf{W}_2^{(l)}\left(e_i^{(l-1)} \odot e_u^{(l-1)}\right)\right), m_{u \leftarrow u}^{(l)} = \mathbf{W}_1^{(l)} e_u^{(l-1)} \tag{6}$$

where $\mathbf{W}_1^{(l)}$, $\mathbf{W}_2^{(l)} \in \mathbb{R}^{d_1 \times d_{1-1}}$ are trainable transformation matrices, where d_l is the size of the transformation; it represents the term generated by the previous message passing step and memorizes information from its $l-1$ hop neighbors. It further helps in representing drug u at layer l. In like manner, the drug v is also acquired with its description at layers l.

After l layers of propagation, we obtain multiple representations of drug u, denoted as $\left\{e_u^{(1)}, \cdots, e_u^{(l)}\right\}$. Because the features acquired at different layers emphasize messages passed through different connections, they contribute differently in reflecting drug relationships. Therefore, this study concatenates them to constitute the high-order overall representation of the drug. Similarly, the same procedure is applied to drug v in this paper, concatenating the representations learned at different layers $\left\{e_v^{(1)}, \cdots, e_v^{(l)}\right\}$ to obtain the final high-order relationship embedding of the graph neural network layer:

$$e_u^{**} = e_u^{(0)} \| \cdots \| e_u^{(l)}, \quad e_v^{**} = e_v^{(0)} \| \cdots \| e_v^{(l)} \tag{7}$$

The ultimate embedding E_u of drug u is obtained by connecting e_u^* and e_u^{**}:
$E_u = e_u^* \| e_u^{**}$.

Therefore, the embedding of drug u not only contains feature information with different weights of neighbours but also includes high-order relationship and structural information. Likewise, the final embedding E_v of drug v can be obtained.

2.3 Binary DDI Prediction

This paper uses a self-attentive mechanism for feature fusion to integrate different potential feature vectors of drug combinations. The self-attention mechanism is a suitable feature fusion approach [11]. It helps the network select important features and assign them high weights for DDI event prediction, as redundant or less important information will be obtained. The calculation formula for multi-head attention [11] is as follows:

$$
\begin{aligned}
X_{\text{attn}} &= \text{Concat}\left(E_u', E_v'\right) W^\circ \\
E_u' &= \text{softmax}\left(\frac{Q_u \times K_u^T}{\sqrt{d_k}}\right) V_u \\
Q_u &= X \times W_u^Q \\
K_u &= X \times W_u^K \\
V_u &= X \times W_u^V \\
X &= \text{Conat}\left(E_u, E_V\right)
\end{aligned}
\tag{8}
$$

After the self-attention layer, vector representations are calculated by re-normalization [12] to improve the "covariate shift" problem and accelerate the convergence of neural network parameters. Simultaneously using residual connections [13] to address the issue of vanishing gradients. Compute the edge loss function as defined in [4]:

$$
\mathcal{L}_b = - \sum_{(u,v)\in\mathcal{G}^+, (u',v')\notin\mathcal{G}^-} \ln\sigma\left(\gamma_2 - X_{\text{attn}}\left(u',v'\right)\right) + \ln\sigma\left(X_{\text{attn}}\left(u,v\right) - \gamma_2\right) \tag{9}
$$

where γ_2 is the margin weight and G^* is interpreted as the drug pair with or without the connection. In addition to the loss caused by knowledge graph embedding (Sect. 2.1), the final loss caused by binary relation prediction is:

$$
\mathcal{L}_B = \mathcal{L}_{\text{KG}} + \mathcal{L}_b + \lambda \|\Theta_1\|_2^2. \tag{10}
$$

2.4 Multi-relational DDIs Prediction

For the task of predicting multiple relationships for DDIs, this paper uses a DNN as the predictor. Given a drug pair (u, v), we use $e^0 = e_u^* \| e_v^*$ as the initial input to the DNN. The result obtained is the probability of each interaction relationship in (u, v) for the drug pair, which is calculated as:

$$
y_m(u, v) = \text{softmax}\left(W_L e^{L-1} + b_L\right) \tag{11}
$$

3 Experimental Setup and Result Analysis

3.1 Experimental Setup

Datasets. (1) The DrugBank [14] database integrates resources from bioinformatics and cheminformatics, including drug chemical substructures, targets, enzymes, pathways, and drug interactions, containing 1317 drugs, of which 198697 are related to 86 types of interactions.(2) The DeepDDI [3] benchmark dataset includes 1710 drugs and 192284 paired DDIs.

Evaluation Metrics and Parameters. This study uses several multi-class evaluation metrics to assess the predictive performance, including ACC, AUPR, AUC, F1, precision, and recall.

In this paper, the appropriate set of hyperparameters is obtained by using nested cross-validation [4]. The maximum number of iterations is set to 100, with 1024 batches used, and Adam algorithm used for learning rate tuning between {0.01, 0.001, 0.0001, 0.00001}.

Baselines: Deepwalk [2]: Learns drug embeddings in the network through random walks to predict drug-drug relationships.

LINE [15]: A neural network-based method that uses first-order approximation (local structure) and second-order approximation (global structure) to model node embeddings directly.

SkipGNN [16]: Predict drug interactions by aggregating information with direct and second-order interactions through two GNNs.

KGNN [17]: Captures the structure of the knowledge graph by mining relationships in the graph with a neural network.

DeepDDI [3]: One of the earliest multi-relation DDI prediction models, based on a DNN framework that uses drug substructure similarity as input to predict the type of interaction between drugs.

DDIMDL [18]: Different drug features were used to construct the similarity matrix, and then a deep neural network pair was used to perform the prediction of drug interactions.

CSMDDI [19]: Obtains drug representations based on a knowledge graph method to predict DDIs in cold start scenarios.

3.2 Results Analysis

During this experiment, we conducted a comparison between KSHFS and several baseline models while also performing a comparison on the dataset. Tables 1 and 2 report the average scores, including AUC, AUPR, and F1, for binary relationship DDI prediction and multi-relationship DDI prediction running on the DrugBank and DeepDDI datasets.

(1) Compared to DeepWalk and LINE, KSHFS has better learning performance. KSHFS can model triplets and integrate edge relationships into the embedding process to learn the graph structure and semantic relationships in the KG, while DeepWalk and LINE only focus on the topological features of the network. This suggests that exploring the features and entities in the knowledge graph during the embedding process is important, and KSHFS can capture more relational information in the DDI network and learn more representative embeddings.

(2) SkipGNN and KGNN outperform DeepWalk and LINE, indicating that using graph neural network models can capture more high-order structures and semantic relationships. The proposed KSHFS combines the KGAT and NGCF models to obtain topological neighbourhood representations of drugs and related entities, achieving better results than SkipGNN and KGNN, demonstrating that KSHFS can better utilize graph neural networks to capture collaborative signals and knowledge propagation in relationship-aware structures.

(3) Compared with the DrugBank dataset, the models with KSHFS and network structure-based models (i.e., KGNN, SkipGNN, and CSMDDI) have marginally lower metrics on the DeepDDI dataset. Because the DeepDDI dataset contains many nodes with simple association relationships (degree <10), it is more challenging for network structure-based models to acquire rich structural information. However, the KSHFS model is always better than others, indicating that the proposed model can obtain rich feature information and embedding information in the DDI network structure. Compared with the DeepDDI model, our method jointly considers the topological neighbourhood structure and related entities in the knowledge graph, which is conducive to improving performance.

In summary, compared to other baseline models, our model achieved an improvement of around 1% on both datasets. We found that using KG significantly improved DDI performance, highlighting the necessity of combining

Table 1. Experimental comparison results of binary DDI prediction on two datasets.

Method	DrugBank dataset			DeepDDI dataset		
	AUC	*AUPR*	*F1*	*AUC*	*AUPR*	*F1*
DeepWalk	70.08	61.60	80.13	75.42	59.63	79.33
LINE	86.20	60.43	78.50	84.36	59.68	79.12
SkipGNN	87.46	81.70	79.23	87.68	85.95	80.30
KGNN	91.78	90.65	89.96	90.37	89.93	87.48
CSMDDI	88.61	86.63	82.65	87.53	85.09	81.51
DeepDDI	94.32	95.20	92.31	94.55	94.16	91.68
Ours	**95.51**	**96.72**	**93.46**	**94.89**	**95.53**	**92.25**
Improvement	**1.26%**	**1.59%**	**1.24%**	**0.35%**	**1.45%**	**0.62%**

knowledge subgraphs with relationship-aware structures to obtain high-order feature information, as it can provide supplementary information for the DDI task. Overall, our KSHFS framework achieved satisfactory performance in DDI prediction. Next, we will further experimentally analyze our model to understand better the contribution of model components to the proposed framework.

3.3   Ablation Study

To demonstrate that each component of KSHFS contributes significantly to the final performance of KSHFS, we conducted a series of ablation studies. KSHFS has three key components, and we studied the performance changes after removing each component:

-G_{sub}: Remove the knowledge graph subgraph extraction module.
-NGCF: Removing the graph neural network layer and using only the graph attention network to obtain first-order neighbour information.

Table 2. Experimental comparison results of multi-relational DDI prediction on two data sets.

Dataset	Method	AUPR	AUC	ACC	F1	Precision	Recall
DrugBank	DeepDDL	90.92	95.48	85.50	74.83	77.92	76.08
	DNN	93.78	96.25	91.27	82.91	85.83	81.70
	DDIMDL(S)	96.33	99.95	92.82	86.36	87.77	85.57
	CSMDDI	96.48	99.93	92.38	90.24	91.14	91.48
	Ours	**97.43**	**99.96**	**94.57**	**91.43**	**92.32**	**92.35**
	Improvement	0.98%	0.03%	1.85%	1.31%	1.29%	0.95%
DeepDDI	DeepDDL	95.87	99.93	92.66	86.72	89.88	88.00
	DNN	94.35	98.75	91.27	85.49	85.98	83.56
	DDIMDL(S)	87.37	99.39	91.44	89.95	91.21	90.43
	CSMDDI	96.03	99.95	93.46	88.21	89.78	87.99
	Ours	**97.74**	**99.96**	**95.18**	**91.59**	**92.65**	**91.58**
	Improvement	1.78%	0.01%	1.84%	1.82%	1.57%	1.27%

DDIMDL(S) indicates that the input to DDIMDL is the chemical substructure.

-Self-attention: Replacing KSHFS's self-attention fusion scheme with the inner product operation of drug pair feature embeddings generated by GCN.

The experimental results presented in Fig. 2 verify the contribution of each component in the KSHFS model to the model's overall performance, indicating that utilizing semantic relationships and high-order feature information of knowledge graph subgraphs in conjunction with topological structures for neighbourhood representation is advantageous to enhance the performance of DDI event prediction. The results show that KSHFS outperforms all three variants on all metrics. Furthermore, we analyzed the three variants, and the -G_{sub} variant reduced the model's performance by 1.64% on Acc, 1.54% on AUC, and

1.73% on AUPR after removing the knowledge subgraph module. The experimental results confirm our motivation - using subgraphs can reduce noise and irrelevant information, effectively utilizing external knowledge.

Comparison of variants -NGCF and KSHFS, where KSHFS contains second-order feature information for drug pairs. -NGCF performs worse than KSHFS, indicating that high-order feature information is essential for predicting drug interactions. This means that high-order neighbourhood information is crucial for predicting interaction links. This discovery identifies high-order feature information as a key driving factor in performance improvement.

The Self-attention variant performs about 1% worse than KSHFS in terms of ACC, AUC, and AUPR, indicating that self-attention fusion is largely superior to the inner product operation. This also confirms our hypothesis that different potential feature vectors have inconsistent predictive effects on drug interaction events, and feature fusion through self-attention modules can learn weight distributions of different features, significantly improving experimental accuracy. In conclusion, we find that all components of KSHFS are necessary for its strong performance.

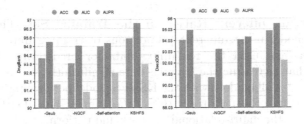

Fig. 2. Results of ablation experiments.

3.4   Parameter Discussion

In this section, we investigate the impact of several key hyperparameters on the proposed KSHFS performance. When studying the impact of one parameter, we fixed the other parameters. Figure 3 reports the ACC, AUC, F1, and AUPR scores on the DrugBank drug dataset.

Influence of the Jump Point of Subgraph k. Figure 3 shows the results of KSHFS as k varies. From the results, we found that for the DrugBank dataset, performance first increases when k is small. However, when k increases from 2 to 3, we observe a slight decrease in the F1 score. This indicates that more significant subgraphs can provide more useful information, but when k is too large, it may also introduce some noise and affect performance. This suggests that two-hop subgraphs provide sufficient information for the DDI task.

Influence of High-Order Embedding Propagation Layers. We found that increasing the value of l in the NGCF layer from 1 to 3 resulted in a decrease in model performance in all metrics due to the introduction of significant noise from the larger l value. Comparing single-layer KSHFS and third-order interaction information, we found that second-order information was more important than third-order information, further confirming the importance of second-order information in biomedical interaction networks, which aligns with the core idea of our proposed model. Experimental results indicate that using an l value of 1 or 2 is typically sufficient for practical purposes.

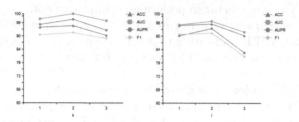

Fig. 3. Parameter experimental results.

Performance at Different Ratios in the Training Set. The experiment trained each model using different proportions of training data (10%–90% of marginal data) and used the remaining data for prediction. Figure 4(a)–(c) demonstrate that increasing the training data significantly improves the predictive performance of all models, with the KSHFS model performing the best across different proportions. Even with only 10% of the training data, it still obtains more information from the multi-relational network and performs well.

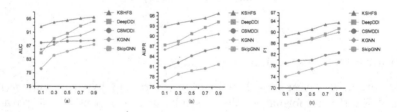

Fig. 4. Comparison of results at different training ratios

4   Conclusion

In this work, a new framework, KSHFS, is designed to model the DDI prediction task in an end-to-end manner explicitly. This study focuses on reducing noise by knowledge subgraph anchoring, enhancing the high-order connectivity of semantic relationships through a high-order feature-aware module, and designing a self-attention mechanism in binary relation prediction to learn the weight distribution of features. Experimental results demonstrate the rationality and effectiveness of KSHFS. In the future, our work can be enriched by combining other heterogenous feature information of drugs to enhance drug representation.

Acknowledgements. This work was supported by the Natural Science Foundation of Shandong Province (ZR2022MF333).

References

1. Qato, D.M., Wilder, J., Schumm, L.P., Gillet, V., Alexander, G.C.: Changes in prescription and over-the-counter medication and dietary supplement use among older adults in the united states, 2005 vs 2011. JAMA Int. Med. **176**(4), 473–482 (2016)
2. Perozzi, B., Al-Rfou, R., Skiena, S.: Deepwalk: online learning of social representations. In: Proceedings of the 20th ACM SIGKDD International Conference on Knowledge Discovery and Data Mining, pp. 701–710 (2014)
3. Ryu, J.Y., Kim, H.U., Lee, S.Y.: Deep learning improves prediction of drug-drug and drug-food interactions. Proc. Natl. Acad. Sci. **115**(18), E4304–E4311 (2018)
4. Hui, Yu., Dong, W.M., Shi, J.Y.: Raneddi: relation-aware network embedding for drug-drug interaction prediction. Inf. Sci. **582**, 167–180 (2022)
5. Sun, Z., Deng, Z.-H., Nie, J.-Y., Tang, J.: Rotate: knowledge graph embedding by relational rotation in complex space. arXiv preprint arXiv:1902.10197 (2019)
6. Wang, X., He, X., Cao, Y., Liu, M., Chua, T.-S. Kgat: knowledge graph attention network for recommendation. In: Proceedings of the 25th ACM SIGKDD International Conference on Knowledge Discovery & Data Mining, pp. 950–958 (2019)
7. Zhao, S., Wei, W., Zou, D., Mao, X.: Multi-view intent disentangle graph networks for bundle recommendation. Proc. AAAI Conf. Artif. Intell. **36**, 4379–4387 (2022)
8. Zhang, M., Chen, Y.: Link prediction based on graph neural networks. Adv. Neural Inf. Process. Syst. **31** (2018)
9. Hamilton, W., Ying, Z., Leskovec, J.: Inductive representation learning on large graphs. Adv. Neural Inf. Process. Syst. **30** (2017)
10. Wang, X., He, X., Wang, M., Feng, F., Chua, T.-S.: Neural graph collaborative filtering. In: Proceedings of the 42nd International ACM SIGIR Conference on Research and Development in Information Retrieval, pp. 165–174 (2019)
11. Guo, S., Wang, Y., Yuan, H., Huang, Z., Chen, J., Wang, X.: Taert: triple-attentional explainable recommendation with temporal convolutional network. Inf. Sci. **567**, 185–200 (2021)
12. Ba, J.L., Kiros, J.R., Hinton, G.E.: Layer normalization. arXiv preprint arXiv:1607.06450 (2016)
13. He, K., Zhang, X., Ren, S., Sun, J.: Deep residual learning for image recognition. In: Proceedings of the IEEE Conference on Computer Vision and Pattern Recognition, pp. 770–778 (2016)
14. Wishart, D.S., et al.: Drugbank: a comprehensive resource for in silico drug discovery and exploration. Nucl. Acids Res. **34**(suppl_1), D668–D672 (2006)
15. Tang, J., Qu, M., Wang, M., Zhang, M., Yan, J., Mei, Q.: Line: large-scale information network embedding. In: Proceedings of the 24th International Conference on World Wide Web, pp. 1067–1077 (2015)
16. Huang, K., Xiao, C., Glass, L.M., Zitnik, M., Sun, J.: Skipgnn: predicting molecular interactions with skip-graph networks. Sci. Rep. **10**(1), 21092 (2020)

17. Lin, X., Quan, Z., Wang, Z.-J., Ma, T., Zeng, X.: KGNN: knowledge graph neural network for drug-drug interaction prediction. IJCAI **380**, 2739–2745 (2020)
18. Deng, Y., Xinran, X., Qiu, Y., Xia, J., Zhang, W., Liu, S.: A multimodal deep learning framework for predicting drug-drug interaction events. Bioinformatics **36**(15), 4316–4322 (2020)
19. Liu, Z., Wang, X.-N., Hui, Yu., Shi, J.-Y., Dong, W.-M.: Predict multi-type drug-drug interactions in cold start scenario. BMC Bioinformatics **23**(1), 75 (2022)

Self-supervised-Enhanced Dual Hierarchical Graph Convolution Network for Social Recommendation

Yixing Guo, Weimin Li[(✉)], Jingchao Wang, and Shaohua Li

School of Computer Engineering and Science, Shanghai University, Shanghai, China
{yx_guo,static,flowingfog}@shu.edu.cn, wmli@t.shu.edu.cn

Abstract. Graph convolution networks (GCNs) have made significant progress in the field of recommendation systems in recent years, and many GCN-based frameworks have applied in social recommendation methods. The essence of social recommendation tasks is modeling user preferences through user social relationships to alleviate the sparsity issue. However, existing GCN-based social recommendation frameworks still have some inherent problems. Firstly, since there are no node attributes available as semantic information in the recommendation task, lightweight graph convolutions that remove feature transformation and non-linear activation function have become widely applied in recommendation task, with the core lies in its message passing mechanism. Social recommendation methods always directly apply existing message passing paradigms, which always have obvious limitations in their message passing mechanisms. Secondly, most existing social recommendation frameworks are limited to pairwise relations and unable to effectively extract implicit inter-graph information. To address these issues, we propose a **S**elf-**S**upervised-Enhanced Dual **H**ierarchical **G**raph **C**onvolution **N**etwork (SSHGCN). In this framework, we first propose a LEWB (Link Encoded-Weight Balanced) message passing paradigm applied to graph convolution network to train user and item representations. Then we explicitly model marginal information between user social graph and user-item interaction graph, item knowledge graph and user-item interactions graph by hypergraph. Finally, we construct hierarchical self-supervised signals and unify self-supervised task and recommendation task for joint training. Extensive experiments on real-world datasets demonstrate that our method outperforms competitive methods. Thorough ablation study verifies the rationality of LEWB message passing paradigm and the effectiveness of the hierarchical self-supervised tasks in our framework.

Keywords: Social Recommendation · Graph Convolution Networks · Self-supervised Learning

1 Introduction

In today's information overload society, recommendation systems are playing an increasingly important role. Collaborative filtering, as a widely used method

B. Luo et al. (Eds.): ICONIP 2023, CCIS 1964, pp. 507–522, 2024.
https://doi.org/10.1007/978-981-99-8141-0_38

in recommendation systems, has also been extensively applied in graph neural networks in recent years [2,5,9]. Furthermore, recommendation methods using social graph [6,12] and knowledge graph [18] to capture self-supervised signals and alleviate the sparsity of user-item interactions have rapidly developed in recent years. Most collaborative filtering-based recommendation methods have data sparsity issue, as user-item interactions are sparse. Social recommendation has been considered a potential solution to alleviate data sparsity issue. It enhances user and item representations by smoothing users with high similarity in social relationships and achieves better performance in recommendation tasks.

Despite the success in graph-based collaborative filtering and graph-based social recommendation, there are still several challenges in current graph-based collaborative filtering for social recommendation. In graph-based collaborative filtering, although methods like LightGCN [5] have been widely applied in GCN-based collaborative filtering models, they still have various limitations. Firstly, existing message-passing mechanism is unable to capture the importance of different relationships, leading to additional noise. Secondly, the existing methods focus on modeling node representations while neglecting the importance of link information in node representation modeling. In social recommendation, there are two issues in current graph-based methods. Firstly, focusing on learning pairwise relations and graph structure information between users, overlooking the complex high-order connectivity information among users and lacking hierarchy [6]. Secondly, information extraction from the user-item graph and user social graph is limited. Some methods only perform graph convolution operations on single graph or separately on two graphs [5,9], which ignores the implicit information that can be obtained from their joint training, referred to as marginal information in this paper. Other methods integrate two graphs and perform convolution operations [16], but this leads to overfitting on training data and cannot reflect the potential of integrated data.

In light of the aforementioned issues, we propose a **S**elf-**S**upervised-Enhanced Dual **H**ierarchical **G**raph **C**onvolution **N**etwork (SSHGCN) for social recommendation. Specifically, in the graph-based collaborative filtering aspect, we introduce a new message-passing paradigm called LEWB(Link Encoded-Weight Balanced). Instead of using the traditional symmetric normalization parameter, we design more reasonable message-passing parameters by approximating the results of infinite-layer graph convolution. Additionally, we incorporate link encoding information as relationship information between nodes to enhance the representation training. As for the social recommendation perspective, we model marginal information between user social graph (user-user graph) and user-item interaction graph, item knowledge graph (item-item graph) and user-item interactions graph explicitly by hypergraph. Concretely, we design several triangle motifs with specific underlying semantics to construct the hypergraph and utilize the hypergraph structure to build hierarchical self-supervised signals after training with graph convolution. Instead of maximizing mutual information between node representations and graph-global representation, we propose a hierarchical mutual information maximization method: We hierarchically maximizes the mutual information among node representations, node-centric sub-hypergraph representations, and graph-global representations based on connected subgraphs. Experimental results demonstrate that our method further reduce uncertainties in local and global structures, preserve more graph structure information, and help to construct more effective self-supervised signals. Finally, we integrate the recommendation task with the hierarchical self-supervised task for jointly optimizing, leading to a substantial improvement in the performance of recommendation task.

The contributions of this paper can be summarized as follows:

- We propose a new message passing paradigm and incorporate link information as relation information between nodes instead of using limited message passing paradigm like LightGCN.
- We leverage hypergraph to model explicitly inter-graph marginal information and design a dual hierarchical self-supervised task. By jointly optimizing the self-supervised task and recommendation task, we further improve the performance of recommendation task.
- Extensive experiments on real-world datasets demonstrate the effectiveness of our proposed method, consistently outperforming state-of-the-art competitive methods.

2   Related Work

Graph-Based Methods in Recommendation. Graph-based methods have achieved great success in recent years. The early research like Pinsage [19] and NGCF [15], which training user and item representations by GCN to captures structural information in user-item graph. However, the feature transformation and non-linear activation in GCN are heavy and redundant for user-item graph lacking node attributes. Therefore, methods like LightGCN [5], LR-GCCF [2], explored lightweight GCN frameworks, which aim to remove unnecessary components and utilize high-order neighbor information. Among them, the removal of feature transformation and non-linear activation components proposed by Light-GCN has become a new paradigm for graph-convolution-based recommendation methods and has been widely adopted in further research [6,10], and other methods such as UltraGCN [9], SVD-GCN [10] further simplify the architecture to improve performance and efficiency.

Social Recommendation. Social relationships can reflect similarity or influence between users, which reflects their potential preferences. Therefore, utilizing social relationships in recommendation systems can alleviate data sparsity problem. In the early research on social recommendation, matrix factorization (MF) are popular methods, which can be categorized into three types: co-factorization methods such as TrustMF [17], ensemble methods such as mTrust [11], and regularization methods such as SR [8]. Recent research such as [7] has also used MF methods. With the development of deep learning, social recommendation with deep learning has emerged as a new direction. Specifically, graph-based methods like GraphRec [3], DiffNet++ [16] and DESIGN [12] used graph neural networks to extract information from user relationships and designed training tasks to integrate this information with the recommendation task, but they all lacked a fine-grained design for more effective information extraction. Besides, there are some methods that learn item trend information instead of user preference, such as TRec [13], that also achieve good performance.

3   Preliminaries

This section defines notations and concepts used in this paper. We denote user set as $U = \{u_1, u_2, ..., u_m\}$, item set as $I = \{i_1, i_2, ..., i_n\}$, node set as V and edge set as E.

Definition 1 User-Item Interaction Graph G_r. The user-item interaction matrix can be defined as $\boldsymbol{R} \in \mathbb{R}^{m \times n}$, when there is an interaction between user u_i and item i_j, $r_{ij} = 1$; otherwise, $r_{ij} = 0$. Based on matrix $\boldsymbol{R}$, we construct user-item interaction graph $G_r = \{V, E\}$.

Definition 2 User Social Graph G_s. The user social matrix can be defined as $\boldsymbol{S} \in \mathbb{R}^{m \times m}$. Based on the matrix $\boldsymbol{S}$, we construct user social graph $G_s = \{V_u, E_u\}$, where edge set E_u is user relationships.

Definition 3 Item Knowledge Graph G_k. The item knowledge matrix can be defined as $\boldsymbol{K} \in \mathbb{R}^{n \times n}$. Based on the matrix $\boldsymbol{K}$, we construct item knowledge graph $G_k = \{V_i, E_i\}$, where edge set E_i is item dependency relationships.

Definition 4 Hypergraph G_H. We define hypergraph as $G_H = \{V_H, E_H\}$, where V_H is the node set with N nodes, and E_H is the hyperedge set with M hyperedges. Each hyperedge $e \in E_H$ can contain an arbitrary number of nodes. The hypergraph can be represented in matrix form as $\boldsymbol{H} \in \mathbb{R}^{N \times M}$, where for any hyperedge $e \in E_H$ containing node $v \in V_H$, $H_{ev} = 1$; otherwise, $H_{ev} = 0$. Clearly, each row of the matrix represents a hyperedge. The node degree matrix and the edge degree matrix can be represented as diagonal matrices $\boldsymbol{D}_v$ and $\boldsymbol{D}_e$, respectively, where element value of $\boldsymbol{D}_v$ and $\boldsymbol{D}_e$ is $D_{vv} = \sum_{e=1}^{M} H_{ev}$ and $D_{ee} = \sum_{v=1}^{M} H_{ev}$. In the following sections, we will describe in detail the construction of enhanced hypergraphs G_{Huu} and G_{Hii} using motifs with specific underlying semantics.

Task Formulation. In this paper, our recommendation task is represented as follows: **Input**: User-item interaction graph G_r, user social graph G_s, item knowledge graph G_k, user-enhanced hypergraph G_{Huu}, item-enhanced hypergraph G_{Hii}. **Output**: A function $F = (u, v | G_r, G_s, G_k, G_{Huu}, G_{Hii}, \Theta)$ that effectively predicts items for which users will have future interactions, where Θ represents model parameters.

4   Methodology

In this section, we present our SSHGCN framework. We provide implementation details of each component, and then describes how to conduct recommendation model training. Figure 1 is a schematic overview of our model.

4.1   Link Encoded-Weight Balanced Message Passing Paradigm

We propose Link Encoded-Weight Balanced (LEWB) message passing paradigm, as illustrated in Fig. 1. The following is the formula form:

$$\boldsymbol{E}^{(k+1)} = \boldsymbol{B}_i \tilde{\boldsymbol{A}} \boldsymbol{B}_j (\boldsymbol{E}^{(k)} + \boldsymbol{L}_T)$$
$$\boldsymbol{E} = (\boldsymbol{E}^{(0)} || \boldsymbol{E}^{(1)} || ... || \boldsymbol{E}^{(K)}) \tag{1}$$

where $\boldsymbol{B}_i$ and $\boldsymbol{B}_j$ are weight-balanced normalized parameter vector inspired by UltraGCN [9]. We explicitly perform the message passing process in this form, which can better capture effective neighbor information and suppress noise, without resulting in an overly simplified model or a loss of high-order neighbor information. Each element β_i and β_j of those two vectors can be calculated by $\beta_i = \frac{\sqrt{d_i+1}}{d_i}$ and $\beta_j = \frac{1}{\sqrt{d_j+1}}$, d_i and d_j is the degrees of nodes i and j, respectively. $\boldsymbol{L}_T$ is link encoding information, which is time information in user-item graph training.

$$\boldsymbol{L}_{T(t_{ij}),2i} = sin(\frac{T(t_{ij})}{10000^{\frac{2i}{d}}})$$
$$\boldsymbol{L}_{T(t_{ij}),2i+1} = cos(\frac{T(t_{ij})}{10000^{\frac{2i+1}{d}}}) \tag{2}$$

where t_{ij} is timestamp, and $T(t_{ij})$ maps the timestamp to a standardized separated time slot, i.e. relative time. $2i+1$ and $2i$ denote the odd and even positions of the elements in $\boldsymbol{L}_{T(t_{ij})}$, respectively. The relative time embedding generated in this way will be regarded as interactive-edge embedding of mapping relative time.

In our work, the message passing in the GCN layer follows the mentioned paradigm for training on different graphs. We train user-item interaction graph to obtain user embeddings $\boldsymbol{E}_U$ and item embeddings $\boldsymbol{E}_I$ by this paradigm. As for hypergraph training, link information is not time-based, we will discuss it in detail in the next section.

4.2   Dual Hierarchical Self-supervised Enhancement Based on Hypergraph

Hypergraph Construction and Hypergraph Convolution Training. We construct hypergraph by several triangle motifs with specific underlying semantics (as shown in Fig. 1). Motifs is specific local structure that describe multiple nodes and is widely used to capture various complex graph patterns. We choose

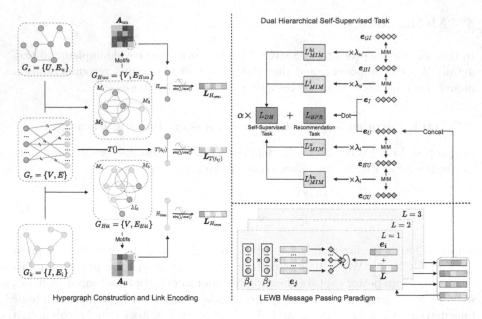

Fig. 1. The proposed framework Self-Supervised-Enhanced Dual Hierarchical Graph Convolution Network (SSHGCN).

triangle motifs because of the ubiquitous triadic closure in social networks [1]. Besides, triangle motifs have precise underlying semantics, they are often be the substructures of more complex structures. Figure 2 shows motifs that we design in our work.

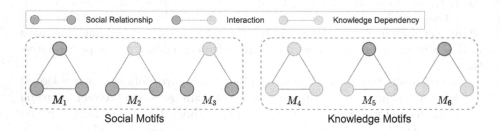

Fig. 2. Our designed motifs. The blue nodes represent users, and the pink nodes represent items. (Color figure online)

M_1-M_3 for integrating user-item interaction graph with user social graph (represented by matrix H_{uu}), and M_4-M_6 for integrating user-item interaction graph with item knowledge graph (represented by matrix H_{ii}), they all have clear semantics. M_1 represents three users who have social relationships with each other. M_2 represents two users with a social relationship who have interactions with the same item. M_3 represents two users without a social relationship who have interactions with the same item. M_4 represents three items that have mutual knowledge dependencies. M_5 represents two items with a knowledge dependency that the same user interacted. M_6 represents two items without a knowledge dependency that the same user interacted. Compared with independent user-item interaction graph and user social graph/item knowledge graph, the hypergraph built by above motifs explicitly models marginal information. Compared with the method of only integrating different graphs for joint training, our motifs have clear underlying semantics integrate different graphs more reasonable.

With reference to the spectral hypergraph convolution proposed in [4], we define hypergraph convolution as follows:

$$E_H^{(k+1)} = D_v^{-\frac{1}{2}} H D_e^{-1} H^\top D_v^{-\frac{1}{2}} E_H^{(k)} \tag{3}$$

where D_v and D_e are node degree and edge degree matrices, respectively, and H is the input hypergraph matrix. The hypergraph convolution in above equation can be regarded as a two-stage message passing of "node-hyperedge-node". Unlike the definition in [4], we define hypergraph convolution by LightGCN paradigm, which removes feature transformation and non-linear activation components. However, it would be costly to construct the hypergraph matrix H. To address this, we refer to the matrix transformation for triangle motifs in [20] and convert the hypergraph convolution to general graph convolution. Table 1 shows matrix transformation in detail.

Table 1. The calculation of the matrix transformation for motifs.

Motif	$A_{M_k} =$
M_1	$(SS) \odot S$
M_2	$(RR^\top) \odot S$
M_3	$RR^\top$
M_4	$(KK) \odot K$
M_5	$(R^\top R) \odot K$
M_6	$R^\top R$

In this table, $\odot$ is hadamard product, S is user social matrix, K is item knowledge matrix, R is user-item interaction matrix, A_{M_k} represents the transformed adjacency matrix for different motifs. The value of $(A_{M_k})_{mn}$ indicates how many times nodes m and n appear in the instance of motif M_k. Without considering self-connections, the sum of A_{M_1} to A_{M_3} is equal to $H_{uu} H_{uu}^\top$, the sum of A_{M_4} to A_{M_6} is equal to $H_{ii} H_{ii}^\top$. Based on above analysis, we reformulate hypergraph convolution as follows:

$$E_H^{(k+1)} = D_v^{-\frac{1}{2}} A D_v^{-\frac{1}{2}} E_H^{(k)} \tag{4}$$

where $\boldsymbol{A}$ represents $\boldsymbol{A}_{uu}$ or $\boldsymbol{A}_{ii}$, and $\boldsymbol{A}_{uu} = \sum_{k=1}^{3} \boldsymbol{A}_{M_k}$, $\boldsymbol{A}_{uu} = \sum_{k=4}^{6} \boldsymbol{A}_{M_k}$. The reformulated hypergraph convolution addresses the issue of constructing hypergraph with high computational cost, however, the hyperedge information is implicitly encoded in the element values of $\boldsymbol{A}_{M_k}$, which prevents the two-stage message passing of "node-hyperedge-node" as in Eq. (3). In order to make the link information fully extracted, we regard the element values in $\boldsymbol{A}_{uu}$ and $\boldsymbol{A}_{ii}$ matrices as link information, encode them with the method of Eq. (2) to obtain hyperedge embeddings $\boldsymbol{L}_H$. We then apply the LEWB message propagation paradigm to our hypergraph training:

$$
\begin{aligned}
\boldsymbol{E}_H^{(k+1)} &= \boldsymbol{B}_{Hm} \boldsymbol{A} \boldsymbol{B}_{Hn} (\boldsymbol{E}_H^{(k)} + \boldsymbol{L}_H) \\
\boldsymbol{E}_H &= (\boldsymbol{E}_H^{(0)} || \boldsymbol{E}_H^{(1)} || ... || \boldsymbol{E}_H^{(K)})
\end{aligned}
\tag{5}
$$

where $\boldsymbol{B}_{Hm}$ and $\boldsymbol{B}_{Hn}$ are weight-balanced normalized parameter vector of hypergraph training.

By training in the aforementioned way, we obtain user hypergraph embedding $\boldsymbol{E}_{HU}$ and item hypergraph embedding $\boldsymbol{E}_{HI}$.

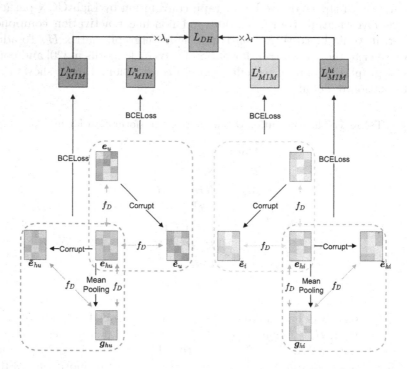

Fig. 3. Construction process of dual hierarchical self-supervised tasks.

Construction of Dual Hierarchical Self-Supervised Task. In representation learning, it is common to use self-supervised tasks as auxiliary tasks to enhance training effectiveness. In our work, we propose a self-supervised auxiliary task based on the Deep Graph Infomax (DGI) framework [14], which learns graph structural information and incorporates it into node representations by self-supervised learning. The core idea of DGI framework is to maximize the mutual information between local and global information. However, this method has limitations in capturing fine-grained structure information and becomes more evident as the graph scale increases. Therefore, our ultimate goal of constructing hypergraph and training its representation is to serve as the input data of the hierarchical DGI framework, and construct a hierarchical self-supervised task. Specifically, the improved DGI framework consists of three levels: nodes, node-centric sub-hypergraphs, and graph-global based on connected subgraphs. With user-user relations and item-item relations as dual centres, we input the corresponding data into DGI framework and maximize the mutual information between adjacent levels, as illustrated in Fig. 3. The following sections provide more details.

We maximize mutual information between node representations and node-centric sub-hypergraph representations, node-centric sub-hypergraph representations and global representations, respectively. In user embeddings E_U and item embeddings E_I, each row represents a specific node representation. Similarly, in the user hypergraph embeddings E_{HU} and item hypergraph embeddings E_{HI}, each row represents the node-centric sub-hypergraph representation. Unlike traditional methods that maximize mutual information between local representations and a single global representation, we utilize the connected subgraph structure of user-item as boundaries for performing average pooling on E_{HU} and E_{HI} to enrich global semantics, generating global representations G_{HU} and G_{HI} at the subgraph level. We use the classifier function f_D as the mutual information estimator, with the dot product between two representations serving as the probability score for sampling pairs. Taking user-user relations as an example, at the node-sub-hypergraph level, we generate positive samples (e_{u_i}, e_{hu_i}) and (e_{i_j}, e_{hi_j}), and also generate negative samples $(\tilde{e}_{u_i}, e_{hu_i})$ and $(\tilde{e}_{i_j}, e_{hi_j})$. At the sub-hypergraph-global level, we generate positive samples (e_{hu_i}, g_{hu_i}) and (e_{hi_j}, g_{hi_j}), as well as negative samples $(\tilde{e}_{hu_i}, g_{hu_i})$ and $(\tilde{e}_{hi_j}, g_{hi_j})$. The negative samples are generated by randomly shuffling the nodes to create misplaced pairs. The same process applies to item-item relations.

We use cross-entropy loss to achieve mutual information maximization and define the loss function for dual hierarchical self-supervised task as follows:

$$L_{MIM} = -\frac{1}{N_{pos} + N_{neg}} \left(\sum_{i=1}^{N_{pos}} t(e_l, e_h) \cdot log\left[f_D(e_l \cdot e_h)\right] \right.$$

$$\left. + \sum_{i=1}^{N_{neg}} t(\tilde{e}_l, e_h) \cdot log\left[1 - f_D(\tilde{e}_l \cdot e_h)\right] \right) \tag{6}$$

$$L_{DH} = \lambda_u(L_{MIM}^u + L_{MIM}^{hu}) + \lambda_i(L_{MIM}^i + L_{MIM}^{hi})$$

where L_{MIM} is the cross-entropy loss for mutual information maximization, N_{pos} and N_{neg} are the numbers of positive and negative samples, respectively. $t()$ is an indicator function that indicates the label values of positive and negative samples. e_l and e_h denote the lower-level representation and higher-level representation, respectively. λ_u and λ_i are hyperparameters that control the weights of different subtasks, which also indirectly affect the weights of recommendation task and auxiliary dual-level hierarchical self-supervised task. Our objective is to minimize L_{DH} in order to maximize mutual information between different levels. Through hierarchical mutual information maximization, we preserve more hypergraph structural information and extract more graph structural information. Meanwhile, we preserve more global semantic information by performing average pooling on the connected subgraph structures to obtain global representations.

4.3   Model Training

In SSHGCN, we employ the Bayesian Personalized Ranking (BPR) loss for recommendation task. It is a pairwise loss that aims to maximize the difference in scores between positive and negative samples:

$$L_{BPR} = \sum_{\substack{i \in N_u \\ i \notin N_u}} -log\sigma(\hat{y}_{ui} - \hat{y}_{uj}) + \lambda||\Theta||^2 \tag{7}$$

where N_u is the observed interactions of user u, i and j are positive and negative samples, respectively. $\hat{y}_{ui}$ and $\hat{y}_{uj}$ are the estimated interaction probabilities for the positive and negative samples, calculated by dot product $e_u^\top e_i$. Θ represents learnable parameters of model, and λ is regularization parameter.

Finally, we integrate recommendation task with auxiliary dual hierarchical self-supervised task for joint training. The overall loss function is:

$$L = L_{BPR} + \alpha L_{DH} \tag{8}$$

The term L_{DH} can be seen as a regularizer for recommendation task, utilizing hierarchical structural information of users and items to enrich representation in recommendation task and achieve better performance.

5   Experiments

In this section, we first describe experimental settings, then compare the proposed SSHGCN method with other state-of-the-art methods. Finally, we perform detailed ablation experiments to verify the rationality of each component in SSHGCN.

5.1   Experimental Settings

Dataset. Our experiments were conducted on two real-world datasets: Epinions and Yelp, both of which are commonly used in GCN-based recommendation models. The statistical information of the datasets is presented in Table 2.

Evaluation Protocols. We employed widely used evaluation metrics in both GCN-based collaborative filtering models and social recommendation models: Normalized Discounted Cumulative Gain (NDCG@N) and Hit Rate (HR@N).

Table 2. Dataset Statistics

Dataset	Epinions	Yelp
Users	18081	43043
Items	251722	66576
Interactions	715821	283512
Density	0.0157%	0.0098%
Social Relations	590641	549451
Social Relations Density	0.1806%	0.0296%
Items Relations	6069106	1847060

Baseline. We compared SSHGCN with the following methods: LightGCN [5], LR-GCCF [2], UltraGCN [9] as graph collaborative filtering methods, and DiffNet++ [16], KCGN [6], GraphRec [3], MHCN [20] as graph social recommendation methods.

Implementation Details. For fair comparison, we referred to the optimal parameter settings as stated in their respective papers for the baseline methods. In our method, we fixed embedding dimension to 64 and set learning rate to 0.001. The batch size was set to 1024, and the hyperparameters λ_u and λ_i were set to 0.1 and 0.001, respectively (adjusted optimal parameters). Training was terminated early when the performance on validation dataset continuously decreased for 15 epochs.

5.2 Performance Comparison

Table 3 presents the performance comparison of SSHGCN with 7 baseline methods on different top-K item ranking metrics. Our model outperforms all the baselines across different top-K values, demonstrating the effectiveness and robust ranking performance of our method. The improvements achieved by SSHGCN can be attributed to the following reasons: (1) LEWB message passing paradigm: It effectively suppresses noise, captures neighbor information more efficiently, and preserves the capability to learn high-order neighbor information. Additionally, the encoding of link information (such as temporal information or hyperedge information) further enriches the node representations. (2) Explicit modeling of marginal information between graphs by utilizing several specific motifs to construct hypergraphs. (3) Construction of dual hierarchical self-supervised auxiliary task: This task extract more graph structure information by fine-grained mutual information maximization. Moreover, the improved global representations preserve more global semantic information.

In the baseline methods, we observed the following results: (1) Among the Graph-based methods, the methods that remove redundant and complex operations generally outperform heavy methods, including LightGCN, UltraGCN, and MHCN. This indicates that for graph neural network recommendation models, a lightweight message propagation paradigm with a simple structure is more effective. (2) MHCN performs better than LightGCN and UltraGCN. This can be attributed to the fact that MHCN is a social recommendation model based on the message passing paradigm of LightGCN, which leverages user social relationships to overcome the sparsity issue. Among heavy Graph-based models, KCGN shows comparable performance to LightGCN. This can be attributed to the fact that KCGN is a social recommendation model that utilizes multi-behavior interaction data and injects knowledge about user relationships and item dependencies into user preference modeling, but the heavy message passing paradigm in this model limits the potential of KCGN. The performance of MHCN and KCGN highlights the inherent advantages of social recommendation over traditional collaborative filtering methods and demonstrates the effectiveness of our approach in constructing fine-grained auxiliary tasks.

Table 3. Overall performance comparison

Dataset	Work	HR@5	NDCG@5	HR@10	NDCG@10	HR@15	NDCG@15
Epinions	LightGCN	0.5792	0.4398	0.7038	0.4824	0.7679	0.4880
	GCCF	0.5408	0.4368	0.6781	0.4788	0.7410	0.4927
	UltraGCN	0.5680	0.4335	0.7070	0.4775	0.7739	0.4988
	DiffNet++	0.5162	0.3821	0.6391	0.4226	0.6981	0.4356
	KCGN	0.5817	0.4430	0.6919	0.4645	0.7654	0.4891
	GraphRec	0.5535	0.4241	0.6866	0.4782	0.7398	0.4689
	MHCN	0.5863	0.4449	0.7012	0.4851	0.7701	0.4974
	SSHGCN	**0.6030**	**0.4636**	**0.7311**	**0.5090**	**0.7906**	**0.5204**
Yelp	LightGCN	0.6645	0.4982	0.8086	0.5358	0.8602	0.5312
	GCCF	0.6218	0.4759	0.7692	0.5190	0.8435	0.5366
	UltraGCN	0.6536	0.4861	0.8015	0.5389	0.8542	0.5495
	DiffNet++	0.6317	0.4628	0.7881	0.5179	0.8633	0.5345
	KCGN	0.6599	0.4885	0.8026	0.5308	0.8687	0.5470
	GraphRec	0.6233	0.4554	0.7605	0.4945	0.8342	0.5139
	MHCN	0.6685	0.5028	0.8074	0.5425	0.8635	0.5458
	SSHGCN	**0.6803**	**0.5121**	**0.8175**	**0.5564**	**0.8774**	**0.5752**

5.3   Ablation Study of SSHGCN Framework

We conducted fine-grained ablation studies from the perspectives of message passing paradigm and self-supervised auxiliary tasks to thoroughly evaluate the effectiveness of each component in our framework.

Ablation Analysis of Message Passing Paradigms. To verify the rationality of LEWB message passing paradigm, we compare it with different message passing paradigms, including: (1) "w/o LE": The encoded link information is not integrated into message passing process, but the normalization method of LEWB is reserved. (2) "w/o WB": The encoded link information is integrated into the message passing process, but the symmetric sqrt normalization method used in LightGCN is applied. (3) "w/o LEWB": Without using LEWB message passing paradigm but using LightGCN message passing paradigm.

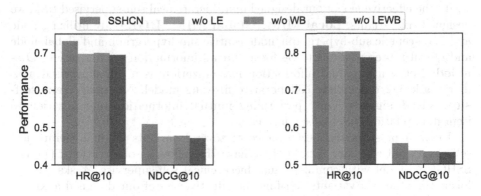

Fig. 4. Ablation Study of LEWB Message Passing Paradigm

Figure 4 presents the performance of several variants of message passing paradigms, and we have the following observations: (1) The performance of SSHGCN using complete LEWB message passing paradigm is significantly better than the other three variants. (2) "w/o LE" and "w/o WB" are variants that partially adopt LEWB message passing paradigm design. Both of them show slightly better performance than "w/o LEWB," which only uses LightGCN message passing paradigm. However, there is still a significant performance gap compared to SSHGCN. The performance improvement achieved by the integration of two components is much greater than the improvement achieved by partial adoption, verifying the rationality of our design.

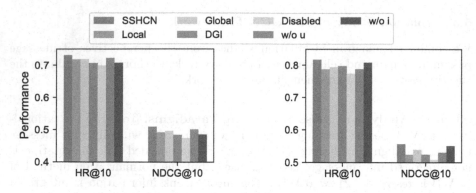

Fig. 5. Ablation study of dual hierarchical self-supervised tasks

Ablation Analysis of Dual Hierarchical Self-Supervised Task. To investigate the effectiveness of our designed dual hierarchical self-supervised task, we designed six variants. **Local MIM, Global MIM, DGI**: Only utilizing node and node-centric sub-hypergraph/node-centric sub-hypergraph and global/node and global representations to perform mutual information maximization. **Disabled**: Not using mutual information maximization as a self-supervised auxiliary task, i.e. a complete collaborative filtering model. **No user/item self-supervised signals**: Only performing mutual information maximization to item-item relations/user-user relations.

Figure 5 presents the performance of several variants of dual hierarchical self-supervised auxiliary task, and we have the following observations: (1) The SSHGCN model with complete dual hierarchical self-supervised tasks outperforms the other six variants, verifing the effectiveness of our designed auxiliary tasks. (2) In different datasets, the Local, Global, and DGI variants show significantly better performance than the Disabled variant, indicating the contribution of self-supervised auxiliary tasks to the overall performance. However, there is no clear advantage among the Local, Global, and DGI variants across different datasets. We attribute this to the varying richness of marginal information in different datasets. (3) The two variants that remove user/item self-supervised signals exhibit significant performance differences in different datasets. We attribute this to the lower number of users in the Epinions dataset, resulting in a lower richness of marginal information and higher noise in constructing user hypergraph. And it is reversed in the Yelp dataset. This issue can be considered as a potential future research direction.

6   Conclusion

In this work, we propose the SSHGCN framework. We first introduce LEWB message passing paradigm for graph convolution training of node and sub-hypergraph representations. We then explicit model inter-graph marginal information by hypergraph to construct a dual hierarchical self-supervised task, capturing graph structure and semantic information in a fine-grained way. Extensive experiments on real-world datasets verify the effectiveness of each component in SSHGCN and demonstrate superior performance compared to state-of-the-art competitive methods.

References

1. Asikainen, A., Iñiguez, G., Ureña-Carrión, J., Kaski, K., Kivelä, M.: Cumulative effects of triadic closure and homophily in social networks. Sci. Adv. **6**(19), eaax7310 (2020)
2. Chen, L., Wu, L., Hong, R., Zhang, K., Wang, M.: Revisiting graph based collaborative filtering: a linear residual graph convolutional network approach. In: Proceedings of the AAAI Conference on Artificial Intelligence, vol. 34, pp. 27–34 (2020)
3. Fan, W., Ma, Y., Li, Q., He, Y., Zhao, E., Tang, J., Yin, D.: Graph neural networks for social recommendation. In: The world wide web conference, pp. 417–426 (2019)
4. Feng, Y., You, H., Zhang, Z., Ji, R., Gao, Y.: Hypergraph neural networks. In: Proceedings of the AAAI Conference on Artificial Intelligence, vol. 33, pp. 3558–3565 (2019)
5. He, X., Deng, K., Wang, X., Li, Y., Zhang, Y., Wang, M.: Lightgcn: simplifying and powering graph convolution network for recommendation. In: Proceedings of the 43rd International ACM SIGIR Conference on Research and Development in Information Retrieval, pp. 639–648 (2020)
6. Huang, C., Xu, H., Xu, Y., Dai, P., Xia, L., Lu, M., Bo, L., Xing, H., Lai, X., Ye, Y.: Knowledge-aware coupled graph neural network for social recommendation. In: Proceedings of the AAAI Conference on Artificial Intelligence, vol. 35, pp. 4115–4122 (2021)
7. Lin, W., Zhang, X., Qi, L., Li, W., Li, S., Sheng, V.S., Nepal, S.: Location-aware service recommendations with privacy-preservation in the internet of things. IEEE Trans. Comput. Soc. Syst. **8**(1), 227–235 (2020)
8. Ma, H., Zhou, D., Liu, C., Lyu, M.R., King, I.: Recommender systems with social regularization. In: Proceedings of the fourth ACM International Conference on Web Search and Data Mining, pp. 287–296 (2011)
9. Mao, K., Zhu, J., Xiao, X., Lu, B., Wang, Z., He, X.: Ultragcn: ultra simplification of graph convolutional networks for recommendation. In: Proceedings of the 30th ACM International Conference on Information & Knowledge Management, pp. 1253–1262 (2021)
10. Peng, S., Sugiyama, K., Mine, T.: Svd-gcn: a simplified graph convolution paradigm for recommendation. In: Proceedings of the 31st ACM International Conference on Information & Knowledge Management, pp. 1625–1634 (2022)
11. Tang, J., Gao, H., Liu, H.: mtrust: Discerning multi-faceted trust in a connected world. In: Proceedings of the fifth ACM International Conference on Web Search and Data Mining, pp. 93–102 (2012)
12. Tao, Y., Li, Y., Zhang, S., Hou, Z., Wu, Z.: Revisiting graph based social recommendation: a distillation enhanced social graph network. In: Proceedings of the ACM Web Conference 2022, pp. 2830–2838 (2022)
13. Tao, Y., Wang, C., Yao, L., Li, W., Yu, Y.: Item trend learning for sequential recommendation system using gated graph neural network. Neural Computing and Applications, pp. 1–16 (2021)

14. Velickovic, P., Fedus, W., Hamilton, W.L., Liò, P., Bengio, Y., Hjelm, R.D.: Deep graph infomax. ICLR (Poster) **2**(3), 4 (2019)
15. Wang, X., He, X., Wang, M., Feng, F., Chua, T.S.: Neural graph collaborative filtering. In: Proceedings of the 42nd International ACM SIGIR Conference on Research and Development in Information Retrieval, pp. 165–174 (2019)
16. Wu, L., Li, J., Sun, P., Hong, R., Ge, Y., Wang, M.: Diffnet++: a neural influence and interest diffusion network for social recommendation. IEEE Trans. Knowl. Data Eng. **34**(10), 4753–4766 (2020)
17. Yang, B., Lei, Y., Liu, J., Li, W.: Social collaborative filtering by trust. IEEE Trans. Pattern Anal. Mach. Intell. **39**(8), 1633–1647 (2016)
18. Yang, Y., Huang, C., Xia, L., Li, C.: Knowledge graph contrastive learning for recommendation. In: Proceedings of the 45th International ACM SIGIR Conference on Research and Development in Information Retrieval, pp. 1434–1443 (2022)
19. Ying, R., He, R., Chen, K., Eksombatchai, P., Hamilton, W.L., Leskovec, J.: Graph convolutional neural networks for web-scale recommender systems. In: Proceedings of the 24th ACM SIGKDD International Conference on Knowledge Discovery & Data Mining, pp. 974–983 (2018)
20. Yu, J., Yin, H., Li, J., Wang, Q., Hung, N.Q.V., Zhang, X.: Self-supervised multi-channel hypergraph convolutional network for social recommendation. In: Proceedings of the Web Conference 2021, pp. 413–424 (2021)

Dynamical Graph Echo State Networks with Snapshot Merging for Spreading Process Classification

Ziqiang Li[1]([✉])[ID], Kantaro Fujiwara[1][ID], and Gouhei Tanaka[1,2][ID]

[1] International Research Center for Neurointelligence, The University of Tokyo, Tokyo 113-0033, Japan
{ziqiang-li,kantaro}@g.ecc.u-tokyo.ac.jp
[2] Department of Computer Science, Graduate School of Engineering, Nagoya Institute of Technology, Nagoya 466-8555, Japan
gtanaka@nitech.ac.jp

Abstract. The Spreading Process Classification (SPC) is a popular application of temporal graph classification. The aim of SPC is to classify different spreading patterns of information or pestilence within a community represented by discrete-time temporal graphs. Recently, a reservoir computing-based model named Dynamical Graph Echo State Network (DynGESN) has been proposed for processing temporal graphs with relatively high effectiveness and low computational costs. Inspired by DynGESN, we propose a novel reservoir computing-based model called the Grouped Dynamical Graph Echo State Network (GDGESN) for dealing with SPC tasks. In this model, a novel augmentation strategy named the snapshot merging strategy is designed for forming new snapshots by merging neighboring snapshots over time, and then multiple reservoir encoders are set for capturing spatiotemporal features from merged snapshots. After those, the logistic regression is adopted for decoding the sum-pooled embeddings into the classification results. Experimental results on six benchmark SPC datasets show that our proposed model has better classification performances than the DynGESN and several kernel-based models.

Keywords: Reservoir Computing · Multiple Reservoir Echo State Networks · Temporal Graph Processing

1 Introduction

The spreading (or dissemination) process is used to describe the spreading of information (e.g. fake news and rumors) or infectious diseases (e.g. Covid-19 and meningitis) within a community. Since spreading patterns of virus strains or different rumors are various, it is hard to recognize them within a relatively short period of time accurately. Based on this background, Spreading Process Classification (SPC) is a highly-demanded technology for experts in relevant fields to distinguish them before carrying out possible interventions and countermeasures.

B. Luo et al. (Eds.): ICONIP 2023, CCIS 1964, pp. 523–534, 2024.
https://doi.org/10.1007/978-981-99-8141-0_39

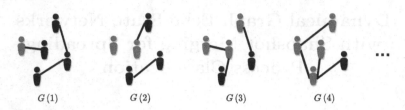

$G(1)$ $G(2)$ $G(3)$ $G(4)$

Fig. 1. An example of epidemic spreading. Icons marked in red and black represent infected and uninfected people, respectively, and a black line between two icons indicates that contact exists between two people at that time step. (Color figure online)

Normally, spreading processes can be represented by temporal graphs with dynamic connections and temporal signals. We show an example of an epidemic spreading in Fig. 1, where $G(t)$ means the t-th snapshot of the temporal graph $\mathcal{G}$. We can notice that uninfected people (marked in black) can be infected probabilistically by contact with infected people (marked in red). Usually, one kind of epidemic has its own basic reproduction number, which leads to different spreading processes.

Generally, SPC can be turned into a Discrete-time Temporal Graph (DTG) classification task. To deal with this task, advanced deep learning models [1,5,10] designed by combining variants of Graph Convolutional Networks (GCNs) [9] with those of Recurrent Neural Networks (RNNs) [2,7] and/or the attention mechanism [20] are widely considered to be ideal choices. The common ground of these models in structure is leveraging multiple graph convolution layers for extracting spatial features and using recurrent layers or attention layers for mining the temporal relationships. In this regard, high computational costs need to be spent to obtain a well-trained complex model. Furthermore, extra efforts for solving gradient explosion and vanishing problems are unavoidable. Another direction is to use some transformation methods to stitch the sequential snapshots into a large-scale static graph and then apply some graph kernel methods (i.e. the Weisfeiler-Lehman graph kernels [17]) to generate the final classification results. Methods following this direction can avoid effects on capturing the temporal dependency in the snapshot sequence, but the large-scale static graphs lead to significantly high computational costs in the calculations of the gram matrix for a support vector machine [6] by using those graph kernels [16].

Reservoir Computing (RC) [13,18] is an efficient framework derived from RNNs, which maps the sequential inputs into high dimensional spaces through a predetermined dynamical system. This characteristic enables the training costs of its derived models to be remarkably lower than those of fully-trained RNNs. The Echo State Network (ESN) [8], as one of the representative models of RC, and its variants have been intensively studied for handling various time series processing tasks [3,12]. Recently, D. Tortorella & A. Micheli successfully extended the standard ESN to a novel RC model called Dynamical Graph Echo State Network (DynGESN) [19], which is capable of dealing with discrete-time

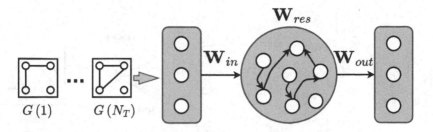

Fig. 2. A schematic diagram of DynGESN [19].

temporal graphs processing tasks. A schematic diagram of DynGESN is shown in Fig. 2. In this model, two fixed weight matrices $\mathbf{W}_{in}$ and $\mathbf{W}_{res}$ are used for mapping original snapshot sequences into vertex embeddings. The trainable weight matrix denoted by $\mathbf{W}_{out}$ is set for transforming the vertex embeddings into desired outputs.

A recent work has demonstrated that DynGESN outperforms some kernel-based methods on a number of SPC benchmark datasets [14]. However, we noticed that only one single dynamical characteristic included in the original temporal graphs is extracted in DynGESN [14]. Obviously, this monotonous strategy may hinder the model from extracting diverse dynamical characteristics extended from the original temporal snapshots.

To solve this problem, a new model, Grouped Dynamical Graph Echo State Network (GDGESN), is proposed for SPC tasks in this study. This model can extract various spatiotemporal features from augmented inputs by group-wise reservoir encoders and generate accurate spreading classification results by a linear classifier efficiently. In this regard, we propose a simple augmentation strategy called snapshot merging to generate multi-timescale temporal graphs and then leverage the multiple-reservoir framework [11] to build the group-wise reservoir encoders. We execute experiments for comparing the classification performances with those of the DynGESN and some kernel-based methods on six benchmark SPC datasets. The experimental results show that the accuracies of our model are higher than those of DynGESN and are close to those of kernel-based methods on some SPC datasets, which manifests the GDGESN owns relatively high effectiveness in dealing with SPC tasks.

The rest of this paper is organized as follows: The preliminary about temporal graphs is introduced in Sect. 2. The proposed method is described in Sect. 3. The analysis about the computational complexity of the proposed model is presented in Sect. 4. The details of the experiments are introduced in Sect. 5. The discussion is given in Sect. 6.

2 Preliminaries

Generally, a discrete-time temporal graph is composed of a sequence of snapshots, which can be denoted by $\mathcal{G} = \{G(t)\}_{t=1}^{N_T}$, where $G(t)$ is the snapshot at

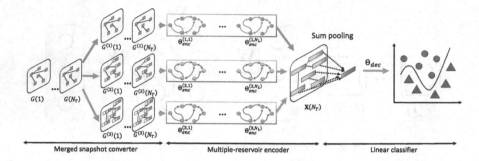

Fig. 3. An example of the proposed model with three groups of reservoir encoders.

time t and N_T is the length of $\mathcal{G}$. The snapshot $G(t) = \{\mathbf{v}(t), \mathbf{A}(t)\}$ contains a time-varying vertex signal vector $\mathbf{v}(t) \in \mathbb{R}^{N_V}$ and the corresponding adjacency matrix $\mathbf{A}(t) \in \mathbb{R}^{N_V \times N_V}$, where N_V is the number of vertices. The state value of the i-th vertex at time t can be represented by $v_i(t) \in \mathbb{R}$. For representing the spreading process, we define that $v_i(t) = 1$ if the i-th person is affected at time t and $v_i(t) = 0$ otherwise. We suppose that each graph is undirected, which can be represented by $A_{i,j}(t) = A_{j,i}(t) = 1$ if there is a contact between the i-th person and the j-th person, $A_{i,j}(t) = A_{j,i}(t) = 0$ otherwise. Moreover, we assume that a spreading process classification dataset has N_S temporal graphs and the corresponding labels, which can be represented by $\{\mathcal{G}_s, \mathbf{y}_s\}_{s=1}^{N_s}$, where $\mathbf{y}_s \in \mathbb{R}^{N_Y}$ is the label represented by the one-hot encoding for $\mathcal{G}_s$.

3   The Proposed Model

A schematic diagram of the GDGESN is shown in Fig. 3. This is a case where a spreading process represented by a discrete-time temporal graph is fed into the GDGESN with three groups of reservoir encoders. We can notice that the model consists of three components, including a merged snapshot converter, a set of multiple-reservoir encoders, and a linear classifier. In the merged snapshot converter, a DTG is transformed into three new DTGs with different window sizes. In the multiple-reservoir encoder, each transformed temporal graph is fed into the corresponding group-wise reservoir encoders for generating various vertex embeddings. In the linear decoder, the aggregated embeddings of the last time step obtained by the sum-pooling operation are collected from all reservoir encoders and then decoded into the classified results. The details about the above-mentioned components are introduced in Sects. 3.1, 3.2, and 3.3, respectively.

3.1   The Merged Snapshot Converter

The merged snapshot converter is proposed to merge several neighboring snapshots into one merged snapshot. To this end, we define a window that slides on the zero-padded snapshot sequence. We denote the size of the sliding window by ω. For simplicity, we fix the stride of this sliding window to be one. In order to keep the length of the merged snapshot sequence the same as that of the original snapshot sequence, we add $(\omega - 1)$ empty snapshots into the beginning of the original snapshot sequence, which can be formulated as follows:

$$\mathcal{P}_s = \left\{ \underbrace{G_{nil}, \ldots, G_{nil}}_{\omega-1}, G(1), G(2), \ldots, G(N_T) \right\}, \tag{1}$$

where $\mathcal{P}_s$ means the s-th snapshot-padded sequence with length $(N_T + \omega - 1)$ and G_{nil} represents the empty snapshot whose signal value of each vertex is zero. We assume that the merged temporal graphs corresponding to N_G different sizes of the sliding windows can be organized into N_G groups. Therefore, we represent the size of the sliding window corresponding to the g-th group as $\omega^{(g)}$ for $g = 1, 2, \ldots, N_G$. Based on the above settings, We can merge snapshots into a new snapshot by executing the logical OR operation within a sliding window with size $\omega^{(g)}$, which can be formulated as follows:

$$\begin{aligned} v_i^{(g)}(t) &= v_i\left(t - \omega^{(g)} + 1\right) \cup v_i\left(t - \omega^{(g)} + 2\right) \cup \cdots \cup v_i(t), \\ A_{i,j}^{(g)}(t) &= A_{i,j}\left(t - \omega^{(g)} + 1\right) \cup A_{i,j}\left(t - \omega^{(g)} + 2\right) \cup \cdots \cup A_{i,j}(t). \end{aligned} \tag{2}$$

From Eq. 2, we can obtain diverse spatiotemporal information with different $\omega^{(g)}$. Figure 4 shows an example of transforming the original snapshot sequence into a merged snapshot sequence by the merged snapshot converter with $\omega = 2$. This example indicates that the merged snapshot converter can produce multi-timescale spatiotemporal inputs with different sizes of sliding windows. The experimental results presented in Sect. 5.4 demonstrate that the merged snapshot converter with various sizes of sliding windows can improve the classification performances of the GDGESN on some SPC datasets.

3.2 The Multiple-Reservoir Encoder

The multiple-reservoir encoder is proposed for extracting spatiotemporal features from merged snapshot sequences. We organize reservoir encoders following the layout described in Ref. [12]. Note that a reservoir encoder denoted by Θ_{enc} contains an input weight matrix $\mathbf{W}_{in} \in \mathbb{R}^{N_R \times N_U}$ and a reservoir matrix $\mathbf{W}_{res} \in \mathbb{R}^{N_R \times N_R}$, where N_R is the size of the reservoir. We add a superscript (g, l) to Θ_{enc} for indicating the encoder located at the l-th layer of the g-th group, which can be formulated by $\Theta_{enc}^{(g,l)} = \left\{ \mathbf{W}_{in}^{(g,l)}, \mathbf{W}_{res}^{(g,l)} \right\}$ for $1 \le g \le N_G$ and $1 \le l \le N_L$, where N_G and N_L are maximal numbers of groups and layers, respectively.

In the encoding process, the vertex embedding matrix at time t, $\mathbf{X}^{(g,l)}(t) \in \mathbb{R}^{N_R \times N_V}$, can be calculated as follows:

$$\begin{aligned} \mathbf{X}_s^{(g,l)}(t) &= \alpha f\left(\mathbf{W}_{in}^{(g,l)} \mathbf{U}_s^{(g,l)}(t) + \mathbf{W}_{res}^{(g,l)} \mathbf{X}_s^{(g,l)}(t-1) \mathbf{A}(t) \right) \\ &\quad + (1-\alpha) \mathbf{X}_s^{(g,l)}(t-1), \end{aligned} \tag{3}$$

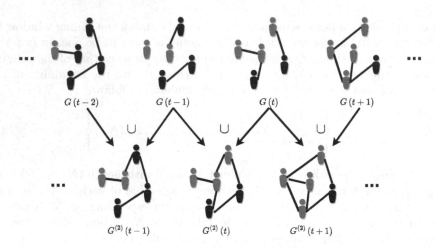

Fig. 4. An example of transforming the original snapshot sequence into a merged snapshot sequence with the size of sliding window $\omega = 2$.

where $\alpha \in (0, 1]$ is the leaking rate, $f(\cdot)$ is an activation function, and $\mathbf{U}^{(g,l)}$ is the input matrix used for receiving the various vertex inputs, i.e.

$$\mathbf{U}_s^{(g,l)}(t) = \begin{cases} \mathbf{v}_s^{(g)}(t) & \text{for } l = 1 \\ \mathbf{X}_s^{(g,l-1)}(t) & \text{for } l > 1 \end{cases}, \tag{4}$$

the element values of $\mathbf{W}_{in} \in \mathbb{R}^{N_R \times N_V}$ are randomly chosen from a uniform distribution with the range of $[-\eta, \eta]$. The element values of $\mathbf{W}_{res}^{(g,l)}$ are randomly assigned from the uniform distribution $[-1, 1]$. In order to ensure the echo state property [19] in each encoder, we keep $\rho\left(\mathbf{W}_{res}^{(g,l)}\right) < 1/\rho\left(\mathbf{A}_s(t)\right)$ at each time step.

3.3　The Linear Classifier

The recognition score of the s-th temporal graph, $\hat{\mathbf{y}}_s \in \mathbb{R}^{N_Y}$, is calculated through a simple linear mapping, which can be formulated as follows:

$$\hat{\mathbf{y}}_s = \mathbf{W}_{out}\mathbf{c}_s + \mathbf{b}, \tag{5}$$

where $\mathbf{b}$ is the bias vector and $\mathbf{c}_s$ is the sum-pooled vector which can be calculated like:

$$\mathbf{c}_s = \left[sp\left(\mathbf{X}_s^{(1,1)}(N_T)\right); sp\left(\mathbf{X}_s^{(1,2)}(N_T)\right); \ldots; sp\left(\mathbf{X}_s^{(N_G,N_L)}(N_T)\right)\right] \in \mathbb{R}^{N_R N_G N_L}, \tag{6}$$

where $[\cdot ; \cdot]$ represents the vertical concatenation and $sp(\cdot)$ acts for the operation of summing N_V column vectors of $\mathbf{X}_s^{(g,l)}(N_T)$ up. The readout matrix $\mathbf{W}_{out} \in \mathbb{R}^{N_Y \times N_R N_G N_L}$ can be calculated as follows:

$$\mathbf{W}_{out} = \mathbf{Y}\mathbf{C}^{\mathrm{T}}\left(\mathbf{C}\mathbf{C}^{\mathrm{T}} + \gamma\mathbf{I}\right)^{-1}, \tag{7}$$

where $\mathbf{C} = [\mathbf{c}_1, \mathbf{c}_2, \ldots, \mathbf{c}_{N_S}] \in \mathbb{R}^{N_R N_G N_L \times N_S}$ is the collected matrix including N_S sum-pooled vectors, $\mathbf{Y} = [\mathbf{y}_1, \mathbf{y}_2, \ldots, \mathbf{y}_{N_S}] \in \mathbb{R}^{N_Y \times N_S}$ is the target matrix, and γ is the regularization parameter. The output for the s-th sample can be determined by the index of the maximum element in $\hat{\mathbf{y}}_s^{(i)}$.

4   The Analysis of the Computational Complexity

We provide an analysis of the computational complexity of training the GDGESN in this section. Since the number of edges in each temporal graph is dynamic, we denote the number of edges for the s-th temporal graph at time t by $E_s(t)$. We define that the sparsity of $\mathbf{W}_{res}$ in each reservoir is $\varphi \in (0, 1]$. The computation in the merged snapshot converter costs $\mathcal{O}\left(\sum_{s=1}^{N_S} \sum_{t=1}^{N_T} E_s(t)\right)$. The computational complexity in each encoder is $\sum_{s=1}^{N_S} \sum_{t=1}^{N_T} \varphi N_R^2 E_s(t)$. The computational complexity of training the linear classifier is $\mathcal{O}\left((N_G N_L N_R)^2 (N_S + N_G N_L N_R)\right)$. It is obvious that the computational complexity of the proposed model in the training phase is mainly determined by the relatively larger part between the cost of running the multiple-reservoir encoder and that of the training decoding module. Therefore, the total computational complexity can be summarized as follows:

$$\max \left(\mathcal{O} \left(N_G N_L \varphi N_R^2 \sum_{s=1}^{N_S} \sum_{t=1}^{N_T} E_s(t) \right), \mathcal{O} \left((N_G N_L N_R)^2 (N_S + N_G N_L N_R) \right) \right).$$
(8)

In this study (see Sect. 5.2), N_G, N_L, and N_R are much smaller than N_S and $\sum_{t=1}^{N_T} E_s(t)$. Therefore, the computational complexity of training the GDGESN can be reduced to $\mathcal{O}\left(\sum_{s=1}^{N_S} \sum_{t=1}^{N_T} E_s(t)\right)$, which is the same with the computational complexity of DynGESN and significantly lower than many kernel-based methods [19].

5   Experiments

5.1   Descriptions of Datasets

Six benchmark spreading process classification datasets released in Ref. [16] were used to evaluate the performances of different models. We present their details in Table 1. For these six datasets, The Susceptible-Infected (SI) epidemic model [15] is used to simulate spreading processes with the corresponding infection probabilities on temporal graphs. Note that there are two categories of infections with probabilities p_1 and p_2 in every dataset, and the spreading pattern corresponding to only one probability (p_1 or p_2) exists in each temporal graph for a dataset. The datasets attached with the suffix '_ct1' indicate that the infection probability of a spreading pattern is $p_1 = 0.5$ or $p_2 = 0.5$ in each temporal graph, and the others show that a spreading pattern with the infection probability $p_1 = 0.2$

or $p_2 = 0.8$ exists in each temporal graph. The goal of the experiment is to test whether a tested model can identify two spreading patterns accurately for each dataset. In this study, we filtered empty adjacency matrices from each temporal graph sequence.

Table 1. Details of six datasets used in the experiment.

	N_S	N_V	N_T	$\sum_{s=1}^{N_S} \sum_{t=1}^{N_T} E^s(t)$
dblp_ct1	755	60	48	835714
dblp_ct2				
highschool_ct1	180	60	205	553013
highschool_ct2				
tumblr_ct1	373	60	91	1039776
tumblr_ct2				

5.2   Tested Models and Experimental Settings

We leverage two categories of models, kernel-based and neural network-based models, for comparison. Kernel-based models can transform a temporal graph into a large-scale static graph and use kernel methods to generate final classification results. A transforming method, the directed line graph expansion (DL) [16], was leveraged to combine with the random walk kernel (RW) [4] and the Weisfeiler-Lehman subtree kernel (WL) [17]. These two combinations are represented by DL-RW and DL-WL, respectively. Since the transformed static graph leads to significantly high computational complexity for these two models [19], a simplified DL-RW method called approximate temporal graph kernel (APPR-$\mathcal{V}$) [16], which can sample k-step random walks starting on only $\mathcal{V}$ vertices of the transformed graph, was used as another tested model. In the experiments, $\mathcal{V}$ was fixed at 250. On the other hand, the prototype of GDGESN, dynamic graph echo state network (DynGESN) [14], is considered as a baseline neural network-based model. Note that the three kernel-based models used supported vector machine [6] rather than the simple linear classifier leveraged by the DynGESN and the proposed GDGESN for generating classification results.

For the proposed GDGESN, the parameter settings are listed in Table 2. We kept values of the spectral radius, the leaking rate, the input scaling, and the regularization factor the same as those of the DynGESN reported in Ref. [19]. We fixed the density of the reservoir connections and the reservoir size to be 1E-3 and 10, respectively. The number of layers and the number of groups were searched in the ranges of $[1, 2, \ldots, 4]$, and $[1, 2, 3]$, respectively. We set the size of the sliding window at $\omega^{(g)} = 2g - 1$ for $g = 1, 2, \ldots N_G$. Note that the target of this study is to show the classification improvement in performances brought about by the merged snapshot strategy in the GDGESN. Therefore we did not

Table 2. Parameter settings of GDGESN.

Parameter	Symbol	Value
Spectral radius	ρ	0.9
Leaking rate	α	0.1
Input scaling	η	1
Density of reservoir connections	φ	1
Regularization factor	γ	1E-3
Reservoir size	N_R	10
Number of layers	N_L	$[1, 2, ..., 4]$
Number of groups	N_G	$[1, 2, 3]$

consider searching the key parameters of encoders and the linear classifier for extreme performances. The computational environment is an Intel (R) Core i9-7900X CPU with 96GB RAM of DDR4 2666MHz.

For the partition of datasets, each dataset was evenly separated into ten parts. We cyclically picked up nine of them for the training set and the rest for the testing set for cross-validation. Based on each partition, we randomly initialized the proposed model 20 times and reported the average performances.

5.3   Evaluation Metrics

The accuracy rate is given by the following evaluations, which can be formulated as follows:

$$\text{Acc} = \frac{\text{The number of correct classified temporal graphs}}{\text{The number of total temporal graphs}} \times 100\%. \qquad (9)$$

5.4   Experimental Results

To investigate the impacts brought about by the merged snapshot converter of the proposed model, we show the average classification performances of the proposed model with different combinations of N_L and N_G on six datasets in Fig. 5. We notice that the GDGESN with $N_G > 1$ outperforms the GDGESN with $N_G = 1$ when varying N_L from one to four on all the SPC datasets except highschool_ct2. Specifically, the classification performances of the GDGESN with $N_G > 1$ obviously surpass those with $N_G = 1$ when $N_L = 1$. These results indicate that multi-timescale spatiotemporal inputs generated by the merged snapshot converter can significantly improve the classification performances of the GDGESN for the tested SPC datasets.

The best average classification performances of the GDGESN and other tested models reported in [19] are listed in Table 3. The best performance of the GDGESN for each dataset is obtained under the best combination of N_P

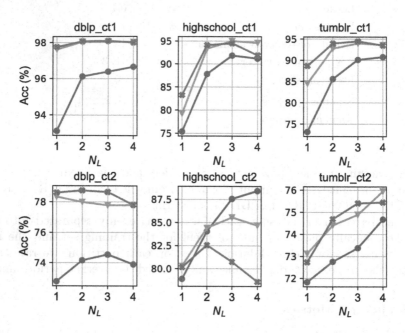

Fig. 5. Average classification performances of the proposed GDGESN with variations of N_L and N_G on six SPC datasets

and N_L shown in Fig. 5. We can see that the GDGESN outperforms the APPR-250 and the DynGESN on dblp_ct1, dblp_ct2, highschool_ct1, and tumblr_ct1. In particular, our model falls behind the DL-RW only on tumblr_ct1. In addition, the GDGESN only has a few inferiorities of classification performances in comparison with the DynGESN on highschool_ct2. Note that the dimension of each vertex embedding for the GDGESN is only 10, whereas that for the DynGESN is 16 [19]. By observing Fig. 5 and Table 3 jointly, we find that our model achieves the highest performances when $N_L < 4$ on all the tested datasets except for highshcool_ct2 and tumblr_ct2, but the DynGESN obtains the best classification performances by setting $N_L = 4$ on all datasets [19].

By observing the performances of kernel-based methods and GDGESN, we can notice that the corresponding performances of GDGESN are still not as good as those of DL-RW and DL-WL in some cases. The DL-RW and the DL-WL can use whole spatiotemporal information of the snapshot sequence, whereas only the sum-pooled vertex embeddings of the last time step are collected for reducing the corresponding training costs in the GDGESN (as shown in Eq. (6)). Since ESP makes the GDGESN has "memory ability" [14], the generated vertex embeddings at the last time step hardly contain spatiotemporal information of the whole snapshot sequences, which could be an important reason why the proposed model underperforms DL-RW and DL-WL.

Table 3. Average classification accuracy rates and standard deviations (%) of the tested models on six datasets.

	DL-RW	DL-WL	APPR-250	DynGESN	GDGESN
dblp_ct1	$98.7 \pm (0.1)$	$98.5 \pm (0.2)$	$97.2 \pm (0.2)$	$97.7 \pm (1.7)$	$98.1 \pm (1.7)$
dblp_ct2	$81.8 \pm (0.9)$	$76.5 \pm (1.0)$	$76.4 \pm (0.9)$	$74.3 \pm (4.7)$	$78.8 \pm (4.2)$
highschool_ct1	$97.4 \pm (0.7)$	$99.2 \pm (0.6)$	$94.0 \pm (1.3)$	$94.4 \pm (5.3)$	$95.0 \pm (5.0)$
highschool_ct2	$93.4 \pm (1.0)$	$89.3 \pm (0.7)$	$90.4 \pm (1.8)$	$92.8 \pm (5.2)$	$88.4 \pm (8.5)$
tumblr_ct1	$95.2 \pm (0.6)$	$94.2 \pm (0.4)$	$92.7 \pm (0.3)$	$93.3 \pm (3.9)$	$94.5 \pm (4.0)$
tumblr_ct2	$77.2 \pm (1.0)$	$78.2 \pm (1.3)$	$78.4 \pm (1.3)$	$76.8 \pm (6.2)$	$77.0 \pm (6.3)$

6 Discussion

We have proposed a new RC-based model for dealing with SPC tasks in this study. The proposed model can transform the original spreading process into various multi-timescale spreading processes and then extract the corresponding spatiotemporal features through fixed group-wise reservoir encoders. These features are decoded into the final classification results by a simple linear classifier. The simulation results show that our proposed model outperforms the DynGESN and even several kernel-based models on some benchmark SPC datasets. In addition, the analysis of computational complexity shows that our model has the same cost as DynGESN in the training process. Based on the above-mentioned contents, we can conclude that the proposed GDGESN can hold relatively high effectiveness and efficiency in dealing with SPC tasks.

It is obvious that the ultimate performances are far from being reached since we only used moderate values of key hyperparameters for multiple reservoirs in our model. We will continue exploring the optimal performances of the GDGESN on various SPC tasks in the future.

Acknowledgements. We thank Bing Wang for valuable comments. This work was partly supported by JST CREST Grant Number JPMJCR19K2, Japan (ZL, FK, GT) and JSPS KAKENHI Grant Numbers 23H03464 (GT), 20H00596 (KF), and Moonshot R&D Grant No. JPMJMS2021(KF).

References

1. Chen, J., Wang, X., Xu, X.: Gc-lstm: graph convolution embedded lstm for dynamic link prediction. arXiv preprint arXiv:1812.04206 (2018)

2. Cho, K., Van Merriënboer, B., Bahdanau, D., Bengio, Y.: On the properties of neural machine translation: encoder-decoder approaches. arXiv preprint arXiv:1409.1259 (2014)
3. Gallicchio, C., Micheli, A., Pedrelli, L.: Deep reservoir computing: a critical experimental analysis. Neurocomputing **268**, 87–99 (2017)
4. Gärtner, T., Flach, P., Wrobel, S.: Graph kernels for chemical informatics. In: Proceedings of the 16th International Conference on Neural Information Processing Systems, pp. 505–512. MIT Press (2003)
5. Guo, S., Lin, Y., Feng, N., Song, C., Wan, H.: Attention based spatial-temporal graph convolutional networks for traffic flow forecasting. In: Proceedings of the AAAI Conference on Artificial Intelligence, vol. 33, pp. 922–929 (2019)
6. Hearst, M.A., Dumais, S.T., Osuna, E., Platt, J., Scholkopf, B.: Support vector machines. IEEE Intell. Syst. Appl. **13**(4), 18–28 (1998)
7. Hochreiter, S., Schmidhuber, J.: Long short-term memory. Neural Comput. **9**(8), 1735–1780 (1997)
8. Jaeger, H.: Tutorial on training recurrent neural networks, covering bppt, rtrl, ekf and the "echo state network" approach (2002)
9. Kipf, T.N., Welling, M.: Semi-supervised classification with graph convolutional networks. arXiv preprint arXiv:1609.02907 (2016)
10. Li, M., Zhu, Z.: Spatial-temporal fusion graph neural networks for traffic flow forecasting. In: Proceedings of the AAAI Conference on Artificial Intelligence, vol. 35, pp. 4189–4196 (2021)
11. Li, Z., Liu, Y., Tanaka, G.: Multi-reservoir echo state networks with hodrick-prescott filter for nonlinear time-series prediction. Appl. Soft Comput., 110021 (2023)
12. Li, Z., Tanaka, G.: Multi-reservoir echo state networks with sequence resampling for nonlinear time-series prediction. Neurocomputing **467**, 115–129 (2022)
13. Lukoševičius, M.: A practical guide to applying echo state networks. In: Montavon, G., Orr, G.B., Müller, K.-R. (eds.) Neural Networks: Tricks of the Trade. LNCS, vol. 7700, pp. 659–686. Springer, Heidelberg (2012). https://doi.org/10.1007/978-3-642-35289-8_36
14. Micheli, A., Tortorella, D.: Discrete-time dynamic graph echo state networks. Neurocomputing **496**, 85–95 (2022)
15. Nowzari, C., Preciado, V.M., Pappas, G.J.: Analysis and control of epidemics: a survey of spreading processes on complex networks. IEEE Control Syst. Mag. **36**(1), 26–46 (2016)
16. Oettershagen, L., Kriege, N.M., Morris, C., Mutzel, P.: Temporal graph kernels for classifying dissemination processes. In: Proceedings of the 2020 SIAM International Conference on Data Mining, pp. 496–504. SIAM (2020)
17. Shervashidze, N., Schweitzer, P., Van Leeuwen, E.J., Mehlhorn, K., Borgwardt, K.M.: Weisfeiler-lehman graph kernels. J. Mach. Learn. Res. **12**(9) (2011)
18. Tanaka, G., et al.: Recent advances in physical reservoir computing: a review. Neural Netw. **115**, 100–123 (2019)
19. Tortorella, D., Micheli, A.: Dynamic graph echo state networks. arXiv preprint arXiv:2110.08565 (2021)
20. Vaswani, A., et al.: Attention is all you need. In: Advances in Neural Information Processing Systems 30 (2017)

Trajectory Prediction with Contrastive Pre-training and Social Rank Fine-Tuning

Chenyou Fan[1], Haiqi Jiang[1], Aimin Huang[1,2], and Junjie Hu[2(✉)]

[1] South China Normal University, Guangdong, China
fanchenyou@scnu.edu.cn
[2] Shenzhen Institute of Artificial Intelligence and Robotics for Society (AIRs),
Shenzhen, Guangdong, China
hujunjie@cuhk.edu.cn

Abstract. This paper focuses on the accurate prediction of pedestrian trajectories in scenarios where individuals walk alone or in social groups, and sometimes alter their paths to avoid collisions. While previous work has improved backbone neural networks to model individual motion patterns, few studies have explicitly addressed the consistency of internal motion patterns or properness of external interactions. To address this, we propose a unified framework consisting of a Contrastive History-Prediction (CHIP) module and a Differentiable Social Interaction Ranking (DSIR) module. The CHIP module utilizes unsupervised contrastive loss to optimize predicted motion patterns consistent with observations, while the supervised DSIR module ensures predicted interactions are compatible with realistic positions. Our analysis and numerical studies demonstrate the effectiveness of our approach, which achieves a 5–10% improvement in positional accuracy and a 3–7% boost in interactive properness. We provide comprehensive visualizations of anticipated trajectories with temporal interactive scores across various scenarios.

Keywords: Trajectory Prediction · Contrastive Learning · Social Interaction

1 Introduction

Predicting the future trajectories of autonomous vehicles is a critical task for safe navigation in dense urban traffic. Recent studies have made substantial progress in developing advanced deep-learning (DL) models to explore human movement patterns, such as using LSTMs [1], GANs [9,20], Transformers [30], and GCNs [17,23]. To model human interactions, these studies have proposed to aggregate neighbors' features with pooling [1,15,31], weighted averaging [17,23], or multi-head attention [30].

Supplementary Information The online version contains supplementary material available at https://doi.org/10.1007/978-981-99-8141-0_40.

In this study, instead of proposing new DL architectures, we focus on two fundamental aspects of trajectory prediction: the consistency of human movement patterns and the properness of human interactions. To this end, we propose a unified, model-agnostic training procedure to explicitly optimize motion consistency and quantify interaction properness. This approach allows us to better explain the interactive mechanisms implied by existing sophisticated DL models and to quantify the properness of the predicted social interactions.

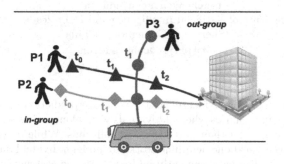

Fig. 1. Demo of pedestrian trajectory prediction. In a same social group, P1 and P2 keep close along the path. P3 is out-group w.r.t. P1 and P2, approaching at some future steps.

Human behaviors have been observed to exhibit stability and predictability, as people often follow consistent patterns in their movements. To investigate this phenomenon, this study focuses on examining the *internal consistency* of human behaviors. Our approach involves optimizing the similarity between observed motion patterns and predicted future patterns for each individual. This is achieved through contrastive learning, which associates a person's historical pattern with their corresponding future pattern. The proposed learning method, named *Contrastive HIstory-Prediction* (CHIP), aims to unsupervisedly ensure that predicted trajectories align with previously observed patterns. To accomplish this, motion embeddings are extracted from historical and future time steps for all individuals in the scene.

Humans are innately social beings who adjust their actions to facilitate appropriate social interactions. Our study identifies two significant types of pedestrian interactions, as in Fig. 1. The first type is called *in-group*, where individuals from the same social group tend to walk together and maintain a close proximity. The second type is *out-group*, where individuals walk separately but anticipate potential collisions in the near future. Consequently, they modify their trajectories using complex dynamics to maintain a comfortable social distance. Thus, the second crucial aspect of this research is to explicitly model the *external social properness* that dictate human behaviors.

In order to capture the *external social properness* of pedestrian behavior, we utilize a pairwise potential energy calculation based on the relative distance between individuals over time. We then create a spatial-temporal ranking of these potentials for all person pairs, reflecting the varying intensities of interactions. This ranking is demonstrated in Fig. 1, where at step t_1, the ranking of $(P1, P3)$ is higher than that of $(P2, P3)$ as they approach, with the order reversing at t_2. During training, the actual ranking is obtained from the ground-truth trajectories. The discrepancy between the predicted and actual ranking is used to determine the properness of predicted interactions, informing the design of a ranking loss which is optimized for the model end-to-end. This process is called Differential Social Interaction Ranking (DSIR), which aims to accurately capture the progression of social potentials in a supervised manner.

In summary, we propose explicitly modeling both *internal* and *external* factors of human behaviors as multi-task learning objectives. Notably, our CHIP and DSIR modules are both model-agnostic and parameter-free. We will demonstrate they can integrate into existing backbones seamlessly and improve prediction accuracy steadily.

In summary, the main contributions of our work include:

- We propose to learn internal movement and external interactive patterns to depict human behaviors in dense traffic;
- We apply unsupervised contrastive learning process to associate observed movements with predicted trajectories for pre-training;
- We design a ranking scheme to describe the dynamic interactions with pairwise potentials based on pedestrians' trajectories, and integrate into model optimization in an end-to-end manner;
- Our approach significantly outperforms existing methods by 5–10% in positional accuracy and 3–7% in interactive properness.

2 Related Work

Trajectory Prediction. In Autonomous Driving (AD) technical stack, trajectory prediction is an important perception task which aims to track mobile agents such as pedestrians and vehicles. Recent approaches commonly use RNNs [1,9,13,27], GANs [9,20] or GNNs [17,23] to encode the history and decode to future trajectories. Recent AD studies [5,7,32] also utilize additional high-definition maps to refine the generation of future coordinates.

Social Interaction. Social-LSTM [1] modeled the interactions by pooling neighboring agents' features. Social-Attention [27] utilized the attention mechanism [26] to model the importance of interactions. PeekFuture [15] additionally modeled the person-scene and person-object interaction with visual contexts. STGCNN [17] built a spatio-temporal graph of the scene with edge weights as the relative distance. SGCN [23] further imposed sparsity constrains on interactions and prune non-influential ones. M2I [25] classified agent relations with heuristics. However, they either implicitly learned the interactions or assumed fixed relations without considering the dynamics. We will consider the dynamic interactions by ranking their potentials temporally. Group detection was extensively studied [16,21,24] as a supervised classification task. However, these approaches become inadequate when handling datasets that lack group labels.

Contrastive Learning with Different Modalities. CLIP [19] model shows contrastive learning effective in large-scale visual concept pre-training. Extended tasks of contrastive learning include object detection [22], text-image retrieval [4], and text-image segmentation [28], etc. In this study, we build a movement pattern embedding space in which the distance of historical and future patterns of a same person is minimized. We formulate our Contrastive History-Future learning as unsupervised pre-training.

3   Our Approach

We begin by outlining the notations and definitions associated with trajectory prediction tasks, followed by our approach description.

Notations. Let T_h be number of historical steps, and T_f be subsequent future steps. For a scene with N persons, their 2-D coordinates are denoted as $\boldsymbol{X}^h$ in history and $\boldsymbol{X}^f$ in future, respectively, as:

$$\begin{aligned} \boldsymbol{X}^h &= \{\boldsymbol{X}_i^h\}_{i=1}^N \ , \ \ s.t. \ \boldsymbol{X}_i^h = \{(x_i^t, y_i^t)\}_{t=1}^{T_h} \ ; \\ \boldsymbol{X}^f &= \{\boldsymbol{X}_i^f\}_{i=1}^N \ , \ \ s.t. \ \boldsymbol{X}_i^f = \{(x_i^t, y_i^t)\}_{t=T_h+1}^{T_h+T_f} \ . \end{aligned} \tag{1}$$

Let $\boldsymbol{G}$ denote a trajectory prediction model. $\boldsymbol{G}$ takes the N-person coordinates $\boldsymbol{X}^h$ as global context of the history and predicts the bi-Gaussian positional parameters (e.g., mean and variance of the XY-coordinates) for T_f future steps such as

$$\begin{aligned} \boldsymbol{Z} &= \{\boldsymbol{z}_i \in \mathcal{R}^{T_f \times 5}\}_{i=1}^N \ , \\ \text{with} \ \ \boldsymbol{z}_i &= \{(\mu_{x,i}^t, \mu_{y,i}^t, \sigma_{x,i}^t, \sigma_{y,i}^t, \rho_i^t)\}_{t=T_h+1}^{T_h+T_f} \ . \end{aligned} \tag{2}$$

For each person i of total N persons in the scene, we extract its D-dim historical feature $\boldsymbol{h}_i$ and its decoded future feature $\boldsymbol{f}_i$ as motion embeddings as:

$$\boldsymbol{H} = \{\boldsymbol{h}_i \in \mathcal{R}^D\}_{i=1}^N \ , \quad \boldsymbol{F} = \{\boldsymbol{f}_j \in \mathcal{R}^D\}_{j=1}^N \ . \tag{3}$$

The collective outputs of $\boldsymbol{G}$ by Eq. 2 and 3 include the parameterized future predictions $\boldsymbol{Z}$, historical features $\boldsymbol{H}$ and future features $\boldsymbol{F}$, as:

$$\boldsymbol{Z}, \boldsymbol{H}, \boldsymbol{F} \leftarrow \boldsymbol{G}(\boldsymbol{X}^h) \ . \tag{4}$$

Depending on the backbone chosen for $\boldsymbol{G}$, $\boldsymbol{H}$ and $\boldsymbol{F}$ can be adaptively collected from the last (or pooled) hidden output of an LSTM, GCN, or Transformer.

3.1   Contrastive History-Prediction Learning

We introduce our Contrastive HIstory-Prediction (CHIP) learning to ensure the internal motion consistency of human behaviours. With historical motion embedding $\boldsymbol{H}$ and future embedding $\boldsymbol{F}$ of Eq. (3), we compute the dot-product for each (i, j) person pair such as $\boldsymbol{Q} = \{q_{ij} = \boldsymbol{h}_i \cdot \boldsymbol{f}_j\}_{i,j=1}^N$.

The concept of internal motion consistency dictates that the expected movement sequence of an individual, denoted as person i, ought to bear greater resemblance to the actual observed pattern compared to the rest of the individuals within the environment.

Consequently, every value along the diagonal of the matrix, represented by q_{ii}, must possess a higher magnitude compared to other entries in the corresponding row and column. Formally, for $i = 1, ..., N$, we have

$$q_{i,i} > q_{i,j} \land q_{i,i} > q_{k,i}, \quad \forall j, k \neq i . \tag{5}$$

To impose above constraints, we formulate our CHIP learning objective as an auxiliary classification task such as

$$L^{chip}(\boldsymbol{Q}) = -\frac{1}{2N} \sum_{i=1}^{N} \left(\log \frac{e^{q_{ii}}}{\sum_{j=1}^{N} e^{q_{ij}}} + \log \frac{e^{q_{ii}}}{\sum_{k=1}^{N} e^{q_{ki}}} \right) , \tag{6}$$

in which q_{ii} gets maximized as a logit. As $\boldsymbol{Q}$ depends on feature embeddings from model outputs, we can optimize the model by minimizing L^{chip} with standard SGD.

Notably, CHIP learning is unsupervised and can serve as a multi-task objective in model training. We describe the details in Sect. 3.4.

3.2   Differentiable Social Interaction Ranking

This study delves into explicitly modeling the properness of social interactions. To achieve this goal, we propose a novel Differentiable Social Interaction Ranking (DSIR) module, which enables the supervised optimization of predicted interactions among all participants to align with their actual positions.

Observations. We highlight two important types of person interactions based on analysis of realistic data. Specifically, Fig. 2a illustrates the *in-group* interaction characterized by individuals walking in close proximity (i.e., less than 1 m) as a same social group. Conversely, Fig. 2b captures the *out-group* interaction where two individuals walk independently but take measures to avoid collision by dynamically adjusting their walking paths while maintaining a comfortable social distance. This finding offers insight into social dynamics and highlights the importance of considering various forms of interpersonal interactions in modeling pedestrian behavior.

We first define the *interaction intensity* of a paired persons (i, j) with the classical Gaussian potential function [2,6,29], depending on their relative distance d_{ij} as

$$\psi(i, j) = \exp\left(-\frac{d_{ij}^2}{2\sigma^2} \right) \in (0, 1], \ \forall i \neq j , \tag{7}$$

in which σ is a constant social distance, d_{ij} is computed based on their positions, and ψ is symmetric ($\psi(i, j) = \psi(j, i)$). We omit $\psi(i, i)$ as we care about interactions with different persons. Thus, in a scene with $N \geq 0$ persons, we have $M = N(N - 1)/2$ unique potentials to consider.

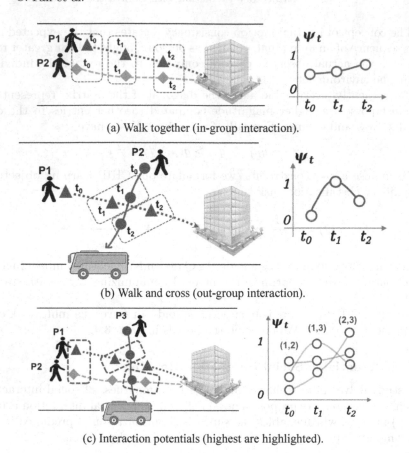

(a) Walk together (in-group interaction).

(b) Walk and cross (out-group interaction).

(c) Interaction potentials (highest are highlighted).

Fig. 2. (a) In-group interaction with high potentials. (b) Out-group interactions with varied potentials at different steps. (c) Ranking of pairwise potentials. The interactions with highest potentials over the time are highlighted in dashed red box. (Color figure online)

Social Descriptors. We propose to describe social interactions in a global view by pairwise potentials of pedestrians depending on their relative distances. The potential ranking over each step provides rich information about the temporal dynamics of human interactions. Figure 2c shows a 3-person scene with complex interactions, i.e., P1 and P2 are in-group while P3 is out-group w.r.t to P1 and P2. Moreover, P3 is expected to interact with P1 and P2 by crossing their paths at subsequent steps. Thus, P1 and P2 need to plan their routes to maintain their intimacy while also keeping a polite distance from P3. As reflected in their ranks, $\psi(1,2)$ remains consistently high across all steps, while $\psi(1,3)$ and $\psi(2,3)$ reach their peaks at t_1 and t_2, respectively, before dropping off in later steps.

To further examine the complexity and appropriateness of human interactions, we propose to learn the predicted ranking of pairwise potentials using trajectory predictions and compare them with the actual order based on the ground truth data.

Task Formulation. Given predicted pairwise potentials, we now formulate the task of ranking them increasingly as an optimization task with a differentiable solution. Thus we can integrate it into our learning procedure.

Let $\psi = [\psi_1, ..., \psi_M]^\top \in \mathcal{R}^M$ be a list of M *unique* pairwise potentials in column form. Let $\mathcal{M} = \{1, 2, ..., M\}$ and $\mathbf{1}_M$ be an all-ones vector of dimension M.

We define an M-element *index array* $\boldsymbol{y}$ as

$$\boldsymbol{y} = [y_j = \frac{j}{M}]_{j=1}^M = \frac{1}{M}[1, \cdots, j, \cdots, M]^\top , \tag{8}$$

in which $y_j \in (0, 1]$ as similar as potential ψ, making the following sorting operations numerically stable.

Let $\boldsymbol{P} = \{p_{ij}\}_{i,j=1}^M$ be an $M \times M$ permutation matrix. $\boldsymbol{P}$ is binary in which each row or column only has one element of 1, otherwise 0. The 1-element p_{ij} ranks j-th element to i-th rank. A sorting permutation matrix $\boldsymbol{P}^*$ permutes ψ in increasing order as $\psi^{\boldsymbol{P}^*}$ as:

$$\psi^{\boldsymbol{P}^*} = \boldsymbol{P}^*\psi = [\psi_1^{\boldsymbol{P}^*}, ..., \psi_M^{\boldsymbol{P}^*}]^\top, \ \forall i < j, \ \psi_i^{\boldsymbol{P}^*} < \psi_j^{\boldsymbol{P}^*}. \tag{9}$$

Sorting process is usually non-differentiable which requires comparing and swapping elements, e.g., QuickSort. We propose to formulate our potential ranking task as a differentiable learning objective and optimize it iteratively.

We first prove that a proper sorting permutation can be obtained by minimizing the following cost function.

Lemma 1. *Let $\boldsymbol{y} = [i/M]_{i=1}^M$ be the index array. Given a vector ψ of M unique elements, the sorting permutation P^* is the unique solution (out of all permutations) which minimizes the following cost*

$$L(\psi^P | \boldsymbol{y}) = \sum_{i=1}^M (y_i - \psi_i^P)^2 . \tag{10}$$

We can perform proof by contradiction by supposing that there exists some non-sorting permutation P' ($P' \neq P^*$) which also minimizes Eq. (10). Then we can show if we further swap the non-sorted pairs we can further lower the cost.

Lemma 1 shows that sorting can be formulated as finding the optimal permutation that minimizes Eq. (10). Based on this, we construct a cost matrix $\boldsymbol{C}_{\psi y}$ as

$$\boldsymbol{C}_{\psi y} = \{c_{ij} = (y_j - \psi_i)^2\}_{i,j=1}^M \in \mathcal{R}^{M \times M}. \tag{11}$$

We formulate sorting operation as a relaxed integer programming as follows:

$$\hat{\boldsymbol{P}}^* = \arg\min \ \langle \boldsymbol{P}, \boldsymbol{C}_{\psi y} \rangle - \lambda H(\boldsymbol{P}) ,$$

$$s.t. \ \&\boldsymbol{P} \geq 0 , \quad \boldsymbol{P}\mathbf{1}_M = \mathbf{1}_M , \quad \boldsymbol{P}^\top \mathbf{1}_M = \mathbf{1}_M ,$$

$$\tag{12}$$

in which $H(P) = -\sum_{i,j} P_{i,j} \log P_{i,j}$ is the entropy term.

The constraints of Problem (12) only limit each row and column of P sums to one and be positive, relaxing the requirement of a binary permutation matrix. This allows soft assignment of ranks, e.g., P_{ij} is interpreted as the weight of assigning element ψ_j to i-th rank.

The solution $\hat{P}^*$ of Problem (12) can be solved in iterative and differentiable way [3].

Lemma 2. *For an $M \times M$ cost matrix C, solving Problem (12) is strictly convex such that there exists a unique minimizer P^* which has the form of $P^* = XAY$, where $A = exp(-\lambda C)$ while $X, Y \in \mathcal{R}_+^{M \times M}$ are both non-negative diagonal matrices which are unique up to a multiplicative factor [3], which can be efficiently solved with the differentiable Sinkhorn algorithm [3].*

Pairwise Ranking Loss. We can estimate the ranks of pairwise potentials with Task 12 as $\hat{R}(\psi) = \{\hat{r}_i\}_{i=1}^M = M \cdot \hat{P}^{*\top} y$. In training stage, we can obtain the actual pairwise potentials ϕ based on true positions X^f, and sort to get their actual ranks $R(\phi) = \{r_i\}_{i=1}^M$ as ground truths.

By comparing the predictions with the ground truths, we develop the social ranking loss L^{dsir} to penalize inconsistent pairwise orders in a supervised manner such that

$$L^{dsir}(\hat{R}|R) = \frac{1}{M^2} \sum_{i=1}^M \sum_{j=1}^M \max(0, -(r_i - r_j)(\hat{r}_i - \hat{r}_j)) . \quad (13)$$

3.3 Bi-Gaussian Regression Loss

We follow previous studies [1,8,9,23] to optimize the predicted trajectories with the bi-Gaussian distribution loss. Let the true future positions be $X^f = (x, y)$ and the parameterized model predictions be $Z = (\mu_x, \mu_y, \sigma_x, \sigma_y, \rho)$ as in Eq. (2). We omit subscript i and t for simplicity. The bi-Gaussian distribution loss follows

$$L^{biG}(Z|X^f) = \frac{1}{2\pi\sigma_x\sigma_y\sqrt{1-\rho^2}} \exp \left(\frac{-1}{2(1-\rho^2)} \right.$$
$$\left. \left[\frac{(x-\mu_x)^2}{2\sigma_x^2} + \frac{(y-\mu_y)^2}{2\sigma_y^2} - \frac{2\rho(x-\mu_x)(y-\mu_y)}{\sigma_x\sigma_y} \right] \right) , \quad (14)$$

which penalizes deviations of predicted (μ_x, μ_y) from the ground-truth (x, y) as well as large variances.

3.4 Two-Stage Multi-task Training Objective

In summary, we can optimize the model by jointly minimizing the bi-Gaussian regression loss in Eq. (14), CHIP learning loss in Eq. (6) and DSIR loss in Eq. (13) as

$$L^{final} = L^{biG} + \alpha_1 L^{chip} + \alpha_2 L^{dsir} , \quad (15)$$

in which α_1, α_2 are scaling factors. Specially, we propose a two-stage best practice of end-to-end model training with the multi-task objective L^{final} with standard SGD.

In Stage-1 *(Pre-training)*, we minimize $L^{pre} = L^{biG} + \alpha_1 L^{chip}$ for fast convergence, i.e., omitting DSIR loss. In Stage-2 *(Fine-tuning)*, we use the full L^{final} in Eq. (15) for fine-tuning trajectory predictions with interaction-aware DSIR loss.

4 Evaluation Metrics

We describe two standard metrics for trajectory prediction, then propose our novel *Intimacy-Politeness Score* to fully evaluate the properness of social interactions.

4.1 Standard ADE and FDE

ADE and FDE are two standard error metrics which measure the deviations from predicted positions to the ground truths. Let $(x_{i,t}, y_{i,t})$ be real position of person i at step t, and $(\hat{x}_{i,t}, \hat{y}_{i,t})$ be the predicted position. Their $L2$-distance is defined as $e_i^t = \sqrt{(\hat{x}_i^t - x_i^t)^2 + (\hat{y}_i^t - y_i^t)^2}$. The *Average Displacement Error* (ADE) [18] calculates the $L2$-distance between predicted future trajectory and ground truth, averaged over all future steps and all N persons in the scene as $\frac{1}{N \cdot T_f} \sum_{i=1}^{N} \sum_{t=T_h+1}^{T_h+T_f} e_i^t$. The *Final Displacement Error* (FDE) [1] computes the $L2$-distance between the predicted position and actual position at the final step as $\frac{1}{N} \sum_{i=1}^{N} e_i^{T_h+T_f}$.

4.2 Surrogate Social Distance Accuracy (SDA)

Public datasets often do not contain labels for social groups, as annotating social interactions can be a time-consuming process.

To address this issue, we suggest using a Social Distance Accuracy (SDA) scoring function to evaluate the quality of social interactions in a weakly-supervised manner. The SDA method uses the pedestrians' predicted distances and compares them with weakly annotated group labels that are based on ground-truth distances. Our research demonstrates that SDA serves as a reliable surrogate measure for assessing the accuracy of trajectory predictions.

Let σ be the minimal distance of social politeness. Let $d_{i,j,t}$ be the actual relative distance between individuals i and j at a given time t, and let $\hat{d}_{i,j,t}$ be the predicted distance. We construct an adaptive in-group distance upper-bound $d_{i,j,t}^+$ and out-group distance lower-bound $d_{i,j,t}^-$, based on $d_{i,j,t}$ and σ. These two bounds define the acceptable social distance range in unsupervised manner without realistic group labels.

We choose a social distance threshold σ as the minimal distance of social politeness. Then we can establish the in-group person triplet set $\mathcal{D}^{In}$ and out-group person triplet set $\mathcal{D}^{Out}$ in unsupervised manner as follows:

$$\mathcal{D}^{In} = \left\{ (i,j,t) \middle| \bar{d}_{i,j,t} \leq \sigma \wedge d_{i,j,t} \leq \sigma \right\} ,$$
$$\mathcal{D}^{Out} = \left\{ (i,j,t) \middle| \bar{d}_{i,j,t} > \sigma \wedge d_{i,j,t} > \sigma \right\} . \tag{16}$$

We construct an adaptive range $[d_{i,j,t}^-, d_{i,j,t}^+]$ based on the actual distance $d_{i,j,t}$ with $\tau \in (0,1)$ such that

$$d_{i,j,t}^+ = (1+\tau) \cdot d_{i,j,t} \text{ and } d_{i,j,t}^- = (1-\tau) \cdot d_{i,j,t} . \tag{17}$$

Now we design the hinge score functions s^I and s^P to measure intimacy and politeness separately, as follows:

$$s_{i,j,t}^I = trunc\left(\frac{d_{i,j,t}^+ - \hat{d}_{i,j,t}}{d_{i,j,t}^+ - d_{i,j,t}}, 0, 1 \right), \ \forall (i,j,t) \in \mathcal{D}^{In} ,$$
$$s_{i,j,t}^P = trunc\left(\frac{\hat{d}_{i,j,t} - d_{i,j,t}^-}{d_{i,j,t} - d_{i,j,t}^-}, 0, 1 \right), \ \forall (i,j,t) \in \mathcal{D}^{Out} , \tag{18}$$

in which $trunc(\cdot, 0, 1)$ clips the value within $[0,1]$. Concretely, s^I rewards a predicted $\hat{d}_{i,j,t}$ to be smaller than $d_{i,j,t}^+$ for in-group triplets in $\mathcal{D}^{In}$, while s^P rewards $\hat{d}_{i,j,t}$ to be larger than $d_{i,j,t}^-$ for out-group triplets in $\mathcal{D}^{Out}$.

Based on the social range, we design a hinge score functions s^I and s^P to measure intimacy and politeness separately, as Eq. 16 shows. In summary, S^I lessens when over-estimating in-group distance d_1, while S^P lessens when under-estimating out-group distance d_2. We use the combined S^I and S^P as the Social Distance Accuracy (SDA).

5   Experiments

We first introduce datasets and benchmark models of existing works. Then we report the performance and carry out ablation studies to show the effectiveness of our methods.

Datasets. We use two widely compared public pedestrian trajectory datasets, i.e., ETH [18] and UCY [14] to evaluate our methods. In particular, ETH dataset contains the ETH and HOTEL scenes, while the UCY dataset contains the UNIV, ZARA1, and ZARA2 scenes. Each data sequence contains observed trajectories extracted from 8 frames (3.2 s) and future trajectories in the next 12 frames (4.8 s). The train/val/test splits are given. Following a standard testing procedure [1, 17], we generate 20 random samples from the predicted distribution for each testing trajectory, then we calculate the minimum ADE and FDE from the predictions to the ground truth. We also calculate the maximum SDA from all samples as our interactive metric value.

Table 1. Min ADE and FDE results on the benchmark ETH and UCY datasets.

Model	Architecture	ETH	HOTEL	UNIV	ZARA1	ZARA2	AVG
Vanilla-LSTM [1]	LSTM	1.09/2.41	0.86/1.91	0.61/1.31	0.41/0.88	0.52/1.11	0.70/1.52
Social-LSTM [1]	LSTM-Pool	1.09/2.35	0.79/1.76	0.67/1.40	0.47/1.00	0.56/1.17	0.72/1.54
Social-GAN [9]	GAN-Pool	0.87/1.62	0.67/1.37	0.76/1.52	0.35/0.68	0.42/0.84	0.61/1.21
Sophie [20]	GAN-Att	0.70/1.43	0.76/1.67	0.54/1.24	0.30/0.63	0.38/0.78	0.51/1.15
Social-BiGAT [11]	GAN-Att	0.68/1.29	0.68/1.40	0.57/1.29	0.29/0.60	0.37/0.75	0.52/1.07
STGCNN [17]	GCN	0.64/1.11	0.49/0.85	0.44/0.79	0.34/0.53	0.30/0.48	0.44/0.75
SGCN [23]	GCN-Att	0.63/1.03	0.32/0.55	0.37/0.70	0.29/0.53	0.25/0.45	0.37/0.65
SGCN+CHIP	Ours	0.59/0.92	0.29/0.51	0.38/0.72	0.28/0.49	0.25/0.45	0.36/0.62
SGCN+CHIP+DSIR	Ours	0.55/0.86	0.28/0.44	0.37/0.69	0.27/0.46	0.23/0.42	**0.34/0.58**

5.1 Our Approaches and Baselines

We construct our learning models based on the state-of-the-art SGCN [23]. **SGCN-CHIP** adds our proposed CHIP learning module to SGCN as a multi-task objective for model regularization. **SGCN-CHIP-DSIR** utilizes both CHIP and DSIR modules to boost training in a two-stage manner as in Sect. 3.4.

We compare our method with various existing methods such as Vanilla LSTM [10], Social-LSTM [1], Social-GAN [9], Sophie [20], Social-BiGAT [12], STGCNN [17] and SGCN [23].

5.2 ADE and FDE Results

We show results evaluated with ADE and FDE (lower is better) in Table 1 and observe the following trends.

- The best performing SGCN-CHIP-DSIR leads SGCN by 8.1% in ADE (0.34 vs. 0.37) and 10.8% (0.58 vs. 0.65) in FDE relatively, establishing new benchmarks for pedestrian trajectory prediction.
- With only CHIP (no DSIR), our SGCN-CHIP still outperforms SGCN by 2.7% in ADE and 4.6% in FDE, showing the effectiveness of contrastive learning for regularizing the model training.
- By additionally performing DSIR, the SGCN-CHIP-DSIR outperforms SGCN-CHIP by 5.6% in ADE (0.34 vs. 0.36) and 6.5% (0.58 vs. 0.62) in FDE, relatively. This shows the effectiveness of potential ranking to explore interactive patterns.

Table 2. SDA results. The higher is the better.

Model	ETH	HOTEL	UNIV	ZARA1	ZARA2	AVG
Van.-LSTM [1]	0.456	0.505	0.536	0.460	0.454	0.482
Soc.-LSTM [1]	0.463	0.497	0.544	0.461	0.450	0.483
Soc.-GAN [9]	0.522	0.542	0.581	0.606	0.640	0.578
Sophie [20]	0.692	0.683	0.740	0.735	0.825	0.735
STGCNN [17]	0.732	0.851	0.679	0.875	0.789	0.785
SGCN [23]	0.783	0.848	0.712	0.873	0.831	0.809
SGCN-CHIP	0.816	0.852	**0.723**	0.886	0.854	0.826
+DSIR	**0.832**	**0.854**	0.721	**0.891**	**0.870**	**0.834**

5.3 SDA Results

We show results of SDA (higher is better) in Table 2 and observe the following
trends. The best performing SGCN-CHIP-DSIR outperforms original SGCN by
3.1% in SDA (0.834 v.s. 0.809). Without DSIR, our SGCN-CHIP still leads
original SGCN 2.1% over original SGCN (0.826 v.s. 0.809). STGCNN and SGCN
have higher SDA with GCN backbones. Social-LSTM/-GAN have low SDA due
to the ineffective social-pooling. Vanilla-LSTM has lowest SDA as it ignores
interaction learning.

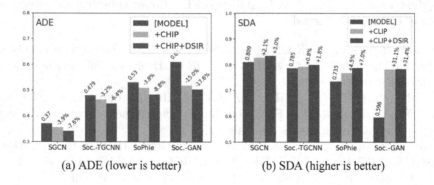

(a) ADE (lower is better)   (b) SDA (higher is better)

Fig. 3. Ablation studies of model compatibility.

5.4 Ablation Study of Model Adaptivity

We study whether our CHIP and DSIR learning modules can adapt to existing
models and boost performance boost upon their original designs. We experiment with several recent studies: Social-GAN [9], Sophie [20], STGCNN [17]

and SGCN [23]. They cover most existing backbones and social learning methods, including LSTM-GAN-Pool, LSTM-GAN-Attention, Transformer, GCN, and GCN-Attention, respectively, as shown in Table 1.

For each existing model, we build two variants as {MODEL}-CHIP and {MODEL}-CHIP-DSIR with one or both our proposed modules respectively. We show the performance boost in ADE and SDA in Fig. 3.

Figure 3a shows that for each model except Social-GAN, our CHIP module lowers ADE by about 3–4%, and CHIP+DSIR together lower ADE by about 6–9%, relative to the original model. Our modules could improve Social-GAN by a substantial 15–18%, indicating the importance of regularizing the path generation process with GANs.

Figure 3b shows that for each model except Social-GAN, our CHIP module can improve SDA about 0.8–4.5%, and CHIP+DSIR together can improve SDA about 1.8–7%. The improvement on Social-GAN as large as 31%.

Compare Fig. 3a and Fig. 3b, we found that SDA and ADE are generally consistent over most methods. This indicates that SDA serves as a trustful evaluation metric for unsupervised social interaction.

In conclusion, our model-agnostic learning modules can integrate into existing models and possibly future state-of-the-art models to achieve performance boosts readily.

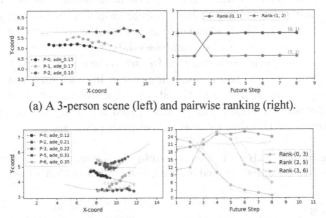

(a) A 3-person scene (left) and pairwise ranking (right).

(b) A multi-group complex scene (left) and pairwise ranking (right).

Fig. 4. Visualize scene-level potential ranking.

5.5   Visualizations of Ranking

In Fig. 4a, we show a 3-person scene on the left and display their potential ranking over 8 future steps on the right. We have excluded pair P-$(0,2)$ due to their significant distance. During steps 1 and 2, Rank-$(1,2)$ is predominant since P-1 (yellow) and P-2 (green) are in closer proximity with higher potential. However, in step 3, P-2 moves away while P-1 continues to follow P-0 (blue) at a closer distance, which leads to a higher rank for P-$(0,1)$. Figure 4b illustrates a scene comprised of five individuals, as well as the potential ranking of three pairs. On the left-hand side, the actual paths of each person are visually represented, while the right-hand side displays the corresponding potential rankings. The shaded lines in both figures show the actual ranking. As P-0 and P-3 move away from each other, the Rank-$(0,3)$ drops over time. On the other hand, Rank-$(2,5)$ remains at the top as P-2 and P-5 approach one another. Rank-$(3,6)$ peaks at step-4, when P-3 and P-6 stand at the closest point before parting ways. Ultimately, our predictions align with the ground truths, demonstrating their accuracy.

6   Conclusion

This study introduces a framework that is model-agnostic and utilizes contrastive learning to ensure motion consistency and potential ranking for tracking interactions. Additionally, a novel metric has been developed to accurately quantify the interactive properness. The performance of our framework surpasses baselines by a significant margin, and can be seamlessly integrated into existing prediction models to enhance their performance. In future work, we aim to extend our methods to dense vehicle traffic scenarios where interactions between cars are constant, but less random due to the stricter constraints of traffic rules. Ultimately, the proposed methods can be integrated into self-driving technology as a crucial module for collision avoidance and path planning.

Acknowledgments. This work is supported by the National Natural Science Foundation of China, Project 62106156, and Starting Fund of South China Normal University.

References

1. Alahi, A., Goel, K., Ramanathan, V., Robicquet, A., Fei-Fei, L., Savarese, S.: Social LSTM: human trajectory prediction in crowded spaces. In: CVPR (2016)
2. Boykov, Y., Veksler, O., Zabih, R.: Markov random fields with efficient approximations. In: CVPR (1998)
3. Cuturi, M.: Sinkhorn Distances: lightspeed computation of optimal transport. In: NeurIPS, pp. 2292–2300 (2013)
4. Dzabraev, M., Kalashnikov, M., Komkov, S., Petiushko, A.: MDMMT: multidomain multimodal transformer for video retrieval. In: CVPRW (2021)
5. Fang, L., Jiang, Q., Shi, J., Zhou, B.: TPNet: trajectory proposal network for motion prediction. In: 2020 IEEE/CVF Conference on Computer Vision and Pattern Recognition (CVPR) (2020)

6. Fathi, A., Hodgins, J.K., Rehg, J.M.: Social interactions: a first-person perspective. In: CVPR (2012)
7. Gao, J., Sun, C., Zhao, H., Shen, Y., Anguelov, D., Li, C., Schmid, C.: VectorNet: Encoding HD maps and agent dynamics from vectorized representation. In: CVPR (2020)
8. Graves, A.: Generating sequences with recurrent neural networks. ArXiv abs/1308.0850 (2013)
9. Gupta, A., Johnson, J., Fei-Fei, L., Savarese, S., Alahi, A.: Social GAN: socially acceptable trajectories with generative adversarial networks. In: CVPR (2018)
10. Hochreiter, S., Schmidhuber, J.: Long short-term memory. Neural Comput. **9**, 1735–80 (1997)
11. Kosaraju, V., et al.: Social-BiGAT: multimodal trajectory forecasting using bicycle-GAN and graph attention networks. In: NeurIPS (2019)
12. Kosaraju, V., Sadeghian, A., Martín-Martín, R., Reid, I.D., Rezatofighi, H., Savarese, S.: Social-BiGAT: multimodal trajectory forecasting using bicycle-GAN and graph attention networks. In: NIPS (2019)
13. Lee, N., Choi, W., Vernaza, P., Choy, C.B., Torr, P.H.S., Chandraker, M.: Desire: distant future prediction in dynamic scenes with interacting agents. In: CVPR (2017)
14. Lerner, A., Chrysanthou, Y., Lischinski, D.: Crowds by example. Comput. Graph. Forum (2007)
15. Liang, J., Jiang, L., Niebles, J.C., Hauptmann, A.G., Fei-Fei, L.: Peeking into the future: predicting future person activities and locations in videos. In: CVPRW (2019)
16. Manfredi, M., Vezzani, R., Calderara, S., Cucchiara, R.: Detection of static groups and crowds gathered in open spaces by texture classification. Pattern Recogn. Lett. **44**, 39–48 (2014)
17. Mohamed, A., Qian, K., Elhoseiny, M., Claudel, C.: Social-STGCNN: a social spatio-temporal graph convolutional neural network for human trajectory prediction. In: CVPR (2020)
18. Pellegrini, S., Ess, A., Schindler, K., van Gool, L.: You'll never walk alone: modeling social behavior for multi-target tracking. In: ICCV (2009)
19. Radford, A., et al.: Learning transferable visual models from natural language supervision. In: ICML (2021)
20. Sadeghian, A., Kosaraju, V., Sadeghian, A., Hirose, N., Rezatofighi, H., Savarese, S.: Sophie: an attentive GAN for predicting paths compliant to social and physical constraints. In: CVPR (2019)
21. Setti, F., Cristani, M.: Evaluating the group detection performance: The GRODE metrics. IEEE Trans. Pattern Anal. Mach. Intell. **41**(3), 566–580 (2019)
22. Shi, H., Hayat, M., Wu, Y., Cai, J.: ProposalCLIP: unsupervised open-category object proposal generation via exploiting clip cues. In: CVPR (2022)
23. Shi, L., et al.: Sparse graph convolution network for pedestrian trajectory prediction. In: CVPR (2021)
24. Solera, F., Calderara, S., Cucchiara, R.: Structured learning for detection of social groups in crowd. In: AVSS (2013)
25. Sun, Q., Huang, X., Gu, J., Williams, B.C., Zhao, H.: M2I: from factored marginal trajectory prediction to interactive prediction. In: CVPR (2022)
26. Vaswani, A., et al.: Attention is all you need. In: NIPS (2017)
27. Vemula, A., Muelling, K., Oh, J.: Social attention: modeling attention in human crowds. In: ICRA (2018)

28. Wang, Z., et al.: CRIS: CLIP-driven referring image segmentation. In: CVPR (2022)
29. Weiss, Y., Freeman, W.: On the optimality of solutions of the max-product belief-propagation algorithm in arbitrary graphs. IEEE Trans. Inf. Theory **47**(2), 736–744 (2001)
30. Yu, C., Ma, X., Ren, J., Zhao, H., Yi, S.: Spatio-temporal graph transformer networks for pedestrian trajectory prediction. In: ECCV (2020)
31. Zhang, P., Ouyang, W., Zhang, P., Xue, J., Zheng, N.: SR-LSTM: state refinement for LSTM towards pedestrian trajectory prediction. In: CVPR (2019)
32. Zhao, H., et al.: TNT: target-driven trajectory prediction. arXiv preprint arXiv:2008.08294 (2020)

Three-Dimensional Medical Image Fusion with Deformable Cross-Attention

Lin Liu[1,2], Xinxin Fan[1,2], Chulong Zhang[1], Jingjing Dai[1], Yaoqin Xie[1], and Xiaokun Liang[1(✉)]

[1] Shenzhen Institute of Advanced Technology, Chinese Academy of Sciences, Shenzhen 518055, China
xk.liang@siat.ac.cn
[2] University of Chinese Academy of Sciences, Beijing 100049, China

Abstract. Multimodal medical image fusion plays an instrumental role in several areas of medical image processing, particularly in disease recognition and tumor detection. Traditional fusion methods tend to process each modality independently before combining the features and reconstructing the fusion image. However, this approach often neglects the fundamental commonalities and disparities between multimodal information. Furthermore, the prevailing methodologies are largely confined to fusing two-dimensional (2D) medical image slices, leading to a lack of contextual supervision in the fusion images and subsequently, a decreased information yield for physicians relative to three-dimensional (3D) images. In this study, we introduce an innovative unsupervised feature mutual learning fusion network designed to rectify these limitations. Our approach incorporates a Deformable Cross Feature Blend (DCFB) module that facilitates the dual modalities in discerning their respective similarities and differences. We have applied our model to the fusion of 3D MRI and PET images obtained from 660 patients in the Alzheimer's Disease Neuroimaging Initiative (ADNI) dataset. Through the application of the DCFB module, our network generates high-quality MRI-PET fusion images. Experimental results demonstrate that our method surpasses traditional 2D image fusion methods in performance metrics such as Peak Signal to Noise Ratio (PSNR) and Structural Similarity Index Measure (SSIM). Importantly, the capacity of our method to fuse 3D images enhances the information available to physicians and researchers, thus marking a significant step forward in the field. The code will soon be available online.

Keywords: Image Fusion · Three-Dimensional · Deformable Attention

1 Introduction

Multimodal image fusion represents a crucial task in the field of medical image analysis. By integrating information from diverse imaging modalities,

---

L. Liu and X. Fan—Contribute equally to this work.

B. Luo et al. (Eds.): ICONIP 2023, CCIS 1964, pp. 551–563, 2024.
https://doi.org/10.1007/978-981-99-8141-0_41

image fusion leverages the complementary characteristics of each technique. Each modality inherently focuses on distinct physiological or pathological traits. Hence, an effective fusion of these attributes can yield a more holistic and intricate image, thereby easing and enhancing physicians' decision-making processes during diagnosis and treatment [1].

For instance, the significance of Magnetic Resonance Imaging (MRI) and Positron Emission Tomography (PET) in image fusion is particularly noteworthy. MRI provides excellent soft tissue contrast and high-resolution anatomical structure information, while PET can present images of metabolic activity and biological processes [2]. The resultant image, derived from the fusion of MRI and PET, provides a comprehensive perspective, encompassing detailed anatomical structures alongside metabolic function data. This amalgamation plays a vital role in early disease or tumor detection and localization and significantly influences subsequent treatment strategies [3].

Traditionally, medical image fusion primarily utilizes multi-scale transformations in the transform domain, which typically comprises three steps. First, the images from each modality are subjected to specific transformations, such as wavelet [4–6] or pyramid transformations [7,8], yielding a series of multi-scale images. Then, these multi-scale images at the same scale level are analyzed and selected to retain the most representative information. Finally, through an inverse transformation, these multi-scale images are amalgamated into a novel image. Beyond these methods, sparse representation has also been applied in image fusion [9–11].

Nevertheless, image fusion tasks face significant hurdles, primarily due to the absence of a "gold standard" or "ground truth" that could encapsulate all modality information-ideally, a comprehensive reference image. Traditional fusion techniques often falter when dealing with high-dimensional data, particularly when encountering noise and complex data distributions across modalities. Moreover, the design of fusion rules remains manual, resulting in suboptimal generalization and unresolved semantic conflicts between different modality images [12]. Deep learning, with its inherent capacity for automatic feature learning and multi-layer abstraction, is poised to mitigate these challenges, enabling more accurate and interpretable image fusion.

Despite the progress, existing medical image fusion techniques primarily focus on two-dimensional (2D) slice fusion, which presents clear limitations. Medical images are predominantly three-dimensional (3D) signals, and 2D fusion approaches often neglect inter-slice context information. This oversight leads to a degree of misinterpretation of spatial relationships crucial for decoding complex anatomical structures. It is in this context that the potential benefits of 3D fusion become salient. By incorporating all 3D of space, 3D fusion can deliver more accurate localization information-a critical advantage in applications requiring precision, such as surgical planning and radiation therapy. Equally important, 3D fusion affords a panoramic view, enabling physicians to inspect and analyze anatomical structures and physiological functions from any perspective, thereby acquiring more comprehensive and nuanced information.

In this paper, we focus on the fusion of PET and MRI medical images, although the proposed methodology is generalizable to other imaging modalities. Our contributions can be delineated as follows: 1) We break new ground by applying a deep learning-based framework for 3D medical image fusion. 2) We introduce a Deformable Cross-Feature Fusion module that adjusts the correspondence information between the two modalities via Positional Relationship Estimation (PRE) and cross-fuses the features of the two modalities, thus facilitating image feature fusion. 3) We evaluate our approach on publicly available datasets, and our method yields state-of-the-art results quantitatively, based on Peak Signal-to-Noise Ratio (PSNR) and Structural Similarity Index Measure (SSIM). Qualitatively, our technique, even within the same 2D slice, surpasses baseline methods focusing on 2D fusion by not only retaining ample PET information but also integrating MRI structural data. The structural information discerned by our approach aligns more closely with the original MRI image.

2   Methods

2.1   Overview of Proposed Method

In this study, we utilized a U-shaped architecture for MRI-PET image fusion, shown in Fig. 1. The network employed a dual-channel input, with MRI and PET 3D image information fed separately. To reduce the parameter count of the 3D network, we applied Patch Embedding using DC2Fusion to both inputs. The resulting patch images were then passed through the Multi-modal Feature Fusion (MMFF) module. MMFF consisted entirely of Cross Fusion Blend (CFB) blocks at different scale levels. Notably, MMFF exhibited a fully symmetrical network structure, allowing the MRI branch to learn PET image information and the PET branch to acquire MRI image information. This symmetrical information interaction was facilitated by the CFB block, which had two input streams: input flow A and input flow B. In the MRI branch, MRI image information served as input flow A, while PET image information served as input flow B. The CFB block enabled the interaction and fusion of features between the two inputs. Consequently, the MRI branch adjusted its own features after perceiving certain characteristics in the PET image and outputted MRI image feature information. When the CFB block operated in the PET branch, input flow A and input flow B reversed their order compared to the MRI branch, enabling the adjustment of PET image features. After passing through the MMFF module, all branches entered a Fusion Layer composed of convolutions to merge the features and reconstruct the Fusion image.

2.2   Deformable Cross Feature Blend (DCFB)

Positional Relationship Estimation. The DCFB Block is a dual-channel input that takes input flow A (I_A) and input flow B (I_B) as inputs, shown in Fig. 2(a). At the Deformable Fusion stage, the two inputs obtain the relative

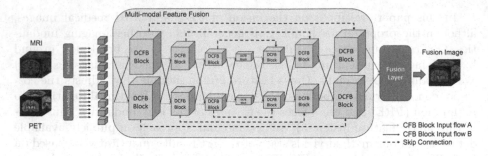

Fig. 1. Overall architecture of the proposed model comprising Patch Embedding and a fully mirrored symmetric Multi-modal Feature Fusion module, followed by a Fusion Layer consisting of fully convolutional operations.

positional deviations (offsets) between I_A and I_B in terms of their corresponding features' absolute positions. To calculate the relative positional deviations between the two images, we introduce depth-wise convolution. Depth-wise convolution partitions the features into groups, prioritizing the calculation of positional deviations within each group and then integrating the positional deviations across multiple groups. This process allows us to obtain the positional deviation of a point in one image with respect to the corresponding point in the other image. The process can be represented as follows:

$$F_I = Concat(F_{I_A}, F_{I_B}) \quad F_{I_A} \in R^{C_1, H, W, D}, F_{I_B} \in R^{C_2, H, W, D} \tag{1}$$

$$\text{inner offset} = Concat(Conv^1(F_I^1), Conv^2(F_I^2), \dots, Conv^{C_1+C_2}(F_I^{C_1+C_2})) \tag{2}$$

$$offset = Conv_{1 \times 1 \times 1}(\text{inner offset}) \tag{3}$$

Here, F_{I_A} and F_{I_B} are feature maps from I_A and I_B, respectively. By concatenating F_{I_A} and F_{I_B}, we obtain the feature map $F_I \in \mathbb{R}^{C_1+C_2, H, W, D}$. We employ depth-wise separable convolution with the number of groups set to $C_1 + C_2$. Based on the calculation of depth-wise separable convolution, the positional relationships between any point and its k-neighborhood (determined by the kernel size of the convolution) can be determined for each feature map. To search for the positional relationships between any two points in a given feature map, we only need to use the inner offset for computation. Currently, the obtained inner offset represents only the relative positional deviations within the feature map. To obtain the relative positional deviations between corresponding points across multiple modalities, we apply a $1 \times 1 \times 1$ convolution along the channel (C) dimension. This allows points $Point_1$ and $Point_2$ in the two feature maps to establish positional relationships based on absolute coordinates, enabling $Point_1$ to locate the position of the corresponding point $Point_1'$ in the other feature map. It is worth noting that $offset$ is a three-channel feature map, where each channel represents the positional deviation in the x, y, and z directions, respectively.

After obtaining the known positional deviation $offset$, we apply it to I_B. I_B is sampled on $offset$ to obtain the position-corrected I_B', expressed as follows:

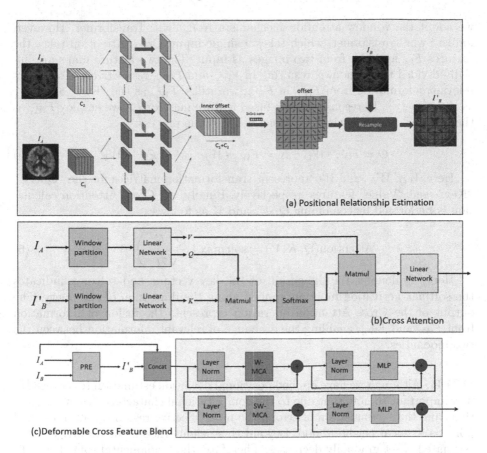

Fig. 2. Overall architecture and specific implementation details of the Cross Feature Blend module: (a) PRE is employed to determine the deviation between corresponding points of the two modalities, optimizing the receptive field shape of Cross Attention. (b) The Cross Attention module learns the correlation between the features of the two modalities and fuses their respective characteristics. (c) The Cross Feature Blend module primarily consists of PRE and Deformable Cross Attention.

$$I'_B = Resample(I_B, offset) \tag{4}$$

This means that a point p on I_B, after undergoing the deviation $offset_p$, obtains a new coordinate point p', and p' represents the same anatomical location on both I_A and I'_B. In other words, the anatomical significance represented by the absolute positions of I'_B and I_A is similar.

Cross Attention. To enable information exchange between I_A and I_B, we introduce the Cross Attention module, as shown in Fig. 2(b). The Cross Attention module is a component commonly used in computer vision to establish connections between different spatial or channel positions. In our implementation, we adopt the window attention mechanism from Swin Transformer. However, unlike Swin Transformer, which takes a single input, Cross Attention takes the features F_{I_A} and F_{I_B} from two images as input. The two feature maps are initially divided into windows, resulting in F_{W_A} and F_{W_B}, respectively. To extract relevant information from F_{W_A} in F_{W_B}, we utilize F_{W_B} as the "Key" and F_{W_A} as the "Query." To reconstruct the fused feature map for I_A, we employ F_{W_A} as the "Value." The following equations describe the process:

$$Q = F_{W_A} \cdot W_Q \quad K = F_{W_B} \cdot W_K \quad V = F_{W_A} \cdot W_V \tag{5}$$

Here, W_Q, W_K, and W_V represent transformation matrices for the "Query," "Key," and "Value" features, respectively. Finally, the Cross Attention calculation is performed by combining Q, K, and V as follows:

$$\text{Attention}(Q, K, V) = \text{softmax}\left(\frac{Q \cdot K^T}{\sqrt{d_k}}\right) \cdot V \tag{6}$$

Here, d_k denotes the dimension of the key vectors, and softmax indicates the softmax activation function applied along the dimension of the query. The output of the Cross Attention operation represents the fusion of information from F_{W_A} and F_{W_B}, enabling the exchange of relevant information between the two modalities.

DCFB. Although we have applied positional deviation estimation to correct I_B, it is important to note that due to the computational characteristics of convolutions and the accumulation of errors, the most effective region for an individual point is its local neighborhood. Beyond this neighborhood, the accuracy of the estimated offset gradually decreases. Therefore, the fundamental role of the offset is to overcome the limited receptive field of the window attention in the Swin Transformer architecture. Moreover, the expansion of the receptive field achieved by the offset follows an irregular pattern. Refer to Fig. 3 for an illustration.

Assuming I_A represents the features extracted from PET images and I_B represents the features from MRI images, our goal is to query the corresponding points in I_B and their surrounding regions based on I_A. However, utilizing the Swin Transformer architecture alone may result in the inability to find corresponding points between the two windows, even with the shift window operation. To address this issue, we propose the Deformable Cross Attention module, which incorporates an offset into the Swin Transformer module. This offset enables the positional correction of I_B beforehand, followed by regular window partitioning on the adjusted I'_B. While employing a similar approach as shown in Fig. 2(c) for I_A and I'_B, the windows on I_B are not regular windows but rather shaped by the features perceived from I_A due to the positional correction applied to I_B.

Loss Functions. In our image fusion algorithm, we utilize three distinct loss functions: Structural Similarity Index (SSIM) $\mathcal{L}_{SSIM}$, Normalized Cross-Correlation (NCC) L_{NCC}, and L1 loss L_1. These loss functions contribute

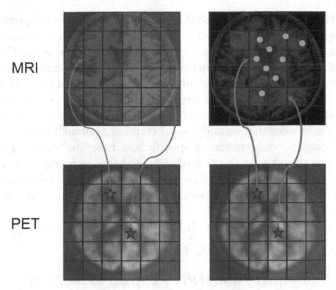

(a) Swin window cross attention   (b) Deformable window cross attention

Fig. 3. Deformable Window Cross Attention achieves equivalent receptive field deformation through PRE, in contrast to Swin Window Cross Attention.

equally to the overall loss with a weight ratio of 1:1:1. Here, we provide a brief explanation of each loss function and their respective contributions to the fusion process. The SSIM ($\mathcal{L}_{SSIM}$) is defined as follows:

$$\mathcal{L}_{SSIM}(I, J) = \frac{(2\mu_I\mu_J + C_1)(2\sigma_{IJ} + C_2)}{(\mu_I^2 + \mu_y^2 + C_1)(\sigma_I^2 + \sigma_J^2 + C_2)} \tag{7}$$

where I and J represent the input and target images, respectively. μ represents the mean, σ represents the standard deviation, and $c1$ and $c2$ are small constants for numerical stability. The Normalized Cross-Correlation ($\mathcal{L}_{NCC}$) is defined as follows:

$$\mathcal{L}_{NCC}(I, J) = \frac{\sum_{x,y,z}\left(I(x, y, z) - \overline{I}\right) \cdot \left(J(x, y, z) - \overline{J}\right)}{\sqrt{\sum_{x,y,z}\left(I(x, y, z) - \overline{I}\right)^2 \cdot \sum_{x,y,z}\left(J(x, y, z) - \overline{J}\right)^2}} \tag{8}$$

where I and J represent the input and target images, respectively. The summation is performed over corresponding image patches. The L1 loss ($\mathcal{L}_1$) is defined as follows:

$$\mathcal{L}_1(I, J) = ||I - J||_1 \tag{9}$$

where I and J represent the input and target images. By combining these three loss functions with equal weights, our image fusion algorithm achieves a balance between multiple objectives. The SSIM loss function focuses on preserving the image structure, the NCC loss function emphasizes structural consistency, and the L1 loss function aims to maintain detail and color consistency. Through the integration of these diverse loss functions, our algorithm comprehensively optimizes the generated image, thereby enhancing the quality of the fusion results.

To avoid biased fusion towards any individual modality solely for the purpose of obtaining favorable loss metrics quickly, we depart from traditional fusion strategies by adopting an end-to-end training approach in our method. To address this concern, we introduce a specific loss function based on the SSIM, which enables us to mitigate the risk of overemphasizing a single modality during the image fusion process solely to optimize loss metrics. This ensures a more balanced fusion outcome and enhances the overall performance of our method:

$$\mathcal{L}_{pair} = ||SSIM(Fusion, MRI) - SSIM(Fusion, PET)|| \tag{10}$$

The formulation of total loss is as follows:

$$\begin{aligned} \mathcal{L} = \mathcal{L}_{SSIM}(Fusion, MRI) + \mathcal{L}_{SSIM}(Fusion, PET) \\ + \mathcal{L}_{NCC}(Fusion, MRI) + \mathcal{L}_{NCC}(Fusion, PET) \\ + \mathcal{L}_1(Fusion, MRI) + \mathcal{L}_1(Fusion, PET) + \mathcal{L}_{pair} \end{aligned} \tag{11}$$

3 Experiments

3.1 Data Preparation and Evaluation Metrics

In this study, we evaluate the performance of our MRI-PET image fusion method on the ADNI-2 dataset [13]. The ADNI-2 dataset consists of 660 participants, each with both MRI and PET images acquired within a maximum time span of three months. We employed the SyN [14] registration algorithm to align the MRI and PET images to the MNI152 standard space, resulting in images with dimensions $182 \times 218 \times 182$. Subsequently, we extracted regions of interest in the form of MRI-PET pairs and resampled them to a size of $128 \times 128 \times 128$. Among the collected data, 528 pairs were used as the training set, while 66 pairs were allocated for validation, and another 66 pairs were designated for testing purposes.

To ensure a fair comparison, we conducted a comparative evaluation between our proposed approach and various 2D image fusion methods. For evaluation, we selected the same slice from the MRI, PET, and the fusion image predicted by our method. Objective metrics, SSIM, PSNR, Feature Mutual Information (FMI), and Normalized Mutual Information (NMI), were used to compare the selected slice with alternative methods. This comprehensive assessment enabled us to evaluate the effectiveness and performance of our proposed method against existing approaches in the field of image fusion.

Table 1. Quantitative results of the MRI-PET fusion task. The proposed methods demonstrate exceptional performance in terms of the SSIM metric and PSNR when employing 2D-based strategies, establishing a new state-of-the-art benchmark.

Method	2D-/3D-	$PSNR \uparrow$	$SSIM \uparrow$	$NMI \uparrow$	$FMI \uparrow$
SwinFuse [15]	2D	14.102±1.490	0.623±0.020	1.275±0.019	0.817±0.009
MATR [16]	-	15.997±1.190	0.658±0.035	1.451±0.009	0.795±0.018
DILRAN [17]	-	19.028±0.821	0.690±0.033	1.301±0.015	0.806±0.019
DC2Fusion(ours)	3D	20.714±1.377	0.718±0.033	1.312±0.012	0.807±0.020

3.2 Implementation Details

The proposed method was implemented using Pytorch [18] on a PC equipped with an NVIDIA TITAN RTX GPU and an NVIDIA RTX A6000 GPU. All models were trained for fewer than 100 epochs using the Adam optimization algorithm, with a learning rate of 1×10^{-4} and a batch size of 1. During training, the MR and PET datasets were augmented with random rotation. Our DC2Fusion model consists of three downsampling stages. To handle the limitations imposed by the image size, we employed a window partition strategy with a window size of 2, 2, 2 at each level. Additionally, the number of attention heads employed in each level is 3, 6, 12, 24.

We conducted a comparative analysis between our method and the 2D image-based fusion methods SwinFuse [15], MATR [16], and DILRAN [17]. Each of these methods was evaluated locally using the provided loss functions and hyperparameter settings by their respective authors. MATR [16] and DILRAN [17] methods employ image pair training, where the input images correspond to slices from MRI and PET, respectively. The SwinFuse method [15] does not provide an image pair end-to-end training approach. Therefore, we adapted its fusion process, as described in the paper, which involved fusing infrared and natural images. During testing, a dual-path parallel approach was employed, where MRI and PET slices were separately encoded. Subsequently, a specific fusion strategy (such as the L1 normalization recommended by the authors) was applied to fuse the features of both modalities. Finally, the fused features were passed through the Recon module for reconstruction.

3.3 Results and Analysis

Figure 4 illustrates the fusion results of DC2Fusion in comparison with other methods. DILRAN, SwinFuse, and our method all exhibit the ability to preserve the highlighted information from PET while capturing the structural details from MRI. However, the MATR algorithm, which lacks explicit constraints on image fusion during training, demonstrates a complete bias towards learning PET image information and severely lacks the ability to learn from MRI information when applied to the ADNI dataset. Consequently, the model quickly achieves high $SSIM(Fusion, PET)$ values during training, thereby driving the overall loss function gradient. In this experiment, our focus is to compare the fusion image quality of DILRAN, SwinFuse, and DC2Fusion. As shown in Fig. 4, DILRAN exhibits less prominent MRI structural information compared to Swin-Fuse and DC2Fusion. Moreover, in terms of preserving PET information, Swin-Fuse and DC2Fusion offer better contrast. For further comparison, Fig. 5 provides additional details. PET images primarily consist of high-signal information with minimal structural details. DILRAN, SwinFuse, and DC2Fusion effectively fuse the structural information from MRI. In comparison to DC2Fusion, both DILRAN and SwinFuse display slightly thinner structures in the gyri region, which do not entirely match the size of the gyri displayed in the MRI. However, DC2Fusion achieves better alignment with the gyri region of the MRI image. Nevertheless, the structural clarity of DC2Fusion is inferior to SwinFuse. This discrepancy arises because the SwinFuse algorithm does not utilize patch embedding operations or downsampling layers, enabling it to maintain high clarity throughout all layers of the model. However, this approach is limited by GPU memory constraints and cannot be extended to 3D image fusion. Therefore, in our method, we made trade-offs in terms of clarity by introducing techniques such as patch embedding to reduce memory usage and accomplish 3D image fusion tasks.

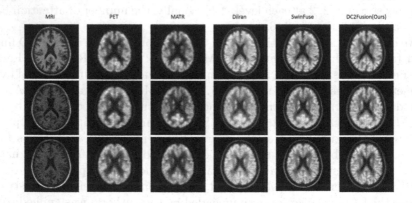

Fig. 4. Comparative images of MRI, PET, and other fusion methods on 3 representative PET and MRI image pairs. From left to right: MRI image, PET image, MATR [16], Dilran [17], SwinFuse [15], and DC2Fusion.

Table 1 presents a summary of the performance metrics for different methods, including 2D-/3D-fusion, namely PSNR, SSIM, NMI, and FMI. Among the evaluated methods, SwinFuse, a 2D fusion technique, achieved a remarkable FMI score of 0.817. However, it is important to note that FMI solely focuses on feature-based evaluation and does not provide a comprehensive assessment of image fusion quality. SwinFuse obtained lower scores in other metrics, namely PSNR (14.102), SSIM (0.623), and NMI (1.275), suggesting potential limitations in preserving image details, structure, and information content. In contrast, our

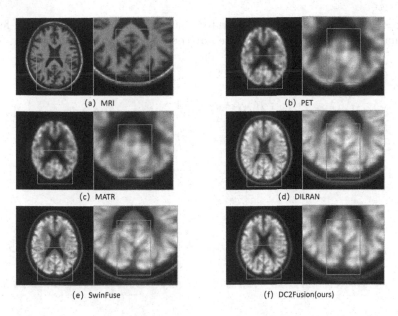

Fig. 5. Qualitative comparison of the proposed DC2Fusion with 3 typical and state-of-the-art methods on a representative PET and MRI image pair: (a) MRI image, (b) PET image, (c) MATR [16], (d) DILRAN [17], (e) SwinFuse [15], (f) DC2Fusion.

method, specifically designed for 3D image fusion, exhibited superior performance. DC2Fusion achieved a notable FMI score of 0.807, indicating consistent and correlated features in the fused images. Moreover, DC2Fusion outperformed other methods in terms of PSNR (20.714), SSIM (0.718), and NMI (1.312), highlighting its effectiveness in preserving image details, structural similarity, and information content. Overall, the results demonstrate the competitive performance of our method in medical image fusion, particularly in 3D fusion tasks. These findings emphasize the potential of DC2Fusion in enhancing image quality and preserving structural information, thereby providing valuable insights for further research in the field of medical image fusion.

As shown in Fig. 6, we present fusion metrics for all samples in our test cases. It is evident that DC2Fusion consistently outperforms other methods in terms of PSNR and SSIM for each sample. However, the NMI metric reveals an anomaly with significantly higher results for MATR. This can be attributed to MATR's fusion results being heavily biased towards the PET modality, resulting in a high consistency with PET images and consequently yielding inflated average values. Nevertheless, these values lack meaningful reference significance. Excluding MATR, our proposed method also achieves better results than other methods in terms of the NMI metric. With regards to the FMI metric, our method does not exhibit a noticeable distinction compared to other methods. Considering Fig. 5, although our results may not possess the same level of clarity as other methods, our approach still obtains relatively high FMI scores in terms of these detailed features. This observation demonstrates that our method preserves the feature information during image fusion, even at the cost of reduced clarity.

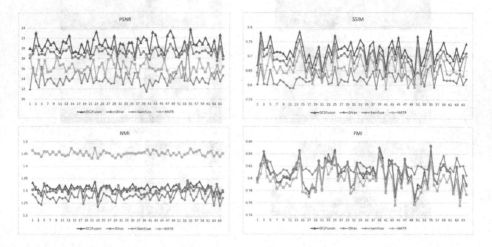

Fig. 6. Illustration of the average fusion metrics (PSNR, SSIM, NMI, and FMI) for each sample in our test cases. These metrics are computed by comparing the fusion images with both the MRI and PET images separately. The reported values represent the average scores across all samples.

4   Conclusion

In this study, we propose a novel architecture for 3D MRI-PET image fusion. Our approach, DCFB, achieves cross-modal PRE and provides a deformable solution for enlarging the receptive field during cross-attention. This facilitates effective information exchange between the two modalities. Experimental results demonstrate the outstanding performance of our proposed 3D image fusion method. Although the image fusion quality of our method is currently limited by GPU constraints, resulting in slightly lower clarity compared to 2D fusion, we have provided a novel solution for 3D medical image fusion tasks. Furthermore, we anticipate applying this framework to various modalities in the future, thereby broadening its applicability across different modal combinations.

References

1. James, A.P., Dasarathy, B.V.: Medical image fusion: a survey of the state of the art. Inf. Fusion **19**, 4–19 (2014)
2. Huang, B., Yang, F., Yin, M., Mo, X., Zhong, C.: A review of multimodal medical image fusion techniques. Comput. Math. Methods Med. 2020 (2020)

3. Yin, M., Liu, X., Liu, Y., Chen, X.: Medical image fusion with parameter-adaptive pulse coupled neural network in nonsubsampled shearlet transform domain. IEEE Trans. Instrum. Meas. **68**(1), 49–64 (2018)
4. Singh, R., Khare, A.: Multimodal medical image fusion using daubechies complex wavelet transform. In: 2013 IEEE Conference on Information & Communication Technologies, pp. 869–873. IEEE (2013)
5. Chabi, N., Yazdi, M., Entezarmahdi, M.: An efficient image fusion method based on dual tree complex wavelet transform. In: 2013 8th Iranian Conference on Machine Vision and Image Processing (MVIP), pp. 403–407. IEEE (2013)
6. Talbi, H., Kholladi, M.K.: Predator prey optimizer and DTCWT for multimodal medical image fusion. In: 2018 International Symposium on Programming and Systems (ISPS), pp. 1–6. IEEE (2018)
7. Du, J., Li, W., Xiao, B., Nawaz, Q.: Union Laplacian pyramid with multiple features for medical image fusion. Neurocomputing **194**, 326–339 (2016)
8. Burt, P.J., Kolczynski, R.J.: Enhanced image capture through fusion. In: 1993 (4th) International Conference on Computer Vision, pp. 173–182. IEEE (1993)
9. Yang, B., Li, S.: Pixel-level image fusion with simultaneous orthogonal matching pursuit. Inf. Fusion **13**(1), 10–19 (2012)
10. Liu, Y., Wang, Z.: Simultaneous image fusion and denoising with adaptive sparse representation. IET Image Proc. **9**(5), 347–357 (2015)
11. Liu, Y., Chen, X., Ward, R.K., Wang, Z.J.: Image fusion with convolutional sparse representation. IEEE Signal Process. Lett. **23**(12), 1882–1886 (2016)
12. Hill, P., Al-Mualla, M.E., Bull, D.: Perceptual image fusion using wavelets. IEEE Trans. Image Process. **26**(3), 1076–1088 (2016)
13. Jack, Jr, C.R.E.A.: The Alzheimer's disease neuroimaging initiative (ADNI): MRI methods. J. magnetic resonance imaging: JMRI **27**(4), 685–691 (2008). https://doi.org/10.1002/jmri.21049. Accessed 11 May 2023
14. Avants, B.B., Epstein, C.L., Grossman, M., Gee, J.C.: Symmetric diffeomorphic image registration with cross-correlation: evaluating automated labeling of elderly and neurodegenerative brain. Med. Image Anal. **12**(1), 26–41 (2008)
15. Wang, Z., Chen, Y., Shao, W., Li, H., Zhang, L.: SwinFuse: a residual swin transformer fusion network for infrared and visible images. IEEE Trans. Instrum. Meas. **71**, 1–12 (2022)
16. Tang, W., He, F., Liu, Y., Duan, Y.: MATR: multimodal medical image fusion via multiscale adaptive transformer. IEEE Trans. Image Process. **31**, 5134–5149 (2022)
17. Zhou, M., Xu, X., Zhang, Y.: An attention-based multi-scale feature learning network for multimodal medical image fusion. ArXiv abs/2212.04661 (2022)
18. Imambi, S., Prakash, K.B., Kanagachidambaresan, G.R.: PyTorch. In: Prakash, K.B., Kanagachidambaresan, G.R. (eds.) Programming with TensorFlow. EICC, pp. 87–104. Springer, Cham (2021). https://doi.org/10.1007/978-3-030-57077-4_10

Handling Class Imbalance in Forecasting Parkinson's Disease Wearing-off with Fitness Tracker Dataset

John Noel Victorino(✉) ⓘ, Sozo Inoue ⓘ, and Tomohiro Shibata ⓘ

Graduate School of Life Science and Systems Engineering, Kyushu Institute of Technology, Kitakyushu, Fukuoka 808-0135, Japan
`jnoelvictorino@brain.kyutech.ac.jp`

Abstract. Parkinson's disease (PD) patients experience the "wearing-off phenomenon", where their symptoms resurface before they can take the following medication. As time passes, the duration of the medicine's efficacy reduces, leading to discomfort among PD patients. Therefore, patients and clinicians must meticulously observe and document symptom changes to administer appropriate treatment.

Forecasting the PD wearing-off phenomenon is challenging due to the class imbalance from the difficulty documenting the phenomenon. This paper compares different approaches for handling class imbalance in forecasting the PD wearing-off phenomenon using the fitness tracker and smartwatch dataset (wearing-off dataset): oversampling, undersampling, and combining the two. Previous studies reported the potential use of commercially-worn fitness tracker datasets to predict and forecast wearing-off periods. However, some participants' high false positives and negatives have been observed with the developed models [16,17].

This paper compares different approaches to handling class imbalance in the wearing-off dataset. First, changes in the data collection process and tools were made during the data collection phase, as the nursing staff struggled with the data collection tool. Second, different existing oversampling and undersampling techniques were used to improve the ratio of wearing-off labels to non-wearing-off instances. Finally, adjustments to forecast probabilities were applied due to the resampling in the second step.

The results showed that an XGBoost model trained with an oversampled dataset using the SMOTE algorithm increased the F1 and AUC-ROC curve scores by 7.592% and 0.754%, respectively, compared to the based XGBoost model. The models in this paper only used the last hour's data to forecast the next hour's wearing-off. The change in the amount of input data also improved the F1 and AUC-ROC curve scores compared with the models that used the current period or the previous day's data [16,17]. These improvements in forecasting wearing-off can help PD patients and clinicians monitor, record, and prepare for wearing-off periods, as we aim to lessen the false positives and negatives in the forecasting results.

© The Author(s), under exclusive license to Springer Nature Singapore Pte Ltd. 2024
B. Luo et al. (Eds.): ICONIP 2023, CCIS 1964, pp. 564–578, 2024.
https://doi.org/10.1007/978-981-99-8141-0_42

Keywords: Parkinson's Disease · Wearing-off Phenomenon · Machine Learning · Class Imbalance

1   Introduction

Parkinson's disease (PD) affects motor and non-motor functions due to the loss of dopamine-producing cells in the brain. Cardinal PD symptoms include tremors, bradykinesia (slow movement), muscle stiffness, and postural instability. Additionally, PD patients suffer non-motor symptoms like mood swings, sleep disturbances, speech problems, and cognitive difficulties. PD, lacking a cure, profoundly impacts daily life and quality of life (QoL) [7].

The "wearing-off phenomenon" is a major source of discomfort for PD patients. Many PD patients rely on Levodopa treatment (L-dopa) to alleviate their symptoms temporarily by increasing dopamine levels in the brain. However, with time, these symptoms resurface earlier than expected, significantly impacting patients. As a result, continuous monitoring and consultation with healthcare providers are essential for adjusting treatment or exploring alternative options [1,4,13,15].

This paper addresses the class imbalance challenge in the wearing-off dataset, which can result from difficulties documenting the phenomenon, its inherent rarity, or unique patterns in some PD patients. We used and compared existing methods, including SMOTE-based and ADASYN oversampling, undersampling, and their combination, to mitigate class imbalance when forecasting the wearing-off phenomenon in PD patients using fitness trackers and smartwatch data. Additionally, we discuss challenges in improving data collection by nursing care staff for PD patients. This paper contributes to answering the following research questions:

1. How can data collection be improved for nursing care staff to increase the number of wearing-off labels in the dataset?
2. Which class imbalance techniques performed well in forecasting wearing-off in the next hour?
3. How does reweighting the forecast probabilities due to resampling affect the forecasting models?

Our results showed that an XGBoost model trained with an oversampled dataset using SMOTE and a resampled dataset using SMOTETomek increased the F1 and AUC-ROC curve scores by 7.592% and 0.754%, respectively. Models in this paper used the last hour's data from the PD patient to forecast the next hour's wearing-off. This paper's best models also improved the F1 score and AUC-ROC curve compared with the models that used the current period's and the previous day's data [18].

The contributions of this paper can be summarized in these two points.

1. **Expanded Data Collection**: We collected additional datasets from new PD patients by optimizing the data collection process to accommodate daily

visits by nursing staff. This adjustment extended the collection period from seven days to two months, resulting in a valuable dataset.

2. **Class Imbalance Handling**: We addressed the inherent class imbalance in the wearing-off dataset, improving the model's forecasting performance for F1 and AUC-ROC scores. Utilizing established oversampling, undersampling techniques, and their combination, our model demonstrated improved scores compared to the base model and prior studies.

These improvements in forecasting wearing-off can help PD patients and clinicians monitor, record, and prepare for wearing-off periods. The results of this paper aim to lessen the false positives and false negatives in forecasting wearing-off among PD patients.

This paper is organized into different sections. Section 2 narrates improvement in the data collection procedure and how the authors managed to increase the wearing-off labels of the dataset. We also provide descriptions of the included techniques to handle class imbalance as well as the primary metrics of this study. Next, Sect. 3 answers the research questions while Sect. 4 highlights the key results of this study. Finally, Sect. 5 summarizes the contributions of this paper, with future opportunities and challenges in this research area.

2   Methodology

This section describes 1) how the nursing staff collected the new datasets from the PD patients, 2) how the data was processed, and 3) how each class imbalance technique was applied before training a forecasting model for wearing-off. This paper compares different approaches to handling class imbalance inherent in the forecasting wearing-off task. The following subsections further explain the data collection and processing and how the other class imbalance techniques are prepared.

2.1   Data Collection

Data collection commenced with a similar approach as previous studies on wearing-off among PD patients [16–18]. Three new PD participants were equipped with Garmin Venu SQ 2 smartwatches through their nursing staff. These smartwatches recorded heart rate, stress score, sleep data, and step count throughout the study. On the other hand, the wearing-off periods were recorded using a smartphone application.

Notable changes were made in this round of data collection. First, the nursing staff recorded wearing-off periods instead of PD patients, with limited input into the smartphone application during the nursing staff's daily visits. Second, the Japanese Wearing-Off Questionnaire (WoQ-9) was changed with a custom form based on the nursing staff feedback. This new form allowed nursing staff to input wearing-off periods and drug intake times by the hour, simplifying data collection during their visits. However, the granularity of wearing-off periods based on specific symptoms was temporarily omitted (Fig. 1).

Fig. 1. The previous smartphone application version collects the Japanese Wearing-off Questionnaire (WoQ-9) responses on whether the PD patient experiences any of the nine symptoms. This version collects one wearing-off period at a time [16,18].

Fig. 2. The adjusted smartphone application collects wearing-off periods by the hour. This version collects multiple wearing-off periods daily, making data collection by nursing staff easier since they only visit the patients once daily.

The datasets obtained from each tool remained the same as in previous studies [16–18], excluding symptoms during the wearing-off periods and the effects of drug intake. The Garmin Health API provides access to the Garmin Venu SQ 2 dataset. Heart rate (HR) in beats per minute (bpm) was recorded every 15 s, while total step count for 15-min intervals was aggregated. Garmin's estimated stress scores were reported every three minutes, with values ranging from 0 to 100, with "−1" and "−2" indicating insufficient data and excessive motion, respectively. Finally, sleep duration data for each sleep stage was provided daily [5,12,14].

The nursing staff utilized a smartphone app to request and record essential information from PD patients, including wearing-off periods, drug intake times, and additional patient details. These patient-specific details encompassed age, sex, Hoehn and Yahr (H&Y) scale, the Japan Ministry of Health, Labor, and Welfare's classification of living dysfunction (JCLD) for the PD stage [2,9], and the Parkinson's Disease Questionnaire (PDQ-8) for self-assessed quality of life [8]. These datasets are summarized in Table 2.

Table 1. Garmin Venu SQ 2 smartwatch datasets via Garmin Health API.

Dataset	Interval	Description
Heart rate	15-s	Beats per minute
Steps	15-min	Cumulative step count per interval, with 0 as the lowest value
Stress score	3-min	Estimated stress score [6] –0–25: resting state, –26–50: low stress, –51–75: medium stress, –76–100: high stress, ——1: not enough data to detect stress, ——2: too much motion
Sleep stages with each sleep duration	Varying interval for each date	Start and end time for each sleep stage: Light sleep, Rapid eye movement (REM) sleep, Deep sleep, Awake [5]

Table 2. The smartphone application collected datasets.

Data Type	Description
Wearing-off periods	Hourly check on whether wearing-off or not
Drug intake time	Hourly check on whether medicine was taken or not
Basic Information	Age and gender
Hoehn & Yahr Scale (H&Y),	
Japan Ministry of Health, Labor, & Welfare's classification of living dysfunction (JCLD)	Participant's PD stage
PDQ-8	PD-specific QoL measurement: 0–100%, with 100% as the worst QoL

The university has conducted an ethical review of this research project. PD patients are eligible for this study if they meet the following criteria.

1. PD patients experience wearing-off periods.
2. PD patients can use the smartwatch.
3. PD patients do not have any serious illnesses or symptoms that could affect them during the data collection.

Upon meeting these conditions, the nursing care facility selected PD patients. Research objectives were communicated to nursing staff, who informed PD patients and obtained their written consent. Subsequently, each nursing staff member received a smartwatch and smartphone for a two-month data collection period, during which PD patients can withdraw voluntarily.

Participants were required to wear the smartwatch continuously, even during bathing and sleep, to record sleep duration data. Except for the noted limitations and recommendations, there were no specific restrictions during the data collection phase.

2.2 Data Processing

The dataset collected from PD patients using the Garmin Venu SQ 2 smartwatch and the wearing-off labels recorded by their nursing staff were processed to produce the combined wearing-off dataset. This section explains the data processing and the important differences with previous studies.

On the one hand, several cleaning and preprocessing steps were implemented for the raw Garmin Venu SQ 2 datasets. Missing values were replaced with "-1," consistent with Garmin's missing value representation [5]. The datasets were then resampled to uniform 15-min intervals to standardize the varying intervals from the Garmin Health API, as depicted in Table 1. Missing values resulting from resampling were filled with the last available data. Each record was matched with its corresponding sleep dataset based on the calendar date. Additional sleep-related features were calculated using equations below [16].

$$\text{Total non-REM duration} = \text{Deep sleep duration} + \text{Light sleep duration}, \quad (1)$$

$$\text{Total sleep duration} = \text{Total non-REM duration} + \text{REM sleep duration} \quad (2)$$

$$\text{Total non-REM percentage} = \frac{\text{Total non-REM duration}}{\text{Total sleep duration}} \quad (3)$$

$$\text{Sleep efficiency} = \frac{\text{Total sleep duration}}{\text{Total sleep duration} + \text{Total awake duration}} \quad (4)$$

On the other hand, the raw smartphone application dataset underwent similar preprocessing. Wearing-off periods were resampled to 15-min timestamps, with "1" indicating the presence of wearing-off and "0" denoting its absence. Drug intake periods were similarly processed to align with the 15-min timestamps.

Finally, the two datasets were merged by matching the 15-min timestamps. The hours of the day and the days of the week were also extracted. For the day of the week, "0" to "6" values were matched with "Monday" to "Sunday." Then, for the hour of the day, sine and cosine transformation was applied. The following features were used to develop the wearing-off forecasting model.

x_1: Heart rate (HR)
x_2: Step count (Steps)
x_3: Stress score (Stress)
x_4: Awake duration during the estimated sleep period (Awake)
x_5: Deep sleep duration (Deep)
x_6: Light sleep duration (Light)
x_7: REM sleep duration (REM)
x_8: Total non-REM sleep duration (NonREMTotal)
x_9: Total sleep duration (Total)
x_{10}: Total non-REM sleep percentage (NonREMPercentage)
x_{11}: Sleep efficiency (SleepEfficiency)
x_{12}: Day of the week (TimestampDayOfWeek)
x_{13}: Sine value of Hour of the day (TimestampHourSin)
x_{14}: Cosine value of Hour of the day (TimestampHourCos)
y: Wearing-off

2.3   Class Imbalance Techniques, Experiments

In this section, we discuss the wearing-off forecasting model. The data split for training and test sets, the metrics used for evaluation, and the different class imbalance techniques considered in this paper are also presented. Data processing and model construction mainly uses Pandas, Scikit-learn, XGBoost, and imbalanced-learn [3,10].

Individualized wearing-off forecasting models were tailored for each PD participant. This approach prioritizes personalized models over general ones, acknowledging the distinct nature of PD experiences among patients. Then, each PD participant's dataset was sequentially partitioned into training and test sets. Each participant's final two days' data constituted the test set, while the remaining data formed the training set. $F1Score$, the area under the Receiver Operator Characteristic curve ($AUC-ROC$), and the area under the Precision-Recall curve ($AUC-PR$) were the main metrics used to evaluate the models, considering the class imbalance challenge addressed by this paper. $Accuracy$, $Precision$, and $Recall$, were also reported.

This paper defined the wearing-off forecasting model as follows.

$$y_{t+1} = f(M(X, y, w)), M = \begin{bmatrix} X_t & y_t \\ X_{t-1} & y_{t-1} \\ \vdots & \vdots \\ X_{t-w} & y_{t-w} \end{bmatrix} \tag{5}$$

In finding the wearing-off in the next hour, y_{t+1}, the model accepts a matrix M of 14 features and wearing-off in previous time steps from the current time until w time step before the current time. This paper used the last hour's data ($w = 1$). The implementation has four records since the data was processed in 15-min intervals. Thus, the matrix, M, has $w * 4$ records or $M_{t-1} \ldots M_t$ to forecast the next hour's wearing-off, y_{t+1}.

This paper compares four existing algorithms for handling class imbalance – SMOTE-based techniques and ADASYN algorithms, where for SMOTE-based techniques, $k\_neighbors$ was set to five.

1. Base XGBoost Model
2. Oversampled SMOTE Model
3. Undersampled+Oversampled SMOTE Model
4. SMOTETomek Resampled Model
5. Oversampled ADASYN Model

After resampling the dataset using the different techniques, we also adjusted the forecasted class probabilities with Eq. 6, which accounts for the real ratio of normal states and wearing-off states [11].

$$adjProb = \frac{a}{a + b}$$

$$a = predProb \cdot \frac{originalProb}{resampledProb}$$

$$b = (1 - predProb) \cdot \frac{1 - originalProb}{1 - resampledProb}$$

where: (6)

$$adjProb := \text{adjusted wearing-off probability}$$

$$predProb := \text{predicted wearing-off probability}$$

$$originalProb := \frac{N_1 + N_0}{N} \text{(original dataset)}$$

$$resampledProb := \frac{N_1' + N_0'}{N'} \text{(resampled dataset)}$$

Finally, the models were trained using a computer with an Intel i7-6700 CPU running at 3.40 GHz, four cores, and 16 GB of RAM. A cross-entropy loss function was used to train the models.

3   Results

This section presents the results of the modified data collection process and the improvements in the wearing-off forecasting models' performance. Three additional PD patients participated in the modified collection process. Despite the improvements in data collection tools and protocols, class imbalance persisted, underscoring its inherent nature in this dataset. Data collection spanned two months, from December 2022 to January 2023, with PD patients initiating data collection at varying times. Consequently, the three new participants contributed different amounts of data, resulting in an average collection duration of 14.31 ± 14.59 days.

Among the new participants, Participant 13, the third oldest at age 70, and with the lowest quality of life outlook (PDQ = 71.88%), contributed data for 61 days. In contrast, Participant 11 provided only 13 days of data due to a delayed start and early conclusion of data collection. The average PD stage remained consistent, with values of 2.69 ± 0.75 on the H&Y scale and 1.69 ± 0.48 on the JCLD scale. These scores correspond to symptoms such as limb shaking, muscle stiffness, mild physical disability, and some daily life inconvenience [2, 9] (Table 3).

Table 3. The participants' demographics. The last three participants were the new PD patients enrolled for this paper. The first 10 PD patients were from the previous data collection [18]

Participant	Age	Gender	H&Y	JCLD	PDQ-8	Number of Collection Days
1	43	Female	2	1	37.50%	9
2	38	Female	3	2	65.63%	6
3	49	Female	3	2	34.38%	10
4	69	Female	3	1	78.13%	10
5	49	Female	2	2	37.50%	8
6	56	Female	2	2	37.50%	9
7	48	Male	3	1	15.63%	6
8	77	Male	3	2	34.38%	11
9	84	Male	4	2	59.38%	11
10	58	Male	3	2	25.00%	10
11	53	Male	3	2	34.38%	13
12	82	Male	1	1	18.75%	22
13	70	Female	3	2	71.88%	61
Average	59.69		2.69	1.69	42.31%	14.31
Std. Dev.	15.16		0.75	0.48	20.04%	14.59

3.1   How Can Data Collection Be Improved for Nursing Care Staff to Increase the Number of Wearing-off Labels in the Dataset?

At the beginning of this study's data collection, the smartphone application utilized the WoQ-9 form, as depicted in Fig. 1. However, the initial number of reported wearing-off labels was notably low, prompting discussions with the nursing staff. These discussions revealed that nursing staff visited PD patients only once during their scheduled visit days, providing essential care and assistance. Consequently, recording detailed wearing-off periods, including precise start and end times, posed an additional burden for them. Moreover, the process of recording each wearing-off period through the smartphone application was time-consuming, as it involved multiple steps.

To address these issues, as illustrated in Fig. 2, nursing staff were tasked with checking wearing-off periods only within the day of their visit. This streamlined the process, reducing the number of steps and simplifying the data collection procedure, resulting in improved wearing-off data collection (Table 4).

Table 4. Wearing-off Percentage. The percentage of wearing-off to the normal state was computed after processing the dataset into 15-min intervals.

Participant	Number of Collection Days	Percentage of Normal Status $(y=0)$	Percentage of Wearing-off Status $(y=1)$
1	9	92.36	7.64
2	6	97.35	2.65
3	10	86.77	13.23
4	10	90.49	9.51
5	8	92.32	7.68
6	9	89.82	10.19
7	6	98.82	1.18
8	11	87.97	12.03
9	11	92.99	7.01
10	10	86.77	13.23
11	13	90.71	9.30
12	22	92.52	7.48
13	61	93.53	6.47

3.2 Which Class Imbalance Techniques Performed Well in Forecasting Wearing-Off in the Next Hour?

Among the experiments that handle class imbalance, the oversampled SMOTE model outperformed the base model and the rest of the models with an F1 Score of 22.31 ± 23.30% and AUC-ROC of 71.92% ± 14.06%. Notably, the oversampled SMOTE model's F1 Score increased by 7.592% compared to the base model. However, it did not have the best AUC-PRC among the experiments (AUC-PRC = 23.49 ± 23.65%). Interestingly, the SMOTETomek Resampled model performed similarly to the SMOTE model regarding the F1 Score, AUC-ROC, and AUC-PRC.

Table 5. Wearing-Off Forecast Model Performance Comparing Different Techniques in Handling Class Imbalance. The reported metric scores were averaged across participants.

Experiment	F1 Score	AUC-ROC	AUC-PRC
Base XGBoost Model	14.72 ± 25.49%	71.16 ± 16.71%	24.19 ± 24.83%
Oversampled SMOTE Model	22.31 ± 23.30%	71.92 ± 14.06%	23.49 ± 23.65%
Undersampled+ Oversampled SMOTE Model	20.09 ± 16.46%	70.35 ± 17.96%	26.01 ± 21.25%
SMOTETomek Resampled Model	22.31 ± 23.30%	71.92 ± 14.06%	23.49 ± 23.65%
ADASYN Oversampled Model	16.52 ± 22.31%	69.85 ± 15.73%	23.69 ± 24.28%
Experiment	Precision	Recall	Accuracy
Base XGBoost Model	14.92 ± 24.93%	16.3 ± 27.63%	91.61 ± 3.87%
Oversampled SMOTE Model	21.46 ± 23.07%	26.65 ± 26.87%	89.75 ± 6.11%
Undersampled+ Oversampled SMOTE Model	12.75 ± 13.59%	87.16 ± 12.44%	41.27 ± 25.43%
SMOTETomek Resampled Model	21.46 ± 23.07%	26.65 ± 26.87%	89.75 ± 6.11%
ADASYN Oversampled Model	16.31 ± 20.71%	20.11 ± 26.61%	90.03 ± 5.34%

3.3 How Does Reweighting the Forecast Probabilities Due to Resampling Affect the Forecasting Models?

Reweighting the forecasted wearing-off probabilities considers the real ratio of wearing-off before applying each resampling technique. The Adj. Oversampled SMOTE and the Adj. SMOTETomek Resampled models still had the highest F1 score and AUC-ROC with 16.28 ± 24.01% and 71.92 ± 14.06%. Although these scores dropped from those in Table 5, the reweighted probabilities outperform the base XGBoost model. In addition, the Adj. The oversampled SMOTE and Adj. SMOTETomek Resampled models had a lower standard deviation across the participants despite having lower AUC-PRC than Adj. ADASYN Oversampled Model (Table 6).

Table 6. Model Performance After Reweighting Forecasted Wearing-off Probabilities. The reweighting takes into account the real ratio of wearing-off from the original dataset 6.

Experiment	F1 Score	AUC-ROC	AUC-PRC
Adj. Oversampled SMOTE Model	16.28 ± 24.01%	71.92 ± 14.06%	23.49 ± 23.65%
Adj. Undersampled+Oversampled SMOTE Model	10.32 ± 22.58%	71.92 ± 14.06%	23.49 ± 23.65%
Adj. SMOTETomek Resampled Model	16.28 ± 24.01%	71.92 ± 14.06%	23.49 ± 23.65%
Adj. ADASYN Oversampled Model	14.76 ± 26.07%	69.85 ± 15.73%	23.69 ± 24.28%
Experiment	Precision	Recall	Accuracy
Adj. Oversampled SMOTE Model	25.47 ± 34.91%	15.66 ± 23.17%	92.26 ± 4.4%
Adj. Undersampled+Oversampled SMOTE Model	13.28 ± 28.35%	8.68 ± 18.93%	93.26 ± 3.52%
Adj. SMOTETomek Resampled Model	25.47 ± 34.91%	15.66 ± 23.17%	92.26 ± 4.4%
Adj. ADASYN Oversampled Model	17.61 ± 28.42%	14.1 ± 24.69%	92.63 ± 4.03%

4 Discussions

The findings of this study have outperformed the performance of the base XGBoost model in terms of F1 Score and AUC-ROC across the 13 participants' forecast models. Both SMOTE and SMOTETomek techniques yielded improved results compared to the base model. The choice between these two methods hinges on the acceptability criteria when utilizing these forecasting models.

To elaborate, oversampling with SMOTE generates synthetic wearing-off records to match the number of non-wearing-off records, achieving a balanced ratio. However, it may not accurately reflect the infrequency of wearing-off periods during the day. Even with an extended data collection period and more wearing-off labels, class imbalance still persisted in the dataset, as evidenced in the new PD patients.

Conversely, SMOTETomek resampling increases the number of wearing-off records while simultaneously reducing the number of non-wearing-off labels in close proximity to the wearing-off records. This approach aligns better with the inherent characteristics of the wearing-off phenomenon.

Lastly, reweighted models offer an alternative for achieving more acceptable forecast probabilities. These models adjust the forecasted wearing-off probabilities based on the original and resampled dataset, enhancing their suitability for real-world applications.

Compared with the previous study, the results presented exhibit better performance. In the last study, an MLP model utilized data from the current time, while a CNN model incorporated one day of data to forecast wearing-off in the next hour. Table 7 shows that the F1 scores have a notable increase of 17.208% for the MLP model and 6.728% for the CNN model. Additionally, the AUC-ROC scores improved by 2.78% and 11.40%, respectively. These improvements suggest that this study's models improved F1 and AUC-ROC scores even when trained with PD patients' last-hour data.

Table 7. Comparison of Wearing-Off Forecast Models with Previous Study. This paper's model accepts the last hour of data, while in the previous study, the MLP model accepts the current time, and the CNN model accepts the last day of data [18].

Model	Input	F1 Score	AUC-ROC
Oversampled SMOTE Model	$M_{t-1} \ldots M_t$	22.31 ± 23.30%	71.92 ± 14.06%
MLP Model	M_t	5.10 ± 11.37%	69.14 ± 10.60%
CNN Model	$M_{t-24} \ldots M_t$	15.58 ± 21.31%	60.52 ± 30.26%
Model	Precision	Recall	Accuracy
Oversampled SMOTE Model	21.46 ± 23.07%	26.65 ± 26.87%	89.75 ± 6.11%
MLP Model	6.25 ± 13.50%	6.15 ± 15.83%	92.49 ± 3.62%
CNN Model	18.61 ± 31.98%	25.06 ± 39.10%	85.09 ± 12.41%

5   Conclusion

This study addressed class imbalance within the wearing-off dataset among PD patients. Two main strategies were used to handle class imbalance. First, The data collection was adjusted to improve data quantity. Second, existing class imbalance techniques were compared with the base model and the previous study's result.

Data collection was adjusted to involve the nursing staff in recording the data, which proved to be more practical than relying on PD patients. Notably, this modification extended the data collection period from seven days in the previous study to up to two months, yielding a more extensive dataset from the new PD patients.

Despite the prolonged data collection period and improved tools, class imbalance in the wearing-off dataset remained apparent, as evident in the percentage of wearing-off instances. Consequently, the second contribution of this paper compared various techniques for managing class imbalance, encompassing the creation of synthetic wearing-off records, reduction of non-wearing-off records, and a combination of both. The experimental results showcased that both an oversampled dataset and a resampled dataset, which involved a combination of oversampling and undersampling, led to a 7.592% increase in F1 and AUC-ROC curve scores compared to the base XGBoost forecasting model. Adjusting forecast probabilities also improved metric scores, albeit with slightly lower scores. When comparing these outcomes to the previous study, they also demonstrated enhancements, suggesting that resampling techniques effectively improved the forecasting outcomes.

These advancements in addressing the class imbalance in the wearing-off dataset can potentially enhance the management of wearing-off and PD among patients. The practical application of the proposed models in real-world settings should be explored to validate their utility. Furthermore, considering alternative decision thresholds below the default 50% might further optimize the F1 scores, providing room for calibration based on data or expert knowledge. Overall, these forecasting models have the potential to significantly improve the management of wearing-off and PD in clinical practice.

Acknowledgements. The authors thank the Kyushu Institute of Technology and the Sasakawa Health Foundation Grant Number 2022A-002 for supporting this study.

References

1. Antonini, A., et al.: Wearing-off scales in Parkinson's disease: critique and recommendations: scales to assess wearing-off in PD. Mov. Disord. **26**(12), 2169–2175 (2011). https://doi.org/10.1002/mds.23875
2. Bhidayasiri, R., Tarsy, D.: Parkinson's disease: Hoehn and Yahr Scale. In: Bhidayasiri, R., Tarsy, D. (eds.) Movement Disorders: A Video Atlas: A Video Atlas, pp. 4–5. Current Clinical Neurology, Humana Press (2012). https://doi.org/10.1007/978-1-60327-426-5_2
3. Chen, T., Guestrin, C.: XGBoost: a scalable tree boosting system. In: Proceedings of the 22nd ACM SIGKDD International Conference on Knowledge Discovery and Data Mining, pp. 785–794 (2016). https://doi.org/10.1145/2939672.2939785, https://arxiv.org/abs/1603.02754
4. Colombo, D., et al.: The "gender factor" in wearing-off among patients with Parkinson's disease: a post hoc analysis of DEEP study (2015). https://doi.org/10.1155/2015/787451
5. Garmin: Garmin vivosmart 4. https://buy.garmin.com/en-US/US/p/605739
6. Garmin: Vívosmart 4 - heart rate variability and stress level. https://www8.garmin.com/manuals/webhelp/vivosmart4/EN-US/GUID-9282196F-D969-404D-B678-F48A13D8D0CB.html
7. Heyn, S., Davis, C.P.: Parkinson's disease early and later symptoms, 5 stages, and prognosis. https://www.medicinenet.com/parkinsons_disease/article.htm
8. Jenkinson, C., Fitzpatrick, R., Peto, V., Greenhall, R., Hyman, N.: The PDQ-8: development and validation of a short-form Parkinson's disease questionnaire. Psychol. Health **12**(6), 805–814 (1997). https://doi.org/10.1080/08870449708406741
9. Kashiwara, K., Takeda, A., Maeda, T.: Learning Parkinson's disease together with patients: toward a medical practice that works with patients, with Q&A. Nankodo, Tokyo, Japan (2013). https://honto.jp/netstore/pd-book_25644244.html

10. Lemaître, G., Nogueira, F., Aridas, C.K.: Imbalanced-learn: a Python toolbox to tackle the curse of imbalanced datasets in machine learning. J. Mach. Learn. Res. **18**(17), 1–5 (2017). https://jmlr.org/papers/v18/16-365.html
11. Matloff, N.: Statistical Regression and Classification: From Linear Models to Machine Learning. Chapman and Hall/CRC, New York (2017). https://doi.org/10.1201/9781315119588
12. Mouritzen, N.J., Larsen, L.H., Lauritzen, M.H., Kjær, T.W.: Assessing the performance of a commercial multisensory sleep tracker. PLoS ONE **15**(12), e0243214 (2020). https://doi.org/10.1371/journal.pone.0243214
13. Stacy, M., et al.: End-of-dose wearing off in Parkinson disease: a 9-question survey assessment. Clin. Neuropharmacol. **29**(6), 312–321 (2006). https://doi.org/10.1097/01.WNF.0000232277.68501.08
14. Stevens, S., Siengsukon, C.: Commercially-available wearable provides valid estimate of sleep stages (P3.6-042). Neurology **92** (2019). https://n.neurology.org/content/92/15_Supplement/P3.6-042
15. Stocchi, F., et al.: Early DEtection of wEaring off in Parkinson disease: The DEEP study. Parkinsonism Relat. Disord. **20**(2), 204–211 (2014). https://doi.org/10.1016/j.parkreldis.2013.10.027
16. Victorino, J.N., Shibata, Y., Inoue, S., Shibata, T.: Predicting wearing-off of Parkinson's disease patients using a wrist-worn fitness tracker and a smartphone: a case study. Appl. Sci. **11**(16), 7354 (2021). https://doi.org/10.3390/app11167354
17. Victorino, J.N., Shibata, Y., Inoue, S., Shibata, T.: Understanding wearing-off symptoms in Parkinson's disease patients using wrist-worn fitness tracker and a smartphone. Procedia Comput. Sci. **196**, 684–691 (2022). https://doi.org/10.1016/j.procs.2021.12.064
18. Victorino, J.N., Shibata, Y., Inoue, S., Shibata, T.: Forecasting Parkinson's disease patients' wearing-off using wrist-worn fitness tracker and smartphone dataset (2023)

Real-Time Instance Segmentation and Tip Detection for Neuroendoscopic Surgical Instruments

Rihui Song[1], Silu Guo[1], Ni Liu[1], Yehua Ling[1], Jin Gong[2], and Kai Huang[1(✉)]

[1] Sun Yat-sen University, Guangzhou, China
huangk36@mail.sysu.edu.cn
[2] Third Affiliated Hospital of Sun Yat-sen University, Guangzhou, China

Abstract. Location information of surgical instruments and their tips can be valuable for computer-assisted surgical systems and robotic endoscope control systems. While real-time methods for instrument segmentation and tip detection have been proposed for minimally invasive abdominal surgeries, the challenges become even greater in minimally invasive neurosurgery due to its narrow operating space and diverse tissue characteristics. In this paper, we introduce a real-time approach for instance segmentation and tip detection of neuroendoscopic surgical instruments. To address the specific requirements of neurosurgery, we design a tailored data augmentation strategy for this field and propose a mask filtering module to eliminate false-positive masks. Our method is evaluated using both a neurosurgical dataset and the EndoVis15' dataset. The experimental results demonstrate that the data augmentation module improves the accuracy of instrument detection and segmentation by up to 12.6%. Moreover, the mask filtering module enhances the precision of instrument tip detection with an improvement of up to 39.51%.

Keywords: Instrument instance segmentation · Neurosurgery · Tip detection · Endoscope image processing

1 Introduction

The locations and tips of surgical instruments are vital for automated endoscope holders [9]. These holders are designed to assist surgeons during surgeries by securely holding the endoscope. In minimally invasive surgery (MIS), surgeons pay close attention to the instruments and their tips. Robotic endoscope control systems aim to automatically adjust the endoscope's position to keep the instruments and their tips within the field of view [15]. Hence, the precise knowledge of the locations and tips of surgical instruments is essential for ensuring accurate automatic adjustments.

Existing research on instrument perception based on endoscopic images can be divided into two categories. The first category involves hardware modifications, where customized marks are added to the instruments and detected

B. Luo et al. (Eds.): ICONIP 2023, CCIS 1964, pp. 579–593, 2024.
https://doi.org/10.1007/978-981-99-8141-0_43

based on their color or shape [23]. The second category utilizes digital image processing or machine learning techniques to directly perceive the instruments without modifying the hardware. Some studies [3] employ hand-crafted features for instrument detection or segmentation. With the availability of datasets like EndoVis for endoscopic images, recent works [17] make use of deep neural networks to achieve higher accuracy in instrument perception, particularly in minimally invasive abdominal surgery scenarios.

There is a need to adapt the segmentation and tip detection techniques used in minimally invasive abdominal surgery to neurosurgical procedures. Recently, there have been significant advancements in endoscope resolution, accompanied by a reduction in diameter. As a result, the endoscope has become a suitable tool for neurosurgery. Furthermore, compared to traditional microscopes, the endoscope provides superior flexibility. Consequently, endoscopy is increasingly used in minimally invasive neurosurgical procedures. Developing instance segmentation and tip detection methods specifically designed for neurosurgical instruments would greatly facilitate the integration of automatic endoscope holders in neurosurgeries.

Developing a real-time instance segmentation and tip detection method for neurosurgical instruments is a challenging task. In comparison to abdominal surgeries, neurosurgeries have a narrow operating space, leading to short distances between the endoscope, instruments, and tissue. This poses several difficulties. Firstly, reflective tissue can appear on the instrument's surface, causing it to resemble actual tissue and potentially leading to identification errors. Secondly, the light emitted from the endoscope creates white spots on both the instruments and the tissue. These areas lacking valid information, such as color and texture, present challenges for instrument perception. Lastly, factors like smoke, fog, water, and blood can easily blur the endoscope images. Overcoming these challenges requires modifications to existing real-time instance segmentation and tip detection methods specifically tailored for neurosurgical instruments.

Being aware of the aforementioned challenges, we propose a real-time method for instance segmentation and tip detection of neurosurgical instruments. Considering the need to run our method on resource-constrained devices, we utilize a deep neural network for instance segmentation and a geometric-based tip detection method inspired by the perception approach used in the endoscope robot [9]. To adapt the method for neurosurgery and improve instrument perception accuracy, we made two key modifications. Firstly, we designed a data augmentation strategy specific to neurosurgical scenarios. Secondly, unlike the approach in Gruijthuijsen et al. [9] where tip detection is viewed as a downstream task of segmentation, we leveraged geometric features from the tip detection method to filter out false positive results from instance segmentation. The performance of our method was evaluated on a neuroendoscopy dataset and the EndoVis15' dataset. The experiments demonstrated that our scheme increased instrument detection and segmentation accuracy by up to 12.6% and improved instrument tip detection precision by up to 39.51%. The main contributions of this work are summarized as follows:

- Design of a specialized data augmentation strategy for neurosurgery.
- Proposal of a module to filter false positive results in instrument instance segmentation and tip detection.

The paper is organized as follows: Sect. 2 provides a review of related work in the literature. Section 3 presents our approach. Experimental results are provided in Sect. 4. Finally, Sect. 5 concludes the paper.

2   Related Work

Surgical instrument perception is a fundamental task for surgical assistant systems, such as robot-assisted systems and clinical augmented reality applications. Various techniques, including traditional digital image processing methods, machine learning approaches with hand-crafted features, and deep learning neural networks, are employed to identify surgical instruments in images.

The methods using digital image processing techniques can be divided into two categories. The first category simplifies tip localization by manually adding markers to the surgical instruments [23]. The second category utilizes color or geometric features without modifying the surgical instruments [1]. Compared with deep learning methods, these approaches typically require fewer computational resources. Considering that our algorithm operates on a resource-constrained robot platform, our tip detection algorithm falls within the domain of digital image processing methods, utilizing the shape characteristics of surgical instruments.

Machine learning approaches utilizing hand-crafted features have been employed for instrument perception. A two-stage framework has been proposed to accomplish segmentation and detection tasks, employing a linear Support Vector Machine (SVM) to globally learn shape parameters [3]. Traditional machine learning methods have also been utilized for surgical tool identification in laparoscopic images [22]. In contrast, our approach does not directly employ machine learning methods with manually designed features for instrument instance segmentation or tip detection. Instead, we introduce a filtering module that employs SVM to eliminate false positive results in the instrument segmentation output.

With the advancement of deep learning techniques, researchers have explored their application in instrument perception as well. One approach combines optical flow tracking with a fully convolutional network (FCN) to achieve real-time segmentation and tracking of laparoscopic surgical instruments [8]. Kurmann et al. [11] addresses the tip localization issue by using U-Net for part segmentation. Another study applies a residual neural network to instrument segmentation during laparoscopy and reports a 4% performance improvement on the EndoVis dataset [16]. The utilization of U-Net for instrument segmentation is also demonstrated by constructing a simulation dataset with virtual reality techniques [17]. Overall, deep learning methods have achieved state-of-the-art performance in instrument segmentation tasks. In our own work, we also employed a deep learning method for the segmentation of surgical instrument instances.

Instrument segmentation and tip detection methods have been integrated into the endoscope robot system. In the study by Gruijthuijsen et al. [9], a robot control system based on surgical instrument tracking is introduced. Their system employs CNN-based real-time tool segmentation networks for instrument segmentation [8], and they propose a tip detection method based on the segmentation results. To mitigate the issue of false-positive tip detection, they apply an instrument graph pruning process assuming a maximum of two tools with two tips each. Our approach also targets endoscopic robots. We propose a filtering module to eliminate false positives in instrument segmentation. The filtering module not only enhances the accuracy of surgical segmentation but also improves the performance of instrument tip detection.

Image data augmentation is a commonly used technique to improve the performance of deep learning methods. It can be classified into two categories: basic image manipulation and deep learning approach [21]. Basic image manipulations involve kernel filters, color space transformations, random erasing, geometric transformations, and image mixing. Deep learning approaches include adversarial training, neural style transfer, and GAN data augmentation. The data augmentation we propose falls under basic image manipulations. We adjust the basic image operations according to the specific characteristics of neurosurgical procedures.

3   Approach

Our objective is to address the challenges posed by the limited operational space in minimally invasive neurosurgery. To tackle this problem, we focus on two aspects. Firstly, we propose an endoscope image data augmentation strategy for neurosurgical scenarios to improve instrument segmentation performance. Secondly, we introduce a filtering module to eliminate false-positive results in instrument segmentation, enhancing both instrument segmentation accuracy and the performance of the instrument tip detection algorithm.

3.1   Framework

Our framework consists of three stages: (1) Instance Segmentation, (2) Mask Filtering, and (3) Tip Detection. The framework is shown in Fig. 1.

Instance Segmentation. The instance segmentation network adopts a similar structure to YOLACT [2] which can be divided into a backbone network, Feature Pyramid Network (FPN), Protonet, prediction head, Non-Maximum Suppression module (NMS), assembly, and post-processing modules. The main modifications include adding convolutional block attention modules (CBAM) to the backbone network and customizing the loss functions. The rest of the modules are consistent with YOLACT.

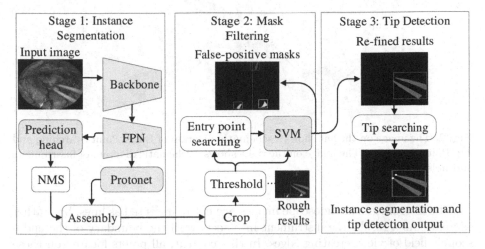

Fig. 1. The architecture of the proposed instrument perception framework is composed of three stages. Modules that can be trained are marked in green. (Color figure online)

The network is trained via end-to-end back-propagation guided by the following loss function:

$$L = \lambda_1 L_{focal} + \lambda_2 L_{loc}, \tag{1}$$

L_{focal} is binary focal loss, representing the loss for each pixel classification. The binary focal loss relaxes the conditions for training the model to make predictions.

$$L_{focal}(p_t) = -\alpha_t (1 - p_t)^\gamma \log(p_t). \tag{2}$$

where p_t refers to the probability of a pixel point being classified as a foreground:

$$p_t = \begin{cases} P, & y = 1 \\ 1 - P, & \text{otherwise} \end{cases}, \tag{3}$$

L_{loc} denotes the localization loss, which is the smooth $L1$ loss [4] between the predicted box and the ground truth box.

Instrument Tip Detection. During surgery, it is possible for the tips of instruments to become obscured by tissue, other instruments, or bloodstains, resulting in their invisibility. To effectively control the endoscope robot, the tip of an instrument is redefined as the end of the visible portion. This means that the tip point and the entry point represent the two ends of the instrument within the field of view. Please refer to Fig. 2 for a schematic diagram illustrating this concept.

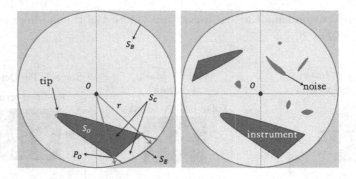

Fig. 2. The image on the left shows a schematic diagram of Entry Point Searching and Tip Point Searching. The image on the right shows a schematic diagram of instruments and noises.

The instruments in endoscope images are not represented as lines, but rather as areas. As a result, the instrument intersects with the boundary of the endoscope's field of view, creating edges. In this context, all points located on these edges are considered entry points. With this definition, the entry point can be described using Eq. (4):

$$S_E = S_C \cap S_B. \tag{4}$$

S_C, S_B, and S_E are three sets of points. The points of S_C and S_B come from the contour of the instrument and the boundary of the view of the endoscope, respectively. Intersection S_E represents the set of entry points.

The outline of each instrument area is represented by S_O. However, due to limitations in segmentation results, the boundaries of the instrument instances may not align perfectly with the actual instrument boundaries, as depicted in Fig. 2. When replacing S_C in Eq. (5) with S_O for operations involving S_B, it often results in an empty set for S_E in most cases. To address this, we approximate the entry point P_E as the point nearest to the boundary of the endoscope's field of view. The approximation for P_E can be described by Eq. (5).

$$P_E \in S_O \wedge \forall P \in S_O, \|OP_E\| \geq \|OP\| \tag{5}$$

The instruments used in minimally invasive surgery are typically long and thin, allowing them to navigate through narrow channels in the body. While the shapes of their ends may vary, the tip points consistently tend to be located near the farthest point from the entry point. Hence, the tip point P_T can be calculated using Eq. (6):

$$P_T \in S_O \wedge \forall P \in S_O, \|P_E P_T\| \geq \|P_E P\| \tag{6}$$

Mask Filtering. In minimally invasive neurosurgery, the narrow surgical space frequently gives rise to various interferences in endoscope images, such as white spots, smoke, and bloodstains. These disturbances make it more susceptible for instrument instance segmentation algorithms to produce false-positive results. These false-positive results adversely impact the accuracy of tip detection algorithms that rely on the segmented masks, ultimately diminishing the precision and stability of robot control. Therefore, we propose a mask filtering module to address this issue.

The areas formed by the misclassified pixels are noise instances in the masks. A schematic diagram is shown in Fig. 2. Since the tip detection algorithm is applied to each instance segmentation result, the noise instance generates false-positive tips. Therefore, noise instances should be removed.

Based on our observations, instruments in neuroendoscopy images are usually of considerable size to ensure clear visibility to the surgeon. Furthermore, due to the limited field of view under the endoscope, there are intersections between the instruments and the boundaries, which act as entry points for the instruments. To distinguish between real instruments and false-positive results, we utilize an SVM (Support Vector Machine) classifier [7]. The area of the instrument region and the distance from the entry point to the center point are utilized as features for this distinction.

The features can be influenced by the transformation of the image size. For instance, if the area of an instrument is initially 200, it becomes 50 when the image is scaled down to half its size. To mitigate the impact of image transformation, we redefine the first feature as the ratio of the number of pixels in the instrument mask to the total number of pixels in the image, rather than solely relying on the area of the instrument region. The first feature can be described using Eq. (7).

$$feature_1 = \frac{Pixel_{tool}}{L_w * L_h} \tag{7}$$

$Pixel_{tool}$ means the number of pixels in the instrument instance. L_w and L_h are the width and height of the image. For the same reason, the second feature is defined as Eq. (8) rather than the distance between the entry point and the center point.

$$feature_2 = \frac{EO}{\sqrt{L_w * L_h}} \tag{8}$$

L_w and L_h are the width and height of the image. EO represents the distance between the entry point E and the center point O. The location of the entry point E is obtained by the entry point searching algorithm.

3.2 Data Augmentation

Due to the limited space in minimally invasive neurosurgery, the distance between the endoscope, instruments, and tissues is significantly reduced. This close proximity often leads to interferences such as white spot, smoke, mist, and bloodstains in endoscope images. Conventional image data augmentation methods have not been able to effectively simulate these interferences in the images.

To address this issue, we propose a novel data augmentation strategy designed specifically for endoscopic images in the field of neurosurgery.

First, we categorized the images in the dataset into two classes: the minority class and the majority class. If an image contained interference, it was assigned to the minority class; otherwise, it was classified as the majority class. The minority class consisted of four situations: Reflection, Overlapping, Multi-tool, and Infrequent Backgrounds. Reflection refers to the phenomenon where a portion of an area in an image reflects the surrounding objects, resulting in a mirror-like effect. Multi-tool indicated images that contained more than two instruments. Overlapping denoted images in which at least one instrument was partially obscured. Infrequent Background encompassed images with rare surgical scenes. For example, if the number of images with the nasal cavity as the background was much smaller than the number of images with the brain as the background in a given dataset, the images with the nasal cavity as the background would be classified as belonging to the Infrequent Background class.

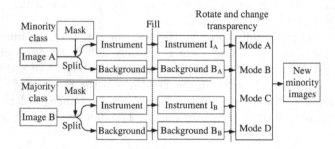

Fig. 3. Overview of data augmentation. Mode A overlays I_A and B_A, mode B overlays I_A, B_A and I_B, mode C overlays I_B and B_A, and mode D overlays I_A and B_B.

The Overview of data augmentation is depicted in Fig. 3. We segment the source images into template images. For instance, Image A and Image B are randomly chosen from the minority class and the majority class, respectively. These images are divided into two parts based on the mask: one contains the background, while the other consists of the instruments. The areas in the instrument-only images that are empty are filled with zeros, whereas the empty areas in the background images are filled with neighboring pixels. From these two images, four template images are generated. Mode A overlays I_A and B_A, mode B overlays I_A, B_A and I_B, mode C overlays I_B and B_A, and mode D overlays I_A and B_B. Prior to overlaying, images I_A and I_B are rotated by an angle of θ and multiplied by a transparency coefficient of λ. The value of θ was chosen randomly from the set of $[0, 90, 180, 270]$. The value of λ was randomly selected from the range of -0.1 to 0.1.

Four modes have the ability to generate images that belong to various types of minority classes. If the transparency coefficient, λ, is not zero, the newly synthesized images from all the modes could potentially be classified as images belonging to the Reflection class, as the background color and texture appear on the instrument's surface. Mode B allows for the synthesis of new overlapping images. Prior to image stitching, when template images are rotated by an angle, θ, there is a possibility of generating overlapping instruments. Additionally, for the Multi-instrument class, new images can be synthesized using Mode B as well. Following the fusion of I_A and I_B, the image may contain three or more instruments. In situations where Image A represents a sample from the Infrequent Background class, new samples synthesized using modes A, B, and C would also fall within the Infrequent Background class. Since white spots, which are indicative of Reflection cases, frequently appear in the images, we randomly add one or two white ellipses of different sizes at various positions on the images with a probability of ρ. In our experiment, the value of ρ was set to 0.1.

4   Results and Discussion

Our method was evaluated using a neurosurgical dataset and the EndoVis15 dataset[1]. We conducted ablation studies to evaluate the impact of different modules. The qualitative and quantitative results are presented and discussed. The source code for our project is publicly available[2].

4.1   Datasets and Metrics

The neurosurgical dataset utilized in our study consisted of 11 neuroendoscopy videos with a resolution of (resolution: 1080×1920, 30 fps) and a frame rate of 30 fps. Videos were sampled at a frequency of 0.2 Hz, resulting in a total of 3,062 images. The images were divided into a ratio of approximately 5:1.25:1, resulting in 2,113 images allocated to the training set, 526 images assigned to the validation set, and 423 images designated for the test set. The images in the training set, validation set, and test set are sourced from different surgical videos.

Table 1. Instrument Segmentation Results on EndoVis

Method	Sensitivity	Specificity	Balance Accuracy
Fcn-8 s [8]	72.2	95.2	83.7
Fcn-8 s+OF [8]	87.8	88.7	88.3
Drl [16]	85.7	98.8	92.3
Csl [12]	86.2	**99.0**	92.6
Cfcm34 [14]	88.8	98.8	93.8
Rnmf(Ours)	**89.1**	98.9	**94.0**

---

[1]   https://endovissub-instrument.grand-challenge.org/.
[2]   https://github.com/RH-Song/Real-time-Neuroendoscopy-Multi-task-Framework.

The dataset for the SVM classifier in the Mask Filtering module was constructed using the following steps. Firstly, we trained an instance segmentation neural network on the training set of the neurosurgical dataset. Next, the trained model was used to infer the images in the training set. The segmentation results were then used to calculate the features described by Eq. (7) and Eq. (8), forming a dataset for the SVM classifier. The labels of the dataset were generated based on the Intersection over Union (IoU) of the masks and segmentation ground truth. If the IoU between a mask and its ground truth was lower than 50%, it was labeled as noise. Otherwise, it was labeled as an instrument.

In our experiment, we employed mean Intersection over Union (mIoU) and mean average precision (mAP) as evaluation metrics to assess the accuracy of instrument instance segmentation [3]. Specifically, we reported the results for mAP_{50} and mAP_{75}, representing the performance when the mean average precision exceeds 50% and 75%, respectively. For instrument tip detection, precision was used as the performance metric. A prediction was considered correct if the distance between the predicted tip and its corresponding ground truth was within σ pixels. To determine a suitable σ, we randomly selected 100 images and obtained five different annotations for each image. By calculating the center of each tip point, we determined that the maximum offset labeled by the annotators was 13.8, which was set as σ for our analysis. To show our results in tables concisely, we call our framework RNMF, which is short for Real-time Neuroendoscopy Multi-task Framework.

4.2   Results

Gruijthuijsen et al. [9] propose a robot control system based on surgical instrument tracking, which utilizes CNN-based real-time tool segmentation networks [8] for instrument segmentation. The CNN-based segmentation model [8] is evaluated using the EndoVis15' dataset. To compare our method with this approach, we also evaluate our method on the same dataset. Table 1 displays the segmentation results of our method and several previous works on the EndoVis15 dataset, including FCN-8 s [8], FCN-8 s with optical flow (FCN-8 s+OF) [8], DRL [16], CSL [12], and CFCM [14]. Our method achieves the highest results in sensitivity and balance accuracy, with the result in specificity being close to the highest result.

The results of instrument instance segmentation and tip detection of images from the neurosurgical dataset are shown in Fig. 4. Five kinds of images are shown in the figure. Object detection results are displayed with bounding boxes. Segmentation of different instances is represented by various colors. Yellow points indicate the results of tip detection. The accuracy of mask edges and tips is high when the image is clear, while the edges of semantic segmentation and the tips of instruments remain mostly correct even in mildly blurred images.

To evaluate the performance of the proposed framework on the instrument detection task, we compare its results of the instrument detection branch against three two-stage methods: Mask R-CNN [10] (MR), Faster R-CNN [19] with DCN [6] (FR+DCN), and Mask R-CNN [10] with DCN [6] (MR+DCN); and three one-stage methods: YOLOv3 [18], Retina-Net [13], and YOLACT [2]. After comparing

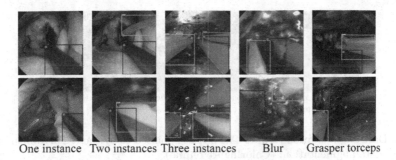

One instance Two instances Three instances   Blur   Grasper torceps

Fig. 4. Instance segmentation and tip detection results of images from the neurosurgical dataset.

Table 2. Instance Segmentation Results on neurosurgical Dataset

Method	Box mAP$_{50}$	Box mAP$_{75}$	Mask mAP$_{50}$	Mask mAP$_{75}$
MR [10]	63.9/67.5	72.2/77.2	66.2/67.5	74.1/77.2
FR [19]+DCN [6]	69.6/69.8	78.9/78.3	–	–
MR [10]+DCN [6]	69.5/71.5	78.4/80.3	68.1/69.7	76.4/78.4
YOLOv3 [18]	55.4/60.4	65.0/69.4	–	–
Retina-Net [13]	63.4/65.6	69.2/73.5	–	–
YOLACT [2]	60.4/67.9	69.6/80.1	63.1/72.8	67.2/80.7
RNMF(Ours)	63.7/72.5	70.2/82.8	67.8/75.7	73.7/83.3

mAP$_1$/mAP$_2$: mAP$_1$ is the result of with data augmentation used in Mask-RCNN [10], and mAP$_2$ is the result of with data augmentation strategy proposed by us.

the results shown in Table 2, our scheme, RNMF with our data augmentation strategy, achieves the highest accuracy for instrument detection. Its box mAP is at least 2.9% higher than the two-stage networks and 9.1% higher than the one-stage networks. The mask mAP of our scheme is 12.6% higher than YOLACT and 7.6% higher than Mask R-CNN with DCN.

Ablation Studies. In the ablation studies, we evaluated the effectiveness of data augmentation, neural network modifications, and mask optimization individually.

Four images generated by data augmentation are shown in Fig. 5. The objects indicated by the red arrows are added through data augmentation. Table 2 presents the results of methods employing different data augmentation strategies. mAP$_1$ represents the result when data augmentation is used in Mask-RCNN [10], while mAP$_2$ represents the result with our proposed data augmentation strategy. Our data augmentation strategy demonstrated the most significant improvement in instrument detection within our method, achieving a 12.6% enhancement in Box mAP$_{75}$. Similarly, in YOLACT [2] for segmentation, there was a 13.5%

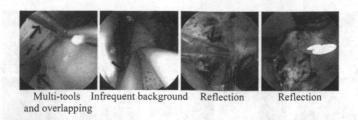

Multi-tools   Infrequent background   Reflection   Reflection
and overlapping

Fig. 5. Data augmentation results. The objects pointed to by the red arrows are added through data augmentation. (Color figure online)

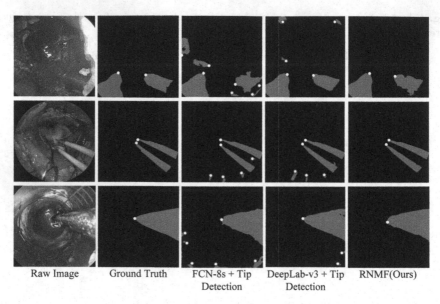

Raw Image   Ground Truth   FCN-8s + Tip   DeepLab-v3 + Tip   RNMF(Ours)
                                     Detection   Detection

Fig. 6. Instrument segmentation results and tip detection results of three challenging scenarios. The red masks are the result of instrument segmentation. The white solid dots are the result of tip detection. (Color figure online)

improvement in Mask mAP_{75}. All models exhibited enhanced performance after training with our data augmentation strategy.

Our framework builds upon the foundation of YOLACT [2]. The key modifications include integrating CBAM into the backbone, refining the loss function, and incorporating a mask filtering module as a post-processing step. Upon comparing the results between YOLACT and RNMF in Table 2, we can observe that these network modifications yield a 3.3% improvement in Box mAP_{75} and a 4.7% enhancement in Mask mAP_{75}, respectively.

To evaluate the performance of the tip detection algorithm, we conducted tip detection using the segmentation results obtained from FCN [20], DeepLab-v3 [5], and our own method. The qualitative results are shown in Fig. 6. Additionally, to assess the effect of the mask filtering module on the tip detection

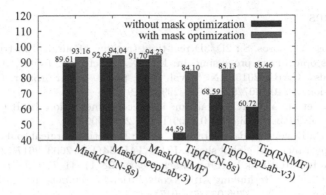

Fig. 7. Instrument segmentation results and tip detection results of the three methods.

algorithm, we performed tip detection using both masks processed by the mask filtering module and masks without any processing. The results of instrument segmentation and tip detection are shown in Fig. 7. It can be observed that the mask filtering module improves the performance of both semantic segmentation and tip detection for all models. The improvement in mIoU for semantic segmentation is smaller than the improvement in the precision of tip detection. The main reason for this phenomenon is that the area of false positive results filtered out by the mask filtering module is small compared to the total area of the instrument mask. Although these false positive mask areas are not large, they are often composed of many small regions. These regions generate false positive detections, significantly reducing the accuracy of tip detection. Therefore, after removing these false positive regions through the mask filtering module, there is a significant improvement in the results of tip detection, with the maximum improvement reaching 39.51%.

5   Conclusion

We have introduced a real-time approach for instance segmentation and tip detection of neuroendoscopic surgical instruments. To address the challenges arising from the restricted operation space in neurosurgery, we have proposed a data augmentation strategy and a mask filtering module. Our method has been evaluated using a neurosurgical dataset as well as the EndoVis15 dataset. Experimental results have demonstrated that both the data augmentation and the mask filtering module contribute to improved performance in instance segmentation and tip detection. In the future, our aim is to develop an end-to-end neural network that can operate in real-time on a resource-constrained robot platform, enabling accurate instrument segmentation and tip detection.

References

1. Agustinos, A., Voros, S.: 2D/3D real-time tracking of surgical instruments based on endoscopic image processing. In: Luo, X., Reichl, T., Reiter, A., Mariottini, G.-L. (eds.) CARE 2015. LNCS, vol. 9515, pp. 90–100. Springer, Cham (2016). https://doi.org/10.1007/978-3-319-29965-5_9

2. Bolya, D., et al.: YOLACT: real-time instance segmentation. In: ICCV, pp. 9157–9166 (2019). https://doi.org/10.1109/ICCV.2019.00925

3. Bouget, D., Benenson, R., Omran, M., et al.: Detecting surgical tools by modelling local appearance and global shape. IEEE TMI **34**(12), 2603–2617 (2015)

4. Caruana, R.: Multitask learning. Mach. Learn. **28**(1), 41–75 (1997)

5. Chen, L.C., et al.: Rethinking Atrous convolution for semantic image segmentation. arXiv preprint: arXiv:1706.05587 (2017)

6. Dai, J., et al.: Deformable convolutional networks. In: ICCV, pp. 764–773 (2017). https://doi.org/10.1109/ICCV.2017.89

7. Fischetti, M.: Fast training of support vector machines with gaussian kernel. Discret. Optim. **22**, 183–194 (2016)

8. García-Peraza-Herrera, L.C., et al.: Real-time segmentation of non-rigid surgical tools based on deep learning and tracking. In: Peters, T., et al. (eds.) CARE 2016. LNCS, vol. 10170, pp. 84–95. Springer, Cham (2017). https://doi.org/10.1007/978-3-319-54057-3_8

9. Gruijthuijsen, C., Garcia-Peraza-Herrera, L.C., Borghesan, G., et al.: Robotic endoscope control via autonomous instrument tracking. Front. Robot. AI **9**, 832208 (2022). https://doi.org/10.3389/frobt.2022.832208

10. He, K., et al.: Mask R-CNN. In: ICCV, pp. 2961–2969 (2017). https://doi.org/10.1109/ICCV.2017.322

11. Kurmann, T., et al.: Simultaneous recognition and pose estimation of instruments in minimally invasive surgery. In: Descoteaux, M., Maier-Hein, L., Franz, A., Jannin, P., Collins, D.L., Duchesne, S. (eds.) MICCAI 2017. LNCS, vol. 10434, pp. 505–513. Springer, Cham (2017). https://doi.org/10.1007/978-3-319-66185-8_57

12. Laina, I., et al.: Concurrent segmentation and localization for tracking of surgical instruments. In: Descoteaux, M., Maier-Hein, L., Franz, A., Jannin, P., Collins, D.L., Duchesne, S. (eds.) MICCAI 2017. LNCS, vol. 10434, pp. 664–672. Springer, Cham (2017). https://doi.org/10.1007/978-3-319-66185-8_75

13. Lin, T.Y., et al.: Focal loss for dense object detection. In: TPAMI, pp. 2980–2988 (2017). https://doi.org/10.1109/TPAMI.2018.2858826

14. Milletari, F., Rieke, N., Baust, M., Esposito, M., Navab, N.: CFCM: segmentation via coarse to fine context memory. In: Frangi, A.F., Schnabel, J.A., Davatzikos, C., Alberola-López, C., Fichtinger, G. (eds.) MICCAI 2018. LNCS, vol. 11073, pp. 667–674. Springer, Cham (2018). https://doi.org/10.1007/978-3-030-00937-3_76

15. Niccolini, M., Castelli, V., Diversi, C., et al.: Development and preliminary assessment of a robotic platform for neuroendoscopy based on a lightweight robot. IJMRCAS **12**(1), 4–17 (2016). https://doi.org/10.1002/rcs.1638

16. Pakhomov, D., Premachandran, V., Allan, M., Azizian, M., Navab, N.: Deep residual learning for instrument segmentation in robotic surgery. In: Suk, H.-I., Liu, M., Yan, P., Lian, C. (eds.) MLMI 2019. LNCS, vol. 11861, pp. 566–573. Springer, Cham (2019). https://doi.org/10.1007/978-3-030-32692-0_65

17. Perez, S.A.H., et al.: Segmentation of endonasal robotic instruments in a head phantom using deep learning and virtual-reality simulation. In: ROBOMEC 2020, pp. 2P2-F01. The Japan Society of Mechanical Engineers (2020)

18. Redmon, J., Farhadi, A.: Yolov3: an incremental improvement. arXiv preprint: arXiv:1804.02767 (2018)
19. Ren, S., He, K., Girshick, R., et al.: Faster R-CNN: towards real-time object detection with region proposal networks. TPAMI **39**(6), 1137–1149 (2016). https://doi.org/10.1109/TPAMI.2016.2577031
20. Shelhamer, E., Long, J., Darrell, T.: Fully convolutional networks for semantic segmentation. TPAMI **39**(4), 640–651 (2017). https://doi.org/10.1109/TPAMI.2016.2572683
21. Shorten, C., Khoshgoftaar, T.M.: A survey on image data augmentation for deep learning. J. Big Data **6**(1), 1–48 (2019)
22. Zappella, L., Béjar, B., Hager, G., et al.: Surgical gesture classification from video and kinematic data. MIA **17**(7), 732–745 (2013)
23. Zhao, Z.: Real-time 3D visual tracking of laparoscopic instruments for robotized endoscope holder. Bio-Med. Mater. Eng. **24**(6), 2665–2672 (2014)

Spatial Gene Expression Prediction Using Hierarchical Sparse Attention

Cui Chen, Zuping Zhang[✉], and Panrui Tang

School of Computer Science and Engineering, Central South University, Changsha 410083, China
{214701013,zpzhang,224711119}@csu.edu.cn

Abstract. Spatial Transcriptomics (ST) quantitatively interprets human diseases by providing the gene expression of each fine-grained spot (*i.e.*, window) in a tissue slide. This paper focuses on predicting gene expression at windows on a tissue slide image of interest. However, gene expression related to image features usually exhibits diverse spatial scales. To spatially model these features, we newly introduced Hierarchical Sparse Attention Network (HSATNet). The core idea of HSATNet is to employ a two-level sparse attention mechanism, namely coarse (*i.e.*, area) and fine attention. Each HSAT Block consists of two main modules: i) An adaptive sparse coarse attention filters out the most irrelevant areas, resulting in the acquisition of adaptive sparse areas. ii) An adaptive sparse fine attention module filters out the most irrelevant windows, leading to the acquisition of adaptive sparse windows. The first module aims to identify similarities among windows within the same area, while the second module captures the differences between different windows. Particularly, after fusing these two modules together, without any additional training data or pre-training, experiments conducted on 10X Genomics breast cancer data show that our HSATNet achieves an impressive PCC@S of **7.43** for gene expression prediction. This performance exceeds the current state-of-the-art model. Code is available at https://github.com/biyecc/HSATNet.

Keywords: Spatial Transcriptomics · Gene Expression Prediction · Deep Learning · Hierarchical Sparse attention · Tissue Slide Image

1 Introduction

Based on the findings in Natural Methods [15], Spatial Transcriptomics (ST) has emerged as the future technique in disease research due to its capability to capture gene expression with high-resolution spatial information. However, low throughput and spatial diversity ST limit the study of gene expression in candidate windows [17]. To effectively and efficiently predict gene expression of each window within slide images (Fig. 1), we propose a novel solution called Hierarchical Sparse Attention (HSATNet).

Previous studies on gene expression prediction have introduced end-to-end neural networks, named STNet [9] and NSL [4]. STNet employs transfer learning by fine-tuning a pretrained DenseNet model to gene expression prediction,

B. Luo et al. (Eds.): ICONIP 2023, CCIS 1964, pp. 594–606, 2024.
https://doi.org/10.1007/978-981-99-8141-0_44

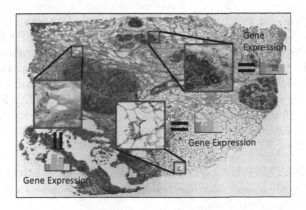

Fig. 1. Overview of fields. Each window of a tissue slide image is with distinct gene expression. Here is an example, we have a tissue slide image with three windows, and each of the windows corresponds with expression of four different gene types. Our goal is to predict the gene expression of each window.

NSL only use convolution operation to map the color intensity of the window to the corresponding gene expression. Although these approaches leverage neural networks and exhibit potential for high throughput analysis, they face two important limitations, lack of local feature aggregation and vulnerable assumption with color intensity of slide image windows. To address these limitations, recent study Exemplar Guided Network (EGN) [25] was proposed. It combined exemplar learning with the vanilla Vision Transformer(ViT) [7]. However, there are two limitations in the selection of exemplars. Firstly, it is challenging to choose the exemplars appropriately. Secondly, there is a lack of consideration for the relationships between different exemplars.

In this work, we tackle these limitations by introducing a novel hybrid hierarchical structure, named HSATNet. As illustrated in (Fig. 2), it flexibly merges coarse level (*i.e.*, area) and fine level (*i.e.*, window) sparse attention, enabling effective and efficient gene expression prediction. HSATNet comprises of four distinct stages, where the input image resolution is decreased through the utilization of a strided convolutional layer, while simultaneously increasing the number of feature maps.

Our contributions are summarised below:

- In this paper, we propose HSATNet , a Hierarchical Sparse Attention network, to effectively and efficiently predict gene expression from the slide image windows. Our approach outperforms state-of-the-art (SOTA) methods, as demonstrated through experiments on two breast cancer datasets.
- At the coarse level, we propose the Adaptive Sparse Area Attention (ASAA) module, which utilizes a sparse-area strategy based on the similarities among different areas. By filtering out the least relevant areas and guiding window-to-area attention, the module focuses on capturing the shared areas of windows

within the same area, which enables us to obtain adaptive sparse coarse-level attention, effectively highlighting the relevant information for feature fusing.

– At the fine level, we propose the Adaptive Sparse Window Attention (ASWA) module which utilizes a sparse-window strategy based on the similarities among different windows. By filtering out the least relevant windows and implementing sparse window-to-window attention, the module focuses on capturing the differences of different windows. This allows us to obtain adaptive sparse fine-level attention, effectively capturing the distinctive characteristics of the windows.

Overall, HSATNet leverages the hierarchical sparse attention approach, incorporating both coarse and fine levels, to enhance gene expression prediction from slide image windows.

2 Related Work

Gene Expression Prediction. Gene expression within the slide image has important biological effects on the properties of the tissue, measuring gene expression is a fundamental process in the development of novel diseases treatments [1]. Deep Learning methods have been introduced to this process, the image-based approaches can be categorized into two branches. The first brunch focuses on bulk RNA-Seq and single-cell RNA [13], which measures gene expression within a large predefined area of up to $10^5 \times 10^5$ in corresponding slide images and the cellular level, respectively. However, both these approach results in the loss of rich spatial information about gene expression, which is crucial when studying tissue heterogeneity [8]. The second brunch is ST, a recently developed technique that utilizes DNA barcodes to distinguish different windows in the tissue and captures spatial information. He et al. designed the STNet [9], which is the first consider integrating the slide image with ST. Dawood et al. [4] propose an NSL, which is "color-aware". Recently, Yang et al. propose exemplar guided deep neural network, named EGN [25].

Vision Transformer. With inspirations drawn from neural activity in biological brains, sparsity of hidden representation in deep neural networks as a tantalizing "free lunch" emerges for both vision and NLP tasks [21,26]. ViT [7] astutely employs attention [20] mechanism, splitting each image into patches and treating them as tokens. It exhibits noteworthy efficiency in various classification tasks [5] but rely on full self-attention mechanism, which leads to quadratic growth in computational complexity as the image size increases. To tackle this dilemma, recent studies focus on sparse attention [6,10,11,24]. Ramachandran et al. have proposed the local window self-attention mechanism [17], along with its shifted or haloed variants. These mechanisms facilitate interactions across different windows and provide a solution to the computational complexity issue. Furthermore, axial self-attention [10] and criss-cross attention [11] have introduced to augment the receptive field. Dong et al. propose [6] both horizontal and vertical stripes window self-attention. Xia et al. [24] trying to make the sparsity adaptive to data. Indeed, it is widely proven that sparse representation also plays a critical role in handling low-level vision problems, such as image deraining [23] and super-resolution [16]. In principle, sparse attention can be categorized into data-based (fixed) sparse attention and content-based sparse attention [3,19]. For data-based sparse attention, several local attention operations are introduced into CNN backbone, which mainly considers attending only to local window size.

3   Method

3.1   Preliminaries

Problem Formulation. There have a dataset collection comprising tissue slide images, where each image comprises multiple windows. Each window within the images is annotated with gene expression information. To be brevity, this dataset collection as pairs consisting of a slide image window $X_i \in \mathbb{R}^{3 \times H_i \times W_i}$ and gene expression $y_i \in \mathbb{R}^M$, ie., $\{(X_i, y_i)\}_{i=1}^N$, where N is the collection size, H and W are the height and width of X_i, and M is the number of gene types. Our goal is train a effective and efficient deep neural network model to predict y_i from X_i.
Attention. The attention function takes input queries $Q \in \mathbb{R}^{N_q \times C}$, keys $K \in \mathbb{R}^{N_{kv} \times C}$, and values $V \in \mathbb{R}^{N_{kv} \times C}$. It computes a weighted sum of the values for each query, where the weights are calculated as normalized dot products between the query and the corresponding keys. This operation can be expressed in a concise matrix form as:

$$\text{Attention}(Q, K, V) = \text{softmax}\left(\frac{QK^\top}{\sqrt{C}}\right) V \tag{1}$$

where the softmax function applied along the query dimension. To prevent concentrated weights and gradient vanishing, a scalar factor of $\sqrt{C}$ is introduced in the attention function. The output of this attention function is a matrix of size $\mathbb{R}^{N_q \times C}$, where each row contains the weighted sum of values for a corresponding query.

3.2   Hierarchical Sparse Attention Network (HSATNet)

In this paper, we propose a Hierarchical Sparse Attention Network (HSATNet). Our key idea is consider coarse-and-fine information and each HSAT Block contains a Adaptive Sparse Area Attention (ASAA) module and a Adaptive Sparse Window Attention (ASWA) module for both coarse-and-fine feature extraction. After combine this two module with LCE, followed by a simple gene prediction layer.
Adaptive Sparse Area Attention (ASAA). Several works [6,10,11,22,24] have proposed various sparse attention mechanisms to address the time and space complexity issues of traditional Transformer. These mechanisms apply only a small number of key-value pairs instead of all pairs. However, existing

approaches either use handcrafted static patterns or share the same subset of key-value pairs among all queries.

To tackle the dilemma of handcrafted approaches, we design a ASAA module enables adaptively sparse area attention through sparse-area strategy. The key idea of strategy is to automatically filter out irrelevant areas in coarse level, resulting in a small subset of routed areas(Note that, the routed areas is the similarity of the windows that belong to query-area, all windows in the query-window will attention with the same windows belong to routing areas). We give a detailed explanation of sparse-area strategy as follows.

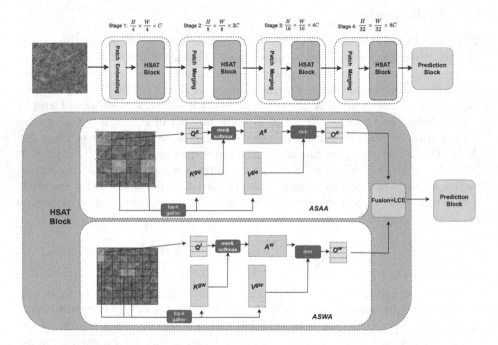

Fig. 2. Network architecture of the proposed HSATNet. Each HSAT Block fusions ASAA and ASWA to the coarse and fine level feature extration. By gathering key-value pairs in top k related areas, ASAA utilizes the sparse-area attention, which find the sameness of the same area. ASWA captures the sparse-window attention, which is the difference of the different windows. Finally, after fusion these two attention and combine with LCE block, HSATNet achieves the gene expression prediction task.

- **Area Partition and Input Projection.** Given a 2D input feature map $X \in \mathbb{R}^{C \times H \times W}$, we begin by partitioning it into non-overlapping areas of size $S \times S$, with each area containing $\frac{HW}{S^2}$ feature vectors. To accomplish this, this step reshape X as $X^a \in \mathbb{R}^{S^2 \times \frac{HW}{S^2} \times C}$ and then use linear projections to obtain the query, key, and value tensors, $Q, K, V \in \mathbb{R}^{S^2 \times \frac{HW}{S^2} \times C}$:

$$Q^i = X^i W^q, \quad K^i = X^i W^k, \quad V^i = X^i W^v. \qquad (2)$$

where $\boldsymbol{W}^q, \boldsymbol{W}^k, \boldsymbol{W}^v \in \mathbb{R}^{C \times C}$ are the projection weights for the query, key, and value, respectively. Specifically, we first construct a directed relationship about per-area, and obtain area-level queries and keys, $\boldsymbol{Q}^a, \boldsymbol{K}^a \in \mathbb{R}^{S^2 \times C}$, by taking the per-area average of $\boldsymbol{Q}^i$ and $\boldsymbol{K}^i$, respectively. To be more specific, We then derive the adjacency matrix, $\boldsymbol{A}^a \in \mathbb{R}^{S^2 \times S^2}$, of area-to-area affinity via matrix multiplication between $\boldsymbol{Q}^a$ and transposed $\boldsymbol{K}^a$:

$$\boldsymbol{A}^a = \boldsymbol{Q}^a (\boldsymbol{K}^a)^T. \tag{3}$$

Entries in the adjacency matrix, $\boldsymbol{A}^a$, measure how much two areas are semantically related.

- **Routing Area.** To filter out irrelevant areas with the query-area, we take advantage of the relation between the query-area and remaining areas(this areas are not query-area). The core step that we perform next is to prune the relation between areas by keeping only top$k$ connections. Specifically, we derive a routing index matrix, $\boldsymbol{I}_a \in \mathbb{N}^{S^2 \times k}$, with the row-wise top$k$ operator:

$$\boldsymbol{I}^a = \text{topkIndex}(\boldsymbol{A}^a). \tag{4}$$

Hence, the ith row of $\boldsymbol{I}^a$ contains k indices of most relevant areas for the ith area.

- **window-to-area attention.** With the area-to-area routing index matrix $\boldsymbol{I}^a$, we can apply external window attention. For each query window in area j, it will attend to all key-value pairs residing in the union of k routed areas indexed with $\boldsymbol{I}^a_{(j,1)}, \boldsymbol{I}^a_{(j,2)}, ..., \boldsymbol{I}^a_{(j,k)}$. To GPU friendly, We gather key and value first, $ie.$,

$$\boldsymbol{K}^{ga} = \text{gather}(\boldsymbol{K}^a, \boldsymbol{I}^a), \quad \boldsymbol{V}^{ga} = \text{gather}(\boldsymbol{V}^a, \boldsymbol{I}^a). \tag{5}$$

where $\boldsymbol{K}^{ga}, \boldsymbol{V}^{ga} \in \mathbb{R}^{S^2 \times \frac{kHW}{S^2} \times C}$ are gathered key and value tensor. We can then apply attention on the gathered key-value pairs as:

$$\boldsymbol{O}^a = \text{Attention}(\boldsymbol{Q}^i, \boldsymbol{K}^{ga}, \boldsymbol{V}^{ga}). \tag{6}$$

Adaptive Sparse Window Attention (ASWA). The window partition is the same as the ASAA module, we get the $\boldsymbol{Q}^i, \boldsymbol{K}^i, \boldsymbol{V}^i$ for each window.

- **Routing Window.** Specifically, we first construct a directed relationship about per-window, and obtain window-level queries and keys, $\boldsymbol{Q}^i, \boldsymbol{K}^i$. To be more specific, We then derive the adjacency matrix, $\boldsymbol{A}^i$ of window-to-window affinity via matrix multiplication between $\boldsymbol{Q}^i$ and transposed $\boldsymbol{K}^i$:

$$\boldsymbol{A}^w = \boldsymbol{Q}^i (\boldsymbol{K}^i)^T. \tag{7}$$

Entries in the adjacency matrix, $\boldsymbol{A}^i$, measure how much two windows are semantically related.

To filter out irrelevant areas with the query-window, we perform next is to prune the relation between windows by keeping only top$k$ connections. Specifically, we derive a routing index matrix, $\boldsymbol{I}^w$, with the row-wise top$k$ operator:

$$\boldsymbol{I}^w = \text{topkIndex}(\boldsymbol{A}^w). \tag{8}$$

Hence, the ith row of $\boldsymbol{I}^w$ contains k indices of most relevant windows for the ith window.

– **window-to-window attention.** With the window-to-window routing index matrix $\boldsymbol{I}^w$, we can apply external window attention. For each query window i, it will attend to all key-value pairs residing in the union of k routed windows indexed with $\boldsymbol{I}^w_{(i,1)}, \boldsymbol{I}^w_{(i,2)}, ..., \boldsymbol{I}^w_{(i,k)}$. To GPU friendly, We gather key and value first, $ie.,$

$$\boldsymbol{K}^{gw} = \text{gather}(\boldsymbol{K}, \boldsymbol{I}^w), \quad \boldsymbol{V}^{gw} = \text{gather}(\boldsymbol{V}, \boldsymbol{I}^w). \tag{9}$$

where $\boldsymbol{K}^{gw}, \boldsymbol{V}^{gw}$ are gathered key and value tensor. We can then apply attention on the gathered key-value pairs as:

$$\boldsymbol{O}^w = \text{Attention}(\boldsymbol{Q}^i, \boldsymbol{K}^{gw}, \boldsymbol{V}^{gw}). \tag{10}$$

Note that, we can parallelize the computation of the attention weights and the attended representations across all tokens in all areas and windows, which makes our approach computationally efficient.

– **Fusion.** We fusion window-to-area attention $\boldsymbol{O}^a$ and internal window-to-window attention $\boldsymbol{O}^w$, we have:

$$\boldsymbol{O}^i = \text{Fusion}(\boldsymbol{O}^a, \boldsymbol{O}^w) \tag{11}$$

there are many Fusion methods of O_i and L_i, This paper choose the simplest way only directly add them. In our extensive experiments, we found that other complex fusion way have little influence on the results of gene expression prediction.

$$\boldsymbol{Y}^i = \boldsymbol{O}^i + \text{LCE}(\boldsymbol{V}^w) \tag{12}$$

Here, we introduce a context enhancement term $\text{LCE}(\boldsymbol{V}^w)$ as in [18]. Function $\text{LCE}(\cdot)$ is parametrized with a depth-wise convolution, and set the kernel size to 5.

Prediction Block. We apply the fusion feature map to prediction. We have:

$$\boldsymbol{y}_i = \text{MLP}_f(\boldsymbol{Y}^i) \tag{13}$$

where MLP_f is a single-layer perception.

Objective. HSATNet is optimized with mean squared loss $\mathcal{L}_2$ and batch-wise PCC $\mathcal{L}_{\text{PCC}}$. The overall objective is achieved by:

$$\mathcal{L}_E = \mathcal{L}_2 + \mathcal{L}_{\text{PCC}}. \tag{14}$$

4 Experiments

4.1 Datasets

We conducted experiments utilizing the publicly available STNet dataset [9] and the 10x Genomics datasets[1]. The STNet dataset contains approximately 30,612

---
[1] https://www.10xgenomics.com/resources/datasets.

pairs of slide image windows and gene expression data, derived from 68 slide images from 23 patients. Following study [9], our objective is to predict the expression levels of 250 gene types exhibiting the highest mean values across the dataset. As for the 10xProteomic dataset, it comprises 32,032 slide image windows and gene expression data obtained from 6 slide images. Similar to the STNet dataset, we employed the same approach to identify the target gene types. To ensure consistency and comparability, we subjected the target gene expression data to both log transformation and min-max normalization.

4.2   Experimental Set-Up

Evaluation Metrics. Following [25], here we also use three metrics to compare: Pearson correlation coefficient (PCC), mean squared error(MSE), and mean absolute error(MAE). All the experimental results are presented 1×10^1. Specifically, to assess the performance, we employ PCC at three different quantiles: PCC@F PCC@S, and PCC@M represent the first quantile, median and mean of PCC. PCC@F reflects the PCC for the least performing model predictions, PCC@S and PCC@M measure the median and mean of correlations for each gene type. Given predictions and Ground Truths(GTs) for all of the slide image windows. Higher values for PCC@F, PCC@S, and PCC@M indicate better performance. Furthermore, MSE and MAE quantify the deviation between predictions and GTs on a per-sample basis for each gene type within each slide image window, lower MSE and MAE value signifies better performance.

Implementation Details. The setting of HSATNet following before study setting. During training, HSATNet is trained from scratch for 100 epochs, with a batch size of 32. We set the learning rate to 5×10^{-4} with a cosine annealing scheduler. To control overfitting, we apply a weight decay of 1×10^{-4}. ASAA is sparse transformer, the patch size is set to 32, the embedding dimension is 1024, the feedforward dimension is 4096, the model consists of 16 attention heads and a depth of 8. All experiments are conducted using 2 NVIDIA A10 GPUs, allowing for efficient processing and training of the model.

4.3   Experimental Results

We have quantitative gene expression prediction comparisons between our HSATNet and with SOTA methods in the STNet dataset and the 10x Genomics datasets (Table 1). All the experimental results are presented 1×10^1 and NBE is Nature Biomedical Engineering. We bold the best results and use '-' to denote unavailable results. Models are evaluated by four-fold cross-validation and three-fold cross-validation in the above datasets. We evaluate the effectiveness of our HSATNet experimentally on a series prior SOTA works, containing models in gene expression prediction [4,9,25] and in ImageNet classification ViT [7], MPViT [12], CycleMLP [2]. Our HSATNet outperforms the SOTA methods in these PCC-related metrics(Note that, PCC-related evaluation metrics are most important in our task.). The results are reported in Table 1, and compare resllts in these two datasets, we find that:

i) Our HSATNet exhibits a notable superiority over the baseline methods when considering the PCC-related evaluations. It is crucial to highlight that a significant proportion of gene types within this first quantile exhibit skewed expression distributions, representing the most difficult aspect of the prediction task. Our method surpasses the second-best approach **2.2%** in terms of PCC@S, indicating a substantial enhancement in capturing correlations for these particularly challenging gene types. This evaluates the performance of the median of correlations for all gene types. AS talking above, PCC-related evaluation metrics are most important in our task. Our HSATNet that uses the Coarse and Fine Attention outperforms them with a reason marginal in PCC-related evaluation metrics, while overall achieving similar MSE and MAE with EGN, which is the SOTA model in gene expression prediction within tissue slide images.

ii) CycleMLP and MPViT, the SOTA methods in the ImageNet classification task, our HSATNet are better then them on every metrics.

To summarize, our HSATNet outperforms the SOTA methods in gene expression prediction task, showcasing a significant improvement in PCC.

Table 1. Comparison to prior works.

Methods	References	STNet Dataset					10x Genomics Dataset				
		MAE	MSE	PCC@F	PCC@S	PCC@M	MAE	MSE	PCC@F	PCC@S	PCC@M
STNet [9]	NBE2020	0.45	1.70	0.05	0.92	0.93	1.24	2.64	1.25	2.26	2.15
NSL [4]	PKDD2021	–	–	-0.71	0.25	0.11	–	–	-3.73	1.84	0.25
ViT [7]	ICLR2021	0.42	1.67	0.97	1.86	1.82	0.75	2.27	4.64	5.11	4.90
CycleMLP [2]	ICLR2022	0.44	1.68	1.11	1.95	1.91	0.47	1.55	5.88	6.60	6.32
MPViT [12]	CVPR2022	0.45	1.70	0.91	1.54	1.69	0.55	1.56	6.40	7.15	6.84
EGN [25]	WACV2023	0.41	1.61	1.51	2.25	2.02	0.54	1.55	6.78	7.21	7.07
Ours		**0.40**	**1.59**	**1.60**	**2.28**	**2.38**	**0.40**	**1.54**	**6.93**	**7.43**	**7.20**

4.4   Ablation Study

The capability of each component by conducting a detailed ablation study in the 10x Genomics datasets.

Effectiveness of the HSATNet Architectures. As one of the key components in our HSATNet, Fusion block merging ASAA with ASWA for better feature aggregation. we compare the proposed method with two modules, one is that 'ASAA only' module and the other is that 'ASWA only'. The results are reported in Table 2. Our findings are as follows:

(i) The 'ASAA only' setting achieves worse performance than the HSATNet. Adding the fine level sparse attention is good for the design of architectures.

(ii) The 'ASWA only' setting achieves worse performance than the HSATNet. Adding coarse level sparse attention is good for the design of architectures.

(iii) We clearly see when the model fusion these two module have the best results in gene expression prediction. It proves that the combination of the two modules is more effective. Note that, here we only the simplest way to combine them, in our extensive experiments, we found that other complex fusion way have little influence on results of gene expression prediction.

Effectiveness of the ASAA. In ASAA, we also have to consider the s and k, here is some considerarion:

(i) s is chosen as similar to SWinTransformer [14], which uses a window size of 7, and it is also void padding. $224 = 7 \times 32$, we use $S = 7$ so that it is a divisor of the size of feature maps in every stage.

(ii) k is chosen as KVT [21] turns out that retaining only about 66% of your attention is enough, we gradually increase k to keep a reasonable number of windows to attend as the area size becomes smaller in later stages. Our goal is that the attention in four stages can be reduced nearly $30\% \sim 40\%$. The k is set to $[1, 4, 4, -2]$ and use 2, 2, 6, 2 blocks for the four stages, non-overlapped patch embedding, set the initial patch embedding dimension $C = 96$ and MLP expansion ratio $e = 4$. Here comparison between ASAA and several existing SOTA sparse attention mechanisms [6, 22, 24] in classification task. The rest of the network stays the same. Following [6], we align macro architecture designs with CSwin [6] for a fair comparison.

Table 2. Ablation study on HSATNet architectures.

Methods	'ASAA only'	'ASWA only'	HSATNet
ASAA	✓		✓
ASWA		✓	✓
MAE	0.58	0.53	**0.40**
MSE	1.64	1.60	**1.54**
PCC@F	6.05	6.78	**6.93**
PCC@S	6.84	7.03	**7.43**
PCC@M	6.57	7.10	**7.20**

Table 3. Ablation study on ASAA.

Methods	DAT [24] CVPR2022	crossformer [22] ICLR2022	CSwin [6] CVPR2022	Ours
MAE	0.58	0.60	0.62	**0.40**
MSE	1.74	1.64	1.89	**1.54**
PCC@F	5.53	6.21	5.46	**6.93**
PCC@S	5.86	7.02	6.05	**7.43**
PCC@M	5.90	6.90	6.30	**7.20**

Table 4. Ablation study on ASWA.

Methods	DAT [24] CVPR2022	crossformer [22] ICLR2022	CSwin [6] CVPR2022	Ours
MAE	0.55	0.45	0.49	**0.40**
MSE	1.65	1.83	1.98	**1.54**
PCC@F	5.76	5.98	6.03	**6.93**
PCC@S	5.62	7.35	6.58	**7.43**
PCC@M	6.16	6.42	6.78	**7.20**

The results are reported in Table 3. In our extensive evaluation, we observed that our sparse-area strategy get the best results in every metric, with all proposed SOTA sparse attention mechanisms, we have the best performance, it proves our sparse-area strategy is strong sparse attention in the coarse level.

Effectiveness of the ASWA. Our ASWA module also uses filter out the most irrelevant feature windows like ASAA. To demonstrate the effectiveness of the proposed ASWA, we verify the effectiveness of like ASAA, only replace ASWA module other sparse attention strategy, the rest of the network stays the same. In Table 4, our findings that our sparse-window strategy get the best results in every metric, with all proposed SOTA sparse attention mechanisms, we have the best performance, it proves our sparse-window strategy is strong sparse attention in the fine level.

5   Conclusion

This paper introduces HSATNet, a novel approach designed to effectively and efficiently predict gene expression from fine-grained spot tissue slide images. HSATNet incorporates both coarse ($i, e., area$) and fine ($i, e., window$) attention through the ASAA (Adaptive Sparse Area Attention) and ASWA (Adaptive Sparse Window Attention) modules, respectively. At the coarse level, HSAT-Net utilizes a sparse-area strategy to acquire sparse window-to-area attention by filtering out the most irrelevant feature areas. This attention mechanism focuses on identifying the same attention areas for windows within the same area. At the fine level, HSATNet employs a sparse-window strategy to acquire window-to-window attention by filtering out the most irrelevant feature windows. This attention mechanism captures the local context and differences between different windows. The obtained coarse and fine attention are combined with a local context enhancement term, LCE, for gene expression prediction. Extensive experiments demonstrate the superiority of HSATNet compared to state-of-the-art (SOTA) methods. HSATNet shows great promise in facilitating studies on diseases and enabling accurate gene expression prediction for novel treatments.

References

1. Avsec, Ž, et al.: Effective gene expression prediction from sequence by integrating long-range interactions. Nat. Methods **18**(10), 1196–1203 (2021)
2. Chen, S., Xie, E., Ge, C., Chen, R., Liang, D., Luo, P.: CycleMLP: a MLP-like architecture for dense prediction. arXiv preprint arXiv:2107.10224 (2021)
3. Correia, G.M., Niculae, V., Martins, A.F.: Adaptively sparse transformers. arXiv preprint arXiv:1909.00015 (2019)
4. Dawood, M., Branson, K., Rajpoot, N.M., Minhas, F.U.A.A.: All you need is color: image based spatial gene expression prediction using neural stain learning. In: Kamp, M., et al. Machine Learning and Principles and Practice of Knowledge Discovery in Databases. ECML PKDD 2021. Communications in Computer and Information Science, Part II, vol. 1525, pp. 437–450. Springer, Cham (2021). https://doi.org/10.1007/978-3-030-93733-1_32
5. Deng, J., Dong, W., Socher, R., Li, L.J., Li, K., Fei-Fei, L.: ImageNet: a large-scale hierarchical image database. In: 2009 IEEE Conference on Computer Vision and Pattern Recognition, pp. 248–255. IEEE (2009)
6. Dong, X., et al.: Cswin transformer: a general vision transformer backbone with cross-shaped windows. In: Proceedings of the IEEE/CVF Conference on Computer Vision and Pattern Recognition, pp. 12124–12134 (2022)
7. Dosovitskiy, A., Beyer, L., Kolesnikov, A., Weissenborn, D., Zhai, X., Unterthiner, T.: Transformers for image recognition at scale. arXiv preprint arXiv:2010.11929 (2020)
8. Gerlinger, M., et al.: Intratumor heterogeneity and branched evolution revealed by multiregion sequencing. N. Engl. J. Med. **366**, 883–892 (2012)
9. He, B., et al.: Integrating spatial gene expression and breast tumour morphology via deep learning. Nat. Biomed. Eng. **4**(8), 827–834 (2020)
10. Ho, J., Kalchbrenner, N., Weissenborn, D., Salimans, T.: Axial attention in multidimensional transformers. arXiv preprint arXiv:1912.12180 (2019)
11. Huang, Z., Wang, X., Huang, L., Huang, C., Wei, Y., Liu, W.: CCNet: criss-cross attention for semantic segmentation. In: Proceedings of the IEEE/CVF International Conference on Computer Vision, pp. 603–612 (2019)
12. Lee, Y., Kim, J., Willette, J., Hwang, S.J.: MPViT: multi-path vision transformer for dense prediction. In: Proceedings of the IEEE/CVF Conference on Computer Vision and Pattern Recognition, pp. 7287–7296 (2022)
13. Li, X., Wang, C.Y.: From bulk, single-cell to spatial RNA sequencing. Int. J. Oral Sci. **13**(1), 36 (2021)
14. Liu, Z., et al.: Swin transformer: Hierarchical vision transformer using shifted windows. In: Proceedings of the IEEE/CVF International Conference on Computer Vision, pp. 10012–10022 (2021)
15. Marx, V.: Method of the year: spatially resolved transcriptomics. Nat. Methods **18**(1), 9–14 (2021)
16. Mei, Y., Fan, Y., Zhou, Y.: Image super-resolution with non-local sparse attention. In: CVPR, pp. 3517–3526 (2021)
17. Ramachandran, P., Parmar, N., Vaswani, A., Bello, I., Levskaya, A., Shlens, J.: Stand-alone self-attention in vision models. In: Advances in Neural Information Processing Systems, vol. 32 (2019)
18. Ren, S., Zhou, D., He, S., Feng, J., Wang, X.: Shunted self-attention via multi-scale token aggregation. In: Proceedings of the IEEE/CVF Conference on Computer Vision and Pattern Recognition, pp. 10853–10862 (2022)

19. Roy, A., Saffar, M., Vaswani, A., Grangier, D.: Efficient content-based sparse attention with routing transformers. Trans. Assoc. Comput. Linguist. **9**, 53–68 (2021)

20. Vaswani, A., et al.: Attention is all you need. In: Advances in Neural Information Processing Systems, vol. 30 (2017)

21. Wang, P., et al.: KVT: k-NN attention for boosting vision transformers. In: Avidan, S., Brostow, G., Cissé, M., Farinella, G.M., Hassner, T. (eds.) Computer Vision - ECCV 2022. ECCV 2022. LNCS, Part XXIV, vol. 13684, pp. 285–302. Springer, Cham (2022). https://doi.org/10.1007/978-3-031-20053-3_17

22. Wang, W., et al.: CrossFormer: a versatile vision transformer hinging on cross-scale attention. arxiv, arXiv preprint arXiv:2108.00154 (2021)

23. Wang, Y., Ma, C., Zeng, B.: Multi-decoding deraining network and quasi-sparsity based training. In: CVPR, pp. 13375–13384 (2021)

24. Xia, Z., Pan, X., Song, S., Li, L.E., Huang, G.: Vision transformer with deformable attention. In: Proceedings of the IEEE/CVF Conference on Computer Vision and Pattern Recognition, pp. 4794–4803 (2022)

25. Yang, Y., Hossain, M.Z., Stone, E.A., Rahman, S.: Exemplar guided deep neural network for spatial transcriptomics analysis of gene expression prediction. In: Proceedings of the IEEE/CVF Winter Conference on Applications of Computer Vision, pp. 5039–5048 (2023)

26. Zhao, G., Lin, J., Zhang, Z., Ren, X., Su, Q., Sun, X.: Explicit sparse transformer: concentrated attention through explicit selection. In: ICLR (2020)

Author Index

A

Akbari, Hesam 439
Ali, Sarwan 215
Ambegoda, Thanuja D. 203

B

Bai, Binhao 176

C

Cai, Yi 312
Chardonnet, Jean-Rémy 325
Charuka, Kaveesh 203
Chen, Cui 594
Chen, Dazhi 453
Chen, Haoran 16
Chen, Jian 74
Chen, Qingliang 260
Chen, Xi 16
Chen, Xiaohe 387
Chen, Yuzhi 3
Chourasia, Prakash 215

D

Dai, Jingjing 551
Deng, Yaoming 312
Deng, Zelin 162
Ding, Qiqi 312
Ding, Yuanzhe 114
Dong, Jia 50

F

Fan, Chenyou 535
Fan, Jun 493
Fan, Xinxin 551
Fang, Xueqing 260
Fang, Yu 3
Fenster, Aaron 125
Fujiwara, Kantaro 523

G

Gan, Haitao 125, 400
Gao, Fei 283
Gao, Qian 493
Gong, Jin 579
Gou, Gang 453
Guo, Silu 579
Guo, Yixing 507

H

Han, Hua 16
He, Pei 162
He, Ruifang 412
He, Yuanhan 28
Hou, Yuexian 412
Hou, Zengguang 50
Hsi, Yuhan 272
Hu, Junjie 535
Huang, Aimin 535
Huang, Jian 347
Huang, Kai 579
Huang, Kun 412

I

Inoue, Sozo 564

J

Jia, Haoze 16
Jiang, Haiqi 535
Jiang, Zekai 260
Jiang, Zihao 228
Jiao, Liqun 50

K

Korani, Wael 439

L

Li, Junli 176
Li, Ning 138, 150

B. Luo et al. (Eds.): ICONIP 2023, CCIS 1964, pp. 607–609, 2024.
https://doi.org/10.1007/978-981-99-8141-0

Li, Shaohua   507
Li, Weimin   507
Li, Xiaomeng   272
Li, Ye   50
Li, Yidong   228
Li, Zhan   260
Li, Zhiqiang   467
Li, Ziqiang   523
Liang, Xiaokun   551
Ling, Yehua   579
Liu, Cheng-Lin   86
Liu, Lin   551
Liu, Ni   579
Liu, Shiqi   50
Liu, Wenqiang   28
Liu, Xiaoming   114
Liu, Xinyu   374
Liu, Xuanming   347
Liu, Yang   125
Liu, Yao   74
Liu, Yongbin   426
Liu, Zhida   244
Long, Jinyu   176
Lou, Weidong   283
Luo, Jichang   50
Luo, Ye   374
Luo, Yiqian   138, 150
lv, Yanan   16

M
Ma, Yan   50
Madhushan, Pasan   203
Merienne, Frédéric   325
Miao, Borui   357
Mondal, Samrat   300
Murad, Taslim   215

O
Ouyang, Chunping   426
Ouyang, Wei   162

P
Palmal, Susmita   480
Pan, Yudong   138, 150
Patterson, Murray   215

Q
Qiu, Wei   138, 150

R
Ran, Longrong   74
Ren, Min-Si   86

S
Saha, Sriparna   300, 480
Sahoo, Pranab   300
Sharma, Saksham Kumar   300
Shi, Jianting   28
Shi, Ming   400
Shibata, Tomohiro   564
Slamu, Wushouer   244
Song, Rihui   579
Sun, Guodong   16
Sun, Shiliang   62
Sun, Yaxuan   189

T
Tan, Chunyu   101
Tanaka, Gouhei   523
Tang, Panrui   594
Tang, Pei   426
Tang, Qiang   162
Tao, Zhiyong   387
Tayebi, Zahra   215
Teng, Lianwei   347
Tian, Yunfei   101
Tong, Tong   467
Tripathy, Somanath   480

V
Victorino, John Noel   564

W
Wang, Annan   189
Wang, Bo   412
Wang, Chengliang   74
Wang, Haonan   272
Wang, Ji   400
Wang, Jialin   357
Wang, Jingchao   507
Wang, Mei   387
Wang, Nana   493
Wang, Qiu-Feng   86
Wang, Shanhu   62
Wang, Siru   125
Wang, Tao   50
Wang, Xinrui   260
Wang, Yicheng   374

Wang, Yuyang 325
Wei, Ping 176
Wen, Jiahui 400
Weng, Libo 283
Wickramanayake, Sandareka 203
Wijesooriya, Dineth 203
Wu, Qiaoyun 101
Wu, Shengxi 189
Wu, Xin 312
Wu, Xing 74

X

Xia, Wei 125
Xiao, Yewei 347
Xie, Xiaoliang 50
Xie, Yaoqin 551
Xiong, Lianjin 138, 150
Xu, Xinyue 272
Xu, Yunfeng 357
Xu, Zhenwei 244

Y

Yang, Zailin 74
Yang, Zhi 400
Yang, Zhibo 176
Yin, Fei 86
Yu, Li 162

Yu, Qing 244
Yuan, Bin 260
Yun, Song 162

Z

Zeng, Jianliang 260
Zhai, Pengjun 3
Zhang, Bairu 50
Zhang, Chulong 551
Zhang, Guifu 28
Zhang, Guiqiang 28
Zhang, Yangsong 138, 150
Zhang, Yan-Ming 86
Zhang, Yi 412
Zhang, Zuping 594
Zhao, Dongming 412
Zhao, Haining 50
Zhao, Jing 62
Zhao, Shaojie 357
Zhou, Hao 162
Zhou, Ran 125
Zhou, Xiaogen 467
Zhou, Xiaohu 50
Zhou, Yaoyong 244
Zhou, Yuanzhen 387
Zhou, Yun 101
Zhu, Aosu 347

Printed in the United States
by Baker & Taylor Publisher Services